Environmental Contaminants in Food

Sheffield Food Technology

Series Editors: P.R. Ashurst and B.A. Law

A series which presents the current state of the art of chosen sectors of the food and beverage industry. Written at professional and reference level, it is directed at food scientists and technologists, ingredients suppliers, packaging technologists, quality assurance personnel, analytical chemists and microbiologists. Each volume in the series provides an accessible source of information on the science and technology of a particular area.

Titles in the Series:

Chemistry and Technology of Soft Drinks and Fruit Juices
Edited by P.R. Ashurst

Natural Toxicants in Food
Edited by D.H. Watson

Technology of Bottled Water
Edited by D.A.G. Senior and P.R. Ashurst

Environmental Contaminants in Food
Edited by C.F. Moffat and K.J. Whittle

Environmental Contaminants in Food

Edited by

COLIN F. MOFFAT
Head of Ecotoxicology and Chemistry
Marine Laboratory Aberdeen

and

KEVIN J. WHITTLE
Director
Torry Research Ltd
Aberdeen

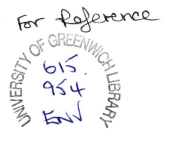

Sheffield
Academic Press

CRC Press

First published 1999
Copyright © 1999 Sheffield Academic Press

Published by
Sheffield Academic Press Ltd
Mansion House, 19 Kingfield Road
Sheffield S11 9AS, England

ISBN 1-85075-921-9

Published in the U.S.A. and Canada (only) by
CRC Press LLC
2000 Corporate Blvd., N.W.
Boca Raton, FL 33431, U.S.A.
Orders from the U.S.A. and Canada (only) to CRC Press LLC

U.S.A. and Canada only:
ISBN 0-8493-9735-9

Printed on acid-free paper in Great Britain by
Bookcraft Ltd, Midsomer Norton, Bath

British Library Cataloguing-in-Publication Data:
A catalogue record for this book is available from the British Library

Library of Congress Cataloging-in-Publication Data:
Environmental contaminants in food / edited by Colin Moffat and Kevin Whittle.
  p. cm. -- (Food technology series)
  Includes bibliographical references and index.
  ISBN 0-8493-9735-9 (alk. paper)
  1. Food contamination. I. Moffat, Colin. II. Whittle, K. J.
(Kevin J.) III. Series.
TX591.E56 1998
664'.117--dc21
                          98-4666
                           CIP

Preface

This volume of the *Sheffield Food Technology* series addresses an important issue of food safety, that of environmental contaminants in food. It is widely acknowledged that in this area of food safety a large difference exists between the high priority assigned by the public perception of risk and the priorities established by risk analysis based on scientific knowledge. Thus it is very important that knowledge is readily available to assist with consumer education. We have taken as broad a view of environmental contaminants in the food chain as is possible within the limited space of this volume, by covering radioactivity, and inorganic and organic chemical contaminants. Veterinary and pesticide residues are included in the latter, although these can be regarded as the result of deliberate use of chemicals in the food chain, rather than as the result of emissions or direct contamination. There is much to be learned from the principle and effectiveness of a withdrawal period in the use of veterinary products, and from the behaviour of the older, persistent pesticides now widespread throughout the global environment—but particularly so in northern latitudes. It is also important to take an holistic view of environmental contaminants, rather than a compartmentalized view which looks at each contaminant group in isolation.

Authors were selected from Asia, Australia, Europe and North America, to contribute to a volume intended to present a state-of-the-art review of the topic and a primary source of reference, world-wide, for food scientists and technologists, suppliers of food ingredients, quality assurance staff, analytical chemists, public analysts, scientists and administrators within the regulatory framework at all levels, and for those at the sharp end of implementation of standards or involved in provision of advice to the public, industry, or government. Contributors range from food and environmental scientists and analysts in academia or in public research institutes who are dealing day-to-day with issues raised by the monitoring or surveillance of specific contaminants or residues, to staff within the regulatory authorities who are continuously assessing the information, data and advice from all sources and are involved in developing appropriate responses, and to consultants, expert in their respective fields, who are often involved in advising industry, government, international organizations (developing standards such as the Codex Alimentarius Commission), consumers and consumer protection organizations.

Development of the contents of this volume took place during the long-awaited debate and consultation in the United Kingdom of the proposed new Food Standards Agency, following a change in government. This proposal was a response to food scares and safety problems in Britain over the last 10 years or so, which not only cost the food industry at all levels and the taxpayer dearly, and nearly wiped out the British beef industry, but also eroded public confidence and, much more importantly, caused unnecessary fatalities. It remains to be seen whether the important lessons have been learnt, the approved legislation proves to be effective, consumer education is improved and public confidence in the food chain is restored. The contents of this volume cover the major environmental contaminants, methodology, sources, occurrence, significance, risk assessment, and standards. Total coverage of all contaminants is of course impossible, due to restrictions of the available space. We hope that the overall combination of contributors and content, bringing together experts from across the world to express their own views—not those of the organizations that employ them—has been timely and has provided another insight into the issue. The learning process can be facilitated by considering the experiences of others with, perhaps, different priorities, thus contributing significantly to furtherance of the debate and helping to focus on the real issues.

We gratefully acknowledge the co-operation of all the authors and those who, in spite of a variety of difficulties such as changes of job, car accidents, ill health, pressures at work, grave financial crises and other problems, produced their manuscripts and dealt patiently with the queries and suggestions of the Editors.

Colin Moffat and Kevin Whittle

Contributors

Dr Peter J. Abbott Australia New Zealand Food Authority, PO Box 7186, Canberra, ACT 2600, Australia

Dr Farid E. Ahmed East Carolina University School of Medicine, Leo W. Jenkins Cancer Center, LSB 114, Greenville, NC 27858, USA

Ms Effie Apostolou Key Centre for Applied and Nutritional Toxicology, RMIT University, City Campus, GPO Box 2476V, Melbourne, Victoria 3001, Australia

Dr Janis Baines Australia New Zealand Food Authority, PO Box 7186, Canberra, ACT 2600, Australia

Dr Jacob de Boer DLO-Netherlands, Institute for Fisheries Research, PO Box 68, 1970 AB IJmuiden, The Netherlands

Dr Simon Brooke-Taylor Australia New Zealand Food Authority, PO Box 7186, Canberra, ACT 2600, Australia

Professor Jana Hajšlová Institute of Chemical Technology, Department of Food Chemistry and Analysis, Technická 3, 16628 Prague, Czech Republic

Mr Wooderck S. Hawer Department of Food Analysis, Korea Food Research Institute, 46-1 Baekheon-dong, Bundang-ku, Songnam, Kyongki-do, 463-420, Republic of Korea

Dr Philipp Hess Marine Laboratory Aberdeen, PO Box 101, Victoria Road, Aberdeen AB11 9DB, Scotland, UK

Mr Peter Howgate 3 Kirk Brae, Aberdeen AB15 9SR, Scotland, UK

Dr Kil-Hwan Kim Department of Food Analysis, Korea Food Research Institute, 46-1 Baekheon-dong,

	Bundang-ku, Songnam, Kyongki-do, 463-420, Republic of Korea
Dr Barrie E. Lambert	Department of Radiation Biology, St Bartholomew's and the Royal London School of Medicine and Dentistry, Charterhouse Square, London EC1M 6BQ, UK
Dr Colin F. Moffat	Marine Laboratory Aberdeen, PO Box 101, Victoria Road, Aberdeen AB11 9DB, Scotland, UK
Dr Katherine J. Mondon	Radiological Safety and Nutrition Division, Joint Food Safety and Standards Group, Ministry of Agriculture, Fisheries and Food, Ergon House, c/o Nobel House, 17 Smith Square, London SW1P 3JR, UK
Dr George Shearer	CSL Food Science Laboratory, Norwich Research Park, Colney Lane, Norwich NR4 7UQ, UK
Dr Luba Tomaska	Australia New Zealand Food Authority, PO Box 7186, Canberra, ACT 2600, Australia
Dr David E. Wells	Marine Laboratory Aberdeen, PO Box 101 Victoria Road, Aberdeen AB11 9DB, Scotland, UK
Dr Kevin J. Whittle	Torry Research Ltd, Cairnview, Westhill, Aberdeenshire AB32 6TX, Scotland, UK
Dr Paul F. A. Wright	Key Centre for Applied and Nutritional Toxicology, RMIT University, City Campus, GPO Box 2476V, Melbourne, Victoria 3001, Australia

Contents

4 Advances in immunochemical methods of analysis for food contaminants **81**
P.F.A. WRIGHT AND E. APOSTOLOU

5 Radioactivity in food **104**
B. E. LAMBERT AND K. J. MONDON

6 Trace metal contaminants in food **146**
F. E. AHMED

7 Pesticides **215**
J. HAJŠLOVÁ

13 Safety standards for food contaminants   500
F. E. AHMED

1 Introduction

Colin F. Moffat and Kevin J. Whittle

1.1 Scope of the volume

The standard approach to environmental contaminants and food safety is to legislate and regulate after identifying and quantifying the risk. This approach, and assessment of exposure by surveillance and monitoring programmes, demands powerful, discriminatory and sensitive quantitative analytical tools and methodology for the production of accurate, precise, assured data. Concentrations are usually in the range of micrograms to nanograms analyte per gram product or less, over the wide array of food matrices that are found spanning raw materials to products prepared for consumption. Recognition of the need for accurate dietary exposure information probably has been the most significant change in risk assessment in recent years (Chapter 12). However, we have not included a specialist chapter dealing with quality assurance in food analysis because the topic was recently reviewed in a companion volume of this series (Wood, 1998), and important specific aspects also have been reviewed recently: development of reference and test materials (Wells, 1998), and proficiency testing in analytical laboratories (Key *et al.*, 1997). Instead, Chapters 2 to 4 introduce and review the most versatile and the most widely used analytical techniques for quantifying trace levels of specific organic environmental contaminants in food and discuss the benefits, limitations and strategies for ensuring data integrity and quality control, and provide a perspective on new developments in immunological methods. For analysis of inorganic contaminants, the reader is referred to inductively coupled plasma–mass spectrometry (ICP-MS), in concert with the necessary procedures to avoid potential interferences, which offers a rapid, reliable and sensitive approach for multielement analysis of a variety of food matrices (Crews *et al.*, 1992), and to recent reviews of the topic (Crews, 1996; Taylor *et al.*, 1998). For measurement of radioactivity in food, the reader is referred to Chapter 5 and the International Atomic Energy Agency guidebook (IAEA, 1989).

Food in the context of this book includes water and beverages unless these are specifically excluded by authors. Environmental contaminants in food have been interpreted as broadly as possible, within the obvious constraints of the size of this volume. Distinctions that are often drawn between environmental as opposed to industrial contaminants, and other ways of categorising contaminants, tend to be arbitrary. In any case, it is the holistic view of toxic chemicals that must prevail in considering the extent of exposure via the diet from food and drink. Succeeding chapters cover radioactivity (Chapter 5), inorganic (Chapter 6) and organic chemical contaminants (Chapters 7–11). The coverage of this volume excludes any consideration of natural toxicants, the subject of a preceding volume in this series, or of microbiological and viral contamination, and prions. We have also excluded consideration of the formation of toxic substances in food within the processing environment. Examples are the nitrosamines, some of which are potent carcinogens, formed by reaction of nitrate with secondary amines during curing and drying processes, and the chloropropanols, also carcinogenic, formed during acid hydrolysis of vegetable proteins. The most common member of this last group of contaminants is 3-monochloropropane-1,2-diol. It is possible that it may migrate into foods from packaging manufactured using epichlorohydrin, and there is also some evidence to suggest that it may occur naturally in cereals.

The contaminants considered are the major substances not intentionally added to food but that can be found in the food as a result of the general and the local environments (e.g. storage and packaging) through which the food progresses, from primary plant or animal production, through the various stages of handling, processing, packaging, storage, distribution and sale, to preparation, cooking and consumption. This, of course, excludes deliberate contamination or sabotage of foods with the obvious intention of causing injury, such as the addition of glass and metal fragments to bakery products, as happened recently in Scotland, or the addition of substances to cause illness or even death, as in an incident in Argentina in which arsenic was added to meat in a butcher's shop. It does include pesticide (Chapter 7) and veterinary residues (Chapter 8). The term pesticides is used generically to include the subdivisions of insecticides, herbicides, fungicides and acaricides. Pesticides are the most intensively studied group of organic chemical contaminants and they can impact food at nearly all the stages from production to consumption. The older and much more persistent pesticides, banned from use in many countries, remain as general environmental contaminants.

Chemical contaminants from industrial, municipal and domestic, atmospheric emissions, from the discharge of pollutants to rivers and seas, from the disposal of industrial waste and sewage sludge on land, and from man's activities in exploiting the earth's natural resources, such as mining, forest clearance, crude oil and gas production and production of energy, including nuclear power, can ultimately contaminate the food chain. In considering such contaminants, it is important to distinguish between these environmental sources and, for example, naturally occurring, high background levels in soils, sediments and water of certain heavy metals such as lead, cadmium and mercury, and the element arsenic, which can lead to accumulation of these elements in offals, fish, shellfish, cereals and some vegetables. There are naturally high levels of background radioactivity in some areas because of the geological nature of the area but they do not usually lead to elevated levels in food. The hydrocarbons and petroleum present a complex picture. They represent both natural products and environmental contaminants (Chapter 10), and some specific compounds and preparations have permitted food usages such as extraction solvents and lubricants. A whole array of hydrocarbons can be deposited on foods that are smoked, either domestically or commercially, as part of the variety of products available to consumers.

The environmentally stable organic contaminants that arise from a variety of sources, rather than a single point source, and accumulate mostly in fatty foods, are ubiquitous food contaminants and probably attract greatest concern. They include the polychlorinated biphenyls (PCBs), polychlorinated dibenzodioxins (PCDDs) and polychlorinated dibenzofurans (PCDFs) (Chapter 9), the polycyclic aromatic hydrocarbons (PAH) (Chapter 10), phthalate esters and organophosphate esters. The last two groups are not the subject of specific chapters but are briefly introduced later in this introduction (sections 1.4 and 1.5).

The food industry and consumers often detect undesirable tastes and flavours in raw materials or food products and an increasing interest has developed in the analysis of unwanted flavours in foods (Henshall, 1991). Chemical analysis is not always possible and reliance has to be placed on sensory assessment. These off-flavours and taints may be due to spoilage or deteriorative changes in storage, or to natural sources such as 'boar' taint in pork or the 'earthy' taints found in freshwater fish. The off-flavours may also be due to contaminants for which much of the available information is anecdotal. Chapter 11 takes a more systematic approach to sensory assessment of tainting and draws on the substantial body of research from the aquatic environment.

Coverage is rounded off appropriately by Chapters 12 and 13, which deal respectively with risk assessment and the translation of this into actual safety standards for food contaminants, attempting to place in perspective the global gathering of the huge amount of quantitative data on trace contaminants.

1.2 Complexity, sources and pathways of contamination

The sources and pathways of input of environmental contamination along the food chain from production in both the terrestrial and the aqueous environments to human consumption of products are complex. Table 1.1, which is by no means exhaustive, lists from around the world some examples of incidents, both major and minor, and as well as concerns that have arisen about chemical environmental contamination over the last 60 years. It includes a wide variety of commodities, the indicative contaminants and the reported sources of contamination. There is current concern that problems with methylmercury in the food chain in a large area of the Amazon basin in Brazil, as a result of uncontrolled gold mining, may well match the scale of human effects of the Minamata Bay incident in Japan, which came to light in the 1950s. Concerns are also being expressed that the more volatile persistent pollutants tend to be transported to and concentrated in the food chain in the northern and sub-arctic latitudes. It has been known for over 20 years that the radionuclide signature of the Sellafield nuclear plant (formerly Windscale) in Cumbria, UK, could be recognised in the Arctic waters inputting to the deep waters of the north Atlantic. Thus, concern about environmental chemical contaminants in the food chain is not misplaced and there is no room for complacency. The complexity of the problem can be shown by a brief look at two examples.

1.2.1 Foods from the aquatic environment

In the aquatic environment, contaminants arise from point and from diffuse sources: industrial and municipal effluents and discharges, accidents in watercourses or at sea (e.g. from bulk carriers), dumping at sea, water run-off, river discharges to coastal waters, atmospheric transport and wet and dry fallout. Accumulation of contaminants by fish and shellfish used for food via the aquatic food web, from primary producers, the phytoplankton, to the species at the top of the food web, is well established. The extent of accumulation depends *inter alia* on location, species, feeding behaviour, position in the food web, solubility and lipophilic properties of the contaminant, chemical stability and persistence in the environment, and on the metabolic and depuration mechanisms of the species.

In the United States, the National Academy of Sciences published a comprehensive report on public health risks associated with seafood (including freshwater resources). Chemically contaminated seafood was ranked fourth in order of importance behind bacteria and viruses in raw molluscan shellfish, naturally toxic fish, and naturally toxic shellfish (Ahmed, 1991). The potential risks from chemical contaminants were generally from freshwater species, but also from specific marine areas and from particular marine species. The risks were highest for subsistence fishermen and sports fishermen who consumed the catch from certain

Table 1.1 Environmental chemical food contamination, some examples of incidents and concerns

Contaminant	Food	Source	Country	Year[a]
Tri-o-cresyl phosphate	Foods cooked in oil	Contaminated oil	Various	1930s
Methylmercury	Fish and shellfish	Industrial chemical waste	Japan	1956
Fluoride	Drinking water	Accidental exposure	Japan	1960
Cadmium	Rice	Mining effluent	Japan	1961
PCB	Rice oil	Contaminated oil	Japan	1968
PCB	Milk	Herbicide carrier	USA	1969
PCB	Milk	Silo sealant	USA	1970
PCB	Chicken	Ground plastic in poultry feed	USA	1970
PCB	Shredded wheat	Recycled paper packaging	USA	1971
PCB	Fish meal	Heat system processing leak	USA	1971
PCB	Meat meal	Heat system processing leak	USA	1971
PCB	Turkey	Not known	USA	1971
Mercury	Bread	Fungicide on grain seeds	Iraq	1971
Arsenic	Drinking water	Contaminated well	Hungary	1971
PBB	Meat, egg, dairy items	Contaminated feed pellets	USA	1973
PCB	Rice oil	Contaminated oil	Taiwan	1979
PCB	Various	Electro-industrial plant	Yugoslavia	1980
Iron	Home-brewed beer	Steel fermenting drums	Sub-Saharan Africa	1980s
Unknown products of aniline denaturation	Cooking oil, fraudulent 'pure olive oil'	Aniline-denatured imported rape-seed oil from France; possibly carried in contaminated tanks	Spain	1981
Lead	Wine	Bottle caps	France	1984
Radioactivity	Animal products, etc.	Chernobyl plant explosion	Europe, various	1986
Pentachlorophenol/dioxin	Animal products	Wood treatment plant fire	USA	1987
Aluminium sulphate	Drinking water	Treated reservoir contamination	UK	1988
Cyanide	Grapes	Not known	Not known	1989
Clenbuterol	Cow's liver	Treated animals	Spain	1989
Lead	Beef offal, milk	Imported rice bran feed pellets	UK	1989
Crude oil	Commercial fisheries	*Exxon Valdez* oil spill	USA	1989

Table 1.1 (Continued)

Contaminant	Food	Source	Country	Year[a]
Methylmercury	Fish, cassava, water	Gold-mining operations	Brazil	1989
Benzene	Bottled water	Process filters	UK	1990
Clenbuterol	Veal liver	Treated animals	France	1990
Lead	Moonshine	Leached from illicit stills	USA	1990
Dioxins	Milk	Municipal incinerators	Netherlands	1990
Dioxins	Milk	Industrial contamination	UK	1990
Penicillin	Chicken	Carcase washing	Spain	1992
PCB	Canned cod liver	Polluted fishing waters	Poland	1992
Crude oil	Fish, shellfish, pasture	*Braer* oil spill	UK	1993
Phthalates	Baby foods	Contaminated edible oils	UK	1995
No. 2 fuel oil	Fish and shellfish	*North Cape* oil spill	USA	1996
Diesel oil	Drinking water	Accidental spillage	UK	1997
Benzene	Carbonated drinks	Contaminated carbon dioxide	UK	1998

[a]Year problem noted, regulated or reported.

areas, pregnant women and children (Ahmed, 1991). The major groups of contaminants considered, with potential for human toxicity, included the elements arsenic, cadmium, lead, mercury and selenium, the PCBs, dioxins, PAHs, chlorinated hydrocarbons and other pesticides. The conclusions were that a small percentage of fish and shellfish were contaminated with potentially hazardous contaminants, but there was a very wide variation with location and species. Generally, marine fish from the open sea contained lower levels of contaminant residues than freshwater fish or shellfish from inshore waters and estuaries. These species, and the particular areas of concern, need to be avoided or closed depending on the severity of the contamination until remedial action results in sufficient improvement to enable fishing for consumption to resume (Price, 1992).

1.2.2 Milk and milk products

On land, food production follows a much more well-defined and controlled food chain from plants and plant products, either directly or through foods of animal origin, to the consumer. In all cases, but particularly in the case of the most persistent and toxic contaminants that accumulate in the food chain or food web, it is most important to protect the consumers at the top of the food chain, the unborn children and infants or young children, because they are most vulnerable to the effects. Consequently, in practical terms, the aim is to keep the persistent and hazardous residues and contaminants at the bottom of the food chain and in basic foods as low as possible. Briefly, we can see what this means for milk and milk products (IDF, 1991).

In most countries, veterinary products have to be licensed, which *inter alia* sets strict withdrawal periods between last application and use for food (the same is true for farmed fish and shellfish). It is essential that only licensed products are used, that they are correctly administered, that usage is contained and that withdrawal periods are observed strictly, because residues in milk will become a safety hazard if these essentials are ignored. Cross-contamination of milk can occur from the bad practice of failing to clean machinery between the milking of treated and untreated cows. Sprays or powders of organophosphates, carbamates or pyrethroids used on cattle against ectoparasites may contaminate milk as a result of inhalation or cutaneous absorption by the animal. In general, residues in milk are low and are present only for a few milkings.

For the last 25 years in many countries, the use of organochlorine pesticides has been banned, or use in the vicinity of milking animals has been restricted, but substitution by less persistent pesticides is not always possible in the tropics where their use can still be essential. Forage, pasture, feed supplements and soil remain important sources of potential contamination, and they are all interlinked. On pasture, cows consume some soil and the plants in the pasture can also be contaminated by contact with and absorption of contaminants from air, water and soil. Similar mechanisms contaminate the plant raw materials made into animal feed supplements, and these pesticides can also be recirculated in animal meal and animal fat. The problem can be controlled by regulating the use of pasture, where possible, and setting tolerances for levels in feedstuffs to achieve as low levels as possible for chlorinated pesticides. Usually, local contamination has been shown to be the source of high levels of organochlorines in milk fat. Overall, over the last 20 years or more, organochlorines in milk and milk products have decreased significantly or shown a tendency to decrease.

Apart from direct application for the control of ectoparasites on dairy cattle, milk and milk products can be contaminated with pesticide residues (in current use) from animal feeds directly treated with insecticides or kept in stores which are treated to control pests, from feeds produced from plant materials treated with insecticides during growth, and from the control of insects in and around milk producing and storage facilities. Milk contamination depends on the stability, type and duration of application, and the rate of degradation of the insecticide by the cow. Residue limits set for milk and milk products depend on the rapid decline of the residues with successive milkings, on the bulking of milk products from different herds and sources before and after processing, and on the veterinary treatment of herds at different times. Consequently, in practice, there is effective dilution leading to low residue levels in commercial milk products, depending on fat content. Nevertheless, good agricultural practices always have to be maintained to keep residues as low as possible.

PCBs were increasingly used by industry from the 1920s to the 1970s. They are persistent in the environment and accumulate particularly in lipid-rich tissues. They are no longer manufactured by many countries but, from use over those 50 years, they became established as global environmental contaminants. Monitoring or surveillance programmes for PCBs from the late 1960s revealed *inter alia* examples of contamination of milk and milk products that were traced to sources such as cattle feeds, silage contaminated by sealants and paints used in silos, the oils and grease used on farms, waste oils, contaminated faeces, sewage sludge spread on fields, and contaminated soil. When the source was removed, total concentrations in milk fat dropped about 10-fold over periods up to about 12 months, depending on previous exposure. Generally, levels in milk and milk products tended to fall in those counties where production and use was halted, but Calabrese (1982) concluded that at that time organochlorine insecticide and PCB residues in human milk in the United States and Canada had not decreased appreciably since restrictions were introduced. Manufacture and use of PCBs should be eliminated everywhere.

Since the 1960s, there has been a growing interest in the dioxins, which are produced as minor by-products of incineration of waste and of industrial, organic, chemical processes using chlorinated products, for example in the paper and oil industries. Other specific inputs were, for example, from impurities in Agent Orange, a defoliant used in the Vietnam war, from use of industrial waste as a dust-binding agent, and from a factory explosion in Seveso, Italy, in 1976. These compounds, which have been produced increasingly for the last 70 years or more as the chemical industry has grown, are very persistent and have become widespread in the environment via air and water. They have accumulated in the food chain, especially in lipid-rich tissues, occurring at low concentrations in some milk and milk products probably as a result of contamination of grass. Another possible source of contamination was chlorine-bleached paper cartons. Contaminated food is probably the main source of dioxins for man, and milk probably contributes about 10% of the burden. Excretion from the body is very slow, so that mother's milk and the infant at the top of the food chain are the most exposed. Possible effects on breast-fed infants have been noted in Holland (Koppe et al., 1991; Pluim et al., 1993). There have been suggestions that the human health risk from dioxins has been exaggerated, although various long-term studies of the possible effects of occupational exposure have yet to be completed.

1.3 Bioavailability, interactions, biomarkers

Accurate measurements of the amounts of most chemical contaminants in foods can be made and the exposure from food, or the amounts consumed with the diet in food and beverages, can be estimated for different groups of consumers with differing diets. Such information is essential to the basic or relatively crude assessment of the safety of food. However, the contaminants are not completely absorbed from the diet. Variable amounts may be excreted unchanged. The variable proportion absorbed is regarded as being bioavailable and is potentially available to exert adverse toxic effects. There are numerous ways of assessing toxicity of contaminants but, because useful information is lacking on the bioavailability of some contaminants, or on the potential interactions between contaminants and between contaminants and other components of the diet, these toxicity assays have limited relevance and usefulness for anything other than crude extrapolation for risk assessment of contaminants in the human diet.

Hoogenboom and Kuiper (1997) recently reviewed the potential of so-called *in vitro* models to assess the presence and safety of contaminant residues and natural toxicants in food products. At the cellular level, using cell lines and tissue slices, these methods can be used for the study of biotransformations, identification of metabolites, and the assessment of toxicity of the contaminants. They can be used to compare and investigate the extrapolation of animal data to humans, in terms of species differences in the bioactivation and detoxification of contaminants. They can be used, for example, as receptor-, enzyme- or cellular-based bioassays to detect the biological activity or activities of contaminants, such as those with dioxin-like activity, including some of the PCBs, those with oestrogenic activity, including some of the phthalates, those with antibiotic activity, agonists and so on. These approaches also enable investigation of interactions between contaminants (Arnold *et al.*, 1996), at least at the human cellular level. The systems are largely developments of methods for screening antibiotic activity, optimising the design of drugs, and testing and screening drugs. Bioavailability is critical to performance for these applications, and it should also be possible to develop similar methods to investigate the absorption and availability of food contaminants. Information on contaminants from *in vitro* studies can be used to address some of the uncertainties in risk analysis that relate to bioavailability, interactions, and extrapolation to humans.

Measures of contaminants or their metabolites, so-called biomarkers, in easily accessible compartments of the human body, blood, breast milk, urine, faeces, breath, and sometimes other tissues, help to assess exposure pathways contributing to the body burden, excretion, and other effects, as a consequence of bioavailability via inhalation, ingestion or dermal absorption from air, water, food, dust and soil. Oxidative damage to DNA is a key aspect of carcinogenesis, ageing, and degenerative disease processes. The food we eat has a large role to play both positively and negatively in the oxidative process. Urinary biomarkers such as thymidine glycol and 8-hydroxydeoxyguanosine can be used to index genetic damage. It has been suggested (Simic and Bergtold, 1991) that such biomarkers can be used to assess the carcinogenic (or anti-carcinogenic) properties of diets, specific foods, or individual components of foods. Whether this approach can be developed to assist in the more sensitive identification of the carcinogenic risk from contaminants within the whole diet context for heterogeneous populations of consumers remains to be seen.

Exposure to contaminants arises from different sources via different pathways, some of which may be dominant, e.g. water, food, a particular food such as a species of freshwater fish, or occupational exposure. Biomarkers can be used to follow exposure to particular contaminants and, by comparison between reference biomarker and exposure data, help to determine whether water, or food, or a particular type of food, for example, is a dominant exposure pathway. Use of the pyrene metabolite 1-hydroxypyrene has confirmed that for the general population diet is the dominant pathway of exposure to PAHs (Buckley and Lioy, 1992; Buckley *et al.*, 1995). An ambitious preliminary study in Texas made biomarker measurements in the urine, blood and breath of subjects, and environmental measurements (air, water, food, dust and soil) in their homes, for persistent and non-persistent pesticides, metals, PAHs and volatile organic compounds, in spring and summer (Akland *et al.*, 1997; Berry *et al.*, 1997a, b; Buckley *et al.*, 1997). The object was to assess the practical difficulties of such an exercise, provide a basis for the design of future studies, and work towards a comparative picture of environmental contamination and associated biomarker levels. Overall, the study did not link seasonal differences or elevated biomarker levels to the various environmental sources owing to low frequency of detection or non-detection of the analyte, and to the relatively small numbers of subjects. However, chlorpyrifos in indoor air and house dust was strongly linked with the metabolite biomarker 3,5,6-TCP in urine, and a sample of carp (taken from a subject's home freezer), caught in a nearby irrigation ditch and intended for the table, was found to have a suprising total PCB concentration of 399 mg/kg, after high PCB concentrations were found in blood serum from the subjects in this household (Berry *et al.*, 1997b).

1.4 Phthalates

Phthalate esters are used widely in industry throughout the world and in household products, particularly as plasticisers; they can slowly volatilise or migrate from plastics, epecially PVC based products, representing a major source and pathway into the environment, from which they can enter the food chain (Seiber, 1990; Gilbert, 1994). Other routes include leaching into water and soil from waste disposal, losses during phthalate production, losses during the manufacture of products to which phthalates are added, and by dispersal from their use as carriers for pesticides, such that they are now global environmental contaminants. Fatty food materials can extract non-polar and highly lipophilic phthalates from plastic tubing, containers and films, but this has been reduced dramatically by the use of polyethylene-based materials. They also pose problems for trace organic analyses in the laboratory. New cabling, plastic ducting and filters in air circulation systems can pose particular, and unwelcome, problems in analytical laboratories. Food chain accumulation can occur in the marine and freshwater environment, but most phthalates are readily metabolised by mammals so that bioaccumulation will be less than in the case of organochlorines and PCBs. They are detectable, usually at low levels, in virtually all foods, but are most abundant in fatty foods. In the United Kingdom, phthalates were detected in all samples in a total diet study and in all samples of infant formula analysed. It is not suprising that phthalates have been found in milk and dairy products, increasingly so for products with increased fat content, ranging from 0.01 mg/kg in milk to over 50 mg/kg in cheese (Sharman *et al.*, 1994); these values are higher than are accountable simply by the increased fat content and suggest significant contamination during processing. They have relatively low toxicity, but some of these compounds appear to have oestrogenic activity and there is some evidence that there may be a strong synergistic interaction between naturally occurring and man-made compounds with oestrogenic activities, even though the structures and other properties may vary considerably (Safe, 1995; Arnold *et al.*, 1996).

1.5   Organophosphate esters

These esters, the trialkyl, triaryl and mixed alkyl/aryl phosphate esters, which have some structural similarity to the organophosphorus ester insecticides, are produced in quantity and widely used in industry as fire-resistant hydraulic fluids, as flame retardants, as plasticisers, and as additives in paints, coatings, adhesives and lubricating oils. Environmental contamination arises from production losses, but the major pathways are probably volatilisation from plastics (e.g. PVC films), accidental leakage to rivers and leaching from solid and liquid wastes (Seiber, 1990; Gilbert, 1994). These volatile and lipophilic compounds are now widely distributed in the environment as a result of atmospheric transport and dry and wet deposition far away from the sources. They are common contaminants in watercourses and sediments, and can be identified in drinking water. Bioconcentration can occur in the food chain, particularly in fatty tissues. Fish caught near industrial sites where these compounds were used can contain tissue concentrations of a few mg/kg. From total diet studies in the United Kingdom, Gilbert *et al.* (1986) concluded that the highest concentrations were consistently found in offal, animal products and nuts. Tri-*o*-cresyl phosphate ester, which caused toxic incidents in the past (Table 1.1), is no longer used and esters in current use are not cholinesterase inhibitors like the related insecticides.

References

Ahmed, F.E. (1991) *Seafood Safety*, Food and Nutrition Board, Institute of Medicine, National Academy of Sciences, National Academy Press, Washington, DC.

Akland, G.G., Schwab, M., Zenick, H. and Pahl, D.A. (1997) An interagency partnership applied to the study of environmental health in the Lower Rio Grande Valley of Texas. *Environment International*, **23** 595-609.

Arnold, S.F.A., Klotz, D.M., Collins, B.M., Vonier, P.M., Guillette, L.J. and McLachlan, J.A. (1996) Synergistic activation of estrogen receptor with combinations of environmental chemicals. *Science*, **272** 1489-1492.

Berry, M.R., Johnson, L.S., Brenner, K.P. and Thomas, K.W. (1997a) Dietary characterizations in a study of human exposures in the Lower Rio Grande Valley: II. Household waters. *Environment International*, **23** 693-703.

Berry, M.R., Johnson, L.S., Jones, J.W., Rader, J.I., Kendall, D.C. and Sheldon, L.S. (1997b) Dietary characterizations in a study of human exposures in the Lower Rio Grande Valley: I. Foods and beverages. *Environment International*, **23** 675-692.

Buckley, T.J. and Lioy, P.J. (1992) An examination of the time course from human dietary exposure to urinary elimination of 1-hydroxypyrene. *British Journal of Industrial Medicine*, **49** 113-124.

Buckley, T.J., Waldman, J.M., Dhara, R., Greenberg, A., Ouyang, Z. and Lioy, P.J. (1995) An assessment of a urinary biomarker for total human environmental exposure to benzo[*a*]pyrene (BaP). *International Archives of Occupational and Environmental Health*, **67** 257-266.

Buckley, T.J., Liddle, J., Ashley, D.L., Paschal, D.C., Burse, V.W., Needham, L.L. and Akland, G. (1997) Environmental and biomarker measurements in nine homes in the lower Rio Grande valley: multimedia results for pesticides, metals, PAHs, and VOCs. *Environmental International*, **23** 705-732.

Calabrese, E.J. (1982) Human breast milk contamination in the USA and Canada by chlorinated hydrocarbon insecticides and industrial pollutants: current status. *Journal of the American College of Toxicology*, **1** (3) 91-98.

Crews, H.M. (1996) ICP-MS for the analysis of trace contaminants in food, in *Progress in Food Contaminant Analysis* (ed. J. Gilbert), Chapman and Hall, London, pp 147-186.

Crews, H.M., Baxter, M.J., Bigwood, T., Burrell, J.A., Owen, L.M., Robinson, C., Wright, C. and Massey, R.C. (1992) Lead in feed incident—multi-element analysis of cattle feed and tissues by inductively coupled plasma–mass spectrometry and co-operative quality assurance scheme for lead analysis of milk. *Food Additives and Contaminants*, **9** (4) 365-378.

Gilbert, J. (1994) The fate of environmental contaminants in the food chain. *Science of the Total Environment*, **143** 103-111.

Gilbert, J., Shepherd, M.J., Wallwork, M.A. and Sharman, M. (1986) A survey of trialkyl and triaryl phosphates in United Kingdom total diet samples. *Food Additives and Contaminants*, **3** 113-122.

Henshall, J.D. (1991) Unwanted flavours in foods, in *Food Contaminants: Sources and Surveillance* (eds C.S. Creaser and R. Purchase), The Royal Society of Chemistry, Cambridge, pp 191-200.

Hoogenboom, L.A.P. and Kuiper, H.A. (1997) The use of *in vitro* models for assessing the presence and safety of residues of xenobiotics in food. *Trends in Food Science and Technology*, **8** 157-166.

IAEA (1989) *Measurement of Radionuclides in Food and the Environment: A Guidebook*, Technical Report Series No. 295, International Atomic Energy Agency, Vienna.

IDF (1991) *Monograph on Residues and Contaminants in Milk and Milk Products*, Special Issue 9101, International Dairy Federation, Brussels.

Key, P.E., Patey, A.L., Rowling, S., Wilbourn, A. and Worner, F.M. (1997) International proficiency testing of analytical laboratories for foods and feeds from 1990 to 1996: the experiences of the United Kingdom Food Analysis Performance Assessment Scheme. *Journal of the Association of Official Analytical Chemists International*, **80** (4) 895-899.

Koppe, J.G., Pluim, H.J., Olie, K. and van Wijnen, J. (1991) Breast milk, dioxins and the possible effects on the health of newborn infants. *Science of the Total Environment*, **106** 33-41.

Pluim, H.J., Koppe, J.G. and Olie, K. (1993) Effects of dioxins and furans on thyroid hormone function in the human newborn. *Chemosphere*, **27** 391-394.

Price, R.J. (1992) Residue concerns in seafoods. *Dairy, Food and Environmental Sanitation*, **12** (3) 139-143.

Safe, S.H. (1995) Environment and dietary estrogens and human health: is there a problem? *Environmental Health Perspectives*, **103** 346-351.

Seiber, J. (1990) Industrial and environmental chemicals in the human food chain, in *Chemicals in the Human Food Chain* (eds C.K. Winter, J.N. Sieber and C.F. Nuckton), Van Nostrand Reinhold, New York, pp 183-219.

Sharman, M., Read, W.A., Castle, L. and Gilbert, J. (1994) Levels of di-(2-ethylhexyl) phthalate esters in milk, butter and cheese. *Food Additives and Contaminants*, **11** (3) 375-385.

Simic, M.G. and Bergtold, D.S. (1991) Dietary modulation of DNA damage in humans. *Mutation Research*, **250** 17-24.

Taylor, A., Branch, S., Halls, D.J., Owen, L.M.W. and White, M. (1998) Clinical and biological materials, food and beverages. *Journal of Analytical Atomic Spectrometry*, **13** (4) 57R-106R.

Wells, D.E. (1998) Development of reference and test materials for organic contaminants in water. *Analyst*, **128** 983-989.

Wood, R. (1998) Quality assurance, in *Natural Toxicants in Food* (ed. D.H. Watson), Sheffield Academic Press, Sheffield, pp 236-264.

2 Gas chromatography and mass spectroscopy techniques

David E. Wells, Philipp Hess and Colin F. Moffat

2.1 Introduction

Gas chromatography (GC), also known initially as gas liquid chromatography (GLC), is a separation technique that relies on the partitioning of molecules between the stationary, liquid phase and the mobile, gas phase. Over the last 45 years the term gas chromatography has become synonymous with an analytical technique that comprises several discrete steps including injection of a mixture of compounds onto a chromatographic column, the separation of the mixture into individual components, the detection of these compounds and their quantification. The technique is well suited to the quantitative determination of volatile and semi-volatile organic environmental contaminants in food products. Chlorinated biphenyls (CBs), phenols, aliphatic hydrocarbons, organohalide pesticides, polychlorinated dibenzo-p-dioxins (PCDDs), polychlorinated dibenzofurans (PCDFs), nitrogen- and phosphorus-containing pesticides, monocyclic aromatic hydrocarbons (MAHs), polycyclic aromatic hydrocarbons (PAHs) and phthalates are often present as highly complex mixtures containing upwards of 40 individual compounds. Most of the components of these mixtures can be resolved and quantified with high-efficiency capillary columns that result in a peak width at half-height (w_h) of 11 seconds for a retention time (t_R) of 70 minutes. A range of injection techniques, universal detectors, highly specific detectors, automation, and ever-improving instrumentation all add to the flexibility of a technique that has found universal acceptance for the quantitative determination of drugs, pesticides, natural products, gases, small molecules, large molecules, polar compounds and non-polar compounds. There are a number of reviews on gas chromatography (Desty, 1958; Ettre, 1975; Gordon, 1990a; Wittkowski and Matissek, 1990; Eiceman et al., 1998) and its applications described in this volume include Chapter 7 (pesticides), Chapter 9 (polychlorinated biphenyls, dioxins etc) and Chapter 10 (PAHs and petroleum products).

This chapter provides an overview on the applicability of GC for the determination of environmental contaminants in foodstuffs. It provides a guide to the selection of the most appropriate columns and detectors based on a review of recent literature. It gives a practical account of the key aspects of quality assurance and control necessary to provide data fit for the purposes of food analysis. A comprehensive range of applications is tabulated with reference to each aspect of the GC instrumentation.

2.2   Historical perspectives and developments

Chromatography has developed over many years, but the concept of gas chromatography and its practical applications started to develop during the early 1950s. Martin and Synge (1941) suggested that 'the mobile phase need not be a liquid but may be a vapour'. This idea was pursued by James and Martin (1952) who described 'the application of the gas–liquid partition chromatogram to the separation of volatile fatty acids'. A 4 ft or 11 ft glass column, held within a vapour jacket, was connected to a titration cell unit used for detection of the acids from formic to dodecanoic. At a very early stage it was realised that the use of a highly sensitive detector was critical to the technique and Martin and James (1956) developed the gas-density meter, which was subsequently used during studies on the separation of methyl esters (James and Martin, 1956; James and Webb, 1957). It was during this time that van Deemter *et al.* (1956) derived the relationship between plate height (h), also called the Height Equivalent to One Theoretical Plate (HETP), and carrier gas velocity (μ) for packed column GC. Over the last 42 years the technique has continued to evolve (Table 2.1). The original packed columns have

Table 2.1 Significant developments in gas chromatography

Date	Development
1941	Basis of the idea put forward
1952	Gas chromatographic separation of volatile fatty acids using a 4 ft × 4 mm glass tube and a silicone/10% stearic acid liquid phase
1950s	Initial development of theory
1958	Flame ionisation detector developed
1959	Glass capillary drawing machine invented
mid-1960s	Commercial instruments readily available
late 1970s	Introduction of the fused-silica capillary column
1977–1979	Uniform oven temperatures and control specifically developed for capillary columns so as to give reproducible retention times
late 1970s	Refined electronics for temperature programming
mid-1980s	Automated data processing packages leading to total PC control of the GC and data processing
late 1980s	Multidimensional GC
late-1980s	Automated on-column injections for capillary columns
early 1990s	Electronic pressure control of gases
1998–	Fast analysis using short columns
	Automatic switching multidimensional GC

been almost totally replaced by the capillary column, glass capillary columns being first reported in 1960 (Grob, 1981). The column length is now more typically 30 m but can be as great as 150 m although in recent times short (3 m) columns have been used for specific applications. The boiler containing a specific liquid to provide a given column temperature (methanol, 65.4°C; water 100°C; ethylene glycol monoethyl ether 137°C; ethylene glycol 200°C) has been replaced with an oven capable of maintaining a temperature from −80°C to +400°C with ramp rates of 0.1–70°C/min. The gas-density balance, a universal detector, has been replaced by a wide range of detectors, some of which are universal while others are highly specific (see section 2.3.3). Manual injections have been replaced, in many applications, with automated injections. These developments have brought lower detection limits, improved compound resolution, greater sample throughput and better precision and reliability. All of these factors contribute to the considerable utilisation of gas chromatography in studies of environmental contaminants in food.

2.3 Instrumentation

2.3.1 Injectors

Injectors are generically described here as a method of introducing the sample to the analytical gas chromatographic column. This may be by vaporised transfer of the sample as a vapour from the injector, or as a liquid by on-column injection or from the primary column (e.g. LC or supercritical fluid extraction). The most commonly used injection techniques in capillary GC are split injection, splitless (non-splitting) injection, direct injection and on-column injection (Grob, 1994). In the first three of these injection techniques there is a vaporisation chamber. This chamber contains a glass or quartz injector liner or insert. During a split injection the liquid samples are vaporised in the heated glass or quartz insert and the resultant vaporised sample is flushed to the column entrance by the carrier gas. At this point the vapours are split, with a small portion entering the column and a larger portion leaving through the split outlet. The split ratio can be adjusted with a needle valve with ratios in the range 10:1 to 200:1 being commonly used. For splitless injections, the split vent is closed during the actual injection and for a set period of time following the injection, with the objective of transferring all the vapours or gaseous products into the column. The direct injection resembles a splitless injection, but the injector does not have a split outlet. The sample liquid is deposited directly into the oven-thermostated column inlet or into an uncoated pre-column, also known as a retention gap, during cold on-column injections. Various applications for the different types of injectors are detailed in Tables 2.2 and 2.3.

Split injection is commonly used for relatively concentrated samples or where the analytes of interest are minor relative to a major component that cannot be removed. An example of the latter situation is the determination of ethyl acetate, methanol, n-propanol and isobutanol in whisky and brandy, where the concentration of these components is extremely small relative to that of ethanol. Splitless injection is often used in trace analysis as, in theory, everything injected should be swept into the column. There is, however, a problem with discrimination in that higher-boiling compounds may not be quantitatively transferred to the column. Non-linear splitting of high-boiling solutes can occur during splitless injections. This can be overcome using on-column injections. The evolution of autosamplers capable of injecting 'on-column' into fused-silica retention gaps has enhanced the use of this technique in trace analysis, but on-column injection is hampered by non-volatile sample by-products accumulating on the retention gap. Splitless injection is commonly used as an alternative for such 'dirty' samples or else the retention gap, used up to the first 3 m prior to the column proper, must be changed at very regular intervals. Splitless, split and direct injections are made at elevated temperatures, but for on-column injections the injector is cooled below the boiling points of the solvent to minimise evaporation inside the syringe needle. Programmed temperature vaporisation (PTV), split, splitless and direct injections represent a modification of the basic injection techniques. In these cases the sample is injected into a cool chamber which can then be rapidly heated.

A limitation of all the above techniques is that the volume that can be injected is generally limited to $\leq 3\,\mu l$. This necessitates the enrichment of trace compounds prior to analysis. Large-volume (up to $140\,\mu l$) injections are, however, now possible using appropriate instrumentation. This technique has been used for the determination of hexachlorobenzene in water (Venema and Jelink, 1996). A 1 ml aliquot of water was mixed with 1 ml of n-pentane and a $140\,\mu l$ aliquot was analysed directly by capillary GC-MS. Grob and Siegrist (1994) used 20–$50\,\mu l$ splitless injections as part of their investigations into mineral hydrocarbon contamination of food associated with transport of the food in jute bags.

Large sample throughput is essential to many industrial and regulatory laboratories. This has been greatly facilitated by the use of automated injection, which can ensure maximum use of costly, capital equipment and can reduce operator errors commonly associated with manual injections.

The principal purpose of the 'injection' is to efficiently deliver a sample into the GC column. As such, the syringe and needle are not the only

Table 2.2 Injection techniques, columns and carrier gases typically used in the determination of environmental contaminants in a range of food products including milk and water

Environmental contaminant[a]	Food product	Injection	Column[b]	Carrier gas	Detector[c]	Detection limits	Reference
Pesticides (diazinon chlorpyrifos)	Apples		Ultra 2;	N_2	NPD	0.3–100 ng linear response range	Asensio et al. (1991)
Organo-phosphorus pesticides	Apples	Split	Ultra 2; 25 m, 0.2 mm, 0.33 μm	N_2	NPD	0.09–60 mg/l linear response range	Barrio et al. (1995)
Pesticides	lettuce	Splitless	DB-1; 20 m, 0.53 mm, 1.5 μm	H_2	ELCD	<90 pg on-column	Fehringer et al. (1992)
	Grapefruit, lemon, orange, cucumber, red salad, mushroom, cherry, asparagus, wheat, soybean	Split/splitless	DB-1; 30 m, 0.25 mm, 1 μm; DB-5; 30 m, 0.25 mm, 1 μm; DB-17; 15 m, 0.55 mm, 1 μm; DB-210; 30 m, 0.25 mm, 0.25 μm; DB-1; 10 m, 0.55 mm, 0.55 μm	He	FPD MS	30 pg	Miyahara et al. (1994)
	Brown rice, potato, cabbage, lettuce, carrot, cucumber, shiitake, apple, strawberry, banana	Splitless	DB-5; 30 m, 0.25 mm, 0.25 μm; DB-210; 30 m, 0.25 mm, 0.25 μm	He	ECD FPD FTD	1 ng/g except organonitrogen at 10 ng/g	Nakamura et al. (1994)
	Onion, red radish, potatoes	Splitless	HP-1; 25 m, 0.32 mm, 0.17 μm; HP-5; 50 m, 0.32 mm, 0.17 μm	He	AED ECD/ NPD	10 ng/g	Stan and Linkerhägner (1993)
	Wine	Programmable pressure splitless	DB-XLB; 30 m, 0.25 mm, 0.25 μm		SIM-MS	0.005–0.01 mg/l	Kaufmann (1997)

Table 2.2 (Continued)

Environmental contaminant[a]	Food product	Injection	Column[b]	Carrier gas	Detector[c]	Detection limits	Reference
Pentachlorophenol	Honey	Split	HP-1; 12 m, 0.2 mm, 0.3 μm	He	SIM-MS	7.2 ng/g	Muiño and Lozano (1991)
Pyrethroids	Apples, oranges, pears, peppers, cabbages, tomatoes	Split/splitless	HP-1; 5 m, 0.65 mm, 2.65 μm	CH_4 (10%)/ Ar (90%)	ECD	3–30 ng/g	Pang et al. (1994)
Insecticide (toxaphene)	Fish, milk	Split	HP SE 52; 25 m	N_2 / He	ECD / MS		Becker et al. (1989)
	Fish		SE-54; 25 m, 0.32 mm; chiral OV1701; 20 m, 0.25 mm	He	MS/MS		Buser and Müller (1994)
	Fish	SPI[d]	DB-5; 30 m, 0.25 mm, 0.25 μm	He	MS/MS	1–3 pg/μl 0.02–0.06 ng/g	Chan et al. (1998)
Insecticide (bromocyclen)	Water fish	On-column SPI	DB-608; 25 m, 0.32 mm, 0.25 μm CP-Sil 8 CB; CP-Sil 19 CB; 60 m, 0.25 mm, 0.25 μm	He / H_2	ECD	0.5 pg/l	Bethan et al. (1997)
		splitless	DB-5; 20 m, 0.25 mm, 0.25 μm	He	EI-MS		
Trichothecenes	Wheat		SE-52; 10 m, 0.32 mm, 0.25 μm	He	CI-MS	10 ng/g	Koistiainen et al. (1989)
OCPs	Cheese	On-column	5% QF-1 on 80–100 mesh Chromsorb W-AW; 2 m×2 mm	Ar-CH_4	ECD		Bentabol and Jodral (1995)
	Milk		DB-1701; 30 m, 0.53 mm, 1.0 μm	N_2	ECD	2–10 ng/g fat	Wong and Lee (1997)
	Milk, butter		Phase for EPA Method 608; 30 m, 0.53 mm, 0.50 μm	N_2	ECD	1–3 ng/g	Waliszewski et al. (1997)
	Butter	On-column from LC	SE-52; 50 m, 0.32 mm, 0.15 μm	H_2	ECD	~1 ng/g	Barcarolo (1990)

Pesticides and CBs	Cow's milk	On-column	Ultra 2; 50 m, 0.32 mm		SIM-MS	0.08–0.20 ng/g fat	Turrio-Baldassarri et al. (1993)
CBs	Human milk	Split	HT-5	H_2	ECD	0.1–1.0 ng/g	Böhm et al. (1993)
	Fish	On-column LC	NB 54;15 m, 0.32 mm, 0.25 μm	He	ECD	1 ng/g	Hyvönen et al. (1991)
	Herring, trout	On-column	CP-Sil 8 CB; 10 m, 0.32 mm HP 22 m, 0.22 mm (GC-GC)		ECD (monitor) SIM-MS (analytical)	0.015–0.80 ng/g	Sippola and Himberg (1991)
	Fish, shellfish	Splitless	CP-Sil 19 CB; 50 m, 0.15 mm, 0.30 μm	H_2	ECD MS (negative CI)		de Boer et al. (1993b)
Chlorobenzenes	Carrot, potato, cabbage, cauliflower, lettuce, onion, broad bean, pea, tomatoe		DB-Wax Ultra 2		ECD	0.002–200 ng/g	Wang and Jones (1994)
PCDD/PCDF	Chicken, eggs, butter, pork, fish, cauliflower, lettuce, cherries, apples, milk, fat	On-column	DB-5; 60 m, 0.32 mm, 0.10 μm CP-Sil 88; 50 m, 0.32 mm, 0.2 μm	He	High-resolution SIM MS	ppq–ppt (pg/g–fg/g)	Beck et al. (1989)
TCDD	Fish	Splitless	Methylsilicone; 25 m, 0.20 mm, 0.33 μm	He+ 1.5% Ar	ECD MI-FTIR	15–45 pg/g	Mossoba et al. (1989)
PAHs	Water	Large volume injector	Rtx-5; 30 m, 0.32 mm, 0.25 μm	He	FID	20–1250 ng/g linearity	Morabito et al. (1993)

Table 2.2 (Continued)

Environmental contaminant[a]	Food product	Injection	Column[b]	Carrier gas	Detector[c]	Detection limits	Reference
PAHs	Fish and shellfish	On-column	Ultra 1; 25 m, 0.2 mm, 0.33 μm	He	MS	0.2 ng/g	Webster et al. (1997a) Topping et al. (1997)
	Seafood	Splitless	95% dimethyl-5% diphenylpoly-siloxane; 30 m, 0.25 mm, 0.5 μm	He	MS	1–5 ng/g	Nyman et al. (1993)
	Meat	Dynamic SFE[e]	DB-1701; 30 m, 0.25 mm, 0.1 μm	He	MS	1 ng/g	Snyder et al. (1996)
	Vegetable oils, smoked fish products, shellfish	On-column	DB-5; 50 m	He	MS (multiple ion detection)	0.1 ng/g	Speer et al. (1990)
	Rougan	Splitless	HP-1; 25 m, 0.32 mm, 0.52 μm	N₂	FID	2.5 ng/g	Wu et al. (1997)
	Mussels	Splitless/split	SPB-5; 15 m, 0.32 mm SPB-20; 15 m, 0.32 mm	He	FID	0.5–2.51 μg/ml	Hernández et al. (1995)
	Milk, eggs, liver, kidney		DB-1		SIM-MS	0.5 ng/g	Husain et al. (1997)
Nitro-PAHs	Meat, vegetables, spices, tea, coffee, paprika, peanuts, olive oil	On-column	HP-5 MS; 30 m, 0.25 mm, 0.25 μm OV-1; 10 m, 0.25 mm, 0.12 μm	He	SIM-MS NPD	5 pg/μl 50 pg/ul	Schlemitz and Pfannhauser (1996)
Aliphatic hydrocarbons	Flat fish	Splitless	OV-1; 25 m, 0.22 mm	N₂	FID		McGill et al. (1987)
Diesel contaminants	Canned fish	Split	SPB-1; 30 m, 0.32 mm, 0.25 μm	He	FID	Semi-quantitative	Newton et al. (1991)

Analyte	Matrix	Sampling	Column	Carrier	Detector	Detection	Reference
n-Alkanes, PAHs pesticides, phenols	Water	On-column from an SPE[f] module	HP-5 MS; 29 m, 0.20 mm, 0.25 μm	He	MS	1–20 ng/l	Louter et al. (1996)
VOCs	Vegetable oil	Headspace	Rtx-Volatiles; 60 m, 0.25 mm, 1.5 μm	He	FID	1 ng/g	McCown and Radenheimer (1994)
	Cooking oils	Headspace/split	DB-5MS; 30 m, 0.25 mm, 0.25 μm	He	MS	~1 ng/ml	Overton and Manura (1995)
	Fruit juices, soft drinks, fruit drinks, milk	On-column/SPME[g]	DB-624; 30 m, 0.32 mm, 1.8 μm	He	ELCD	0.002–1.5 ng/g	Page and Lacroix (1993)
	Butter, margarine, peanut butter, mayonnaise, coffee, flour, pastry mix, beer, lemonade, citrus juices	On-column from purge and trap	DB-624; 30 m, 0.32 mm, 1.8 μm	He	ELCD	0.04–0.24 ng	Page and Lacroix (1995)
	Vegetable oils	From purge and trap	DB-1; 30 m, 0.25 mm, 0.25 μm	He	SIM-MS	0.4–7.6 ng/g	Thompson (1994)
	Fish	From purge and trap	Rtx-502.2; 60 m, 0.32 mm, 1.8 μm	He	SIM-MS	5–80 pg/g	Roose and Brinkman (1998)
Organotin Organolead	Water	On-column	DB-5; 30 m, 0.32 mm, 0.25 μm DB-1; 30 m, 0.32 mm, 0.25 μm	He	AED	0.5 ng/l	Sadiki and Williams (1996)
Methyl bromide	Wheat, brown rice, raisins, cocoa beans	Headspace	DB-624; 60 m, 0.25 mm, 1.8 μm	He	ECD	10 ng/g	Norman et al. (1995)

Table 2.2 (Continued)

Environmental contaminant[a]	Food product	Injection	Column[b]	Carrier gas	Detector[c]	Detection limits	Reference
Methylmercury	Fish and shellfish	On-column	5% diethylene glycol succinate polyester; 2 m, 2 mm	N_2	ECD	1.6 ng/g	Pan et al. (1995)
	Fish	On-column from headspace analyser	10% AT-1000 on Cromosorb WAW (80–100 mesh); 1 m × 3 mm	Ar	Microwave-induced plasma atomic emmission	25–100 ng/ml linearity range	Lansens et al. (1991)

[a]CBs, chlorobiphenyl congeners; OCPs, organochlorine pesticides; PAHs, polycyclic aromatic hydrocarbons; PCDDs, polychlorinated dibenzo-p-dioxins; PCDFs, polychlorinated dibenzofurans; TCDD, 2,3,7,8-tetrachlorodibenzo-p-dioxin; VOCs, volatile organic compounds.

[b]Column description; column dimensions—length, internal diameter, film thickness.

[c]AED, atomic emission detector; ECD, electron capture detector; ELCD, electrolytic conductivity detector; FID, flame ionisation detector; FPD, flame photometric detector; FTD, flame thermionic detector; MI-FTIR, matrix isolation Fourier transform infrared spectroscopy; MS, mass spectrometry; CI, chemical ionisation, EI, electron ionisation, SIM, selective ion monitoring; NPD, nitrogen phosphorus detector.

[d]SPI, programmable septum injector (Chan et al., 1998) or septum-equipped programmable injector (Bethan et al., 1997).

[e]SFE, supercritical fluid extraction.

[f]SPE, solid-phase extraction.

[g]SPME, solid-phase microextraction.

Table 2.3 Typical GC injectors, columns, carrier gases and detectors used in the determination of polycyclic aromatic hydrocarbons (PAHs), monocyclic aromatic hydrocabons (MAHs), chlorobiphenyls (CBs), polychlorinated dibenzo-p-dioxins (PCDDs) and dibenzofurans (PCDFs), tetrachlorinated dibenzo-p-dioxins (TCDDs) and dibenzofurans (TCDFs). Several of these examples include comparisons of different column phases and parameters

Analyte	Injector	Column[a]	Carrier gas	Detector[b]	Detection limits[c]	Reference
PAHs	Splitless	Ultra 1; 25 m, 0.22 mm, 0.33 μm	He	MS	2.4 pg	Brindle and Li (1990)
	On-column	Rtx-200; 15 m, 0.28 mm, 0.10 μm	H$_2$	FID		Bemgård et al. (1993)
	On-column, temperature programmable	Smectic liquid-crystalline; 30 m, 0.2 mm, 0.15 μm	He	MS		Berset and Holzer (1994)
Sulphur-containing PAHs	On-column	CP-Sil 8; 25 m, 0.25 mm, 0.25 μm	H$_2$	FPD	0.1–0.2 ng	Morel et al. (1991)
Nitrophenols	Splitless	PTE-5; 30 m, 0.32 mm, 0.25 μm	He	NPD FPD	10–30 pg/μl	Wennrich et al. (1995)
MAHs	Purge and trap	DB-5; 30 m, 0.32 mm, 0.25 μm	He	FID	2 ng/g	Komolprasert et al. (1994)
Aliphatic hydrocarbons Biomarkers	Splitless	DB-5; 30 m, 0.32 mm, 0.25 μm	He	FID		Wang et al. (1994)
Aliphatic hydrocarbons	On-column	DB-1; 30 m, 0.25 mm, 0.25 μm	H$_2$	FID		Morel et al. (1991)
CBs	Splitless PTV[d]	CP-Sil 8; 50 m, 0.25 mm, 0.26 μm HT-5; 25 m, 0.22 mm, 0.10 μm	He	ECD MS (full scan EI and SIM)	25–50 pg (full scan MS)	Bøwadt and Larsen (1992)

Table 2.3 (Continued)

Analyte	Injector	Column[a]	Carrier gas	Detector[b]	Detection limits[c]	Reference
CBs	Splitless	CP-Sil 3, CP-Sil 19; 50 m, 0.15 mm, 0.30 μm CP-Sil 12	H_2	ECD	0.2 ng/g	de Boer et al. (1993a)
			He	MS		
	Splitless	OV-1; 20 m, 0.25 mm, 0.30 μm	He	MS	9–10 pg on-column	Blanch et al. (1996)
	Splitless	CP-Sil 8 CB; 50 m, 0.15 mm	H_2	ECD		de Boer and Dao (1989)
	Cool on-column	Smectic liquid-crystalline; 30 m, 0.2 mm, 0.15 μm	He	ECD		Berset and Holzer (1994)
		SE-54; 25 m, 0.32 mr, 0.25 μm and OV-210; 25 m, 0.32 mm, 0.25 μm	H_2	ECD	0.01 pg	Kannan et al. (1993)
	LC-on-column	DB-5MS; 27 m, 0.32 mm, 0.5 μm	He	ECD	0.1 ng/ml	Gort et al. (1997)
	On-column	CP-Sil 8 CB; 10 m, 0.32 mm, 0.25 μm DB-FFAP; 30 m, 0.25 mm, 0.25 μm	He	ECD SIM-MS	7–70 pg	Himberg and Sippola (1990)
	LC-on-column	CP-Sil 5 CB; 50 m, 0.32 mm, 0.13 μm	N_2	ECD	1–330 pg injected	Maris et al. (1988)
Brominated DDs and DFs		DB-5; 30 m, 0.25 mm, 0.25 μm	1.01% Ar in He	MI-IR	1 ng/μl on-column	Childers et al. (1992)

Compound	Injection	Column	Carrier gas	Detector	Notes	Reference
TCDDs		CP-Sil 88; 50 m, 0.23 mm, 2 μm		MI- IR	Low ng range	Grainger et al. (1989)
PCDDs		CP-Sil 88; 50 m, 0.23 mm, 2 μm		FTIR		
PCDDs/ PCDFs	Split/splitless	CP-Sil 88; 50 m, 0.25 mm	He	MS and FT-IR	10 ng PCDD, 20 ng PCDF by FT-IR	Sommer et al. (1997)
TCDD/TCDF	Splitless	SP-2250, DB-1701, DB-225 (30 m), SP-2401, DB-Wax; 60 m, 0.25 mm, 0.25 μm	H_2	ECD		Harden et al. (1989)
TCDD/TCDF	Splitless	DB-530; ?? m, 0.32 mm, 0.25 μm SB-Smectic; 25 m, 0.20 mm, 0.15 μm	He	ECD (monitor detector) SIM-MS (analytical detector)	4–2000 pg on-column linearity	Lamparski et al. (1990)
Toxaphene	On-column	DB-5; 60 m, 0.25 mm, 0.25 μm	N_2	ECD and FID	10 ng/g fat	Lach and Parlar (1991)
	Splitless	HP-1; 25 m, 0.2 mm, 0.33 μm	He	MS (EI and NCI)		
	Splitless	CP-Sil 2; 50 m, 0.25 mm, 0.25 μm β-BSCD[e]; 30 m, 0.2 μm	He	SIM-MS (EI and NCI)		Vetter et al. (1997)
	On-column	DB-5; 30 m, 0.25 mm, 0.25 μm	He	MS (scanning and SIM) CI		Lau et al. (1996)

Table 2.3 (Continued)

Analyte	Injector	Column[a]	Carrier gas	Detector[b]	Detection limits[c]	Reference
Pesticides	Splitless	BP-5; 25 m, 0.22 mm, 0.25 μm BP-5; 50 m, 0.22 mm, 0.25 μm BP-10; 50 m, 0.22 mm, 0.25 μm DB-17; 30 m, 0.22 mm, 0.25 μm OV-1701; 25 m, 0.25 mm, 0.25 μm CP-Cyclodextrin-B-2,3,6-M-19; 30 m, 0.25 mm, 0.25 μm	He	ECD	5–191 pg	Picó et al. (1994)
	On-column/SPME[f]	DB-5; 15 m, 0.25 mm, 0.25 μm	He	ELCD	0.3–10 ng/l	Page and Lacroix (1997)
PAHs Phthalates Pesticides	Splitless	Rtx-5, 30 m, 0.32 mm, 1.0 μm and Rtx-1701; 30 m, 0.32 mm, 1.0 μm		IR MS		Krock and Wilkins (1996)

[a]Column description; column dimensions—length, internal diameter, film thickness.
[b]ECD, electron capture detector; ELCD, electrolytic conductivity detector; FID, flame ionisation detector; FPD, flame photometric detector; FTD, flame thermionic detector; MI-FTIR (MI-IR), matrix isolation Fourier transform infrared spectroscopy; MS, mass spectrometry, CI, chemical ionisation, EI, electron ionisation, NCI, negative chemical ionisation, SIM selective ion monitoring; NPD, nitrogen phosphorus detector.
[c]This information was not always provided in the reference.
[d]PTV, programmed temperature vaporization.
[e]β-BSCD, 25%-t-butyldimethylsilylated β-cyclodextrin in 85%-dimethyl–15%-diphenylpolysiloxane, a chiral phase.
[f]Solid-phase microextraction.

delivery systems available to the analyst. Solid-phase microextraction (SPME) has been used by Shirey (1995) and Nilsson *et al.* (1995) for the determination of volatile organic compounds (VOCs), including the volatile aromatic hydrocarbons, benzene, toluene, ethylbenzene and the xylenes (BTEX compounds), in drinking water. The SPME device is a fused-silica fibre coated with a layer of polydimethylsiloxane and housed in a stainless steel needle. The fibre is exposed to the water within a headspace vial and then transferred to the GC injector, where the SPME fibre is pushed out of the needle and thermally desorbed into the carrier gas stream within the injector. The heated split/splitless injector, septum programmable injector (SPI) or the on-column port of the gas chromatograph can be used for the thermal desorption of the analytes from the fibre. Desorption takes from a few seconds up to several minutes depending on the class of compound. VOCs are cryo-focused prior to being chromatographed (Eisert and Levsen, 1996).

Volatile compounds, such as benzene, the other BTEX compounds, chloroform, trichloroethane, trichloroethylene, etc., are commonly analysed using static headspace, dynamic headspace or purge and trap techniques. Commercial headspace samplers are available and can be directly coupled to the GC where the injector can be operated in the split mode (Jickells *et al.*, 1990). Alternatively, the volatiles can be trapped and then desorbed into a cryofocusing module that is connected to the analytical column (Roose and Brinkman, 1998). Other variations on this theme exist, but ultimately this technique is limited to volatile compounds.

Direct 'injection' of less volatile contaminants in food is a considerable challenge. Schulzki *et al.* (1995) used on-line LC-GC to determine the concentration of radiolytic hydrocarbons in Camembert cheeses. Although radiolytic hydrocarbons are not strictly environmental contaminants, aliphatic hydrocarbons present in mineral oils, and transferred to food products during transport in jute bags, can be classified as such and have been studied by LC-GC (Grob *et al.*, 1991). In this case the hydrocarbons were isolated from extracts of foods by HPLC using a 10 cm × 2 mm i.d. LC column packed with silica gel and then transferred directly onto a PS-225 (25 m, 0.32 mm, 0.6 μm film thickness) capillary column. Products investigated included bread, chocolate and bonbons (a sweet candy). LC-LC-GC has been applied to the determination of aliphatic and aromatic hydrocarbons in vegetable oils. The first LC column was used to remove the fat and the second LC column was used to fractionate the hydrocarbons. These fractions then entered a vaporiser and from there went directly onto the GC column (Moret *et al.*, 1996). Uncoated, and usually deactivated, GC capillary pre-columns are generally used in LC-GC to accommodate the LC effluent while it vaporizes. In addition, the pre-column provides for the solute reconcentration by phase-ratio focusing, cold trapping and the so-called 'solvent effect' (Davies *et al.*, 1989). Microcolumn HPLC, using a 0.32 mm × 400 mm microcapillary LC column, has been used to deliver CBs, PCDDs and PCDFs to a 3 m pre-column ahead of a DB-5 (30 m, 0.25 mm) analytical column with mass spectrometric (MS) detection (Welch and Hoffman, 1993).

Supercritical fluid extraction (SFE)-GC was used by Snyder *et al.* (1996) to study PAHs in fire-exposed meat. The compounds were extracted from the meat using static SFE and then transferred directly onto the capillary column (DB-1701 30 m, 0.25 mm, 0.1 μm film thickness) by dynamic SFE, the column being held at $-50°C$ cryogenically with liquid nitrogen.

2.3.2   Columns

Capillary columns, and more specifically wall-coated open tubular (WCOT) columns, are today used almost exclusively for the determination of environmental contaminants. The WCOT columns contain the stationary phase as a thin film coated or chemically bonded to the wall of a fused-silica capillary. The original capillary columns were made of glass, but since the 1980s they are almost exclusively made from fused silica coated with a polyamide to provide protection and flexibility. A wide range of column dimensions, film thicknesses and phases have been utilised (Tables 2.2, 2.3 and 2.4). Furthermore, a range of speciality columns are now prepared commercially, often designed to meet the needs of a specific EPA method of analysis (Table 2.5).

Where the resolution of a single column is insufficient to obtain the required separation of the compounds of interest, then it may be necessary to apply a multidimensional capillary GC (MDGC) approach. In this technique two columns of differing polarity are sequentially connected, with a switching mechanism between them. The sample is chromatographed on the first column and when the compound(s) of interest elute from the first column they are switched or *heart-cut* onto the second column to provide further separation (Duebelbeis *et al.*, 1989a, b; Stan, 1992; Himberg and Sippola, 1993; de Boer, 1995; Hess *et al.*, 1995).

2.3.3   Detectors

Capillary GC requires detectors with the lowest noise, highest sensitivity, maximum baseline stability and the highest selectivity possible (Yang and Cram, 1981). The combined effect of using detectors with these

Table 2.4 Similar GC phases supplied by various companies[a]

Polarity	Phase composition	Column specific description
Nonpolar	100% Dimethylpolysiloxane	BP-1, DB-1, HP-1, Ultra 1, Rtx-1, ZB-1, SPB-1, OV-1, PE-1, CP-Sil 5 CB, SE-30
	5% Diphenyl– 95%-dimethylpolysiloxane	BP-5, DB-5, HP-5, Ultra 2, Rtx-5, ZB-5, SPB-5, OV-5, PE-2, CP-Sil 8 CB, SE-54
Low to midpolar	6% Cyanopropylphenyl– 94% dimethylsiloxane	BP-624, DB-1301, DB-624, HP 1301, Rtx-624, ZB-624, SPB-1301, CP-624, CP-1301, AT-624
Intermediate	35% Diphenyl– 65% dimethylsiloxane	DB-35, HP-35, Rtx-35, ZB-35, SPB-35, PE-11, AT-35
	50% diphenyl– 50% dimethylsiloxane	DB-17, HP-50 +, Rtx-50, ZB-50, SPB-50, OV-17, PE-17, CP-Sil 24 CB, AT-50
	7% Cyanopropyl– 7% phenyl– 86% dimethylsiloxane	DB-1701, HP-1701, Rtx-1701, ZB-1701, SPB-1701, OV-1701, PE-1701, CP-Sil 19 CB, AT-1701
Intermediate to polar	50% Cyanopropylphenyl– 50% dimethylsiloxane	BP-225, DB-225, HP-225, Rtx-225, OV-225, CP-Sil 43CB
Polar	Poly(ethylene glycol)	BP-20, DB-Wax, HP-INNO-Wax, Stabilwax, ZB-Wax, Supelwax-10, Carbowax-20M, PE-CW, CP-Wax 52 CB, AT-Wax
	Cyanopropylpolysiloxane	CP-Sil 88

[a]BP, SGE; DB/SE, J&W; HP/Ultra, Hewlett-Packard; Rtx, Restek; ZB, Phenomenex; SPB, Supelco; OV, Ohio Valley; PE, Perkin Elmer, CP-Sil, Chrompack; AT, Alltech.

Table 2.5 Speciality columns and associated EPA[a] methods

Column	Environmental contaminant	EPA methods
HP-608 DB-608, SPB-608, BP-608	Chlorinated pesticides	608, 508, 8080, 8081, 8150, 8151, 505
HP-624 DB-624, Rtx-Volatiles, Rtx-624, ZB-624, BP-624	Volatile halogenated compounds	501.3, 503.1, 524.2, 601, 602, 603, 624, 8260
HP-VOC, Rtx-502, DB-502.2, DB-VRX, Rtx-Volatiles	Purgeable organics	502.2, 524.2, 601, 602, 8024, 8260
HP PAS 5	Pesticides, CBs, herbicides	608, 508, 8080, 8081, 505, 515, 8150
HP-5 MS	Semivolatiles	625, 525, 8250, 8270

[a]EPA, US Environmental Protection Agency (http://www.epa.gov/epahome/standards.html).

characteristics and the high resolving power of capillary GC means that extremely complex mixtures of organic compounds can now be characterised at very low concentrations. Many environmental studies require the analysis of such complex mixtures, whether it be aliphatic hydrocarbons, PAHs, VOCs, CBs or pesticides.

Table 2.6, giving an overview of the detector types commonly used for the measurement of environmental contaminants in food, provides the key information to selecting the most appropriate instrument.

A less commonly used detector is the sulphur chemiluminescence detector (SCD), which has been used to study the sulphur components of gasoline in process streams where a detection limit of 50 ppb (ng/g) per sulphur component was obtained (Di Sanzo et al., 1994). Another less used detector is the plasma or ion mobility detector, which has been used in the past for the determination of PCBs (Karasek, 1971).

2.3.4   Gases

Gases are a fundamental aspect of the technique since they form the mobile phase. Gases are, however, also used as the fuel in the flames of the FID, FPD or NPD, and are used as 'make-up' gas, which is an additional inert gas added to improve the performance of, for example, an FID. This relates to the fact that the response of an FID is dependent not only on the amount of hydrogen and air but on the amount of carrier gas, which can help to stabilize the flame, resulting in a reduction of noise. The highest sensitivity can be obtained with approximately 30 ml/ min nitrogen.

In theory, any inert gas could be used as the carrier gas, but normally only nitrogen, helium or hydrogen are used (Tables 2.2 and 2.3). A small number of workers use an argon (90%):methane (10%) mixture as a carrier gas, particularly for ECD detection (Table 2.2). The performance of these three gases varies quite markedly owing to their reducing viscosity going from N_2 to He to H_2 (Wittkowski, 1990). Nitrogen produces a minimum height equivalent to a theoretical plate (h/mm) at the slowest average linear velocity of the three gases, as derived from the van Deemter plot (Gordon, 1990b; Wittkowski, 1990). In contrast, hydrogen shows a minimum at the greatest average linear velocity. Furthermore, after the minimum, the steepest increase in h with average linear velocity is observed for N_2, while H_2 produces the shallowest curve. Hydrogen is the easiest to optimise, provides the shortest retention times for the same temperature programme and column, and is available at a relatively lower cost for a high-purity product. With adequate safety valves on most modern GC instruments, the risk of an explosion with hydrogen is very low; at least 4% H_2 in the atmosphere is required to

Table 2.6 Detectors typically used for the determination of environmental contaminants (Fürst, 1990a, b; Vieths, 1990; Gordon, 1990b)

Detector	Abbreviation(s)	Common application	Introduced	Advantages	Limitations
Flame ionisation detector	FID	Phenols, BTEX compounds, PAHs, aliphatic hydrocarbons, free fatty acids, polychlorinated biphenyls, monomer residues	1958	Good linearity, highly sensitive	Not specific
Electron capture detector[a]	ECD	Fungicides, organochlorine pesticides, trihalomethanes, toxaphene, polychlorinated biphenyls, polybrominated biphenyls, halogenated solvents and volatile halocardons, phthalates, disinfectants and cleansing agents	Early 1960s	High sensitivity for nitro- and thiophosphate-containing phosphoric acid esters	Variation in response relating to number and location within molecule Interference from sulphur-containing compounds when analysing the likes of onions and garlic Interference from phthalates
Flame photometric detector	FPD	S in natural gas, dimethyl sulphide	1966[b]	Most selective of the common detectors (ECD, NPD, FPD) used in pesticide analysis	Limited linearity, especially with in the sulphur mode
Nitrogen–phosphorus detector	NPD, TID, PND, N-FID	Nitrogen/phosphorus-containing pesticides, triazine herbicides, nitrogen-containing pharmaceutically active substances	1966	Highly selective, limited sample clean-up required	Demands considerable fine tuning Require to perform analysis on two columns of different polarity Response relatively low to pesticides containing only one nitrogen atom
Thermal conductivity detector	TCD	CO_2, O_2, N_2, CH_4, CO, H_2S, NH_3	1954[c]	Particularly useful for gases	Sensitivity less than other common detectors

Table 2.6 (Continued)

Detector	Abbreviation(s)	Common application	Introduced	Advantages	Limitations
Photoionisation detector	PID	Pesticides, CBs, PAHs, VOCs		Nondestructive detector, wide linear dynamic range, high sensitivity	Not universal
Electrolytic conductivity detector	ELCD	Halogen-, sulphur- and nitrogen-containing compounds	1965	Selective, i.e. the nitrosamine-selective detector	
Atomic emission detector	AED	Organotin, organomercury, organolead, hydrocarbons, CBs, PBBs[e], PBDEs, organophosphorus pesticides, sulphur-containing pesticides, thiols	Late 1970s[d]	High element selectivity, high sensitivity, useful for volatile and semivolatile compounds, reduces the probability of false positive signal interpretation	Chemical structure (i.e. the nature of the bond to the heteroatom) influences the response to the elements
Infrared detector	IRD	Lubrication oil, industrial solvents, TCDD, brominated DDs		Provides complementary information when used with MS	Sensitivity, thermally sensitive
Mass spectrometer	MS, MSD, ITD	Dioxins, semivolatile organic compounds, herbicides, insecticides, fungicides, aliphatic hydrocarbons, polychlorinated biphenyls, tetrachlorobenzyltoluenes	1964	Full mass spectra for structure elucidation Sensitivity and selectivity when using selected ion monitoring (SIM) or multiple ion detection (MID) Permits the use of labelled internal standards for increasing accuracy and precision	Sensitivity of MS to CBs decreases with increasing chlorine number

[a]Based on a modification to the argon ionisation detector which was introduced in 1958 (Ettre, 1975).
[b]The FPD was first presented in 1962 and developed for used in GC in 1966 (Blomberg, 1976).
[c]The initial TCD was the katharometer.
[d]First commercial instrument. GC-AED was first reported in 1965 (Pedersen-Bjergaard and Greibrokk, 1996).
[e]PBBs, polybrominated biphenyls.

create an explosive mixture, and hence a serious, major leak. Thus, hydrogen is the preferred carrier gas for capillary GC, but helium and nitrogen are frequently used.

Impurities in the carrier gas should be minimised. This, in part, can be achieved by using a high-purity gas. Stan and Linkerhägner (1993) used helium of 99.999% purity when using an ECD, NPD and AED for the analysis of pesticides. In addition to the use of high-purity gas, however, gas scrubbers or absorbers are used. The most common are charcoal filters for hydrocarbons etc., water traps and oxygen traps. Sommer *et al.* (1997), for example, used ultrapure helium and cleaned it with an O_2 absorber, ANOXY-CIL, and a water filter when analysing PCDDs and PCDFs by GC-MS-FTIR.

2.3.5 Derivatisation

The primary prerequisite for a compound to be chromatographed by GC is that it is volatile or semivolatile and will partition between the gas and the stationary liquid phase over a temperature range between ambient and $\sim 350°C$. Highly polar or ionic compounds tend to be less volatile so, in addition to being more difficult to match the chromatographic requirements of the determinand with the performance of the column, they are less likely to pass readily through the column. Many of the environmental contaminants identified in foods are non-polar and do not need to be derivatised prior to analysis by gas chromatography. There are, however, occasions where derivatisation of the isolated contaminant is a part of the analytical procedure. Generally, derivatisation of a compound is done to improve volatility, to improve separation and/or resolution, and to improve the detection of the non-derivatised (parent) compound by a specific detector. PAHs are generally determined directly by GC-MS, GC-FID (or HPLC-fluorescence) as detailed in Chapter 10. As an alternative to this, bromination of the PAHs, using Br_2 with an iron catalyst, has been used so as to allow the use of an ECD, which greatly improved the detection limits relative to the FID (Seym and Parlar, 1991). Ting and Lee (1995) derivatised triclopyr, a systemic herbicide used for the control of woody plants and broad-leaved weeds in grasslands, uncultivated lands and rice fields, also with the objective of enhancing detector response. Additionally, however, they were attempting to improve compound resolution.

With the increase in the thermal stability and coating of the polar GC liquid phases it has been possible to chromatograph highly polar compounds directly without derivatisation, e.g. 2-ethyl hexanol (Webster *et al.*, 1997b). Such developments have generally reduced the necessity to include a derivatisation step in the analysis. Derivatisation was generally hampered by the need to carefully optimise the reaction conditions and the purity of the reagents. Although the methods could be automated, the sensitivity and the shelf-life of the reactants invariably required the system to be well maintained with the fresh preparation of the reagents on the day of use.

2.4 Mass spectrometry

To say that a mass spectrometer is simply a detector for gas chromatography would cause offence to many people and is, of course, not factually correct since it is now an LC detector! By the same token, to say that GC is simply a sample delivery system for a mass spectrometer would be equally wrong. The two techniques do, however, complement each other. A high-efficiency gas chromatographic column can separate a complex mixture into individual components that can then be detected in such a way as to produce a positive electron ionisation mass spectrum for structural confirmation, a molecular ion and, by using high-resolution accurate mass measurement, a molecular formula. The sensitivity can be enhanced by using selective ion monitoring (SIM), and moving to high-resolution SIM further improves sensitivity such that detection of 0.015 ng/g CBs in a food matrix can be achieved (Sippola and Himberg, 1991). Mass spectrometry is unique in as much as it is not only a quantitative detector but is equally powerful as a qualitative detector, providing detailed structural information. For environmental contaminants in food, however, the technique is very much used quantitatively. The variety of techniques available for quantitative analysis, and covered by the general heading of 'mass spectrometry', is large. A small number of examples are cited below.

- Low-resolution quadrupole instruments including the ion trap detector for bromocyclin (Bethan *et al.*, 1997), the MSD for PAHs (Brindle and Li, 1990; Louter *et al.*, 1996) or other quadrupole instruments for CBs and PCDD/Fs (de Boer *et al.*, 1993a).
- Quadrupole MS for high molecular mass PAHs such as dibenzo[*a,j*]coronene, which has a molecular mass of 400 Da, and other 9-ring PAHs with molecular masses of up to 426 Da (Bemgård *et al.*, 1993).
- High-resolution (10 000) mass spectrometry using a double focusing instrument for the determination of 2,3,7,8-TCDD (Brown *et al.*, 1989), PCDD and PCDF (Beck *et al.*, 1989) and CBs (Böhm *et al.*, 1993).

- Low-resolution mass spectrometry can be used for the analysis of chlorinated dibenzo-*p*-dioxins in conjunction with high-resolution GC-GC (Lamparski *et al.*, 1990).
- MS/MS for toxaphene (Buser and Müller, 1994).
- Ion trap for MS/MS of toxaphene (Chan *et al.*, 1998).
- For a specific contaminant there may be several options for the MS technique, as for example with toxaphene where Lau *et al.* (1996) compared electron capture negative ion (ECNI) mass spectrometry with electron ionisation high-resolution mass spectrometry (EI-MS).

The application of MS to environmental contaminants is considerable and the provision of more detailed information is outwith the scope of this chapter, but further reading is recommended, specifically the reviews presented in *Analytical Chemistry*, the most recent of which is by Burlingame *et al.* (1996).

2.5 Quantification

Once the gas chromatographic elution time of each determinand has been characterised by comparison with the retention index of a standard solution of known composition and/or by reference to the mass spectral library (RSC, 1991; NIST, 1998), then a method to quantify that compound can be established by direct comparison of the instrumental response of the known concentration to that of a characterised standard solution of known concentration.

2.5.1 External standards and calibration solutions

External standards are primarily used to calibrate the detector of the GC on a regular basis to confirm the response factor of each determinand prior to making the measurement of the unknown concentration of that determinand. These external standards are normally prepared as calibration solutions from crystalline solids or liquids of high purity (> 99%). In some cases these compounds are available as Certified Reference Material (CRMs) (National Research Council of Canada (NRCC) http://www.nrc.ca; Food and Drug Administration (FDA) http://www.fda.gov/ora/science_refs/lpm/-; Environmental Protection Agency (EPA) http://www.epa.gov) or as Standard Reference Materials (SRMs) from the National Institute for Standards and Technology (NIST, http://ts.nist.gov/srm).

The preparation and control of the calibration solutions are essential to ensure that the bias and precision of the analysis are minimised and that there is comparable between-laboratory agreement for these determinations (Wells *et al.*, 1992). Stock solutions should be checked either against the previous calibration solutions or against an independent solution of known, documented origin. Where neither of these options is possible, a duplicate set of primary standards should be prepared by two separate analysts and compared. Only in such a way can the errors associated with the preparation of the calibrants be kept to a minimum. Details of the preparation, control and storage of the calibrants are available for most environmental contaminants (Wells *et al.*, 1992; Wells, 1993a, b).

The concentration of the external standards should also be prepared and controlled by weight. This enables the solutions to be prepared easily in mixed solvents, where necessary, and eliminates the need for any temperature corrections. The following key information and documentation on these calibrants is required.

- Where there is no certified or well-documented information, then the purity of the solid calibrant should always be checked.
- The shelf-life and storage conditions of the calibrant are particularly important for photosensitive compounds like PAHs or thermally and hydrolytically unstable materials such as organophosphorus (OP) pesticides. For such materials the *expiry date* of the solutions should be clearly stated. This is particularly important for the dilute working standard which may, at best, only be sufficiently stable over a period of a few weeks. In some cases, for selected OPs it is necessary to prepare fresh working standards prior to the calibration of the instrument and analysis of each batch of prepared samples. The purity of the solvent used to prepare the standards should be known and checked by concentrating ×100 and analysing by ECD and FID.
- The GC detector should be calibrated with sufficient frequency to ensure that the response factor for each determinand is within ~±5% or better of the actual response at any time. Some detectors, such as the ECD, require frequent calibration. This may be as often as with each batch of 10–20 samples. In contrast, other detectors may only require re-calibration after maintenance. The frequency of the calibration is primarily a function of the cleanliness of the samples that are analysed. Insufficient clean-up in the sample preparation stages can seriously affect both the resolution of the capillary column and the sensitivity of the detector.

Since many capillary GC analyses can take up to 1–2 hr per sample to complete, it is essential to minimise the time taken to calibrate the GC and run the system checks (e.g. blanks).

2.5.1.1   Single point calibration

The different GC detectors have widely differing dynamic and linear ranges. Single point calibration is therefore rarely appropriate for any GC detector. All injector/column/detector systems have a finite limit of detection and therefore the calibration is unlikely to pass through the origin. Therefore the origin cannot automatically be used as one of the calibration points. Single point calibration can be used when the response factor of the calibration solution is within ±5% of that of the compound in the sample. This approach is valid where a screening or semiquantitative measurement is required and the concentration of the determinand in the samples is relatively constant. It is also essential that the dynamic range of this type of calibration is restricted to the linear detector response. The main advantage in this approach is that it overcomes the need to curve-fit the detector response. However, the calibration is always made retrospectively since the calibrant response is required to closely match that of the sample. This calibration process rapidly becomes time consuming when there are a large number of determinands in the sample that occur over a wide concentration range.

2.5.1.2   Bracketing concentration calibration

The next type of calibration curve uses a set of bracketing standards that are set at the upper and lower limits of the selected working range of the detector. The response factor of the detector for each compound must be known and must be constant over the range selected. The determinand is measured only if the detector response lies within the upper and lower masses injected from the bracketing standards. A minimum number of calibrants is used to cover a wide concentration range. However, the linearity of the detector must be verified over the range used *and* these linearity checks must be verified with time and use.

2.5.1.3   Multipoint calibration

The more usual type of calibration uses between three and ten separate calibration points to describe the detector response curve. The upper and lower masses injected provide the dynamic range and the distribution of the calibration points is placed over the range to allow a curve-fit of the detector response. The types of curve fitting usually depend on the type of algorithms used in the data analysis software of the GC. Usually, the software allows a number of options including cubic spline and quadratic fitting. It is also possible that a *point to point* linear fit can be used. One advantage of the latter is that if the range of the determinand concentration is very narrow then the calibration can revert to the previous style of bracketing standards.

2.5.2   Internal standards

Internal standards are added to the sample or the sample extract to reduce the effect of the variance or the bias of either the whole method during extraction, clean-up, chromatographic separation and detection or the variability of the instrumentation at the final detection. Ideally, the internal standard should reflect the physicochemical characteristics of the determinand. The compounds that most clearly reflect these requirements are labelled (^{13}C, ^{2}H, ^{17}O) homologues. These homologues have all been used to provide labelled compounds that are detected using single or multiple ion monitoring by mass spectrometry. The main disadvantages of this approach are the availability and cost of the labelled compounds and the need to have the MS as the detector. This is a particular problem when there are a large number of individual compounds in any one single analysis, e.g. PAHs, CBs, PCDDs and PCDFs.

The variability of instrumental analysis at the ultra-trace level ($<$ pg/kg) is usually very high due to adsorption losses. It is practically mandatory to use labelled homologues to reduce this variability (Rymen *et al.*, 1992). In such cases the most reliable methods recommend using ^{2}H-labelled homologues with high resolution (HR) GC interfaced with HRMS.

It is not always possible or convenient to use a labelled analogue for every compound determined, owing to the lack of availability or the cost of the materials. The next group of internal standards that have been used are close structural analogues or isomers to the determinand, such as *o,o'*-DDE for the *p,p'*-DDE or ε-HCH for the α-,β-,γ-HCH pesticides. However, there are often a limited number of suitable options with structural isomers and their addition requires adequate chromatographic separation in what is often a crowded chromatogram. The separation must be made not only between the analyte and the internal standard but also from other co-eluting compounds. The addition of such analogues is particularly applicable when the analytes in the group are similar; for example, CB 53 has been used as an internal standard for chlorobiphenyls. Attempts to use internal standards to correct for recovery for quite different compound types can lead to a greater variance being introduced into the analysis than if there were no recovery correction at all. Many of the organochlorine pesticides (OCPs) are determined together in one chromatographic fraction, but owing to the considerable differences in the physicochemical parameters it is not good practice to use a single representative compound, e.g. ε-HCH to correct for all compounds present as given in Table 2.7 (D. Pirie, personal communication). In such cases the most suitable alternative is to undertake a series of recovery tests using a spiked addition of the external standard for each of the determinands and obtain a level of the recovery at the 10% level of the range and at ∼75% of the range.

Most of the variability of the measurement still comes from the accuracy of the calibration of the GC and the calibration compounds. It is possible to reduce the effects of the instrumental variability with the addition of specific markers that can be used both as an internal standard and as retention index markers (Wells *et al.*, 1985; Morosini and Ballschmiter, 1994). The 2,4-dichlorobenzyl alkyl ethers (Wells *et al.*, 1985) and the 2,4,6-trichlorophenyl alkyl ethers (Morosini and Ballschmiter, 1994) are detectable by FID, ECD and AED as well as having a suitable intense base m/z for SIM.

2.5.3 Methods of data analysis

2.5.3.1 Manual methods

The quantification is made by the direct comparison of the calibrated response of the detector with the response of the same determinand of unknown concentration. Most GCs now have integrated data systems and software that controls and provides all data acquisition, integration and data output as well as controlling the GC programme conditions. It is often easy to allow the system to control all decision making, which can result in the analyst (i) being unaware of the actual numerical manipulations and (ii) not developing or maintaining the necessary skills

Table 2.7 Comparison of percentage recovery of organochlorine pesticide (OCP) residues relative to ε-HCH and historical data from spiked addition of the actual OCP

Compound in order of GC elution time	Relative to ε-HCH	Historical recovery
α-HCH	97	101
γ-HCH	103	100
ε-HCH (IS)[a]	100	100
2,4-DDE	111	96
Dieldrin	116	101
2,4'-DDD	122	105
2,4'-DDT	128	100
4,4-DDT	132	99

[a]IS, Internal standard

to make the appropriate checks and, where necessary, corrections. Prior to the development of the computerised data systems, the output from the detector was made by chart recorder. From these manual methods came the integrator, which gave a printout of the chromatogram as well as producing the integrated table of results. The results from the chart recorder were calculated by hand, either by using the measured peak height or by cutting and weighing the area of paper under each peak. Although it would be a highly retrograde step to revert to such practices, it is important that the analyst not only is aware of the details of the calculation but also has a programme which checks the numerical manipulations as well as making a series of manual checks. This is still currently a requirement for laboratories that wish to be accredited for GC analytical methods by the United Kingdom Accreditation Service (UKAS) to confirm the instrumental operation. Many of the gross errors (e.g. $\times 1000$, 100, 5, and 2) that occur in the results from External Quality Assurance Schemes such as the Food Analysis Performance Assessment Scheme (FAPAS) and the Quality Assurance of Information in Marine Environmental Monitoring in Europe (QUASIMEME) (Wells and Cofino, 1995) occur as result of incorrect reporting units or calculation errors.

All automatic integration of complex chromatograms should be inspected visually. The main source of errors come from incorrect setting of the sensitivity of the peak start and, more importantly, the peak finish positions. The latter is affected by incomplete compound separation and peak tailing from column degradation. Often the quality of the chromatography is different when going from standard to sample, owing to inadequate clean-up and the presence of co-eluting compounds that can interfere with the measurement in the real sample. Shoulders and unresolved peaks cause additional difficulty, both in deciding the shape of the area under the peak and the repeatability of the peak shape containing unresolved compounds. Decisions whether to drop perpendicular lines or tangents to the shoulder can be undertaken by the algorithm of the data system, but the main difficulty is in the reproducibility of the peak shape which, for actual samples, is often affected by the state of the capillary column, which is modified by the previous samples. Some of these problems can be overcome by a more rigorous clean-up procedure, either by tailoring the sample preparation to the specific determinand rather than having a multiresidue scheme (Wells, 1993b) or by using multidimensional GC to eliminate the co-extracted interference. If these options are not available, then it is also possible to make a spiked addition of the standard to have a similar integration for the sample and the standard.

2.5.3.2 *False positives and false negatives*

For most GC detectors, with the exception of MS in the total ion mode or multiple ion monitoring, the presence of a response is not immediately confirmatory that the specific compound is present. One of the weaknesses of these detectors is that false positives can easily be recorded if additional qualitative examinations are not made on alternative column phases or by MS. Alternative confirmatory analyses are now mandatory requirements for the characterisation and certification of reference materials and are given as clear recommendations in protocols for Laboratory Performance Studies. Impurities in the sample, solvent or GC system, e.g. from the septum during injection, can provide false positive values either in addition to the determinand or in place of the determinand. This problem can occur at all levels including CRMs. In one instance, a fish oil certified for CB 138 was retrospectively reassessed as being a composite of CB 138 and CB163 (de Boer and Dao, 1991).

False negatives, or signal reduction, can also occur. Co-eluents, if present in sufficient quantities, can reduce the signal of the determinand. This has been observed during the measurement of organochlorine residues in mineral oil when there is sufficient hydrocarbon present from incomplete clean-up to reduce the ECD signal even though the hydrocarbons themselves were not detected. A similar situation can occur during single ion monitoring MS if the quenching co-extracts do not have a measurable fragment at the mass(es) that are being monitored (Griepink *et al.*, 1990).

2.5.4 *Performance criteria*

The GC performance data are an essential part of the validation of any method. Although the actual details of the performance data will be specific to each application and purpose, there are general categories common to most procedures.

The range of the GC instrument's capability is primarily dictated by the capacity of the stationary phase and the detector. Highly efficient thin-film columns (ca $< 0.1\ \mu m$) have been developed for maximum resolution (see manufacturers' catalogues for details) and can set the upper limit of the GC working range. The lower limit will be set by the intrinsic sensitivity of the detector. The limit of detection (LOD) is defined as 3σ of the baseline signal (Taylor, 1987) and the limit of the determination of quantification is set at 10σ. Other criteria have been used (Funk *et al.*, 1995), these being slightly more pragmatic, and include 'What is the minimum mass that can be detected with a stated precision?'. The magnitude of the precision of the measurement increases significantly at the lower detection signals. It is, therefore, usual to provide a *reporting limit* at a concentration below which the uncertainty of the measurement becomes unacceptable for the purposes of the analysis. The lower limits may be extended for the method by providing a larger sample mass. However, additional clean-up steps may also be required to eliminate the increased mass of the co-extractants. Lower final volumes of the sample extract also have finite limits. Final sample volumes have typically been ~1 ml. The introduction of micro-vials has allowed these volumes to be reduced to ~50–100 μl, with an increase in sensitivity of ×10–20. Although further reduction in the solvent volume is possible, the impurities in the solvent have been concentrated ×100–1000 and there is little to be gained in additional solvent reduction for quantitative analysis.

2.5.4.1   Precision
As a minimum, the precision of the instrumentation should be recorded at 10% and the 90% level of the working range. Although the precision may initially be obtained in the short term, within 1–2 sample batches, the long-term (months) instrument precision should also be obtained and updated over the lifetime of the equipment. Although most users provide precision data based on standard solutions, it is more realistic to have these data generated by actual sample extracts, since the presence of trace co-extractants often has a considerable influence on the column and detector performance.

2.5.4.2   Bias
The bias of an instrument calibrated against itself is close to zero. The effect(s) of sample extracts can be determined by spiked addition into different sample matrices. The bias of the whole method is measured by analysing samples of known concentration using either Certified Reference Materials (CRMs) or, in the absence of appropriate CRMs, a spiked addition of a field blank sample.

2.5.4.3   Accuracy
This is a combination of the precision (σ) and the bias (β), given by the formula

$$c \pm (|\beta| + 1.645\,\sigma_a^2)$$

and is discussed in detail in relation to trend analysis of environmental chemical contaminants by Nicholson (1994).

2.6 Quality control

Once an analytical procedure has been developed and applied in the laboratory and the performance criteria have been established, then it is necessary to maintain the validation by routine quality control (QC) procedures (Dux, 1990; Günzler, 1996) and appropriate external quality assurance (QA) schemes. Many of the applications for the measurement of environmental contaminants in food have been documented and published as standard methods by the AOAC (http://www.aoac.org/pubs/pubcat16.htm) and EPA (http//www.epa.gov/epahome/standards.html) (Table 2.5). The advantage of implementing a standard method is that there is a common point of reference and data should be directly comparable with a probability of a better between-laboratory agreement than if widely different methodologies are used. The main drawbacks are that it tends to stifle progress and improvement in methodology. New instrumentation, equipment and analytical techniques are not naturally included if only the standard methods are implemented, with any innovation being reserved for specific reference or research laboratories. Equally, it is possible that an intrinsic bias may go undetected or develop if laboratories only compare data on the basis of one method alone. Provided that an alternative method to any reference or standard method is fully validated, then its inclusion is likely to produce a more robust approach while allowing developments in methodology to be incorporated into routine laboratory measurements. There are recognised training programmes for analysts to develop valid analytical methods (VAM), organised by the Laboratory of the Government Chemist (UK).

In-house quality control over the precision of a method can be provided by the use of Shewhart charts. These charts typically use data from repeated analysis of well-characterised laboratory reference materials (LRMs). Laboratory reference materials should be well-characterised in terms of their homogeneity, stability and concentration of analytes (Dux, 1990). Where a CRM or an LRM is not available, replicates and spiked replicates can provide data for the R Charts (Dux, 1990) to act as a QC measure and a record of laboratory analytical performance. Control, evaluation and, where necessary, corrective improvement are essential aspects of a working quality system for the laboratory and its clients.

The organisation of an institute's policy that underpins the analytical procedures and laboratory records is formally embodied within the national Accreditation Standards. A common standard is now provided between national accreditation bodies in Europe through the European Cooperation for Accreditation of Laboratories (EAL POB 29152,

3100 GD, Rotterdam, The Netherlands). In addition to the in-house QC programmes, most accreditation bodies now require evidence of participation in an appropriate proficiency testing scheme (PT) or laboratory performance study. These external QA schemes provide suitable test materials to match the determinands, matrices and range of concentrations required for analysis, and a normalised assessment of the laboratory performance following the analyses of these blind test samples. Such schemes are organised centrally for the food sector by the Food Analysis Performance Assessment Scheme (FAPAS) operated in the United Kingdom by the Ministry of Agriculture Fisheries and Food (MAFF). Participation in a PT scheme should be part of the laboratory QA procedures, where successful participation should be expected rather than anticipated. PTs are *not* part of quality control, but provide the independent evidence for the overall quality assurance information for the analysis. The control and validation of procedures must come from within the laboratory with the independent conformation provided by the independent assessment .

2.7   Benefits, limitations and future perspectives

The key advantages of GC for the determination of environmental contaminants in food include:

(1) a wide range of volatiles and semivolatiles;
(2) different selective detectors, e.g. NPD, ECD, MS, which can easily be interfaced by direct coupling;
(3) high chromatographic resolution, high-efficiency columns, high thermal stability of a wide range of stationery phases of differing polarity;
(4) easy of automation, high reliability and reproducibility, cost effectiveness.

High-performance liquid chromatography (HPLC) provides a highly complementary technique that is particularly well suited to the determination of polar and ionic compounds, e.g. aflatoxins and paralytic shellfish poisons (PSP) and diarrhoetic shellfish poisons (DSP). This highlights one of the main limitations of GC in that it is only amenable to the separation of volatile and semivolatile compounds or those that can be derivatised to increase their volatility. Although this includes a large number of environmental contaminants, in itself it is only a small number of the total possible that are a potential hazard both to the environment and to human health from consuming contaminated food.

By comparison to GC, some other chromatographic techniques such as thin-layer chromatography (TLC) are at best semiquantitative, with problems associated with plate edge effects and a resolution synonymous with packed column technology. TLC is relatively quick and visual for general screening purposes, whereas GC, with its array of detector choice, provides a more definitive qualitative and quantitative analysis. Capillary electrophoresis (CE) currently has a limited application, primarily for more polar, ionisable determinands, and is hampered by the small quantities of material it can cope with.

Although the major developments in GC would appear to be historical and the technique is now a stable work-horse for many sectors, particularly in food analysis, there are still some valuable developments that will secure the future of this technique beyond the year 2000. Multidimensional GC (MDGC) provides a mechanism for a radical improvement in resolution and analysis time. Short columns (ca $< 10 \, m$) of differing polarity may be coupled in the one GC oven to provide switching of selected areas of the chromatogram of the first column to the second, more polar, column in a quantitative transfer both of the analytes and of any internal standards (J. de Boer, personal communication).

The power of the computer to control systems and interrogate data has led to development of electronic pressure control of the GC with retention time locking. The independent development of retention indices as a more definitive tool for peak identification can be coupled with the mass spectral library to provide an integrated approach to identifying each selected component in the chromatogram.

Today the development leaders in the field of GC analysis are less likely to be academic researchers but to be found among the manufacturers who have, over the last two decades, listened to the requirements in industry and provided the necessary tools in a rapid flow of developments. With the reduction in the time from innovation to final routine product, it is now more cost effective for these improvements to occur in one place.

One of the final frontiers is to match chemical detection and the measurement of the biological effect, not at the acute level but at the chronic and sub-chronic, long-term environmental effect level . Analytical methodology is currently adequate from the % to ppt (pg/g)/ppq (pg/kg) level. Most acute effects are rare, but prevalent at the 0.1% level or above. Chronic effects for known toxins extend to the ppb (ng/g) range. It is, therefore essential to keep the ability for specific chemical measurements in step with the developments in biological effects studies for both environmental and human health surveillance.

References

Asensio, J.S., Barrio, C.S., Juez, M.T.G. and Bernal, J.G. (1991) Study of the decay of diazinon and chlorpyrifos in apple samples, using gas chromatography. *Food Chemistry*, **42** 213-224.

Barcarolo, R. (1990) Coupled LC/GC: A new method for the on-line analysis of organochlorine pesticide residues in fat. *Journal of High Resolution Chromatography*, **13** 465-469.

Barrio, C.S., Asensio, J.S., Medina, M.P., Clavijo, M.P. and Bernal, J.G. (1995) Evaluation of the decay of malathion, dichlofluanid and fenitrothion pesticides in apple samples, using gas chromatography. *Food Chemistry*, **52** 305-309.

Beck, H., Eckart, K., Mathar, W. and Wittowski, R. (1989) PCDD and PCDF body burden from food intake in the Federal Republic of Germany. *Chemosphere*, **18** 417-424.

Becker, F., Lach, G. and Parlar, H. (1989) Analysis of insecticide 'Toxaphene' in fish products. *Toxicological and Environmental Chemistry*, **20–21** 203-208.

Bemgård, A., Colmsjö, A. and Lundmark, B.-U. (1993) Gas chromatographic analysis of high-molecular-mass polycyclic aromatic hydrocarbons. II. Polycyclic aromatic hydrocarbons with relative molecular masses exceeding 328. *Journal of Chromatography*, **630** 287-295.

Bentabol, A. and Jodral, M. (1995) Determination of organochlorine pesticide residues in cheese. *Journal of the Association of Official Analytical Chemists*, **78** 94-98.

Berset, J.D. and Holzer, R. (1994) Separation of important PCBs and PAHs on a prototype smectic liquid-crystalline polysiloxane stationary phase capillary column. *Chemosphere*, **28** 2087-2099.

Bethan, B., Bester, K., Hühnerfuss, H. and Rimkus, G. (1997) Bromocyclen contamination of surface water, waste water and fish from northern Germany, and gas chromatographic chiral separation. *Chemosphere*, **34** 2271-2280.

Blanch, G.P., Glausch, A., Schurig, V., Serano, R. and Gonzales, M.J. (1996) Quantification and determination of enantiomeric ratios of chiral PCB 95, PCB 132, and PCB149 in shark liver samples (*C. coelolepis*) from the Atlantic Ocean. *Journal of High Resolution Chromatography*, **19** 392-396.

Blomberg, L. (1976) Gas chromatographic separation of some sulphur compounds on glass capillary columns using flame photometric detection. *Journal of Chromatography*, **125** 389-397.

Böhm, V., Schulte, E. and Thier, H.-P. (1993) Polychlorinated biphenyl residues in food and human milk. *Zeitschrift für Lebensmittel Untersuchung und Forschung*, **196** 435-440.

Bøwadt, S. and Larsen, B. (1992) Improved congener-specific GC analysis of chlorobiphenyls on coupled CPSil-8 and HT-5 columns. *Journal of High Resolution Chromatography*, **15** 377-380.

Brindle, I.D. and Li, X.-F. (1990) Investigation into the factors affecting performance in the determination of polycyclic aromatic hydrocarbons using capillary GC/MS with splitless injection. *Journal of Chromatography*, **498** 11-24.

Brown, R.S., Pettit, K. and Jones, P.W. (1989) A GC/GC-MS study of 2,3,7,8-TCDD using a DB-225 column and selected decomposition monitoring as a quantitative tool. *Chemosphere*, **19** 171-176.

Burlingame, A.L., Boyd, R.K. and Gaskell, S.J. (1996) Mass spectrometry. *Analytical Chemistry*, **68** 599R-651R.

Buser, H.-R. and Müller, M.D. (1994) Isomer- and enantiomer-selective analyses of toxaphene components using chiral high-resolution GC and detection by MS/MS. *Environmental Science and Technology*, **28** 119-128.

Chan, H.M., Zhu, J. and Yeboah, F. (1998) Determination of toxaphene in biological samples using high resolution GC coupled with ion trap MS/MS. *Chemosphere*, **36** 2135-2148.

Childers, J.W., Wilson, N.K., Harless, R.L. and Barbour, R.K. (1992) Characterization of brominated and bromo/chloro dibenzo-*p*-dioxins and dibenzofurans by gas chromatography/matrix isolation-infrared spectrometry. *Chemosphere*, **25** 1285-1290.

Davies, I.L., Markides, K.E., Lee, M.L., Raynor, M.W. and Bartle, K.D. (1989) Applications of coupled LC/GC: a review. *Journal of High Resolution Chromatography*, **12** 193-207.

de Boer, J. (1995) Analysis and biomonitoring of complex mixtures of persistent halogenated micro-contaminants. PhD Thesis, Vrije Universiteit Amsterdam, IJMUIDEN, January, 1995.

de Boer, J. and Dao, Q.T. (1989) The analysis of individual chlorobiphenyl congeners in fish extracts on 0.15 mm i.d. capillary columns. *Journal of High Resolution Chromatography*, **12** 755-759.

de Boer, J. and Dao, Q.T. (1991) Analysis of seven chlorobiphenyl congeners by multidimensional gas chromatography. *Journal of High Resolution Chromatography*, **14** 593-596.

de Boer, J., Dao, Q.T., Daudt, M.J.M. and Wester, P.G. (1993a) Determination of mono-*ortho* substituted chlorobiphenyls by GC/ECD, GC/NCI-MS and multidimensional GC. *15th International Symposium on Capillary Chromatography, Riva Del Garda, Italy*, **2** 782-789.

de Boer, J., Stronck, C.J.N., Traag, W.A. and Meer, J. van der. (1993b) Non-*ortho* and mono-*ortho* substituted chlorobiphenyls and chlorinated dibenzo-*p*-dioxins and dibenzofurans in marine and freshwater fish and shellfish from the Netherlands. *Chemosphere*, **26** 1823-1842.

Desty, D.H. (1958) *Gas Chromatography*, Butterworth Scientific, London.

Duebelbeis, D.O., Kapila, S., Clevenger, T., Yanders, A.F. and Manahan, S.E. (1989a) A two-dimensional reaction gaschromatographic system for isomer-specific determination of polychlorinated biphenyls. *Chemosphere*, **18** 101-108.

Duebelbeis, D.O., Pieczonka, G., Kapila, S., Clevenger, T.E., Yanders, A.F. and Wilson, J.D. (1989b) Application of a dual column reaction chromatography system for confirmatory analysis of polychlorinated biphenyl congeners. *Chemosphere*, **19** 143-148.

Dux, J.P. (1990). *Handbook of Quality Assurance for the Analytical Chemistry Laboratory*, Van Nostrand Reinhold, New York.

Eiceman, G.A., Hill, H.H. and Gardea-Torresday, J. (1998) Gas chromatography. *Analytical Chemistry*, **70** 321R-339R.

Eisert, R. and Levsen, K. (1996) Solid-phase microextraction coupled to gas chromatography: a new method for the analysis of organics in water. *Journal of Chromatography A*, **733** 143-157.

Ettre, L.S. (1975) The development of gas chromatography. *Journal of Chromatography*, **112** 1-26.

Fehringer, N.V., Gilvydis, D.M., Walters, S.M. and Poole, C.F. (1992) Optimization of an electrolytic conductivity detector for the determination of toxic nitrogen-containing food contaminants separated by open tubular column gas chromatography. *Journal of High Resolution Chromatography*, **15** 124-127.

Fürst, C. (1990a) Pesticide analysis, in *Capillary Gas Chromatography in Food Control and Research* (eds R. Wittkowski and R. Matissek), Technomic, Lancaster, PA, pp 207-238.

Fürst, C. (1990b) Contaminant analysis, in *Capillary Gas Chromatography in Food Control and Research* (eds R. Wittkowski and R. Matissek), Technomic, Lancaster, PA, pp 239-276.

Funk, W., Dammann, V. and Donnevert, G. (1995) *Quality Assurance in Analytical Chemistry*, VCH, Weinheim.

Gordon, M.H. (ed.) (1990a) *Principles of Gas Chromatography in Food Analysis*, Ellis Horwood, Chichester.

Gordon, M.H. (1990b) Principles of gas chromatography, in *Principles of Gas Chromatography in Food Analysis* (ed. M.H. Gordon), Ellis Horwood, Chichester, pp 11-58.

Gort, S.M., Hoff, R. van der, Baumann, R.A., Zoonen, P. van, Martin-Moreno, J.M. and Veer, P. van't. (1997) Determination of *p,p′*-DDE and PCBs in adipose tissue using high-performance liquid chromatography coupled on-line to capillary GC—the EURAMIC study. *Journal of High Resolution Chromatography*, **20** 138-142.

Grainger, J., Patterson, D.G. Jr and Presser, D. (1989) Structure/retention time correlations for the 22 tetrachlorodibenzodioxin isomers by gas chromatography/matrix isolation Fourier transform infrared spectroscopy. *Chemosphere*, **19** 1513-1520.

Griepink, B., Maier, E.A., Muntau, H. and Wells, D.E. (1990) *The certification of the content of CBs in waste mieral oil CRM 420*, Community Bureau of Reference, Rue De La Loi 200, B-1049 Brussels, BCR/139/90.

Grob, K. (1981) Twenty years of glass capillary columns, in *Recent Advances in Capillary Gas Chromatography* (eds W. Bertsch, W.G. Jennings and R.E. Kaiser), Dr. Alfred Hüthig Verlag, Heidelberg, pp 83-96.

Grob, K. (1994) Injection techniques in capillary GC. *Analytical Chemistry*, **66** 1009A-1019A.

Grob, K., Artho, A., Biedermann, M. and Egli, J. (1991) Food contamination from hydrocarbons from lubricating oils and release agents: determination by coupled LC-GC. *Food Additives and Contaminants*, **8** 437-446.

Grob, K. and Siegrist, C. (1994) Determination of mineral oil on jute bags by 20–50 µl splitless injections onto a 3 m capillary column. *Journal of High Resolution Chromatography*, **17** 674-675.

Günzler, H. (ed.) (1996) *Accreditation and Quality Assurance in Analytical Chemistry*, Springer, Berlin.

Harden, L.A., Garrett, J.H., Solch, J.G., Tiernan, T.O., Wagel, D.J. and Taylor, M.L. (1989) Results of comparative evaluations of various fused silica GC columns for retention of tetrachlorinated dibenzo-*p*-dioxins and dibenzofurans. *Chemosphere*, **18** 85-91.

Hernández, J.E., Machado, L.T., Corbella, R., Rodriguez, M.A. and Garcia Montelongo, F. (1995) *n*-Alkanes and polynuclear hydrocarbons in fresh-frozen and precooked-frozen mussels. *Bulletin of Environmental Contamination and Toxicology*, **55** 461-468.

Hess, P., Boer, J.de, Cofino, W.P., Leonards, P.E.G. and Wells, D.E. (1995) Critical review of the analysis of non- and mono-*ortho*-chlorobiphenyls. *Journal of Chromatography A*, **703** 417-465.

Himberg, K.K. and Sippola, E. (1993) Multidimensional gas chromatography–mass spectrometry with NCI detection and 13C-labelled internal standards: a novel approach to the separation and determination of coplanar PCB congeners. *Chemosphere*, **27** 17-24.

Husain, A., Naeemi, E., Dashti, B., Al-Omirah, H. and Al-Zenki, S. (1997) Polycyclic aromatic hydrocarbons in food products originating from locally reared animals in Kuwait. *Food Additives and Contaminants*, **14** 2952-2999.

Hyvönen, H., Auvinen, T., Riekkola, M.-L. and Himberg, K. (1991) Two-dimensional separation of PCB compounds in baltic Herring by on-line HPLC-HRGC. *13th International Symposium on Capillary Chromatography, Riva Del Garda, Italy*, 1289-1297.

James, A.T. and Martin, A.J.P. (1952) Gas–liquid partition chromatography: the separation and micro-estimation of volatile fatty acids from formic acid to dodecanoic acid. *Biochemical Journal*, **50** 670-690.

James, A.T. and Martin, A.J.P. (1956) Gas–liquid chromatography: the separation and identification of the methyl esters of saturated and unsaturated acids from formic acid to *n*-octadecanoic acid. *Biochemical Journal*, **63** 144-152.

James, A.T. and Webb, J. (1957) The determination of the structure of unsaturated fatty acids on a microscale with the gas–liquid chromatogram. *Biochemical Journal*, **66** 515-520.

Jickells, S.M., Crews, C., Castle, L. and Gilbert, J. (1990) Headspace analysis of benzene in food contact materials and its migration into foods from plastics cookware. *Food Additives and Contaminants*, **7** 197-205.

Kannan, B., Petrick, G., Schultz-Bull, D.E. and Duinker, J.C. (1993). Chromatographic techniques in accurate analysis of chlorobiphenyls. *Journal of Chromatography*, **642** 425-434.

Karasek, F.W. (1971) Plasma chromatography of polychlorinated biphenyls. *Analytical Chemistry*, **43** 1982-1986.

Kaufmann, A. (1997) Fully automated determination of pesticides in wine. *Journal of the Association of Official Analytical Chemists*, **80** 1302-1307.

Koistiainen, R., Rizzo, A. and Hesso, A. (1989) The analysis of trichothecenes in wheat and human plasma samples by chemical ionization tandem mass spectrometry. *Archives of Environmental Contamination and Toxicology*, **18** 356-364.

Komolprasert, V., Hargraves, W.A. and Armstrong, D.J. (1994) Determination of benzene residues in recycled polyethylene terephtalate (PETE) by dynamic headspace-gas chromatography. *Food Additives and Contaminants*, **11** 605-614.

Krock, K.A. and Wilkins, C.L. (1996) Qualitative analysis of contaminated environmental extracts by multidimensional gas chromatography with infrared and mass spectral detection (MDGC-IR-MS). *Journal of Chromatography A*, **726** 167-178.

Lach, G. and Parlar, H. (1991) Comparison of several detection methods for toxaphene residue analysis. *Toxicological and Environmental Chemistry*, **31–32** 209-219.

Lamparski, L.L., Nestrick, T.J., Janson, D. and Wilson, G. (1990) Development of an HRGC-HRGC-LRMS system: instrument design and performance data. *Chemosphere*, **20** 635-645.

Lansens, P., Leermakers, M. and Baeyens, W. (1991) Determination of methylmercury in fish by headspace gas chromatography with microwave-induced-plasma detection. *Water, Air, and Soil Pollution*, **56** 103-115.

Lau, B., Weber, D. and Andrews, P. (1996) GC/MS analysis of toxaphene: a comparative study of different mass spectrometric techniques. *Chemosphere*, **32** 1021-1041.

Louter, A.J.H., Ramalho, S., Vreuls, R.J.J., Jahr, D. and Brinkman, U.A.Th. (1996) An improved approach for on-line solid-phase extraction-gas chromatography. *Journal of Microcolumn Separations*, **8** 469-477.

Maris, F.A., Noroozian, E., Otten, R.R., Dijk, R.C.J.M. van, Jong, G.J. de, and Brinkman, U.A.Th. (1988) Determination of chlorinated biphenyls in sediment by on-line narrow-bore column liquid chromatography/capillary gas chromatography. *Journal of High Resolution Chromatography*, **11** 197-202.

Martin, A.J.P. and Synge, R.L.M. (1941) A new form of chromatogram employing two liquid phases. 1. A theory of chromatography 2. Application to the micro-determination of the higher mono-amino-acids in proteins. *Biochemical Journal*, **35** 1358-1368.

Martin, A.J.P. and James, A.T. (1956) Gas–liquid chromatography: the gas-density meter, a new apparatus for the detection of vapours in flowing gas streams. *Biochemical Journal*, **63** 138-143.

McCown, S.M. and Radenheimer, P. (1994) An equilibrium headspace gas chromatographic method for the determination of volatile residues in vegetable oils and fats. *LC-GC*, **7** 918-924.

McGill, A.S., Mackie, P.R., Howgate, P. and McHenery, J.G. (1987) The flavour and chemical assessment of dabs (*Limanda limanda*) caught in the vicinity of the Beatrice oil platform. *Marine Pollution Bulletin*, **18** 186-189.

Miyahara, M., Okada, Y., Takeda, H., Aoki, G., Kobayashi, A. and Saito, Y. (1994) Multiresidue procedures for the determination of pesticides in food using capillary gas chromatographic, flame photometric, and mass spectrometric techniques. *Journal of Agriculture and Food Chemistry*, **42** 2795-2802.

Morabito, P.L., McCabe, T., Hiller, J.F. and Zakett, D. (1993) Determination of PAHs in samples using large volume on-column injection capillary gas chromatography. *Journal of High Resolution Chromatography*, **16** 90-94.

Morel, G., Samhan, O., Literathy, P., Al-Hashash, H., Moulin, L., Saeed, T., Al-Matrouk, K., Martin-Bouyer, M., Salier, A., Paturel, L., Jarosz, J., Vial, M., Combet, E., Fachinger, C.

and Suptil, J. (1991) Evaluation of chromatographic and spectroscopic methods for the analysis of petroleum-derived compounds in the environment. *Fresenius Journal of Analytical Chemistry*, **339** 699-715.

Moret, S., Grob, K. and Conte, L.S. (1996) On-line high-performance liquid chromatography–solvent evaporation–high-performance liquid chromatography–capillary gas chromatography–flame ionisation detection for the analysis of mineral oil polyaromatic hydrocarbons in fatty foods. *Journal of Chromatography*, **750** 361-368.

Morosini, M. and Ballschmiter, K. (1994) 2,4,6-Trichlorophenyl alkyl ethers as retention index markers in capillary gas chromatography with electron capture and mass spectrometric detection. *Analytica Chimica Acta,* **286** 451-456.

Mossoba, M.M., Niemann, R.A. and Chen, J.-Y.T. (1989) Picogram-level quantitation of 2,3,7,8-TCDD in fish extracts by capillary gas chromatography/matrix isolation/Fourier transform infrared spectrometry. *Analytical Chemistry,* **61** 1678-1685.

Muiño, M.A.F. and Lozano, J.S. (1991) Mass spectrometric determination of pentachlorophenol in honey. *Analytica Chimica Acta,* **247** 121-123.

Nakamura, Y., Tonogai, Y., Sekiguchi, Y., Tsumura, Y., Nishida, N., Takakura, K., Isechi, M., Nakamura, M., Kifune, M., Yamamoto, K., Tersawa, S., Oshima, T., Miyata, M., Kamakura, K. and Ito, Y. (1994) Multiresidue analysis of 48 pesticides in agricultural products by capillary gas chromatography. *Journal of Agriculture and Food Chemistry*, **42** 2508-2518.

Newton, J.M., Rothman, B.S. and Walker, F.A. (1991) Separation and determination of diesel contaminants in various fish products by capillary gas chromatography. *Journal of the Association of Official Analytical Chemists*, **74** 986-990.

Nicholson, M.D. (1994) QA/QC in trend analysis of chemical contaminants, in *Analysis of Contaminants in Edible Aquatic Resources* (eds J.W. Kiceniuk and S. Ray), VCH, Weinheim.

Nilsson, T., Pelusio, F., Montanarella, L., Larsen, B., Facchetti, S. and Madsen, J.O. (1995) An evaluation of solid-phase micro-extraction for analysis of volatile organic compounds in drinking water. *Journal of High Resolution Chromatography*, **18** 617-624.

NIST (1998) National Institute of Standards and Technology, Standard Refernce Data Program, MS Windows NIST Mass Spectral Search Programme for the NIST/EPA/NIH Mass Spectral Library, Version 1.1a, National Institute of Standards and Technology, Gaithersburg, USA [www.nist.gov/srd/nist1a.htm].

Norman, K.N.T., Scudamore, K.A., Matthews, W.A. and Wilson, M.F. (1995) Determination of methyl bromide residues in stored foods using automated headspace gas chromatography. *Pesticide Science,* **44** 309-316.

Nyman, P.J., Perfetti, G.A., Joe, F.L. and Diachenko, G.W. (1993) Comparison of two clean-up methodologies for the gas chromatographic/mass spectrometric determination of low nanogram/gram levels of polynuclear aromatic hydrocarbons in seafood. *Food Additives and Contaminants*, **10** 489-501.

Overton, S.V. and Manura, J.J. (1995) Analysis of volatile organics in cooking oils by thermal desorption–gas chromatography–mass spectrometry. *Journal of Agriculture and Food Chemistry*, **43** 1314-1320.

Page, B.D. and Lacroix, G. (1993) Applications of solid-phase microextraction to the headspace analysis of halogenated volatiles in selected foods. *Journal of Chromatography A*, **648** 199-211.

Page, B.D. and Lacroix, G.M. (1995) On-line steam-distillation/purge and trap analysis of halogenated, nonpolar, volatile contaminants in foods. *Journal of the Association of Official Analytical Chemists International*, **78** 1416-1428.

Page, B.D. and Lacroix, G. (1997) Application of solid-phase microextraction to the headspace gas chromatographic analysis of semi-volatile organochlorine contaminants in aqueous matrices. *Journal of Chromatography A*, **757** 173-182.

Pan, T.-M., Wu, S.-J., Huang, K.-C., Wang, R.-T., Wang, C.-T. and Pan, T.-C. (1995) Determination of methylmercury in fish and shellfish. *Japanese Journal of Toxicology and Environmental Health*, **41** 364-366.

Pang, G.-F., Fan, C.-L., Chao, Y.-Z. and Zhao, T.-S. (1994) Rapid method for the determination of multiple pyrethroid residues in fruits and vegetables by capillary column gas chromatography. *Journal of Chromatography A*, **667** 348-353.

Pedersen-Bjergaard, S. and Greibrokk, T. (1996) Environmental applications of capillary gas chromatography coupled with atomic emission detection—a review. *Journal of High Resolution Chromatography*, **19** 597-607.

Picó, Y., Manes, J. and Font, G. (1994) Optimisation of experimental conditions for the identifications of pesticide mixtures on six GLC columns. *Journal of Chromatographic Science*, **32** 386-392.

RSC (1991) *Eight Peak Index of Mass Spectra*, 4th edn, Royal Society of Chemistry, Cambridge.

Roose, P. and Brinkman, U.A.Th. (1998) Determination of volatile organic compounds in marine biota. *Journal of Chromatography A*, **799** 233-248.

Rymen, T., Hinschberger, J., Maier, E.A. and Griepink, B. (1992) *The quantitative determination of PCDD and PCDF: improvement of the analytical quality up to a level acceptable for the certification of certified reference materials*, EUR-Report 14357-EN, CEC, Luxembourg.

Sadiki, A. and Williams, D.T. (1996) Speciation of organotin and organolead compounds in drinking water by gas chromatography–atomic emission spectrometry. *Chemosphere*, **32** 1983-1992.

Sanzo, F.P.di, Bray, W. and Chawla, B. (1994).Determination of the sulphur components of gasoline streams by capillary column gas chromatography with sulphur chemiluminescence detection. *Journal of High Resolution Chromatography*, **17** 255-276.

Schlemitz, S. and Pfannhauser, W. (1996) Monitoring of nitropolycyclic aromatic hydrocarbons in food using gas chromatography. *Zeitschrift für Lebensmittel Untersuchung und Forschung*, **203** 61-64.

Schulzki, G., Siegelberg, A., Bögl, K.W. and Schreiber, G.A. (1995) Detection of radiation-induced hydrocarbons in camembert irradiated before and after the maturing process—Comparison of Florisil column chromatography and on-line coupled liquid chromatography–gas chromatography. *Journal of Agriculture and Food Chemistry*, **43** 372-376.

Seym, M. and Parlar, H. (1991) Sensitive GC/ECD analysis of selected PAHs by bromination. *Toxicological and Environmental Chemistry*, **31–32** 227-233.

Shirey, R.E. (1995) Rapid analysis of environmental samples using solid-phase micro-extraction (SPME) and narrow bore capillary columns. *Journal of High Resolution Chromatography*, **18** 495-499.

Sippola, E. and Himberg, K. (1991) Determination of toxic PCB congeners in biological samples by multidimensional gas chromatography–mass spectrometry (GC/GC/MS). *Fresenius Journal of Analytical Chemistry*, **339** 510-512.

Snyder, J.M., King, J.W. and Nam, K.-S. (1996) Determination of volatile and semivolatile contaminants in meat by supercritical fluid extraction/gas chromatography/mass spectrometry. *Journal of the Science of Food and Agriculture*, **72** 25-30.

Sommer, S., Kamps, R., Schumm, S. and Kleinermanns, K.F. (1997) GC/FT-IR/MS spectroscopy of native polychlorinated dibenxo-*p*-dioxins and dibenzofurans extracted from municipal fly-ash. *Analytical Chemistry*, **69** 1113-1118.

Speer, K., Steeg, E., Horstmann, P., Kühn, T. and Montag, A. (1990) Determination and distribution of PAHs in native vegetable oils, smoked fish products, mussels and oysters, and Bream from the river Elbe. *Journal of High Resolution Chromatography*, **13** 104-111.

Stan, H.J. (1992) Two-dimensional capilary GC for residue analysis, in *Emergency Strategies for Pesticide Analysis* (eds Cairns and Sherma), pp 175-211.

Stan, H.-J. and Linkerhägner, M. (1993) Capillary gas chromatography—atomic emission detection: a useful instrumental method in pesticide residue analysis of plant foodstuffs. *Journal of High Resolution Chromatography*, **16** 539-548.

Taylor, J.K. (1987) *Quality Assurance of Chemical Measurements*. Lewis Publishers, Michigan.

Thompson, D.W. (1994) Determination of volatile organic contaminants in bulk oils (edible, injectable, and other internal medicinal) by purge-and-trap gas chromatography/mass spectrometry. *Journal of the Association of Official Analytical Chemists International*, **77** 647-654.

Ting, K.-C. and Lee, C.-S. (1995) Gas chromatographic determination of triclopyr in fruits and vegetables. *Journal of Chromatography A*, **690** 119-129.

Topping, G., Davies, J.M., Mackie, P.R. and Moffat, C.F. (1997) The impact of the Braer spill on commercial fish and shellfish, in *The Impact of an Oil Spill in Turbulent Waters; The Braer* (eds J.M. Davies and G. Topping), The Stationary Office, Edinburgh, pp 121-143.

Turrio-Baldassari, L., Domenico, A.D., Fulgenzi, A., Iacovella, N., Bocca, A. and Larsen, B.R. (1993) GC-MS determination of HCB, DDT, DDE, and selected PCB congeners in cow's milk. *Fresenius Environmental Bulletin*, **2** 370-374.

van Deemter, J.J., Zuiderberg, F.J. and Klinkenberg, A. (1956) Longitudinal diffusion and resistance to mass transfer as causes of nonideality in chromatography. *Chemical Engineering Science*, **5** 271-289.

Venema, A. and Jelink, J.T. (1996) Automated trace analysis at ppt level, using large volume injection. *Journal of High Resolution Chromatography*, **19** 234-236.

Vetter, W., Krock, B., Klobes, U. and Luckas, B. (1997) Enantioselective analysis of a heptachlorobornane isolated from the technical product melipax by gas chromatography/mass spectrometry. *Journal of Agriculture and Food Chemistry*, **45** 4866-4870.

Vieths, S. (1990) Headspace gas chromatography of highly volatile compounds, in *Capillary Gas Chromatography in Food Control and Research* (eds R. Wittowski and R. Matissek), Technomic, Lancaster, PA, pp 277-324.

Waliszewski, S.M., Pardio, V.T., Waliszewski, K.N., Chantiri, J.N., Aguirre, A.A., Infanzon, R.M. and Rivera, J. (1997). Organochlorine pesticide residues in cow's milk and butter in Mexico. *The Science of the Total Environment*, **208** 127-132.

Wang, M.J. and Jones, K.C. (1994) Occurrence of chlorobenzenes in nine UK retail vegetables. *Journal of Agriculture and Food Chemistry*, **42** 2322-2328.

Wang, Z., Fingas, M. and Li, K. (1994) Fractionation of a light crude oil and identification and quantitation of aliphatic, aromatic and biomarker compounds by GC-FID and GC-MS, Part I. *Journal of Chromatographic Science*, **32** 361-366.

Webster, L., Angus, N., Topping, G., Dalgarno, E.J. and Moffat, C.F. (1997a) Long-term monitoring of polycyclic aromatic hydrocarbons in mussels (*Mytilus edulis*) following the Braer oil spill. *Analyst*, **122** 1491-1495.

Webster, L., Mackie, P.R., Hird, S., Munro, P.D., Brown, N.A. and Moffat, C.F. (1997b) Development of analytical methods for the determination of synthetic mud base fluids in marine sediments. *Analyst*, **122** 1485-1490.

Welch, K.J. and Hoffman, N.E. (1993) Analysis of polychlorinated biphenyls, dibenzodioxins and dibenzofurans by on-line coupled microcolumn HPLC capillary GC/MS. *Journal of Liquid Chromatography*, **16** 307-313.

Wells, D.E. (1993a) Current developments in the analysis of polychlorinated biphenyls (PCBs) including planar and other toxic metabolites in environmental matrices, in *Environmental Analysis: Techniques, Applications and Quality Assurance* (ed. D. Barcelo), Elsevier Science, Amsterdam, pp 113-148.

Wells, D.E. (1993b) Extraction, clean-up and recoveries of persistent trace organic contaminants from sediment and biota samples, in *Environmental Analysis: Techniques, Applications and Quality Assurance* (ed. D. Barcelo), Elsevier Science, Amsterdam, pp 79-109.

Wells, D.E. and Cofino, W.P. (1995) A holistic structure for quality management: a model for marine environmental monitoring, in *Quality Assurance in Environmental Monitoring, Sampling and Sample Pretreatment*, VCH, Weinheim, pp 255-290.

Wells, D.E., Gillespie, M.J. and Porter, A.E.A. (1985) Dichlorobenzyl alkyl ether homologues as retention index markers and internal standards for the analysis of environmental samples using capillary gas chromatography. *Journal of High Resolution Chromatography and Chromatographic Communications*, **8** 443-449.

Wells, D.E., Maier, E.A. and Griepink, B. (1992) Calibrants and calibration for chlorobiphenyl analysis. *International Journal for Environmental and Analytical Chemistry*, **46** 255-264.

Wennrich, L., Efer, J. and Engewald, W. (1995) Gas chromatographic trace analysis of underivatized nitrophenols. *Chromatographia*, **41** 361-366.

Wittkowski, R. (1990) Fundamental aspects of capillary gas chromatography, in *Capillary Gas Chromatography in Food Control and Research* (eds R. Wittkowski and R. Matissek), Technomic, Lancaster, PA, pp 1-49.

Wittkowski, R. and Matissek, R. (eds) (1990) *Capillary Gas Chromatography in Food Control and Research*, Technomic, Lancaster, PA.

Wong, S.-K. and Lee, W.-O. (1997) Survey of organochlorine pesticide residues in milk in Hong Kong 1993–1995. *Journal of the Association of Official Analytical Chemists*, **80** 1332-1335.

Wu, J., Wong, M.K., Lee, H.K., Shi, C.Y. and Ong, C.N. (1997) Determination of polycyclic aromatic hydrocarbons in *Rougan*, a traditional Chinese barbecued food, by capillary gas chromatography. *Environmental Monitoring and Assessment*, **44** 577-585.

Yang, F.J. and Cram, S.P. (1981) Characteristics and performance of gas chromatographic detectors with glass capillary columns, in *Recent Advances in Capillary Gas Chromatography* (eds W. Bertsch, W.G. Jennings and R.E. Kaiser), Dr. Alfred Hüthig Verlag, Heidelberg, pp 57-82.

3 Instrumentation and application of HPLC for the analysis of food contaminants

Wooderck S. Hawer and Kil-Hwan Kim

3.1 Introduction: historical perspective

Liquid chromatography represents the origin of the present-day so called chromatographic kingdom. In some respects, liquid chromatography (LC) is less developed than gas chromatography (GC). However, LC has many advantages not only in the analysis of organic or non-organic components but also in preparative clean-up of samples for further analysis by GC, GC coupled with mass spectrometry (GC-MS), high-performance (or high-pressure) liquid chromatography (HPLC) and other analytical and identification methods. This chapter provides a review of the essential features of liquid chromatography rather than a detailed presentation of the analytical methods for contaminants in food. The appropriate methods are included in succeeding chapters. This chapter presents an historical perspective, and deals with instrumentation, derivatization, clean-up and isolation techniques, benefits and limitations.

The history of classical chromatography dates back to 1906. Michael Tswett (1872–1919) was the originator of chromatography as it is practised today. His first paper (Tswett, 1906), described the separation of plant pigments as a 'chromatogram' and the corresponding methodology as a 'chromatographic method'. However, in the strictest sense, the term 'chromatography' is a misnomer these days. Most of the materials chromatographed are colourless or, even if they are coloured, cannot be seen inside the separation system with the naked eye. Consequently, chromatography should be defined as a separation technique in which a mixture is separated by ion-forces, differential sorption, the partition ratio between stationary and mobile phases, or the molecular size of the mixture.

The chromatographic kingdom may be divided into three families according to the mobile phase (Figure 3.1). The oldest is LC, which was followed in 1952 by GC (Chapter 2), and the most recent development is supercritical fluid chromatography (SFC). GC is subdivided into gas–solid chromatography (GSC) and gas–liquid chromatography (GLC). SFC was considered a promising new technique in analytical chemistry. However, it is not yet in widespread routine use in laboratories for analysis of environmental food contaminants, although supercritical fluid

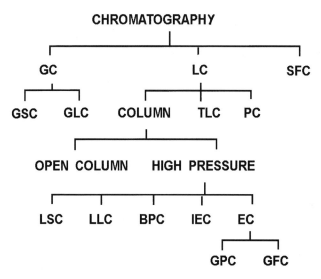

Figure 3.1 The classification of chromatography. LC, liquid chromatography; GC, gas chromatography; SFC, supercritical fluid chromatography; GSC, gas–solid chromatography; GLC, gas–liquid chromatography; TLC, thin-layer chromatography; PC, paper chromatography; LSC, liquid–solid chromatography; LLC, liquid–liquid chromatography; BPC, bonded-phase chromatography; IEC, ion exchange chromatography; EC, exclusion chromatography; GPC, gel permeation chromatography; GFC, gel filtration chromatography.

techniques have been applied, for example, to the multiresidue analysis of pesticides (Chapter 7) in foods and to phthalate plasticizers (Marin *et al.*, 1998). Liquid chromatography is much more diverse than the other two families. It may be divided into column chromatography, paper chromatography (PC) and thin-layer chromatography (TLC). Column chromatography is further subdivided into open column and HPLC. Nowadays, LC is often used as a synonym for HPLC. Liquid–liquid chromatography (LLC) is partition chromatography. The analytes in the sample are retained to a greater or lesser extent by differential partitioning between the mobile and stationary phases. In liquid–solid chromatography (LSC), the solid phases, such as silica gel, alumina, molecular sieves or porous glass, are packed in the column to adsorb the components in the sample. In bonded phase chromatography (BPC), an organic stationary phase is chemically bonded to silica. In this type, various functional groups such as nitrile, amino and ester groups are bonded to achieve special selectivity. The most popular is the octadecyl group, which consists of a C_{18} hydrocarbon covalently bonded to the surface. Ion exchange chromatography (IEC) is commonly used to analyse amino acids and other charged molecules. In this system, zeolite and synthetic organic or inorganic resins perform separation by an exchange of ions between the sample and the resins. Consequently, compounds that have different affinity to the resin can be separated. In exclusion chromatography (EC), highly porous polymers are used to separate the sample by molecular size. The smaller molecules are retained as a result of their entering into the polymer network, while the large molecules, which are excluded from entering the network, are swept out more rapidly by the eluent; the largest molecules are eluted first, medium size molecules next, and the smallest last.

HPLC is not restricted to analytical separation. It is frequently used in the preparative separation of a substance from a mixture, in the study of reaction kinetics, in molecular structural investigations, and so on. Recently, clean-up by gel permeation chromatography (GPC) has been applied widely to extracts of food and environmental matrices. The method is cost-effective; it has the potential to provide reliable data; detection limits are usually adequate for most purposes; and maintenance and calibration costs can be minimized. Successful use of LC depends on choosing the right combination of operating conditions: the type of column packing, solvents for elution, the length, diameter, and operating pressure of the column, separation temperature, sample size, flow rate, and so on. This in turn requires a basic understanding of the various factors that should be controlled in LC separation.

Following Tswett's early work, chromatography was applied to the separation of other small molecules, most notably to amino acids (Martin and Synge, 1941). Tiselius made great progress in the development of LC, carrying out very important research into the principles of frontal analysis, elution analysis and displacement development (Tiselius, 1941). IIe also introduced the continuous monitoring of column effluent by measuring refractive index and developed an automatic recording system. Consden et al. (1944) developed modern paper chromatography. They described the technique, its application for amino acids analysis, and the underlying theory. The possibility of gas chromatography was also described by Martin and Synge in 1941, but the first gas–liquid chromatograph was developed by James and Martin in 1952. The theory of chromatography was greatly advanced by Van Deemter et al. (1956). They developed rate theory, expressing column efficiency as a function of the flow velocity of the mobile phase.

In spite of the rapid development of gas chromatography, in the field of liquid chromatography it was not until the late 1950s that Spackman et al. (1958) developed the amino acid analyser. This work was further developed by Hamilton (1963) and Giddings (1965) with their contributions to the basic theory of HPLC. In the late 1960s, wide-ranging research on instruments, columns and theory was carried out by many research groups. Many reviews have traced the development of modern liquid chromatography, theory and practice—for example, Kirkland (1971), Scott (1983) and Gilbert (1987).

The application of chromatographic techniques to the separation of macromolecules did not begin to gain acceptance until the 1950s, with the introduction of a universally applicable gel medium for gel filtration by Porath and Flodin (1959). In this system, separation is based on molecular size. The technique has now become the method of choice for the separation of macromolecules. Sephadex and agar gels are hydrophilic and are used for aqueous applications, while hydrophobic polystyrene gels have been developed for non-aqueous applications. The first GPC instrument, described by Moore (1968), revolutionized the analysis of macromolecules by enabling separation by relative molecular mass from thousands to millions.

3.2 Instrumentation for HPLC

For a better understanding and interpretation of HPLC separation, it is advisable to understand the basic principles and instrumentation. In an HPLC system the arrangement and combination of its several components is very flexible and versatile. The components for a modern HPLC system are very different from the relatively simple and unsophisticated equipment used for the early, classical, liquid chromatographic separations. The components for modern LC can be assembled in various combination from items that are available from a variety of manufacturers or suppliers. The more complex systems use two or three pumps with a similar number of detectors. However, it is also possible to purchase completely assembled packaged systems from various manufacturers.

The liquid chromatograph consists of a solvent reservoir, pump(s), sample injector, chromatographic column, detector(s) and data-handling equipment (Figure 3.2).

3.2.1 Solvents and solvent reservoir

Solvents are used as the mobile phase in LC. In GC, the mobile phase contributes little to the elution of the components in the sample. However, in LC the effects of the mobile phase are so powerful that the elution order can be changed by changing the solvent, which can range from organic solvents to aqueous buffer solutions, from non-polar to polar solvents, and from strongly acidic to basic solutions.

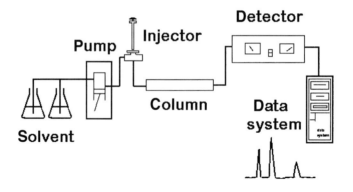

Figure 3.2 The components of a liquid chromatograph.

Particular care is required to select a solvent of sufficient purity, because some solvents are stabilized by small amounts of certain reagents that may affect the separation. Solvents employed in HPLC must be compatible with the detectors and columns, be safe to use, be relatively unreactive and be of the required purity. The reservoir generally holds 0.5 to 1.0 litre of mobile phase. To remove fines or dust particles suspended in the solvent, a porous, stainless steel frit is placed at the reservoir end of the Teflon tubing delivering solvent to the pump.

Dissolved gas in the mobile phase can interfere with the chromatographic procedure by de-gassing and bubbling in the pump and detectors. Some reservoirs are designed so that the mobile phase may be de-gassed in the system. Helium sparging is the most popular de-gassing method. Helium de-aeration is very simple and effective but rather expensive. Nearly 90% of the dissolved air in the mobile phase can be removed within 10 min. Vacuum de-gassing is the second most popular method; however, special attention should be paid not to disturb the equilibrium of solvents presaturated with water or organic stationary phases. An on-line de-gasser using special fluoropolymer membrane tube to remove O_2 and N_2 under vacuum (Figure 3.3) has become available. Ultrasonic de-gassing can also be used to de-gas the mobile phase. It is less effective compared with helium or vacuum de-gassing. Sometimes, ultrasonic de-gassing may actually increase the air levels in the solvent, so that sonic de-gassing should not be used in conjunction with an electrochemical detector that is sensitive to oxygen. Ultrasonic de-gassing is often combined with vacuum de-gassing. Heating is another possibility for de-gassing, but it is the least effective and the most inconvenient approach.

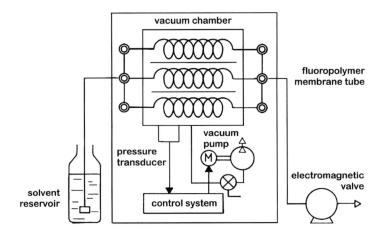

Figure 3.3 Block diagram of JASCO de-gasser.

3.2.2 Pumps

The function of the pump in an HPLC system is to supply a constant and reproducible flow of the mobile phase through the system without any pulse or fluctuation of the flow. The column is packed with small particles of micrometre size, which cause such a considerable resistance to flow that high-pressure pumps are required. The maximum pressure used is about 34.5 MPa (5000 psi). Typical flow rates range between 0.1 and 10 ml/min for analytical columns and up to 1000 ml/min for preparative separation. The most important specifications for pumps are reproducibility, precision, noise, drift and accuracy.

Currently, many types of pump design are in use. However, three major pump types can be distinguished: pneumatic amplifier, syringe, and reciprocating piston. The pneumatic amplifier pump is somewhat like a hydraulic jack. Gas pressure is applied to one end of a piston with a large surface area and the pressure is measured by a Bourdon type of pressure gauge or a strain-gauge pressure transducer. The much smaller surface area at the other end of the piston displaces solvent from a sealed reservoir. The simple pneumatic amplifier is low in cost. It can deliver a pulse-free flow of solvent that is ideal for quantitative analysis, but gradient elution is difficult because the solvent reservoir is sealed under high pressure. The syringe pump consists of a large cylinder in which a piston is driven slowly forward by a lead screw and stepping motor to displace solvent. This type of pump is considered to deliver a high pressure and pulse-free constant flow. However, a critical disadvantage is the limited solvent capacity, which requires the pumps to be stopped for refilling of the reservoir. No check valves are needed and the system is convenient to operate. Probably most of the pumps in use today are based on the reciprocating piston pump of fixed frequency and variable stroke. The main advantage of this type of pump is that solvent delivery is continuous and the flow rate is constant regardless of solvent viscosity and column permeability. A small piston is motor-driven rapidly back and forth in a hydraulic chamber. By means of check-valves, the piston draws in solvent from a reservoir on the backward stroke, and pushes out solvent to the column on the forward stroke.

3.2.3   Sample injection systems

The injection system is designed to inject reproducibly a precise volume of the solution to be analysed onto the column, i.e. to the centre of the top of the column packing. It is important to inject the sample in a narrow plug to maintain the maximum separation performance of the column. Sample introduction is performed by syringe injectors, injection valves, stop-flow injectors and automatic injection devices. Injection volumes usually range between 1 and 250 µl. It is very important to load the sample onto the column properly to obtain a good chromatogram. When the sample is introduced onto the column in a diffused state, the band of analyte separated at the end of the column will be wider and the separation efficiency will be lower.

The injection devices used in HPLC are more sophiscated than those in GC because of the extremely high pressure in the system, and because the system cannot be stopped to release the pressure for injection. In order to inject directly into the high-pressure system, most devices incorporate a sample loop system for applying the sample to the column. Recently, autosamplers that permit fully automated, unattended operation of an LC system have enabled increased labour saving and round-the-clock operation.

3.2.3.1   Syringe injectors

The simplest type of sample introduction is syringe injection to the injection port. The sample solution taken in the microsyringe is injected through a septum accommodated in an injection port. It is possible to load the sample directly onto the inlet spot of the column to produce a narrow band of sample. However, the system has an inherent a critical disadvantage. The sample is loaded with a syringe designed to withstand high pressure up to 10.3 MPa (1500 psi) through a septum made of elastomeric silicone. Special attention should be given to solvents such as chloroform and dichloromethane which denature the elastomers. After repeated injections, the needle will make a hole in the septum, causing leakage.

3.2.3.2 *Microsampling valve*

A six-way valve consisting of a rotating seal and fixed body is the basic structure of a microsampling valve (Figure 3.4). The sample solution is introduced into the fixed-volume loop by syringe, the solvent is pumped to the column and the sample in the loop can be swept onto the column by rotating the lever to injection mode. The loop volume can be adjusted by simply changing the loop from 1 μl to several millilitres. The main advantages of this type of injector are that it is simple and reproducible, and it can be operated at high pressure up to 34.5 MPa (5000 psi).

To guarantee that a precise sample volume is injected, sufficient volume of sample solution should be flushed through the loop to ensure that the loop is completely full. However, partial-loop injection, in which a volume of sample less than the loop volume is injected, is also available. In this technique, the rest of the loop remains filled with the solvent. When a sample is injected by partial-loop injection, the volume of the sample should not exceed 50% of the loop volume. Without any precautions, of course, the sample at the front end will be diluted with solvent in the loop so as to make a broad injection. To prevent this dilution, it is recommended to introduce a tiny bubble of air after taking the exact volume of sample solution into the syringe. When an injection is made into the loop, the air bubble will be introduced first to make a separate seal between the solvent and sample solution. Generally, partial-loop injection is less precise than full-loop injection.

3.2.3.3 *Automatic sample injector*

An autosampler is a sample injection valve designed to inject the sample automatically under electronic control. This type of injector has revolutionized laboratory efficiency and reduced labour costs. Currently, a fully automated HPLC system can be left unattended for round-the-clock operation, even though the samples may be perishable, because most autosamplers are designed to control the temperature of the sample in the tray. The mechanism of operation of autosamplers varies from manufacturer to manufacturer. However, fixed or variable injection

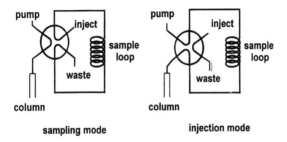

Figure 3.4 Typical six-port microsampling valve.

volume, injection order, flushing times and volumes, precolumn mixing for derivatives, dilution, urgent intercept, self-diagnosis, and so on can be controlled by entering these parameters into the automatic control system.

3.2.4   Columns and packing

The most important consideration in LC is selection of the appropriate column to provide the resolving power for separation of the analytes. The tubing material of the column must withstand the pressures to be used and be inert to any chemical reaction with the solvent or the sample. Currently, HPLC columns range from 100 to 250 mm in length and from 2 mm (microcolumns) to 8 mm in bore and are generally made from stainless-steel tubing. However, heavy-walled glass columns are sometimes employed for low-pressure work up to 4.1 MPa (600 psi) and other applications requiring an inert surface.

Packings made from microparticulate silica are held in the column by a retainer that may be a fine porosity metal frit or a disk of stainless steel gauze. Porous plugs are also used at both ends of the column. Columns have a limited useful analytical life and are expendable. This means that they must be easy to replace. There are several designs for column fitting. Connections between the injector, column top, column outlet and the detector in series should be made by zero-dead-volume fittings, otherwise the fittings act as mixing vessels and cause extra broadening of the analyte bands in the sample. The most popular connections are $\frac{1}{16}$-inch Swagelok.

Typical packings are based on porous silica particles offering a fixed porosity or on cross-linked organic polymer gels that swell when suspended in a solvent. Spherical and angular particles ranging in size from 3 to 10 μm are available for both types. Columns packed with 5–10 μm particles usually contain 40 000 to 50 000 theoretical plates. The most important properties of the packing that specify the performance of the column are mean particle size and particle size distribution. A column packed with 3–5 μm particles will have some 100 000 plates.

The chromatographic retention involves a dynamic equilibrium distribution of the components between the stationary and mobile phases. Hence, the characteristics of the packings relate to the types of analysis, such as LSC, LLC, IEC, BPC and size exclusion chromatography (SEC). Currently, most packings for LC columns are made from porous silica particles, which are prepared by agglomerating silica particles to increase the surface area and produce highly uniform size. The surface of the silica particles can be modified chemically or physically to provide specific sites to bind selective functional groups. Columns from different manufacturers show significant variations in separation, pressure required to overcome resistance to flow, reproducibility of resolution, useful life-span, and so on. The major types of column are listed in Table 3.1. To keep an HPLC column functioning properly and operating for as long a time as possible, a short guard column is employed just prior to the analytical column. The guard column removes particles and contaminants in the solvent and samples. The packing of the guard column should be similar to the packing of the analytical column, otherwise the component to be analysed may be retained on the guard column packing. After clean-up, the sample should be completely dissolved in the solvent mixture that is to be applied for the separation. Particular attention should be paid to the adjustment of pH when working with aqueous-organic or buffered eluents to ensure reproducibility.

Before carrying out a separation with a new column or when changing the mobile phase, the system must be brought into equilibrium. Equilibrium can be achieved by pumping the eluent at a low flow rate for sufficient time. The state of equilibrium can easily be monitored by repeated injection of a test compound and checking the chromatogram for consistent retention time and detector peak shape. For practical use of an LC system, it is recommended to set up a chromatographic system for one type of analysis. This will improve both the efficiency and the management of the system. Optimum flow rate is very important in chromatography. As the rate theory of Van Deemter *et al.* (1956) predicts, the flow rate has to be set to a certain range for the highest efficiency. In practice, the flow rate for a 4 mm ID column ranges from 0.5 to 2.0 ml/min.

3.2.5 Detectors

Detectors for LC are another critical component. The function of the detector is to monitor continuously the mass or concentration of the

Table 3.1 Functional groups of the column packing surface in different LC modes

Mode	Functional group
Reversed-phase	Amino, cyano, dimethyl, diphenyl, octadecyl, octyl, perfluorodecyl, phenyl, polystyrene, trimethyl
Normal phase	Amino, cyano, silica
Ion exchange	Bonded-phase ion exchangers, polystyrene
Size exclusion	Silica, polystyrene

solutes eluted from the column. The ideal detector should provide high sensitivity. Sensitivity is defined as the strength of the detector signal relative to the unit mass or concentration of the solute in the eluent. Hence, more sensitive detectors produce larger peaks for the same concentration. The next requirement is to respond to all relevant solutes but, of course, this is non-selective; if the detector has a high response to all solutes, it is non-specific. Ideal detectors should not be affected by changes in temperature, flow rate and composition of the mobile phase. In addition, the detector should generate a signal that is proportional to mass per unit time (g/s) or concentration (g/ml). None of the available detectors is universally applicable to all types of LC analysis. This has led to the appearance of a variety of special detectors for LC. Hence, a satisfactory detector (or detectors) for a particular application must be selected from those available.

The most common detectors in use are the UV–visible, refractive index, fluorescence, electrochemical and light-scattering detectors. Photometric detectors measure the variation of a property such as optical absorbance, fluorescence and refractive index of both the sample and solvent, whereas, electrochemical and conductivity detectors are sensitive only to changes in the electrical characteristics of the eluent. The last group of detectors are based on solute–eluent separation, such as the hot wire detector, light-scattering detector or liquid chromatography coupled to mass spectrometry (LC-MS). Table 3.2 compares the most popular detectors for LC by several criteria.

3.2.5.1 Ultraviolet and visible spectrophotometric detector

The detectors most widely used for LC are probably spectrophotometric detectors based on ultraviolet/visible absorption. They are relatively inexpensive and tend to be one of the first detectors in train in an LC system. The detectors measure the absorbance changes in both the UV

Table 3.2 Comparison of LC detectors

Detectors	Specificity	Detection limit	Gradient	Limitation
Ultraviolet	Selective	10^{-10} g/ml	Yes	Non-UV-active solvent
Refractive index	Universal	10^{-7} g/ml	No	Low sensitivity, precise temperature control required
Fluorescence	Selective	10^{-11} g/ml	Yes	Limited dynamic range
Electrochemical	Selective	10^{-11} g/ml	No	Compound adsorption, no electroactive solvent
Light scattering	Universal	10^{-9} g/ml	Yes	Organic solvent or pure water eluent

and visible regions. Some detectors are designed to provide absorbance measurements at a fixed wavelength, but variable wavelengths are also available in most detectors. They are relatively insensitive to flow and temperature changes but, in general, have good selectivity and a high sensitivity in the analysis of specific components. They can frequently be used in gradient elution systems. Quantitative analysis by peak height or peak area using a UV–visible photometric detector is very reliable, with high precision in gradient or isocratic separation. Since they are based upon absorption of ultraviolet or visible radiation, they are particularly useful for detection of many organic components. Most organic components contain $-C=C-$, $-C=O$, $-N=O$ or $-N=N-$ functional groups that absorb UV radiation between 254 and 280 nm. The detector cell is a cuvette through which the solvent eluted from the column is passed. UV/visible light passed through the cuvette is detected by photomultiplier tube. To increase the sensitivity, several types of efficient flow cell (Figure 3.5) combining low volume and long light path are available. A variation of the UV detector is the photodiode-array detector. In this system, a photodiode array is used to detect multiple wavelengths simultaneously in order to obtain a spectrum for each component over a selected wavelength range.

3.2.5.2 Refractive index detector

Refractive index (RI) detectors measure changes in the refractive index of the column eluent. This device continuously monitors the difference in refractive index of the mobile phase containing components as it elutes at the end of the column compared with the pure mobile phase as reference. Since these detectors respond to all solutes but have a low sensitivity, they are commonly used for preparative-scale chromatography or size

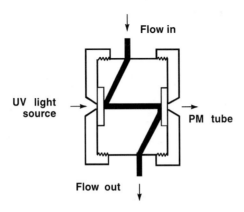

Figure 3.5 Cross-section of flow cell.

exclusion chromatography. Refractive index is very sensitive to changes in temperature, bubbles in the solvent, and flow rate of the solvent, requiring close control of the operating conditions in the system.

There are three types of refractive index detector (two are shown in Figure 3.6). The Fresnel refractometer is based on Fresnel's law of reflection. The law states that the percentages of light reflected at a glass–liquid interface will vary with the angle of incident light and the refractive indices of the solutions. In the Fresnel RI detector, a very small flow cell is accommodated and the cell is cleanly swept by mobile phase. However, it requires clear cell windows and the linear detection range is limited. In addition, two different prisms are required to cover the useful refractive index range. In the deflection refractometer, a cell with sample and reference compartment is mounted in the optical path. The compartments are separated by a diagonal piece of glass. The main advantages of this detector are its wide range of linearity and excellent versatility. However, the volume of the flow cell is rather bigger than in the Fresnel-type, so that band spreading can be expected. In the interferometric refractometer, polarized light from the lamp is passed through a beam splitter to cells of sample and reference. The advantages of this detector are a minimized, small-volume flow cell resulting in less band broadening, and sensitivity that is 10–100 times higher than that of other RI detectors. Refractometers are reliable and convenient; however, they are impractical for application to gradient analysis. Despite low sensitivity, instability to temperature change and impracticability for gradient analysis, the refractometer is the second most widely used detector in liquid chromatography.

3.2.5.3 Fluorescence detector

Fluorescence detectors are another of the main types of detector used in LC. These detectors are less sensitive than UV detectors but are more

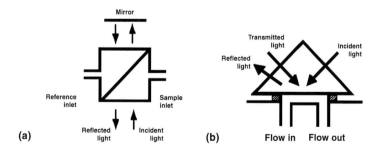

Figure 3.6 (a) Deflection and (b) Fresnel-type optical cells.

selective. In fluorescence detectors, the light source, sample and photomultiplier tube are usually arranged at right angles, not in a straight path as in the UV detector arrangement. Ultraviolet light from the light source is filtered to allow a specific wavelength to excite the compounds to emit fluorescence. The fluorescence is filtered again to remove incident radiation and detected with a photomultiplier tube. Compounds that do not fluoresce naturally may be detected by forming a fluorescent complex with a derivatizing reagent. To induce fluorescence in a non-fluorescent compound, a derivatizing reagent is supplied to the column eluent via a specially designed pump. Since the reaction takes some finite time to complete, the mixture of eluent and reagent is introduced to a reaction coil or a reactor where the temperature is controlled for optimum reaction (Figure 3.7). This is very effective for the analysis of amino acids and peptides.

3.2.5.4 Light-scattering detector

Light scattering is a relatively recent development in detectors for LC (Figure 3.8). The effluent from the chromatographic column is sprayed by an inert gas stream. At the outlet of the nebulizer, the aerosol is introduced into a chamber to eliminate large droplets by means of an evacuating siphon. The fine homogeneous mist of particles is passed towards an evaporating tube, a heated tube that ensures the evaporation of the solvent. Thus, only a fine mist of microparticles of non-volatile analytes arrives at the detecting head. The detector is designed to be able to work with organic solvents or with pure water as eluents. In some cases, it is also possible to use mobile phases containing sublimable salts.

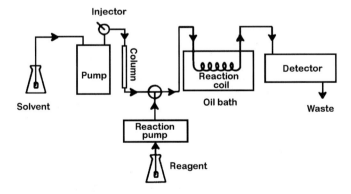

Figure 3.7 Schematic diagram of a post-column derivatization reaction system.

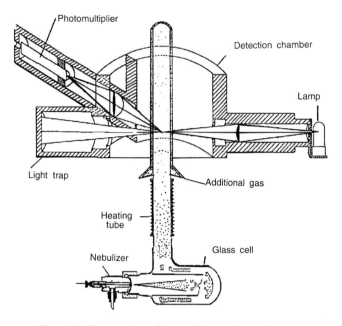

Figure 3.8 Cross-section of Sedere light-scattering detector.

An inlet of additional gas forms a concentric shield that surrounds the mist that enters the light beam, avoiding contamination of the cell.

The photomultiplier is located at 120° to the light beam, and collects the light scattered by the particles. The signal is proportional to the mass of analyte. This detector is now being used instead of the refractive index detector because of it has higher sensitivity and can be applied to solvent-programmed analysis.

Discussion here has been limited to UV, RI, fluorescence and light-scattering detectors. Besides these, there are several other types of special detectors available for specific applications (Parriott, 1993).

3.2.6   Recorder and data-handling systems

The signal produced by detectors requires a fast and accurate monitoring device such as a high-speed recorder. For routine operations, digital electronic integrators and computers are available. The most recent integrated HPLC instruments include a microprocessor system to control the operating variables and operate as a data-handling centre to provide various outputs such as the chromatogram, operating conditions, quantitation and sample identification, etc.

3.3 Application for the analysis of contaminants in food

3.3.1 Direct application of HPLC to analysis of contaminants in food

The application of LC for the analysis of contaminants in food has many advantages over other analytical methods. In LC, components that are not readily volatilized can be separated. In addition, this approach can be applied to highly polar, high molecular mass, strongly ionic or thermally unstable components. Thus, derivatization of the analyte is not required so often as in gas-chromatographic analysis. However, pre-column or post-column derivatization for LC has received special attention in the analysis of contaminants such as pharmaceuticals, drugs and hormones to achieve higher detection limits.

Derivatization involving fluorogenic labelling and fluorescence detection for the analysis of glyphosate and its metabolite, aminomethylphosphonic acid was described by Wigfield and Lanouette (1991). Miles and Zhou (1991) reported a LC method with post-column reaction for detection of a fungicide, nabam. Nabam was recovered in high yield from the fortified samples. Aflatoxins are one of the most popular components analysed by LC. Aflatoxins are also derivatized with trifluoroacetic acid prior to injection (Tarter et al., 1984). The proposed method for determination of aflatoxins in peanuts, peanut butter and other nuts was accurate, precise and sensitive. Post-column iodine derivatization and fluorescence detection was used for quantitation by Dorner and Cole (1988) and Beaver and Wilson (1990).

Vanillin post-column derivatization was adopted for quantitation of monensin in chicken tissues (Moran et al., 1994) and cattle rations (Bridges et al., 1996). Monensin is a feed additive for chickens and turkeys to prevent coccidiosis, and for cattle to improve feed efficiency. The method showed improved specificity, analysis time and precision compared with microbiological methods. Vanillin has also been applied to the analysis of other contaminants. Ericson et al. (1994) adopted a post-column derivatization with vanillin for determination of semduramicin sodium in poultry tissues to develop a sensitive and reliable LC method. The mean recovery of the drug was 95% and the coefficient of variation was less than 10%.

HPLC systems are often used in the analysis of miscellaneous pesticides because of the advantages of simplicity of sample treatment, good potential resolution compared with capillary GC columns and high sensitivity compared with the electron-capture detector of GC. Gas chromatographic analysis has been the preferred analytical technique for organochlorine pesticides and organophosphate insecticides. However, HPLC has been used in the analysis of organophosphate residues in food because of its advantages of simple clean-up steps and minimized interferences. Martinez *et al.* (1992) reported a rapid and sensitive method for the determination of organophosphorus pesticides in fruits and surface waters by HPLC with UV detection within 10 minutes. Ioerger and Smith (1993) separated 17 different organophosphates and their metabolites by HPLC. They analysed the compounds in a reversed-phase column using a step-gradient mobile phase with increasing acetonitrile content in water from 55% to 65% within 11 minutes, and the compounds were identified using a diode-array detector by matching with standards in the spectral library.

The common functional carbamate group (NH_2COO^-) in carbamate pesticides is esterified to a variety of aliphatic and aromatic groups. The pesticides are not easily analysed by GC without derivatization because they are not volatile and are heat labile to some extent. The most common method for carbamate analysis is the HPLC method developed by Krause (1985a,b), who separated the carbamate insecticides using a Zorbax C8 column and gradient mobile phase system with a linear increase of acetonitrile in water.

Benzimidazoles are another group for which analysis by HPLC is the method of choice because they are typically non-volatile. Once they are extracted with acidified methanol, they are partitioned into methylene chloride (Long *et al.*, 1989). Quantitation of antimicrobial drugs is another application for HPLC. Smedley and Weber (1990) separated 10 sulphonamides under isocratic conditions and Zomer *et al.* (1992) separated 12 sulphonamides and *p*-aminobenzoic acid in milk. Smallidge *et al.* (1988) reported sulphamethazine and sulphathiazole determination in swine feed by reversed-phase liquid chromatography with post-column derivatization.

Many liquid chromatographic methods have been described for determination of antibiotics. The major advantage of LC over other chromatographic approaches for antibiotic residue analysis is that frequently little or no sample preparation is required. Another advantage is that procedures can be partially or completely automated. Many of these methods are intended for determination of residues in formulations, fermentation broths, or biological fluids for clinical applications, but considerable progress has also been made in development of LC methods for residues of some antibiotics in tissues and foods. Louis *et al.* (1982) reported liquid chromatographic determination of natamycin in cheese at residue levels, and antibiotic residues in salmon tissue were analysed by ion-pair reversed-phase liquid chromatography using an electrochemical detector (Luo *et al.*, 1996).

Sponholz and Lamberty (1980) quantified styrene in wines by a reversed-phase HPLC method. The styrene contaminant came from improperly cleaned polyester tanks and led to impaired flavour and aroma; it was detected at 0.38 and 0.87 mg/l.

3.3.2 Solid phase extraction

The principle of liquid chromatography is applied to many analytical problems in various ways. Liska *et al.* (1989) reviewed solid phase extraction for the concentration of organics from water matrices. Solid phase extraction is now one of the most convenient and common sample clean-up methods used in environmental analysis, and many methods that include solid phase extraction clean-up have been described for analysis of contaminants in food. Haagsma and Water (1985) developed a cation exchange, solid phase extraction for rapid, simultaneous quantitation of sulphonamides in swine tissue by HPLC. Kohler and Su (1986) used solid phase extraction for rapid sample preparation of organochlorine pesticides from seafood. They compared recoveries by solid phase extraction for seven pesticides with recoveries by conventional liquid–liquid extraction. Slates (1988) determined bensulphuron methyl residues in rice grain and straw by HPLC. He extracted the herbicide with methylene chloride and separated the bensulphuron methyl residues from interfering components by solid phase extraction; recoveries averaged 97% for rice grain. Fungicide residues were also analysed using solid phase extraction by Newsome and Collins (1989). Thereafter, the use of this extraction technique became more common.

Molto *et al.* (1990) reported a combination of reversed-phase and adsorption chromatography for sample clean-up with a gas chromatographic procedure for analysis of organochlorine pesticides. Manes *et al.* (1990) described the advantages of solid phase extraction; recoveries were 60–105% for 19 spiked pesticides and the method was faster and easier to use than conventional solvent extraction. In addition, solvent consumption was far less, and the use of glass microcolumns avoided contamination of the samples by plasticizers. Farran *et al.* (1991) used a XAD-2 resin for a solid phase preconcentration step for organophosphorus pesticides coupled to a continuous, unsegmented-flow HPLC separation system. The influence of different parameters on the retention and recovery of the pesticides was also studied. Recoveries higher than 95% were obtained and the method was successful for application to spiked river water samples. Lagana *et al.* (1995) analysed eight triazine herbicides in milk by the reversed-phase mode with UV detection. Solid phase extraction was performed using a non-specific adsorbent together with a cation exchanger. Another combination of double trapping sample clean-up was described by Parrilla and Martinez-Vidal (1996) for the simultaneous determination of 21 pesticides in vegetables. Chee *et al.*

(1995) described a two-level, orthogonal-array design to optimize solid phase extraction of organochlorine pesticides from water. A fast and reliable method for monitoring of organophosphorus pesticides in water was reported by Driss and Bouguerra (1996). They used C-18 bonded silica cartridges for extraction and enrichment of 16 pesticides. The recoveries by solid phase extraction were 70–98%, except for dimethoate, which was exceptionally low at 52–60%. Nau *et al.* (1996) described an efficient analysis of pesticide and herbicide residues in food including meat, fish, fruits, vegetables and processed foods. Schenck (1996) reported the coefficients of variation (CV) for six participating laboratories in an interlaboratory trial of a solid phase extraction clean-up method for determination of organochlorine pesticides in non-fatty seafood products; the within-laboratory CV ranged from 4.2% to 8.5% and the between-laboratory CV ranged from 10.9% to 26.5%.

3.3.3   GPC clean-up system

HPLC is also widely used to clean up the sample solution prior to analysis for contaminants in food. Analysis can be divided into three steps. The first step is extraction of the components to be analysed from the sample; the next step is the clean-up procedure, followed by qualitative and quantitative analysis. Of the three steps, the clean-up procedure is the most laborious step and the one consuming the most solvent/reagent within the total analytical procedure.

Separation in gel permeation chromatography is based on differences in molecular size and has revolutionized the analysis and preparative separation of macromolecules. It is also widely used as a sample clean-up procedure and has proved to be a most versatile and convenient sample clean-up technique for a wide range of sample matrices including food products, tissues, grains, plants and environmental samples such as soil, sludge and hazardous wastes. The main advantage of the technique is that samples cleaned up once on GPC can be introduced to GC or LC without additional clean-up procedures. Currently, the automated GPC methods are officially accepted by the US government agencies. The technique is highly recommended as a step in the preparation of environmental samples for analysis in EPA SW-846, method 3460 (Figure 3.9). The Food and Drug Administration accepted the technique for pesticide analysis in Pesticide Analytical Manual, Volume 1, Section 221, FDA. USDA accepted the GPC clean-up technique in the Official Methods of Analysis 14th edition, in section 29.037-.043.

The introduction of gel permeation chromatographic clean-up techniques for pesticide residue analysis was made by Stalling *et al.* in 1972. They reported higher than 95% recoveries from fish extracts for

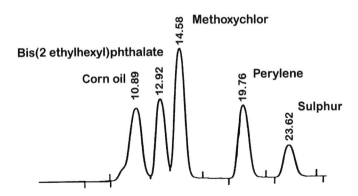

Figure 3.9 Typical chromatogram of EPA calibration standard for GPC using Envirosep-ABC.

seven non-ionic chlorinated pesticides, polychlorinated biphenyls, malathion and parathion. The technique was automated by Tindle and
Stalling (1972). With this system, 23 samples were cleaned up completely
unattended. Two years later, Griffitt and Craun (1974) indicated that
their GPC clean-up system provided better recoveries of pesticides than
the Florisil clean-up system. They used Bio-Beads SX-2 and cyclohexane
for the GPC system. David Stalling, the originator of GPC clean-up, now
leads the research programme at O.I. Analytical, which continues to
develop applications and refine techniques for GPC clean-up. Autovap
AS-2000 (Figure 3.10), is an automated GPC clean-up system that is now
integrated with on-line, unattended, sample concentration to provide up
to 23 processed samples in series, transferred to sealed vials at a low final
volume.

The application of an improved gel–solvent system to the clean-up of
pesticide residues by GPC was investigated by Johnson *et al.* (1976).
Results were in excellent agreement with those from accepted manual
partitioning methods and were achieved with significant savings of both
labour and chemicals. Later, Hopper and Griffitt (1987) evaluated an
automated GPC clean-up and evaporation system for determining
pesticide residues in fatty samples, and showed that the system is
valuable for automation of the analysis of fatty food samples for pesticide
and industrial chemical residues. It is particularly useful for the analysis
of foods not requiring additional clean-up prior to GC analysis.

Ogawa *et al.* (1997) reported a rapid determination of multiple
pesticide residues in agricultural products by GPC clean-up and GC-MS.
They extracted 95 pesticide residues by solvent extraction, followed by
GPC separation for sample clean-up. GPC profiles of pesticides and
major components in agricultural products were studied with various

Figure 3.10 Automated GPC clean-up system, Autovap AS-2000.

organic solvents, and toluene was found to be the most suitable mobile phase for GPC with Bio Beads S-X3. Detection limits for the 95 pesticides analysed by this method were satisfactory for practical use.

Analysis of a selective herbicide, ethofumesate, in plants, soil and water has been described by Alawi (1990). Samples were extracted with dichloromethane and followed by clean-up with GPC combined with a mini silica gel column, and analysed. Recovery of ethofumesate added to water was 90.2–103.3% and the method was found to be applicable also to other types of plant matrices.

There are several literature references for the successful application of gel permeation chromatography to the clean-up of samples for the analysis of semi-volatile organic pollutants and polychlorinated biphenyls (PCBs). Czuczwa (1989) described a repetitive system for GPC processing of waste oil samples that was shown to be successful for determination of PCBs in waste oils. Pietsch *et al.* (1996) determined aliphatic and alicyclic amines in water by GC and LC after derivatization with chloroformates.

3.4 Benefits and limitations

Analysis is relatively simple by injection into a system. It takes usually less than 20 minutes for an analysis. LC offers high resolution, which means that very complex sample mixtures can be resolved. Quantitative analysis is possible with high precision and accuracy, and sensitivity down to nanogram levels. Fully automatic systems are available that perform analysis unattended round-the-clock. These offer substantial labour saving and have revolutionized and increased laboratory efficiency.

The application of LC to the analysis of contaminants in food has many advantages over other analytical methods. Components that are not readily volatilized can be separated. In addition, very polar, high molecular mass, strong ionic or thermally unstable components can be separated more easily. Thus, on the one hand, derivatization of the analyte is not required so often as in gas-chromatographic analysis; on the other hand, derivatization offers both further analytical options and improved sensitivity. The versatility is clearly illustrated by the various types of components that can be separated. These include pesticides, drugs, mycotoxins, antibiotics, plasticizers, and polycyclic aromatic hydrocarbons. These provide an impressive list, but constitute only a fraction of the wide variety of samples amenable to analysis by HPLC.

The principle of liquid chromatography is applied in analytical chemistry and biochemistry in various ways. Solid phase extraction is now one of the most convenient and common sample clean-up methods available and there are many applications of this technique in most areas of analysis. The GPC clean-up technique is the another proven universal clean-up method for removing lipids, polymers, resins and other high molecular mass compounds from a wide range of tissue, food matrix, soil and sediment sample extracts. Removal of these interfering compounds improves analytical recoveries while extending the life of columns and reducing downtime of analytical instruments.

The limitations of LC include expensive instrumentation and operation. For many laboratories, high-purity solvents for LC are a major expenditure. However, the critical disadvantage is the lower resolution of the columns compared to capillary GC. Hence, the columns cannot separate some complex mixtures effectively, resulting in poor qualitative and quantitative analysis. At the present time, there is no universal and sensitive detector and, finally, the consumable supplies for operation, such as replacement columns, are moderately expensive.

References

Alawi, M.A. (1990) A high performance liquid chromatographic approach for monitoring ethofumesate in plant, soil and water. *Analytical Letters*, **23** (9) 1695-1709.

Beaver, R.W. and Wilson, D.M. (1990) Comparison of postcolumn derivatization–liquid chromatography with thin-layer chromatography for determination of aflatoxins in naturally contaminated corn. *Journal of the Association of Official Analytical Chemists*, **73** (4) 579-581.

Bridges, E.A., Roth, D.M., Cleveland, C.M., Moran, J.W. and Coleman, M.R. (1996) Determination of monensin in high-moisture cattle rations by liquid chromatography with postcolumn derivatization. *Journal of the Association of Official Analytical Chemists International*, **79** (6) 1255-1259.

Consden, R., Gordon, A.H. and Martin, A.J.P. (1944) Qualitative analysis of proteins: a partition chromatographic method using paper. *Biochemical Journal (London)*, **38** 224-232.

Chee, K.K., Wong, M.K. and Lee, H.K. (1995) Optimization by orthogonal array design of solid phase extraction of organochlorine pesticides from water. *Chromatographia*, **41** (3/4) 191-196.

Czuczwa, J.M. (1989) Optimized gel permeation chromatographic cleanup for soil, sediment, wastes, and oily waste extracts for determination of semivolatile organic pollutants and PCBs. *Journal of the Association of Official Analytical Chemists*, **72** (5) 752-759.

Dorner, J.W. and Cole, R.J. (1988) Rapid determination of aflatoxins in raw peanuts by liquid chromatography with postcolumn iodination and modified minicolumn cleanup. *Journal of the Association of Official Analytical Chemists*, **71** (1) 43-47.

Driss, M.R. and Bouguerra, M.R. (1996) Solid phase extraction of organophosphorus pesticides from water using capillary gas chromatography with thermionic specific detection. *International Journal of Environmental Analytical Chemistry*, **65** (1/4) 1-10.

Ericson, J.F., Calcagni, A. and Lynch, M.J. (1994) Determination of semduramicin sodium in poultry liver by liquid chromatography with vanillin postcolumn derivatization. *Journal of the Association of Official Analytical Chemists International*, **77** (3) 577-582.

Farran, A., Pablo, J. de. and Hernandez, S. (1991) Continuous flow determination of organophosphorus pesticides using solid phase extraction coupled on-line with high performance liquid chromatography. *International Journal of Environmental Analytical Chemistry*, **45** (4) 245-253.

Giddings, J.C. (1965) *Dynamics of Chromatography*, Marcel Dekker, New York.

Gilbert, M.T. (ed.) (1987) *High Performance Liquid Chromatography*, IoP Publishing, Bristol.

Griffitt, K.R. and Craun, J.C. (1974) Gel permeation chromatographic system: an evaluation. *Journal of the Association of Official Analytical Chemists*, **57** 168-172.

Haagsma, N. and Water, C. Van De (1985) Rapid determination of five sulphonamides in swine tissue by high performance liquid chromatography. *Journal of Chromatography*, **333** (1) 256-261.

Hamilton, P.B. (1963) Ion exchange chromatography of amino acids: recent advances in analytical determination. *Advances in Chromatography*, **2** 3-62.

Hopper, M.1. and Griffitt, K.R. (1987) Evaluation of an automated gel permeation clean-up and evaporation systems for determining pesticide residues in fatty samples. *Journal of the Association of Official Analytical Chemists*, **70** (4) 724-726.

Ioerger, B.P. and Smith, J.S. (1993) Multiresidue method for extraction and detection of organophosphate pesticides and their primary and secondary metabolites from beef tissue using HPLC. *Journal of Agricultural and Food Chemistry*, **41** 303-307.

James, A.T. and Martin, A.J.P. (1952) Gas–liquid partition chromatography: the separation and micro-estimation of volatile fatty acids from formic acid to dodecanoic acid. *Biochemical Journal (London)*, **50** 679-690.

Johnson, L.D., Walts, R.H., Ussary, J.P. and Kaiser, F.E. (1976) Automated gel permeation chromatographic clean-up of animal and plant extracts for pesticide residue determination. *Journal of the Association of Official Analytical Chemists*, **59** (1) 174-187.

Kirkland, J.J. (ed.) (1971) *Modern Practice of Liquid Chromatography*, Wiley Interscience, New York.

Kohler, P.W. and Su, S.Y. (1986) Analysis of some organochlorine pesticides in seafood samples by solid phase extraction–gas chromatography. *Chromatographia*, **21** (9) 531-537.

Krause, R.T. (1985a) Liquid chromatographic determination of N-methyl carbamate insecticides and metabolites in crops. I. Collaborative study. *Journal of the Association of Official Analytical Chemists*, **68** (4) 726-733.

Krause, R.T. (1985b) Liquid chromatographic determination of N-methyl carbamate insecticides and metabolites in crops. II. Collaborative study. *Journal of the Association of Official Analytical Chemists*, **68** (4) 734-741.

Lagana, A., Mario, A. and Fago, G. (1995) Evaluation of double solid-phase extraction system for determining triazine herbicides in milk. *Chromatographia*, **41** (3/4) 178-182.

Liska, I., Krupcik, J. and Leclercq, P.A. (1989) The use of solid sorbents for direct accumulation of organic compounds from water matrices—a review of solid-phase extraction techniques. *Journal of High Resolution Chromatography*, **12** 577-590.

Louis, G.M., Tunistra, T.H. and Traag, W.A. (1982) Liquid chromatographic determination of natamycin in cheese at residue levels. *Journal of the Association of Official Analytical Chemists*, **65** (4) 820-822.

Long, A.R., Hsieh, L.C., Malbrough, M.S., Short, C.R. and Barker, S.A. (1989) Multiresidue method for isolation and liquid chromatographic determination of seven benzimidazole anthelmintics in milk. *Journal of the Association of Official Analytical Chemists*, **72** (5) 739-741.

Luo, W., Hansen, E.B., Jr. Ang, C.Y.W. and Thompson, H.C., Jr. (1996) Determination of lincomycin residues in salmon tissues by ion-pair reversed-phase liquid chromatography with electrochemical detection. *Journal of the Association of Official Analytical Chemists*, **79** (4) 839-843.

Manes, J., Pico, Y., Molto, J.C. and Font, G. (1990) Solid phase extraction of pesticides from water samples. *Journal of High Resolution Chromatography and Chromatography Communications*, **13** (12) 843-845.

Marin, M.L., Lopez, J., Sanchez, A., Vilaplana, J. and Jimenez, A. (1998) Analysis of potentially toxic phthalate plasticizers used in toy manufacturing. *Bulletin of Environmental Contamination and Toxicology*, **60** 68-73.

Martin, A.J.P. and Synge, R.L.M. (1941) A new form of chromatogram employing two liquid phases. *Biochemical Journal (London)*, **35** 1358-1368.

Martinez, R.C., Gonzalo, E.R., Moran, M.J. and Mendez, J.H. (1992) Sensitive method for the determination of organophosphorus pesticides in fruits and surface waters by high-performance liquid chromatography with ultraviolet detection. *Journal of Chromatography*, **607** 37-45.

Molto, J.C., Albelda, C., Font, G. and Manes, J. (1990) Solid-phase extraction of organochlorine pesticides from water samples. *International Journal of Environmental Analytical Chemistry*, **41** (1/2) 21-26.

Miles, C.J. and Zhou, M. (1991) Determination of nabam fungicide in crops by liquid chromatography with postcolumn reaction detection. *Journal of the Association of Official Analytical Chemists*, **74** (2) 384-388.

Moore, J.C. (1968) Gel permeation chromatography: its inception, in *Analytical Gel Permeation Chromatography* (eds J.F. Johnson and R.S. Porter), *Journal of Polymer Science, Part C, Polymer Symposium* (**21**), Interscience, New York, pp 1-3.

Moran, J.W., Rodewald, J.M., Donoho, A.L. and Coleman, M.R. (1994) Determination of monensin in chicken tissue by liquid chromatography with postcolumn derivatization. *Journal of Association of Official Analytical Chemists International*, **77** (4) 885-890.

Nau, D., Hung, N. and Pocci, R. (1996) Efficient analysis of pesticide and herbicide residues in food. *European Food and Drink Review*, (Winter) 63-69.

Newsome, W.H. and Collins, P. (1989) Multiresidue method for fungicide residues in fruit and vegetables. *Journal of Chromatography*, **472** (2) 416-421.

Ogawa, M., Sakai, T., Ohkuma, K., Matsumoto, T., Hisamatsu, Y. and Nakazawa, H. (1997) Rapid determination of multiple pesticide residues in agricultural products by GPC clean-up and GC/MS-SIM. *Journal of the Food Hygienic Society of Japan*, **38** (2) 48-61.

Parrilla, P. and Martinez-Vidal, J.L. (1996) HPLC determination of pesticides in green bean samples after SPE clean-up. *Chromatographia*, **43** (5/6) 265-270.

Parriott, D. (ed.) (1993) *A Practical Guide to HPLC Detection*, Academic Press, Orlando, FL.

Pietsch, J., Hampel, S., Schmidt, W., Brauch, H.J. and Worch, E. (1996) Determination of aliphatic and alicyclic amines in water by gas and liquid chromatography after derivatization by chloroformates. *Fresenius*, **355** (2) 164-173.

Porath, J. and Flodin, P. (1959) Gel filtration: a method for desalting and group separation. *Nature (London)*, **183** 1657-1659.

Schenck, F.J. (1996) Screening of nonfatty fish for organochlorine pesticide residues by solid phase extraction clean-up: interlaboratory study. *Journal of the Association of Official Analytical Chemists International*, **79** (5) 1215-1219.

Scott, R.P.W. (1983) Small-bore columns in liquid chromatography. *Advances in Chromatography*, **22** 247-294.

Slates, R.V. (1988) Determination of bensulfuron methyl residues in rice grain and straw by HPLC. *Journal of Agricultural and Food Chemistry*, **36** (6) 1207-1211.

Smedley, M.D. and Weber, J.D. (1990) Liquid chromatographic determination of multiple sulfonamide residues in bovine milk. *Journal of the Association of Official Analytical Chemists*, **73** (6) 875-879.

Smallidge, R.L. Kentzer, E.J., Stringham, K.R., Kim, E.H., Rodger, C.L., Stringham, R.W. and Mundell, E.C. (1988) Sulfamethazine and sulfathiazole determination at residue levels in swine feeds by reverse-phase liquid chromatography with post-column derivatization. *Journal of the Association of Official Analytical Chemists*, **71** (4) 710-717.

Spackman, D.H., Stein, W.H. and Moore, S. (1958) Automatic recording apparatus for use in the chromatography of amino acids. *Analytical Chemistry*, **30** 1190-1206.

Sponholz, W.R. and Lamberty, P. (1980) Determination of styrene in wines. *Zeitschrift für Lebensmittel Untersuchung and Forschung*, **171** (6) 451-452.

Stalling, D.L., Tindle, R.C. and Johnson, J.L. (1972) Cleanup of pesticide and polychlorinated biphenyl residues in fish extracts by gel permeation chromatography. *Journal of the Association of Official Analytical Chemists*, **55** 32-38.

Tarter, E.J., Hanchay, J.P. and Scott, P.M. (1984) Improved liquid chromatographic method for determination of aflatoxins in peanut butter and other commodities. *Journal of the Association of Official Analytical Chemists*, **67** (3) 597-600.

Tindle, R.C. and Stalling, D.L. (1972) Apparatus for automated gel permeation clean-up for pesticide residue analysis. Applications to fish lipids. *Analytical Chemistry*, **44** (11) 1768-1773.

Tiselius, A. (1941) Adsorption analysis of amino acids and peptides. *Arkiv för Kemi, Mineralogi och Geologi*, **15B** (6) 1-5.

Tswett, M. (1906) Physikalisch-chemische Studien über das Chlorophyll. Die Adsorptionen. *Berichte der Deutschen Botanischen Gesellschaft*, **24** 316-323.

van Deemter, J.J., Zuiderweg, F.J. and Klinkenberg, A.A. (1956) Longitudinal diffusion and resistance to mass transfer as causes of non-ideality in chromatography. *Chemical Engineering Science*, **5** 271-289.

Wigfield, Y.Y. and Lanouette, M. (1991) Residue analysis of glyphosate and its principal metabolite in certain cereals, oilseeds and pulses by liquid chromatography and postcolumn fluorescence detection. *Journal of the Association of Official Analytical Chemists*, **74** (5) 842-847.

Zomer, E., Saul, S. and Charm, S.E. (1992) A method for confirmation and identification of antimicrobial drugs, by using liquid chromatography with micorbial receptor assay. I. Sulfonamides in milk. *Journal of the Association of Official Analytical Chemists International*, **75** (6) 987-993.

4 Advances in immunochemical methods of analysis for food contaminants

Paul F.A. Wright and Effie Apostolou

Abbreviations

AFM$_1$, aflatoxin M$_1$; anti-id, anti-idiotype antibodies; BSA, bovine serum albumin; Cd-EDTA-KLH, keyhole limpet haemocyanin–thioureido-L-benzylethylenediamine tetraacetic acid–cadmium; CT, cholera toxin; DDT, dichlorodiphenyltrichloroethane; 2,4-D, 2,4-dichlorophenoxyacetic acid; EDTA, ethylenediaminetetraacetic acid; ELISA, enzyme-linked immunosorbent assay; FB$_1$, fumonisin B$_1$; GC, gas chromatography; GC-MS, gas chromatography–mass spectrometry; HCB, hexachlorobenzene; HPLC, high-performance liquid chromatography; HPTLC, high-performance thin-layer chromatography; KLH, keyhole limpet haemocyanin; LIA, liposome immunoaggregation assay; MeIQx, 2-amino-3,8-dimethylimidazo[4,5-f]quinoxaline; NMC, N-methyl carbamate; OC, organochlorine; PhIP, 6-phenyl-2-amino-1-methylimidazo[4,5-f]pyridine; PBS, phosphate buffered saline; PCB, polychlorinated biphenyl; PCP, pentachlorophenol; ppb, parts per billion (ng/g); ppt, parts per trillion (pg/g); RIA, radioimmunoassay; TLC, thin-layer chromatography.

4.1 Historical perspectives

Immunochemical methods have become very popular since their introduction in the 1960s. The first radioimmunoassay publication, by Yalow and Berson (1960), for the measurement of serum insulin, was a significant event that generated great interest in immunochemical methods owing to their sensitivity, specificity, simplicity and almost universal application. When enzyme immunoassays were first reported a decade later (Engvall and Perlman, 1971), their principles were similar to those of radioimmunoassays. The advantage of these methods lay in the use of 'labels' other than radioisotopes, such as fluorophores or enzyme activities measured spectrophotometrically, fluorometrically or by chemiluminescence (Edwards, 1985). During the early 1970s, much work was done to improve labelling techniques, methods of separating bound and free label, and antibody production (Abraham, 1977).

Owing to an initial lag in the development of suitable reagents, it was not until the 1980s that the food industry began to appreciate the benefits of using immunochemical methods for food monitoring of pesticide and mycotoxin residues, as well as trace amounts of other contaminants in foods and animal feeds (Marks, 1985). In the past few years, there have been many new developments in immunochemical methods. The trend has been to move away from liquid phase assays and the use of radioisotopic labels towards the use of sensitive and rapid homogeneous solid phase assays (Nakamura, 1992).

4.2 Methodology

4.2.1 Choice of method

The method chosen for a given application depends on the following factors (Barker and Walker, 1992):

(1) the sample size required to achieve the desired limit of detection for the analytical instrument available;
(2) the nature of the matrix and the ability of existing methods to isolate target compounds from the matrix;
(3) the specificity required from the isolation technique, such as the ability to detect a single mycotoxin from a group of related mycotoxins;
(4) the number of samples and time limitations; and
(5) the cost of the method, including consumables, labour, and the cost of the instrumentation used.

4.2.2 Antibody production

Monoclonal antibodies are antibodies that specifically recognise one region (or 'epitope') on the surface of an antigen. They are produced through the fusion of a specific antibody-producing splenic B-lymphocyte with a mutant myeloma cell line, to produce a hybridoma. Although the production of monoclonal antibodies is not technically difficult, it is quite demanding in terms of time, effort and money and ideally requires the attention of a small research team and specialist facilities (Siddle, 1985).

Recently, a comprehensive method for the production of monoclonal antibodies has been described by Blake *et al.* (1996) for cadmium–EDTA complexes. In this study, the cadmium chelate complex of interest was too small to elicit an immune response, so a bifunctional EDTA derivative known to be immunogenic in mice was loaded with cadmium and covalently conjugated to a large carrier protein (keyhole limpet haemocyanin, KLH). This conjugate was then injected into the peritoneal cavity of three mice at two-week intervals, using 50 µg of the KLH–thioureido-L-benzylethylenediaminetetraacetic acid–cadmium (Cd-

EDTA-KLH) conjugate emulsified in Ribi adjuvant. The third and fourth injections were given 30 and 40 days after the second boost. Blood was then collected from the tail vein, and the antibody response was determined by an indirect enzyme-linked immunosorbent assay (ELISA, see section 2.4). A final boost of the Cd-EDTA-KLH conjugate in phosphate buffered saline (PBS) was given intraperitoneally 4 days prior to fusion of mouse spleen cells to myeloma cells. The presence of antibodies to the cadmium-chelate complex was confirmed by competitive ELISA (see section 2.4). In this study, only one mouse showed a preferential binding of the cadmium-loaded conjugate, so only the spleen from this mouse was used for the preparation of hybridoma cells.

Polyclonal antibodies are a mixture of antibodies that recognise more than one epitope on the antigen, and are produced by immunisation of animals with an antigen that elicits a humoral immune response (or 'immunogen'). When the desired antibody titre is achieved, blood is collected and the serum is harvested, aliquoted and stored at $-80°C$ (Lu et al., 1995). The rabbit is the most common animal species utilised in polyclonal antibody production, but pigs, mice and rats are also used. In the production of antibodies to mycotoxins, an oil-based adjuvant containing non-viable Mycobacterium, such as Freund's complete adjuvant, is co-injected to allow slow release of the immunogen, and also to non-specifically stimulate the immune response. Recently, cholera toxin (CT) has been used as an adjuvant with several advantages. There was a rapid yield of quality antibodies without impairment of animal health as compared to Freund's adjuvant, which typically gives rise to abscesses, ulcers or granulomas at the injection site. CT would also be valuable when immunogen availability is limited, as only relatively low doses of immunogen are required to induce a rapid and strong antibody response (Pestka et al., 1995). Azcona-Olivera et al. (1992) used CT as the carrier-adjuvant to generate antibodies for fumonisins, where fumonisin B_1 (FB_1) was conjugated to CT by glutaraldehyde to form an immunogen. For this protocol, mice were immunised with three doses of 7.5 µg of FB_1-CT intravenously in the lateral tail vein at days 0, 10 and 16. Mice were then bled at day 21, and serum titre and antibody specificity were determined by indirect ELISA. Three weeks after the last dose, a 4 µg intravenous dose of immunogen was given, and the fusion of splenic B-lymphocytes with myeloma cells was performed 4 days later.

Many food contaminants, such as most mycotoxins, heavy metals, polychlorinated biphenyls, pesticides, herbicides, medicines and poly-cyclic aromatic hydrocarbons, are not immunogenic and therefore need to be conjugated to a protein prior to immunisation, e.g. attached to ovalbumin, bovine serum albumin (BSA) or KLH. In the case of heavy metals, a metal ion chelator is first conjugated to the protein, and the metal–chelator–protein complex is used as an immunogen to obtain antibodies, which have been found to be highly specific for the bound target metal (Blake *et al.*, 1996; Khosraviani *et al.*, 1998).

It is usually necessary to introduce a reactive group on the molecule of interest, prior to conjugation of the contaminant to the protein (Chu, 1996). Generally, when a carbonyl group in the molecule is available, an *O*-carboxymethyl oxime derivative can be prepared, while a reactive hydroxyl group can be derivatised (using bifunctional acid anhydrides) to a hemisuccinate or hemiglutarate (Chu, 1986). For chemicals with an amino reactive group, such as some antibiotics and sulphonamides, glutaraldehyde is used as the coupling agent (Märtlbauer *et al.*, 1994). In most cases, the same conjugation techniques used for immunogen preparation can also be used to link the analyte to enzyme markers, as long as the reaction conditions do not denature the enzyme (Pestka *et al.*, 1995). It is critical to optimise the molecular ratio of antibody or antigen to the detection enzyme for each system. Furthermore, purification of the conjugate to remove unbound antibody, antigen or detection enzyme will prevent a loss in sensitivity (Rittenburg, 1990).

An example of antigen conjugation is the immunoassay for tissue residues of ivermectin, an antiparasitic avermectin that is frequently administered to livestock (Crooks *et al.*, 1998). Ivermectin has been measured in animal tissues using polyclonal antibodies raised against 5-*O*-succinoylivermectin conjugated to apo-transferrin, via carbodiimide activation. Horseradish peroxidase was similarly conjugated to the succinoylated ivermectin and was used as the detection enzyme in a direct competitive ELISA procedure (Crooks *et al.*, 1998).

Anti-idiotype antibodies (anti-id) are antibodies which are considered to be an 'internal image' of the antigen, i.e. they contain regions that bear a structural resemblance to the antigen epitopes that are recognised by other antibodies (Nisonoff, 1991). Anti-id antibodies act by treating the original antibody as the immunogen, thus suppressing its immuno-reactivity by blocking antigen binding. Anti-id antibodies that mimic the epitopes of large molecules are used routinely for clinical diagnosis and immunotherapy and have now also been generated against a number of small molecular mass haptens, including some mycotoxins and herbicides (Chu, 1996). Polyclonal anti-id antibodies are usually generated from mouse ascites, after immunisation with a particular purified rabbit polyclonal antibody that is directed against the target antigen. ELISA analyses have shown that anti-id antibodies bind specifically to the original polyclonal antibody, and not to other types of monoclonal antibodies or normal mouse IgG (Hsu and Chu, 1995). Owing to the structural similarity to antigen epitopes, an anti-id antibody can be used as a substitute for a purified antigen when the antigen is not readily available or is prohibitively expensive. It is also useful for low molecular mass antigens, for which immobilisation to a solid phase can frequently mask epitopes and render the bound antigen immunologically unreactive (Kricka, 1994).

4.2.3 Sample clean-up

The sensitivity of an immunoassay method can be improved by a simple clean-up procedure after extraction (Chu, 1986). Immunoaffinity cartridges are commercially available and are routinely used for the clean-up of extracts during the analysis of food samples contaminated with environmental contaminants, such as mycotoxins (Figure 4.1). These cartridges contain antibodies specific for a certain chemical (or related group of chemicals), which are immobilised on a suitable solid phase, such as Sepharose. The extract, usually diluted in phosphate buffer, is forced through the cartridge while the mycotoxin is left bound to the recognition site of the antibody specific for that molecule. After washing to remove impurities, the mycotoxin is then eluted from the column with an elution solvent such as methanol or acetonitrile, for quantitative determination by immunoassay or high-performance liquid chromatography (HPLC) (van Egmond, 1991). Such columns have also been used for the isolation of dietary heterocyclic amine carcinogens (Vanderlaan et al., 1993) and residues of antibiotics in foods of animal origin (Shaikh and Moats, 1993; Godfrey et al., 1997). Similarly, immunoassay microtitre plate wells can be coated with antibodies that can specifically bind the analyte in a sample or standard. The wells can then be washed to remove interfering material and the bound analyte further purified or analysed (Lucas et al., 1995).

An immunoaffinity column can be prepared as previously described by Ueno et al. (1991). Briefly, monoclonal antibodies specific for the mycotoxin ochratoxin A, were coupled with bromium cyanide-activated Sepharose 4B and then 0.5 ml of the gel is packed into a small glass column containing 0.5 mol/l NaCl PBS. The sample was loaded onto the column, which was then washed with water, and subsequently the ochratoxin A was eluted with 50% dimethyl sulphoxide–40 mmol/l phosphate buffer (pH 5.0, 1:1, v/v).

4.2.4 Detection and quantitation

Various methods are used in the determination of food contaminants, including immunochemical methods such as ELISA and radioimmunoassay (RIA). Generally, an 'immunoassay' is an assay in which the amount of

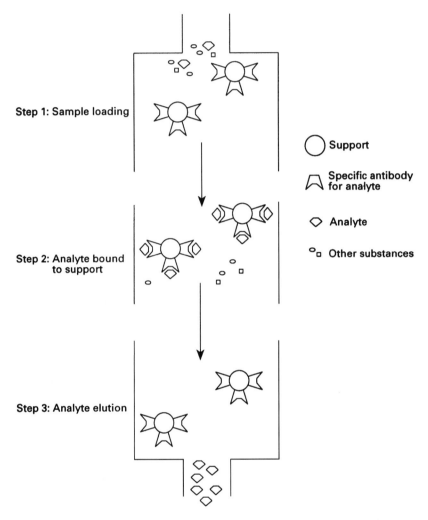

Figure 4.1 Schematic diagram of an immunoaffinity column for the purification of analytes.

available antibody is limited, whereas in an 'immunometric' assay the antibody is in excess and the amount of antigen is limited. A 'heterogeneous' assay is an assay in which the interaction between the antigen and the antibody does not affect the detection-enzyme activity, while in a 'homogeneous' assay the interaction between the antigen and the antibody does alter the activity of the detection enzyme (Edwards, 1985). In competitive immunoassays, the theoretical maximum sensitivity can only be achieved if highly specific detection enzymes are used in the labelling procedure, such as horseradish peroxidase and alkaline phosphatase.

The competitive RIA involves the incubation of a specific antibody with a solution of either a known standard or an unknown sample, and also a known amount of radiolabelled contaminant. The radiolabelled and unlabelled molecules of the contaminant then compete for a limited number of antibody binding sites, so the amount of radiolabelled contaminant bound to the antibody is inversely proportional to the concentration of the contaminant present in the sample. The affinity of the antibody must be the same for the free and labelled antigen, as higher or lower affinities for the labelled antigen will reduce the sensitivity of the assay (Märtlbauer et al., 1994). A variation is the immunoradiometric assay, where the antibody is radiolabelled and present in excess, so that the amount of bound radiolabelled antibody is directly proportional to the amount of contaminant present in the food sample.

For the analysis of mycotoxins, pesticides, heavy metals or other environmental contaminants in foods, one of two types of heterogeneous competitive ELISAs is usually used. These are: (a) the direct competitive ELISA, which involves the use of a toxin–enzyme conjugate, and (b) the indirect competitive ELISA, which involves the use of a protein–toxin conjugate and a secondary antibody to which an enzyme has been conjugated (Figure 4.2). Generally, the ELISA is 10–100 times more sensitive than a RIA when purified mycotoxins are used (Chu, 1986). ELISAs are available in a variety of formats, e.g. as microwell, tube, card, cup, probe and test strip assays (Scott, 1995).

It is important to note that small batch-to-batch differences may occur in commercial immunochemical test kits with regard to the quality of the antibodies and the number of available binding sites immobilised on the support. As the stability of these kits is limited (usually around 6 months), it is essential that they are not used after their expiry date. Each test series or microtitre plate should include a negative (blank) control sample; a positive control sample, which is an artificially contaminated (or 'spiked') sample to indicate false negatives; and possibly a blind control sample (Keukens et al., 1992). An example of the layout of a 96-well plate for ELISA is shown in Figure 4.3. In this 'alternative' plate layout, the outer wells are used only as background wells to avoid problems due to dehydration (as all wells are typically used in the 'standard' plate layout). Replicates (of at least two) can be arranged vertically, horizontally or randomly according to the operator's preference. However, some methods specify the number of wells required for the preparation of a standard curve and the number of replicates required for standards and samples, as well as the number of dilutions of these samples (Märtlbauer et al., 1994).

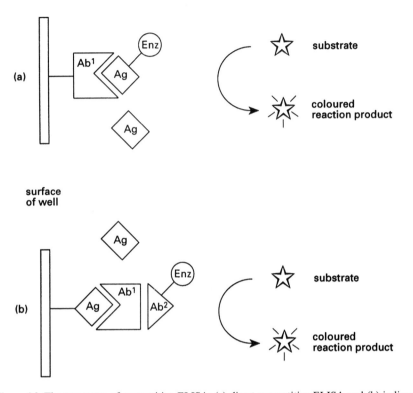

Figure 4.2 The two types of competitive ELISA: (a) direct competitive ELISA and (b) indirect competitive ELISA: Ag, antigen; Ab[1], primary antibody directed towards the Ag; Ab[2], secondary antibody directed toward Ab[1]; Enz, conjugated detection enzyme. In the direct competitive ELISA, the antibody is bound to the well of a plate and to this is added a solution containing the antigen extracted from the food sample and a known concentration of the relevant antigen bound to the detection enzyme. The plate is washed to remove unbound antigen–enzyme conjugate; a substrate for the detection enzyme is then added and the resulting colour is measured spectrophotometrically. The indirect competitive ELISA differs from the competitive ELISA in that it is the antigen that is bound to the plate and a mixture of antigen from the food and the relevant antibody that is added. The plate is washed to remove excess antigen and unbound antibody and a second antibody, conjugated with a detection enzyme, is added. The excess antibody–enzyme conjugate is removed prior to the addition of the enzyme substrate.

4.3 Typical analytes

With over 5 million registered chemicals and 1000 new chemicals being registered every year, the opportunity for the exposure of humans to a large number of chemicals is great. Despite the imposition by governments of regulations designed to control the use of pesticides and

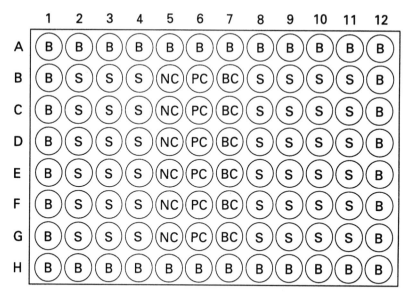

Figure 4.3 An example of the layout of a standard 96-well microtitre plate for ELISA: B, background wells; S, sample wells; NC, negative (blank) control wells; PC, positive control wells; BC, blind control wells (optional).

insecticides in agriculture, high levels of mercury (0.5–10 ppb (ng/g) in Wylie *et al.*, 1991; see Chapter 6), polychlorinated biphenyls (PCBs; see Chapter 9), dichlorodiphenyltrichloroethane (DDT; see Chapters 7 and 9) and other contaminants are found in our waterways and in the fish that inhabit them. The commercial production of foods, involving intensive farming methods, factory production and storage, is also a major source of contamination by compounds such as pesticides (Chapter 7), herbicides, mycotoxins and bacterial toxins. Bushway and Fan (1995) have reviewed the main commercially available immunoassay techniques used for the detection of pesticides and drug residues in food. Below is a short description of some food contaminants that can be detected using immunochemical methods.

4.3.1 Heavy metals

Heavy metals are widely used in industry and discharged into the environment. At least 20 are known to be toxic and many of these, including arsenic, cadmium, copper, lead, mercury, nickel, silver, selenium and zinc, are present in the environment at levels that pose a risk to human health. Immunoassays for metals can utilise a monoclonal antibody that recognises a specific metal–EDTA–protein complex, but not free EDTA or EDTA chelated with other metals (Khosraviani et al., 1998). This method was first applied to indium, which was measured in concentrations as low as 5 ppt (pg/g), using a test system designed for waste water (Chakrabarti et al., 1994).

Cadmium has also been measured by this method using EDTA conjugated to KLH (Blake et al., 1996; see section 2.2) or BSA (Khosraviani et al., 1998). The latter procedure employed an indirect ELISA format with a detection limit of 7 ppb, with little or no interference from a range of cations commonly found in water samples (i.e. Ca^{2+}, Na^+, K^+, Fe^{3+}, Mg^{2+} and Pb^{2+}). Other cations, including Zn^{2+}, Ni^{2+}, In^{3+} and Mn^{2+}, had minimal effect on the assay, but Hg^{2+} caused false positive results at concentrations greater than 1 µmol/l.

Mercury and most of its derivatives are extremely toxic and ubiquitous in the biosphere (WHO, 1990). An indirect ELISA has recently been developed by Wylie et al. (1991) that can detect mercury at concentrations of 0.5 ppb and above in water, as an alternative to atomic absorption spectrometry, which is the current method of choice. This assay has been found to be as sensitive as cold-vapour atomic absorption for mercury detection with only 100 µl of sample (Wylie et al., 1991). At present, the only commercially available metal ion immunoassay is based on this method (i.e. BiMelyze from BioNebraska, for detecting inorganic mercury in soils), and employs a reactive sulphydryl surface to trap ionic mercury and form a mercury–glutathione complex that is recognised by a specific monoclonal antibody (Wylie et al., 1991).

4.3.2 N-*methylcarbamate pesticides (NMCs)*

Carbaryl, carbofuran, aldicarb, methomyl and oxymyl are a few of the most common N-methylcarbamate (NMC) pesticides on the market. There is a magnetic particle-based ELISA, where antibodies are attached to magnetic particles, and commercial ELISA kits are available for the determination of NMCs and their metabolites in water, soil, fruits, vegetables and other crops (Yang et al., 1996).

4.3.3 Organochlorine pesticides (OCs)

The bioaccumulation of OCs in the food chain and the relatively high concentrations found in breast milk (Stevens et al., 1993) are indicative of the persistent nature of OCs in the environment. In the study by Stevens et al. (1993), hexachlorobenzene (HCB) was detected at a median level of 0.1 mg/kg in both breast milk and adipose tissue of nursing mothers in

Western Australia, despite being deregistered in 1972. Although many classes of OCs have been banned in several countries, there is continued detection of traces of DDT and some cyclodienes in food and water (ANZFA, 1996).

Immunoassay kits have been used to detect OCs in a wide range of food types. Assay kits, such as Res-I-Mune (ImmunoSystem Inc., Scarborough, ME, USA), are used for the qualitative screening of food samples for cyclodiene residues, with a detection limit of 10–30 μg/g of endosulfan, endrin and dieldrin in apples, and 10 and 20 μg/g of endosulfan in tomatoes and lettuce, respectively (Wigfield and Grant, 1992). These analyses took less than 20 minutes and required only a UV spectrophotometer, in contrast to the considerable time and specialised instrumentation required for conventional multiresidue chromatographic methods.

4.3.4 Polychlorinated biphenyls (PCBs)

Many people are exposed to PCBs through the consumption of contaminated fish (see Chapter 9), although in some places significant amounts are also accumulated from milk, meat and eggs and by inhalation from the air. The aquatic environment is, however, the major route of entry of PCBs to the food chain (Millichap, 1993). Roberts and Durst (1995) have developed two liposome-based immunomigration techniques for the detection of PCBs. The liposome immunocompetition assay format measures the competitive reaction between analyte-tagged liposomes and the sample analyte for immobilized antibodies and can detect 0.4 nmol of PCB in less than 8 minutes. The more sensitive liposome immunoaggregation assay (LIA) detects the inhibition of immunospecific liposome aggregation in solution and can detect 2.6 pmol of PCB in less than 23 minutes. Both formats utilize capillary action to transport liposome-containing solutions along strips of nitrocellulose. Measurement of colour intensity is then carried out visually or with a desktop scanner (Roberts and Durst, 1995).

4.3.5 Pentachlorophenol (PCP)

Exposure to PCP for the general population is through drinking water and food (up to 40 μg/kg in composite food samples). Exposure to hexachlorobenzene and related compounds in foods, which are then biotransformed to PCP, may also be another important source of exposure (WHO, 1987). Rapid immunoassay methods have been developed for on-site screening of soil samples for PCP, such as Penta

RISc (EnSys Inc., Research Triangle Park, NC, USA), which has a detection limit of 500 ppb and a sample processing time of less than 20 minutes (Mapes *et al.*, 1992). A similar immunoassay compared very well with the standard chromatographic methods for screening water samples for the presence of PCP (as described in section 4.4) (Van Emon and Gerlach, 1992).

4.3.6 Triazine herbicides

Atrazine is one of the most widely applied herbicides and is used to control annual grasses and broad-leaf weeds in a range of crops, including corn, pineapple, sugarcane and macadamia orchards. Many other crops, however, are exposed as a result of soil carry-over in rotational crop programmes (Bushway *et al.*, 1989). Immunoassays have been used to measure atrazine in a range of liquid and solid food samples, with a detection limit of 50 pg/assay tube (Bushway *et al.*, 1989). Simazine, another triazine compound, has also been measured in soil samples, with a detection limit of 2–5 ng/g, using a monoclonal antibody (Goh *et al.*, 1992).

4.3.7 Heterocyclic amines

Heterocyclic amines are a family of carcinogens that are produced through the cooking of meats. 2-Amino-3,8-dimethylimidazo[4,5-*f*]quinoxaline (MeIQx) is responsible for most of the bacterial mutagenicity of well-done cooked beef, although the weaker mutagen 6-phenyl-2-amino-1-methylimidazo[4,5-*f*]pyridine (PhIP) is more common. Immunoaffinity columns containing immobilised monoclonal antibodies to MeIQx and PhIP have been used to purify these mutagens from cooked beef and the urine of consumers (Vanderlaan *et al.*, 1993).

4.3.8 Mycotoxins and fungi

Mycotoxins are a very diverse group of highly toxic secondary metabolites produced by a wide range of filamentous fungi (Blunden *et al.*, 1991). Throughout the world, aflatoxin B_1, zearalenone, ochratoxin A and deoxynivalenol are among the mycotoxins most commonly found following the fungal colonisation of cereal grains (Abouzied *et al.*, 1991). Other commonly affected foods are corn, nuts, rice, soybeans and animal feeds and, consequently, animal tissues and products.

El-Nezami *et al.* (1995) have described an indirect competitive ELISA for the detection of aflatoxin M_1 (AFM_1) in breast milk, with a detection limit of 0.5 pg AFM_1/ml (using a 5 ml sample extracted and reconstituted in 0.5 ml PBS). Both ELISA and RIA have been used in the analysis of trichothecene mycotoxins and it has been found that, although there is good correlation between these methods, values from the RIA method tend to be higher owing to some cross-reaction of the antibody (Park and Chu, 1996). Fungi also produce unique exoantigens for which species-specific rabbit polyclonal antibodies can be produced and utilised for the discriminative monitoring of particular fungal species in food (Lu *et al.*, 1995).

4.3.9 *Veterinary drugs*

The use of veterinary drugs for food-producing animals, such as poultry, cows, swine and farmed fish, can affect public health because of the presence of drug residues or their metabolites in edible products such as milk, eggs and body tissues (Aerts *et al.*, 1995; see Chapter 8). Commonly used veterinary drugs include the sulphonamides, chloramphenicol, tetracyclines, antiparasitic drugs (e.g. avermectins) and tranquillisers. An example of an immunochemical test is the Cite Probe test (Idexx Corporation, Portland, ME, USA), which is available in a kit form and screens for residues of chlortetracycline, oxytetracycline and tetracycline in milk (Barker and Walker, 1992). It is a competitive immunoassay with an assay time of 5 minutes and a detection limit in milk of 20–40 ng/ml. As detection is by visual comparison of the standard and test spots, it is very easy to use on-site (Barker and Walker, 1992). Other immunochemical test kits are available that can also be used to test meat for tetracyclines, such as the La Carte test kit, which contains polyclonal antibodies directed against the analyte immobilised in wells on the card and a solution of enzyme-labelled analyte, a negative control buffer solution and a colour substrate solution (Transia-Biocontrol, Waddinxveen, The Netherlands) and Quik card (Environmental Diagnostics, Burlington, VT, USA) (Keukens *et al.*, 1992).

4.4 Benefits and limitations

The need to control the level of environmental contaminants in food has prompted many countries to pass legislation that stipulates maximum tolerance limits for concentrations of contaminants. If these regulations are to be implemented efficiently, simple, sensitive and rapid methods of analysis are required. The extensive development of methodologies for the production of specific monoclonal and polyclonal antibodies against many food contaminants has resulted in the development of ELISA and RIA methods for the rapid screening of a large number of food samples. These methods are simple, fast, specific, sensitive and inexpensive, and are therefore suitable for the routine diagnostic application of food contaminant detection (Ramakrishna *et al.*, 1990; also see examples described in sections 4.3.1, 4.3.3, 4.3.4, 4.3.6 and 4.3.9).

A typical example of the direct comparisons that have been made between immunoassays and chromatographic methods is a study that compared the PCP immunoassay produced by Westinghouse Bio-Analytic Systems (Rockville, MD, USA) and the standard gas chromatography (GC) Method 604 prescribed by the US Environmental Protection Agency for the detection of PCP in water samples (Van Emon and Gerlach, 1992). Although the GC analysis was more accurate and had a lower limit of detection, it was considerably more expensive and time consuming (Table 4.1). Furthermore, the immunoassay had the additional advantages of not requiring sample extraction and the capability of on-site analysis, and was therefore ideal for the purpose of screening water samples for PCP (Van Emon and Gerlach, 1992).

Immunoassays can be used as confirmation methods for nondestructive classical analyses, such as HPLC and thin-layer chromatography. They can also be used as the method of detection for analytes not amenable to routine chromatographic methods owing to cost and time limitations, or a requirement for rapid on-site analyses without the use of complex analytical equipment, as long as specific antibodies are available and confirmation is performed by GC or gas chromatography–mass spectro-

Table 4.1 Comparison between immunoassay and chromatographic methods for detecting pentachlorophenol (PCP) in water samples

Performance parameters	Immunoassay[a]	Gas chromatography[b]
Detection limit (ppb)	30–40	1–15
Linear range (ppb)	30–400	1–200
Precision (%)	20–30	10–20
Accuracy (%)	±40–50	±30
Analysis time based on sample load	2.5 h/40 samples	0.5 h/sample
Extraction required	No	Yes
Cost per sample[c] ($)	2.50	100–300
Rapid on-site analysis capability	Yes	No
Total analysis time	5 h/40 samples	4.5 hr/sample

Adapted from Van Emon and Gerlach, 1992.
[a]Immunoassay for PCP produced by Westinghouse Bio-Analytic Systems (Rockville, MD, USA).
[b]Standard gas chromatography Method 604 prescribed by the US Environmental Protection agency for the detection of PCP in water samples.
[c]US dollars as estimated by Van Emon and Gerlach (1992).

metry (GC-MS) (Lucas *et al.*, 1995). Therefore, immunoassays, such as ELISA, are very useful for on-site analysis in the field, e.g. the measurement of fumonisin mycotoxins in beer (Scott, 1995; Scott *et al.*, 1997) and of PCP (Mapes *et al.*, 1992) and inorganic mercury in soils (Wylie *et al.*, 1991).

The main disadvantage of immunoassays is that, owing to interfering substances in the sample matrix, most methods require a clean-up step before analysis, and generally different food commodities require different clean-up methods (Chu *et al.*, 1987). The analyte in question may remain bound to sample proteins, or exist in complexes that do not allow them to bind with their specific antibody, leading to false negative results (Barker and Walker, 1992). Interfering substances that bind to the antibody can also lead to false-positive results. For example, both monoclonal and polyclonal antibodies directed to specific sulphonamides exhibit up to 10% cross-reactivity (Aerts *et al.*, 1995) owing to the structural similarity of this group of antibiotics, e.g. sulphamethazine and sulphamerazine differ by a single methyl group (Märtlbauer *et al.*, 1994). However, there are several commercial immunoassays available that are quite specific and quantitative for individual compounds within a chemical class. For example, a number of ELISA kits are available for a range of mycotoxins (and also for NMC and pesticide analysis) that are specific for a compound, or group of compounds. These kits tend to reduce the matrix-dependent interference, as well as the need for sample clean-up and concentration (Yang *et al.*, 1996).

Kits can be used for rapid qualitative screening or for quantitation (Scott, 1995). The typical format for quantitative evaluation is the microtitre plate assay. Quick tests, such as tube tests and dipsticks, are used for qualitative or semiquantitative testing. In a dipstick immunoassay, the nylon membrane that is used as a solid support is pretreated to covalently bind proteins (such as antibodies), which are pipetted onto the membrane in a buffer solution. This test will provide qualitative or semiquantitative results in less than an hour. Compared to microtitre plate assays, this test is up to two orders of magnitude less sensitive, mainly because visual evaluation is less precise and sensitive than measurement of sample absorbance with a spectrophotometer. However, many dipstick immunoassays are less subject to interferences from sample matrices and solvent composition of the extract than are microtitre plate assays (Märtlbauer *et al.*, 1994).

The advantage of noncompetitive immunometric methods is that the amount of analyte-bound sites is determined directly, instead of an indirect determination of the proportion of unbound sites. However, these techniques cannot presently be applied to small molecular mass analytes, such as drugs and pollutants, as these compounds are not large enough to bind simultaneously more than one antibody independently (Self and Cook, 1996).

In pesticide analysis, immunoassays cannot compete with HPLC and GC in terms of the information obtained about the sample composition. The main limitation is that only one compound can be determined at a time and an erroneous result can be produced owing to matrix effects or cross-reactivity (Torres *et al.*, 1996). Nevertheless, immunoassays are very useful for screening a large number of samples for a single analyte in a short time.

The advantage of using monoclonal antibodies over polyclonal antibodies in an enzyme immunoassay is that monoclonal antibodies tend to have higher affinity and specificity. Polyclonal antibodies are simpler to produce, but usually can only be produced in small amounts, while hybridomas are an immortal source of monoclonal antibodies (Candlish *et al.*, 1988). The inherent variability of polyclonal antibodies also makes it difficult to use them in commercial kits with defined performance characteristics (Pestka *et al.*, 1995). Furthermore, the commercial availability of antibodies is currently limited to the most frequently measured food contaminants and drugs.

4.5   Quantification and accuracy

Enzyme immunoassays for antimicrobial drugs in food of animal origin have been shown to have the advantage of speed, sensitivity and simplicity. Cross-reactivity can produce false positive results. Thus, confirmation of immunoassay results by HPLC and GC is required for all legal and statutory purposes (Märtlbauer *et al.*, 1994). A screening method should not produce false negatives, but a limited number of false positives is acceptable. However, the confirmation method must not give false positive results and must reliably quantify the amount of analyte when a maximum residue level has been set (Keukens *et al.*, 1992).

In the case of polychlorinated contaminants, analyses are subject to various uncertainties, such as the disappearance of analytes known to be present (to cause false negatives) as well as the appearance of artefacts (resulting in false positives). Absolute proof of the presence of false positives can only be approached by interlaboratory testing. Within a laboratory, assay replication is not independent because the same variables are usually operative, e.g. instruments, reagents, standards, calibration curve, operations and time (Horwitz and Albert, 1996).

Factors that may contribute to immunoassay variability include pipetting errors, bound/free analyte separation, and signal measurement. Other factors include changes in calibration materials, reagent ageing, the effect of different operators, and antibody/analyte carry-over (the last factor is known to be trivial) (Sadler *et al.*, 1997). Operator effects have been shown to have an important influence on the total error in RIA; however it should be noted that this is not solely a feature of manual assays (Sadler *et al.*, 1997). Automated instruments are also subject to operator effects, such as failure to warm reagents to recommended temperatures and failure to properly follow priming, calibration and maintenance procedures (Sadler *et al.*, 1997).

A crucial point to note about the accuracy of immunoassays is the specificity of the antibodies used. As cross-reaction is possible, it is good laboratory practice to confirm positive immunoassay results with another method of analysis (van Egmond, 1991). Chromatographic techniques are commonly used for confirmation as they have been shown to have better specificity than immunochemical methods (Ferrara *et al.*, 1994).

The precision of an immunoassay is affected mainly by two factors. The first is the location of the absorbance values on the calibration curve: as the calibration curve is nonlinear, the values nearest to 50% binding give the most precise results. The second factor is the number of replicates used for each sample, as obviously the method is more precise if the number of replicates is increased (Märtlbauer *et al.*, 1994).

Current opinion suggests that the total analytical error should not exceed one-half of the intra-individual biological variation. Thus, increase in biological variation due to analytical errors is generally limited to around 10% (Sadler *et al.*, 1997).

4.6 Quality control

There are no internationally recognised reference methods for highly sensitive immunoassays, possibly because these methods are very dependent on the quality of the reagents used and the contaminant in question is sometimes insufficiently characterised to allow preparation of adequate reference materials (Bergamaschi, 1995).

In order for a method to be certified, a number of internal and external quality assurance measures must be included in the procedure. These are described by Keukens *et al.* (1992) for the analysis of chloramphenicol residues in meat by immunochemical methods, HPLC and GC-MS. In short, these measures include:

- the inclusion of recovery samples;
- the inclusion of blank samples;
- the inclusion of known (internal) or unknown 'blind' (external) samples that are naturally contaminated control or reference materials;
- steps to prevent cross-contamination;

- use of at least two replicates;
- the establishment of a maximum allowable deviation of individual results from the average recovery; and
- involvement in quality-controlled collaborative studies.

Currently, there is a need for such (inter)national collaborative studies, as well as certified reference materials, to test the accuracy of these methods (Aerts *et al.*, 1995). Control over pre-analytical factors is important, as even a well-planned investigation is useless if samples are not handled correctly. Therefore, care must be taken during the collection, storage and transportation of samples (Bergamaschi, 1995).

4.7 Relationship to chemical analysis

A combination of the sensitivity of an immunochemical assay, such as an ELISA, with the selectivity of a physicochemical method, such as HPLC, results in a highly reliable analysis. However, owing to time and cost limitations, the chromatographic techniques are generally restricted to confirmation analyses, while other more rapid and inexpensive test formats are utilised for first-action screening (Märtlbauer *et al.*, 1994). This trend is not surprising as the costs of chemical analyses are ever increasing owing to the expense of solvents and of their disposal, and also the need for trained personnel and expensive equipment (Bushway and Fan, 1995). In pesticide analysis, results from ELISA kits were found to be comparable to those from other methods such as HPLC (Yang *et al.*, 1996). In the analysis of animal products for drug residues, such as antibiotics, immunoassays have often been found to lack specificity and also to be subject to false positive and false negative results; however, they are still used for rapid screening in food monitoring programmes to determine which samples should be subjected to more elaborate analytical methods (Barker and Walker, 1992).

Immunoassays and chemical analyses are not mutually exclusive methods, but rather can be used in conjunction in the determination of analytes. An ELISAgram has been devised to separate structurally related mycotoxins, combining the sensitivity and specificity of a competitive ELISA with the capacity of high-performance thin-layer chromatography (HPTLC) (Pestka *et al.*, 1995). In this method, the mycotoxins are separated by HPTLC and the TLC plate is blotted with nitrocellulose coated with a mycotoxin-specific monoclonal antibody. The nitrocellulose is then incubated with a mycotoxin–enzyme conjugate, and the bound conjugate (which identifies the unreacted antibody binding sites) is detected with a precipitating substrate, either visually or densitometrically, as inhibition bands (Pestka *et al.*, 1995). The authors believe that this method could be widely applicable to the simultaneous quantitation and confirmation of multiple analytes with a single cross-reactive antibody.

In the detection of veterinary drugs such as tetracyclines, the development of simple isocratic HPLC analysis methods (which can simultaneously detect most compounds) reduces the analytical time-frame required per sample. A combination of immunochemical detection and screening techniques, coupled with multiresidue detection and quantitation through rapid HPLC analyses, provides the necessary speed and accuracy required for the regulation of these drugs in foods of animal origin (Barker and Walker, 1992).

4.8 Future perspectives

There is currently an increase in the utilisation of immunochemical methods for screening and quantitation of food contaminants. Present methods are being replaced by antibody–antigen-based systems that simultaneously perform extraction, isolation, detection and, in some cases, quantitation (Barker and Walker, 1992). However, the lack of reference methods continues to be a problem for many immunoassays (Sadler *et al.*, 1997). There is a great need, therefore, for the validation of immunochemical methods by regulatory agencies.

There is also a need to develop immunoaffinity columns and disks with specific antibodies for a wider range of food contaminants. Progress in the development of more specific antibodies will continue to be made, as well as improvements in the use of anti-idiotype antibodies, and also structural modification (Chu, 1996). If a food contaminant is too small to provide recognition sites for the antibody, it can be derivatised to a product that is often better recognised by the antibody (Lucas *et al.*, 1995).

Continued progress with immunoassays has led to the development of immunosensors. Immunosensors are analytical devices that detect the binding of an antigen to its specific antibody by coupling the immunochemical reaction to the surface of a transducer device (Gizeli and Lowe, 1996). They have the advantage over current immunoassay methods of providing continuous 'on-line' measurements. In piezoelectric immunosensors, a piezoelectric crystal acts as a microbalance and changes in its mass due to an antigen–antibody reaction on the crystal surface produce a decrease in resonance frequency. However, early prototypes of this immunosensor have been found to bind non-specifically to proteins in the sample (Kricka, 1994).

Optical immunosensors have been used for the detection of the pesticide atrazine. In this competitive assay system, an atrazine derivative immobilised on the device is exposed to the test sample, which has previously been incubated with a monoclonal anti-atrazine antibody. However, the lower detection limit of this method (0.25 ppb) is higher than that imposed by the European Union regulations (0.1 ppb) (Brecht et al., 1995).

In an alternative approach, an amperometric (electrochemical) immunosensor that employs a screen-printed electrode as an ampero-metric transducer, combined with monoclonal antibodies against 2,4-dichlorophenoxyacetic acid (2,4-D) in a competitive assay format, has been successfully used to detect 2,4-D with a detection limit of 0.1 ppb (Kal and Sklad, 1995).

Gizeli and Lowe (1996) have suggested that an ideal immunosensor system should possess all of the following properties:

(1) the ability to detect and quantify antigen within the required concentration range and within a reasonable time-frame (i.e. a few seconds);
(2) the ability to transduce the binding of antigen to antibody without additional reagents;
(3) use of a reversible immunochemical reaction, i.e. a capacity to repeat the measurement; and
(4) the ability to detect the specific binding of the antigen in naturally contaminated samples.

Currently, there are very few immunosensor systems, if any, that possess all of these properties. However, owing to rapid developments in the field of immunosensor technology, some of these difficulties may soon be overcome.

References

Abouzied, M.M., Azcona, J.I., Braselton, W.E. and Pestka, J.J. (1991) Immunochemical assessment of mycotoxins in 1989 grain foods: evidence for deoxynivalenol (vomitoxin) contamination. *Applied and Environmental Microbiology*, **57** 672-677.

Abraham, G.E. (1977) *Handbook of Radioimmunoassay*. Marcel Dekker, New York.

Aerts, M.M.L., Hogenboom, A.C. and Brinkman, U.A.Th. (1995) Analytical strategies for the screening of veterinary drugs and their residues in edible products. *Journal of Chromatography B*, **667** 1-40.

ANZFA (Australia New Zealand Food Authority). (1996) *The 1994 Australian Market Basket Survey*. AGPS, Canberra.

Azcona-Olivera, J.I., Abouzied, M.M., Plattner, R.D. and Pestka, J.J. (1992) Production of monoclonal antibodies to the mycotoxins fumonisins B1, B2, and B3. *Journal of Agricultural Food Chemistry*, **40** 531-534.

Barker, S.A. and Walker, C.C. (1992) Chromatographic methods for tetracycline analysis in foods. *Journal of Chromatography*, **623** 195-209.

Bergamaschi, E. (1995) Quality assurance for chemical methods. *Toxicology Letters*, **77** 205-208.

Blake, D.A., Chakrabarti, P., Khosraviani, M., Hatcher, F.M., Westhoff, C.M., Goebel, P., Wylie, D.E. and Blake II, R.C. (1996) Metal binding properties of a monoclonal antibody directed toward metal–chelate complexes. *Journal of Biological Chemistry*, **271** 27677-27685.

Blunden, G., Roch, O.G., Rogers, D.J., Coker, R.D., Bradburn, N. and John, A.E. (1991) Mycotoxins in food. *Medical Laboratory Sciences*, **48** 271-282.

Brecht, A., Piehler, J., Lang, G. and Gauglitz, G. (1995) A direct optical immunosensor for atrazine detection. *Analytica Chimica Acta*, **311** 289-299.

Bushway, R.J. and Fan, T.S. (1995) Detection of pesticide and drug residues in food by immunoassay. *Food Technology*, **49** 108-115.

Bushway, R.J., Perkins, B., Savage, S.A., Lekousi, S.L. and Ferguson, B.S. (1989) Determination of atrazine residues in food by enzyme immunoassay. *Bulletin of Environmental Contamination and Toxicology*, **42** 899-904.

Candlish, A.A.G., Stimson, W.H. and Smith, J.E. (1988) Determination of ochratoxin A by monoclonal antibody-based enzyme immunoassay. *Journal of the Association of Official Analytical Chemists*, **71** 961-964.

Chakrabarti, P., Hatcher, F.M., Blake II, R.C., Ladd, P.A. and Blake, D.A. (1994) Enzyme immunoassay to determine heavy metals using antibodies to specific metal–EDTA complexes: optimization and validation of an immunoassay for soluble indium. *Analytical Biochemistry*, **217** 70-75.

Chu, F.S. (1996) Immunochemical methods for fumonisins. *Advances in Experimental Medicine and Biology*, **392** 123-133.

Chu, F.S. (1986) Recent studies on immunochemical analysis of mycotoxins, in *Mycotoxins and Phycotoxins* (eds P.S. Steyn and R. Vleggar), Elsevier Science, Amsterdam, pp 277-292.

Chu, F.S., Fan, T.S.L., Zhang, G.-S., Xu, Y.-C., Faust, S. and McMahon, P.L. (1987) Improved enzyme-linked immunosorbent assay for aflatoxin B1 in agricultural commodities. *Journal of the Association of Official Analytical Chemists*, **70** 854-857.

Crooks, S.R.H., Baxter, A.G., Traynor, I.M., Elliott, C.T. and McCaughey, W.J. (1998) Detection of ivermectin residues in bovine liver using an enzyme immunoassay. *Analyst*, **123** 355-358.

Edwards, R. (1985) *Immunoassay. An Introduction*. William Heinemann Medical Books, London.

El-Nezami, H.S., Nicoletti, G., Neal, G.E., Donohue, D.C. and Ahokas, J.T. (1995) Aflatoxin M_1 in human breast milk samples from Victoria, Australia and Thailand. *Food and Chemical Toxicology*, **33** 173-179.

Engvall, E. and Perlman, P. (1971) Enzyme-linked immunosorbent assay (ELISA). Quantitation of immunoglobulin G. *Immunochemistry*, **8** 871-874.

Ferrara, S.D., Tedeschi, L., Frison, G., Brusini, G., Castagna, F., Bernardelli, B. and Soregaroli, D. (1994) Drugs-of-abuse testing in urine: statistical approach and experimental comparison of immunochemical and chromatographic techniques. *Journal of Analytical Toxicology*, **18** 278-291.

Gizeli, E. and Lowe, C.R. (1996) Immunosensors. *Current Opinion in Biotechnology*, **7** 66-71.

Godfrey, M.A., Luckey, M.F. and Kwasowski, P. (1997) IAC/cELISA detection of monensin elimination from chicken tissues, following oral therapeutic dosing. *Food Additives and Contaminants*, **14** 281-286.

Goh, K.S., Spurlock, F., Lucas, A.D., Kollman, W., Schoenig, S., Braun, A.L., Stoddard, P., Biggar, J.W., Karu, A.E. and Hammock, B.D. (1992) Enzyme-linked immunosorbent assay (ELISA) of simazine for Delhi and Yolo soils in California. *Bulletin of Environmental Contamination and Toxicology*, **49** 348-353.

Horwitz, W. and Albert, R. (1996) Reliability of the determinations of polychlorinated contaminants (biphenyls, dioxins, furans). *Journal of the Association of Official Analytical Chemists International*, **79** 589-621.

Hsu, K.-H. and Chu, F.S. (1995) Anti-idiotype and anti-anti-idiotype antibodies genearated from polyclonal antibodies against aflatoxin B_1. *Food and Agricultural Immunology*, **7** 139-151.

Kal, B.T. and Sklad, L.P. (1995) A disposable amperometric immunosensor for 2,4-dichlorophenoxyacetic acid. *Analytica Chimica Acta*, **304** 361-368.

Keukens, H.J., Aerts, M.M.L., Traag, W.A., Nouws, J.F.M., De Ruig, W.G., Beek, W.M.J. and Den Hartog, J.M.P. (1992) Analytical strategy for the regulatory control of residues of chloramphenicol in meat: preliminary studies in milk. *Journal of the Association of Official Analytical Chemists International*, **75** 245-256.

Khosraviani, M., Pavlov, A.R., Flowers, G.C. and Blake, D.A. (1998) Detection of heavy metals by immunoassay: optimisation and validation of a rapid, portable assay for ionic cadmium. *Environmental Science and Technology*, **32** 137-142.

Kricka, L.J. (1994) Selected strategies for improving sensitivity and reliability of immunoassays. *Clinical Chemistry*, **40** 347-357.

Lu, P., Marquardt, R.R. and Kierek-Jaszczuk, D. (1995) Immunochemical identification of fungi using antibodies raised in rabbits to exoantigens from *Aspergillus ochraceus*. *Letters in Applied Microbiology*, **20** 41-45.

Lucas, A.D., Gee, S.J., Hammock, B.D. and Seiber, J.N. (1995) Integration of immunochemical methods with other analytical techniques for pesticide residue determination. *Journal of the Association of Official Analytical Chemists International*, **78** (3) 585-591.

Mapes, J.P., McKenzie, K.D., McClelland, L.R., Movassaghi, S., Reddy, R.A., Allen, R.L. and Friedman, S.B. (1992) Penta RISc^{TM} soil—a rapid, on-site screening test for pentachlorophenol in soil. *Bulletin of Environmental Contamination and Toxicology*, **49** 334-341.

Marks, V. (1985) Uses of immunoassay, in *Alternative Immunoassays* (ed. W.P. Collins), Wiley, Chichester, pp 1–5.

Märtlbauer, E., Usleber, E., Schneider, E. and Dietrich, R. (1994) Immunochemical detection of antibiotics and sulfonamides. *Analyst*, **119** 2543-2548.

Millichap, J.G. (1993) *Environmental Poisons in Our Food*, PNB Publishers, Chicago.

Nakamura, R.M. (1992) General principles of immunoassays, in *Immunochemical Assays and Biosensor Technology for the 1990s* (eds R.M. Nakamura, Y. Kasahara and G.A. Rechnitz), American Society for Microbiology, Washington, DC, pp 3-21.

Nisonoff, A. (1991) Idiotypes: concepts and applications. *Journal of Immunology*, **147** 2429-2438.

Park, J.J. and Chu, F.S. (1996) Assessment of immunochemical methods for the analysis of trichothecene mycotoxins in naturally occurring moldy corn. *Journal of the Association of Official Analytical Chemists International*, **79** 465-471.

Pestka, J.J., Abouzied, M.N. and Sutinko. (1995) Immunological assays for mycotoxin detection. *Food Technology*, **49** 120-128.

Ramakrishna, N., Lacey, J., Candlish, A.A., Smith, J.E. and Goodbrand, I.A. (1990) Monoclonal antibody-based enzyme immunosorbent assay of aflatoxin B1, T-2 toxin and ochratoxin A in barley. *Journal of the Association of Official Analytical Chemists*, **73** 71-76.

Rittenburg, J.H. (1990) Fundamentals of immunoassay, in *Development and Application of Immunoassay for Food Analysis* (ed. J.H. Rittenburg), Elsevier Applied Science, New York, pp 29-57.

Roberts, M.A. and Durst, R.A. (1995) Investigation of liposome-based immunomigration sensors for the detection of polychlorinated biphenyls. *Analytical Chemistry*, **67** 482-491.

Sadler, W.A., Smith, M.H., Murray, L.M. and Turner, J.G. (1997) A pragmatic approach to estimating total analytical error of immunoassays. *Clinical Chemistry*, **43** 608-614.

Scott, P.M. (1995) Mycotoxin methodology. *Food Additives and Contaminants*, **12** 395-403.

Scott, P.M., Yeung, J.M., Lawrence, G.A. and Prelusky, D.B. (1997) Evaluation of enzyme-linked immunosorbent assay for analysis of beer for fumonisins. *Food Additives and Contaminants*, **14** 445-450.

Self, C.H. and Cook, D.B. (1996) Advances in immunoassay technology. *Current Opinion in Biotechnology*, **7** 60-65.

Shaikh, B. and Moats, W.A. (1993) Liquid chromatographic analysis of antibacterial drugs in food products of animal origin. *Journal of Chromatography*, **643** 369-378.

Siddle, K. (1985) Properties and applications of monoclonal antibodies, in *Alternative Immunoassays* (ed. W.P. Collins), Wiley, Chichester, pp 13-37.

Stevens, M.F., Ebell, G.F. and Psaila-Savona, P. (1993) Organochlorine pesticides in Western Australian nursing mothers. *Medical Journal of Australia*, **158** 238-241.

Torres, C.M., Picó, Y. and Mañes, J. (1996) Determination of pesticide residues in fruit and vegetables. *Journal of Chromatography A*, **754** 301-331.

Ueno, Y., Kawamura, O., Sugiura, Y., Horiguchi, K., Nakajima, M., Yamamoto, K. and Sato, S. (1991) Use of monoclonal antibodies, enzyme-linked immunosorbent assay and immunoaffinity column chromatography to determine ochratoxin A in porcine sera, coffee products and toxin-producing fungi, in *Mycotoxins, Endemic Nephropathy and Urinary Tract Tumours* (eds M. Castegnaro, R. Pletina, G. Dirheimer, I.N. Chernozemsky and H. Bartsch), International Agency for Research on Cancer, Lyon, pp 71-75.

Vanderlaan, M., Hwang, M. and Djanegara, T. (1993) Immunoaffinity purification of dietary heterocyclic amine carcinogens. *Environmental Health Perspectives*, **99** 285-287.

van Egmond, H.P. (1991) Methods for determining ochratoxin A and other nephrotoxic mycotoxins, in *Mycotoxins, Endemic Nephropathy and Urinary Tract Tumours* (eds M. Castegnaro, R. Pleština, G. Dirheimer, I.N. Chernozemsky and H. Bartsch), International Agency for Research on Cancer, Lyon, pp 57-70.

Van Emon, J.M. and Gerlach, R.W. (1992) Evaluation of a pentachlorophenol immunoassay for environmental water samples. *Bulletin of Environmental Contamination and Toxicology*, **48** 635-642.

Wigfield, Y.Y. and Grant, R. (1992) Evaluation of an immunoassay kit for the detection of certain organochlorine (cyclodiene) pesticide residues in apple, tomato, and lettuce. *Bulletin of Environmental Contamination and Toxicology*, **49** 342-347.

World Health Organisation (1987) *Environmental Health Criteria 71—Pentachlorophenol*, United Nations Program, Finland.

World Health Organisation (1990) *Environmental Health Criteria 101—Methylmercury*, WHO, Geneva.

Wylie, D.E., Carlson, L.D., Carlson, R., Wagner, F.W. and Schuster, S.M. (1991) Detection of mercuric ions in water by ELISA with a mercury-specific antibody. *Analytical Biochemistry*, **194** 381-387.

Yalow, R.S. and Berson, S.A. (1960) Immunoassay of endogenous plasma insulin in man. *Journal of Clinical Investigation*, **39** 1157-1175.

Yang, S.S., Goldsmith, A.I. and Smetena, I. (1996) Recent advances in the residue analysis of *N*-methylcarbamate pesticides. *Journal of Chromatography A*, **754** 3-16.

5 Radioactivity in food

Barrie E. Lambert and Katherine J. Mondon

5.1 Introduction

Radioactivity occurs naturally and as a consequence of the use of radiation in nuclear power, nuclear weapons and medicine and may expose people to radiation through a number of pathways—direct exposure, inhalation or ingestion (Figure 5.1). Under normal conditions, contemporary waste discharges to the atmosphere and to rivers and the sea from the nuclear power industry do not now represent a serious threat in terms of the contamination of food. This has not, however, always been so and, in addition, accidents have added substantially to both local and continental pollution. Together with the radioactive fallout from atmospheric nuclear weapons tests (now banned), these sources have resulted in significant food contamination in the past. Consequently, the need for control of the contamination of food has been recognised and standards have been developed internationally. This chapter gives an overview of radioactivity in food and concludes that only in exceptional circumstances does the contamination of the diet from human endeavours now result in significant risk.

5.2 Radiation units and biological effects

The unit of radioactivity is the becquerel (Bq), where one Bq is one radioactive disintegration per second. A disintegrating radionuclide will emit γ-rays and/or subatomic α- or β-particles (all collectively referred to as radiation), which are capable of causing ionisations in the material through which they pass. As radiation interacts with matter, it deposits energy, called the absorbed dose, measured in grays (Gy), where 1 Gy is equal to 1 joule/kg. Dose for dose, α-particles are more damaging to living tissues than β-particles or γ-rays and hence, for the purposes of comparing doses from different radiations, the absorbed dose is multiplied by a factor that takes account of how a particular radiation distributes energy in tissues. It is then called *equivalent dose*, measured as sieverts (Sv), where

$$1\,\text{Sv} = 1\,\text{Gy} \times W_r \tag{5.1}$$

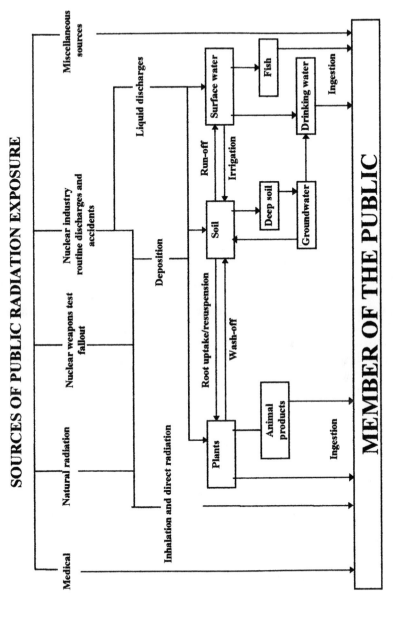

Figure 5.1 Major pathways of public exposure to radioactivity in the environment.

W_r = radiation weighting factor for the radiation under consideration;
    W_r = 1 for γ-rays and β-particles;
    W_r = 20 for α-particles.

Organs in the body have different susceptibilities to radiation. To sum the effects of radiation on the whole body, the effects on different organs are weighted (by a factor W_t) and the resultant dose is known as the *effective dose*, also in Sv. Finally, the kinetics of the turnover of a radionuclide taken into the body depends on (a) the physical half-life ($T_{1/2}$), which is the time taken for the radioactivity to halve by decay; and (b) the biological half-life (T_{biol}), which is governed by the behaviour of the radionuclide once in the body and usually follows the same pattern as for stable chemical analogues.

The overall residence time in the body is described by the effective half-life (T_{eff}), where

$$\frac{1}{T_{eff}} = \frac{1}{T_{1/2}} + \frac{1}{T_{biol}} \tag{5.2}$$

The dose from an intake will be received over a period of time, that is, after intake one is 'committed' to receiving a radiation dose. The total dose is, therefore, calculated at the time of intake, as the committed effective dose, given as sieverts. All references to doses as sieverts in this chapter are to the committed effective dose and therefore take all the above complications into consideration. The sievert is a very large unit of dose, and therefore most routine dose calculations from the ingestion pathway are expressed as millisieverts (mSv) or microsieverts (μSv). Current annual average doses to adults from all sources are given in Table 5.1(a).

Present knowledge suggests that all radiation exposure increases the risk of cancer, although acute doses over about 2–3 Sv can also cause more immediate death from the so-called 'radiation sickness' syndrome. Likely doses due to the ingestion of artificially contaminated food will be of the order of a few μSv per year, the only effects of which will be a very, very small increase in the risk of cancer. As the normal or spontaneous risk of cancer over a lifetime in all populations is about 1 in 4, the risk from ingestion of food contaminated at current levels (apart from those near Chernobyl, see section 5.9.2) would only be detectable by epidemiological studies on vast populations. As the consequences of radiation exposure at low levels are considered to be probabilistic, the effects on an individual would be practically impossible to ascribe in terms of causation. However, there is no doubt that the ingestion of food containing normal levels of naturally occurring radionuclides is

Table 5.1 Current annual average doses (μSv) to adults

Source	μSv/y
(a) Doses from all sources	
Natural sources	
Cosmic rays	380
Cosmogenic radionuclides	10
Terrestrial radiation	
external exposure	460
internal exposure[a]	229
radon + daughters	1270
Total	2349
Artificial sources	
Medical uses (developed countries)	1000–2000
Weapons testing	5
Nuclear power[b]	1–200
Occupational doses	1000–10000
(b) Doses from ingestion of naturally occurring radionuclides	
Radionuclide	
^{14}C	12
Uranium isotopes	0.28
Thorium isotopes	0.79
Radium isotopes	7.7
$^{210}Pb/^{210}Po$	43
^{40}K	165
Total (rounded)	229

From UNSCEAR 1993.
[a] Excludes radon; data for individual isotopes presented in part (b).
[b] Most highly exposed people living near nuclear installations.

responsible for some fraction (approximately 3–5%) of normal or spontaneous cancers.

The biological effects of radiation are regularly reviewed by expert groups, in particular the United Nations Scientific Committee on the Effects of Atomic Radiation (UNSCEAR, 1993) and the Biological Effects of Ionising Radiation Committee (BEIR, 1989) in the United States. Recommendations on radiation protection standards for both workers and the public are made by the International Commission on Radiological Protection (ICRP), and most countries use ICRP publications as the basis for their own national guidelines. In its latest publication on dose limits (ICRP, 1991) ICRP has recommended an annual dose limit to the public of 1 mSv excluding medical uses and doses from natural sources. It is generally accepted that the risk of cancer induction is linear with exposure, with no threshold, and is currently estimated at 5% per Sv for fatal cancer for members of the public. This means that if 100 people each received 1 Sv, 5 of them would be expected to die of cancer as a consequence of their exposure—this is the same as

100 000 people each receiving 0.001 Sv (1 mSv). Hence the 1 mSv limit can be interpreted as a risk of 5×10^{-5} (5 in 100 000) per year. This is judged to be an 'acceptable' risk for a member of the public by comparison with other risks faced by society (Royal Society, 1992; HSE, 1992). Apart from recommending dose limits, the ICRP's main principles of radiation protection include the concepts of justification (i.e. no practice involving radioactivity shall be adopted unless it produces a net benefit) and optimisation (this assumes all doses carry some risk and therefore exposures should be kept as low as reasonably achievable, the ALARA principle). The ALARA principle, in particular, dominates the way in which the nuclear industry is regulated so that the public dose limit is not treated as a ceiling that is reached during normal operations. Thus the dose limit is considered as the border of acceptability (the third ICRP principle: limitation).

5.3　Assessing the dose from ingestion of radionuclides

Table 5.2 lists those radionuclides of potential importance to the ingestion pathway. If a radionuclide is ingested, the dose it will deliver is determined from its gut uptake, its effective half-life (T_{eff}, see section 5.2) and the type and energy of its emitted radiations. Gut uptake factors, known as the fractional absorption (f_1) and expressed as the fraction of activity taken up into the blood from the gut, are mainly a function of chemistry but are often complicated by active biological transport and the amount and type of material in the gut. For example, isotopes of iodine appear to be completely absorbed across the gut regardless of dietary sources, whereas plutonium has an extremely low gut uptake factor (see Table 5.2 for f_1 values). In contrast, the uptake of radiostrontium is strongly influenced by that of its chemical congener, calcium, which is a homeostatically controlled essential element (Russell, 1966); there is a roughly inversely proportional relationship between strontium uptake and calcium concentration in the diet. Although some radionuclides may have low gut uptake factors, they may significantly irradiate the gut and other body organs as they pass through the body because of the energy and penetrating ability of their radioactive emissions. An example of such a radionuclide is ruthenium-106 (^{106}Ru) which has an energetic β emission. These points can be illustrated by considering the examples of both a very short-lived fission product[1], such

---

[1] Fission products are the radioactive products of fission, or splitting of the atomic nucleus of an element such as uranium or plutonium.

Table 5.2 Radionuclides of potential importance to the ingestion pathway

Radionuclide	Physical half-life[a]	Most important routes of ingestion	f_1[b]	Ingestion dose coefficient (nSv/Bq[c])		
				1-year-old	10-year-old	Adult
Pure β Emitters						
^{3}H	12.3 y	All food (tritiated water)	1	0.05	0.02	0.02
^{3}H	12.3 y	All food (organically bound tritium)	1	0.12	0.06	0.04
^{14}C	5730 y	All food	1	1.6	0.80	0.58
^{89}Sr	50.5 d	Milk/seafood/cereals	0.3	18	5.8	2.6
^{90}Sr	29.1 y	Milk/seafood/cereals	0.3	73	60	28
^{99}Tc	213 000 y	Crustaceans	0.5	4.8	1.3	0.64
β/γ Emitters						
^{60}Co	5.27 y	Molluscs	0.1	27	11	3.4
^{65}Zn	244 d	Molluscs	0.5	16	6.4	3.9
^{106}Ru	1.01 y	Offal/crustaceans	0.05	49	15	7
^{129}I	15.7×10^{6} y	Milk/crops	1	220	190	110
^{131}I	8.04 d	Milk/crops	1	180	52	22
^{134}Cs	2.06 y	Milk/crops/animal products/seafood	1	16	14	19
^{137}Cs	30 y	Milk/crops/animal products/seafood	1	12	10	13
^{210}Pb	22.3 y	Offal/molluscs/water/reinder meat	0.2	3600	1900	690
α Emitters						
^{210}Po	138 d	Offal/molluscs/water/reindeer meat	0.5	8800	2600	1200
^{226}Ra	1600 y	Milk/animal products/crops/brazil nuts/water	0.2	960	800	280
^{228}Ra	6.75 y	Milk/animal products/crops/brazil nuts/water	0.2	5700	3900	690
^{228}Th	1.91 y	Crops/water	0.0005	370	150	72
^{232}Th	14×10^{9} y	Crops/water	0.0005	450	290	230
^{235}U	700×10^{6} y	Crops/water	0.02	130	71	47
^{238}U	4.5×10^{9} y	Crops/water	0.02	120	68	45
^{238}Pu	87.8 y	Offal/molluscs	0.0005	400	240	230
^{239}Pu/^{240}Pu	24 100 y	Offal/molluscs	0.0005	420	270	250
^{241}Am	432 y	Offal/molluscs	0.0005	370	220	200

[a] y = year; d = day.
[b] Source: ICRP (1996). f_1 is the fractional absorption (gut uptake factor).
[c] 1nSv = 10^{-9} Sv.

as iodine-131 (^{131}I) with a $T_{1/2}$ of 8 days, and a very long-lived activation product[2], such as plutonium-239 (^{239}Pu) which has a $T_{1/2}$ of 24 000 years. Iodine-131 sublimes and is therefore very volatile and would be one of the most significant radionuclides following an accidental release from a reactor, because it enters the body readily through inhalation, and via the foodchain. Once in the body, ^{131}I concentrates in the thyroid, where it remains for some time, having a T_{biol} of 80–120 days. Its T_{eff} is obviously dominated by its $T_{1/2}$, and is about 7.5 days. It is a potent causative agent in thyroid cancer—11 years after the Chernobyl accident, the only quantifiable effect so far on the exposed population has been a marked increased in childhood thyroid cancers (OECD, 1995). Stable iodine can 'block' the thyroid and protect it from radioiodine. It is for this reason that emergency plans associated with most nuclear reactors include the issue of iodate tablets to the immediate population in the event of a release of volatile fission products. Plutonium-239 is an α emitter. If ingested, it is hardly absorbed by the gut, but once in the body it concentrates in the skeleton and liver, where it has a T_{biol} of 100 years and 40 years respectively, with an overall T_{eff} exceeding the lifetime of any person (200 years). Thus, even though plutonium is poorly absorbed from the diet, it presents a significant risk because of its persistence in the body and its α emissions.

Fortunately, the factors discussed above have been taken into account in tables of dose coefficients, in units of sieverts of dose per becquerel ingested, for different age groups (ICRP, 1996) and these are reproduced in Table 5.2 for a number of radionuclides. In theory, therefore, all that is needed to calculate a radiation dose from ingestion of a radionuclide is the concentration in the food or water, the intake of the food or water over the period of interest, and the relevant dose coefficient for the radionuclide.

$$\text{Annual dose (Sv)} = \begin{matrix} \text{Radioactivity} \\ \text{in the food} \\ \text{(Bq/kg)} \end{matrix} \times \begin{matrix} \text{Food} \\ \text{consumed} \\ \text{(kg/year)} \end{matrix} \times \begin{matrix} \text{Dose coefficient} \\ \text{for the radionuclide} \\ \text{(Sv/Bq)} \end{matrix}$$

(5.3)

To calculate the total dose from ingestion, this process needs to be reiterated for all nuclides and all foods in the diet.

In practice, determination of the total dose from ingestion is not quite this simple. First, there is the problem of data at the limit of detection (LOD, see section 5.7). Zero concentration could be assumed, or an actual concentration at the LOD. More usually, a final dose is recorded

---

[2] Activation products result from capture of neutrons by the nucleus of an element.

as 'less than' a certain amount obtained using a LOD concentration. Secondly, appropriate consumption rates need to be obtained. In many cases, dose assessments are undertaken to establish the dose to those most at risk from particular sources of anthropogenic radioactivity, such as nuclear facilities (called the critical group). Usually, food consumption rates are needed for different age groups and for a wide range of foods since different radionuclides concentrate in different foods. A further choice is needed in the use of average rates of consumption or elevated rates, e.g. 97.5 percentiles, or some combination thereof.

The choice of method for dose estimation will depend on the intended use of the data, and the availability of site-specific information. In any case, the methodology should be explained clearly for those using the information. It can therefore be seen that a dose calculated from ingestion of radionuclides can have a number of uncertainties associated with it. These uncertainties are associated with, *inter alia*, the methods of obtaining data on ingestion rates. For instance, using a duplicate diet study to measure intake directly depends on the honesty of participants in providing a complete portion of what they have eaten. This method has been shown to result in about a 20% underestimate of food intakes compared with other methods (Bull and Wheeler, 1986). The assumption of intakes using the 97.5 centile gives an assessment of intake that is about three times the mean. This is conservative and implies an overestimate for all but 2.5% of those studied.

5.4   Standards for radioactivity in food and water

Accidents involving radioactivity have highlighted the vital importance of having pre-set standards for permitted concentrations in foods that can be rapidly implemented. The European Union has regulations (Council Regulations No. 3954/87 and 2218/89), binding on all member states, giving a range of maximum permissible radionuclide limits in foods (Table 5.3) that would come into force following a nuclear accident with transboundary consequences. They were derived by an expert group after the Chernobyl accident, in the light of the very wide range of standards across Europe at that time. They are based on the concept of 10% of a person's diet being contaminated at the maximum permissible level for a year. Standards for use after accidents in the rest of the world are largely taken from the IAEA's International Basic Safety Standards (IAEA,1996; Table 5.3).

For routine risk assessments, two approaches are possible for interpretation of data on levels of radioactivity in food. One is the calculation of a dose to an individual using specific annual consumption

Table 5.3 Limits of radioactivity in foods after accidents (Bq/kg, wet weight)

EU limits[a]

	Baby foods	Dairy and liquid foods	Other foods
Radiostrontium, notably ^{90}Sr	75	125	750
Radioiodine, notably ^{131}I	150	500	2000
α Emitters, notably ^{239}Pu and ^{241}Am	1	20	80
All other radionuclides with half-lives > 10 days, notably ^{134}Cs and ^{137}Cs	400	1000	1250

International basic safety standards[b]

	Foods destined for general consumption	Milk, infants' foods and drinking water
^{134}Cs, ^{137}Cs, ^{103}Ru, ^{106}Ru, ^{89}Sr	1000	1000
^{131}I	1000	100
^{90}Sr	100	100
^{241}Am, ^{238}Pu, ^{239}Pu, ^{240}Pu, ^{242}Pu	10	1

[a]From European Union Council Regulations 3954/87 and 2218/89.
[b]From IAEA (1996).

rates and the comparison of this with 1 mSv, or some other limit. Another method is to use a 'derived level', which relates amounts of radioactivity in foodstuffs to a theoretical dose using generic consumption rates. Generalised derived levels, for example, are sometimes used in the United Kingdom. These represent the concentration of a radionuclide in a food which, if maintained for a year, would theoretically deliver an individual dose of 1 mSv (Attwood *et al.*, 1996).

For drinking water, the World Health Organisation has published guidelines of 0.1 Bq/l for total α activity and 1 Bq/l for β activity (WHO, 1993).

5.5   Collective doses

Although individual doses are usually calculated for the purposes of assessing the risk from a particular source of radioactivity, it is sometimes useful to assess the risk to a population. This is called the collective dose and is expressed as man-sieverts. The collective dose is simply the product of the (usually very small) doses received by individuals and the (usually vast) number of individuals in the group. Estimates of collective dose are used in particular for comparing the potential radiological impact of different waste treatment regimes at a nuclear plant, or the collective risk from widescale consumption of radioactively contaminated food. However, there are drawbacks to the use of collective dose if the scenario that produces the dose has to be projected far into the future or if models have to be used to predict environmental contamination over large areas. There is no doubt that the collective dose can be used as a surrogate for collective or absolute risk if the individual doses can be prescribed, for example after the consumption of contaminated food. Nevertheless, because the method of calculation will maximise any imprecisions in the dose, the resultant risk estimate is nearly always subject to large, undefined, uncertainties.

5.6 Radionuclide movement through the food chain

The movement of radioactivity through the food chain is a complex subject and the interested reader is referred to some other publications for general introductions and for information on the plethora of mathematical models that are used to predict concentrations in the environment (Carter, 1988; Eisenbud and Gesell, 1997; Russell, 1966; Simmonds et al., 1995). The most important concepts will be considered in this section.

As many radionuclides are isotopes of biologically essential stable elements, they behave according to biochemical expectations. Tritium (^{3}H) and carbon-14 (^{14}C), for example, follow the same metabolic pathways as their stable analogues. Others behave similarly to essential elements that are in the same chemical group, radiostrontium behaving like calcium and radiocaesium like potassium. Such similarities have been extremely useful in radioecological studies. For example, radiostrontium was an important contaminant in cows' milk following fallout from weapons testing (see section 5.9.3). The uptake of calcium in ruminants has been found to be dependent on the calcium requirement of the animal, which in turn is influenced by its age, growth rate and milk yield. For a given calcium requirement, the absorption of calcium is inversely proportional to the amount of calcium in the diet, and a similar relationship between calcium intake and the transfer of radiostrontium from diet to cows' milk has been derived (Beresford et al., 1998).

Some radionuclides that occur in plants and animals largely as a result of pollution from human activity, including uranium, technetium and the actinides, have no obvious chemical analogues. Data on their environmental behaviour have, therefore, been derived empirically.

5.6.1   Radionuclides in the atmosphere

The main sources of atmospheric radioactivity are weapons testing, accidents and discharges from nuclear plants, together with naturally formed ^{14}C and ^{3}H. In the case of artificial radionuclides emitted to the atmosphere, dispersion is affected mainly by the effective release height (e.g. height of explosion of a weapon, or stack plus buoyancy of the plume from heat and velocity of exhaust gas) and meterological conditions. Other important factors are the topography, presence of buildings, surface roughness of the environs, and the physical state of the release, i.e. whether gaseous or as an aerosol. Deposition from the atmosphere is largely governed by washout in rain and by particle size. For example, after the Chernobyl accident (section 5.9.2) a wide range of radionuclides were released from the reactor, but the majority of countries receiving fallout were mainly affected by volatile radionuclides, in particular ^{131}I in the short term, and ^{134}Cs and ^{137}Cs. Within 30 km of the accident site, though, additional problems were caused by isotopes of strontium and plutonium that were present as particulates and hence were deposited sooner.

In the case of atmospheric weapons testing, where highly buoyant plumes were generated, radioactivity reached the stratosphere (above 30 000 m) and circulated round the globe, resulting in worldwide exposures (see section 5.9.3).

Once deposited on the ground, radioactivity may be ingested directly (from crops), or may enter the food chain via animals grazing contaminated pastures or through root uptake from activity that has entered soil (Figure 5.1). Foliar contamination via resuspension or 'rain splash' may also result in significant intake. After an accidental release of radioactivity to the atmosphere, the relative importance of these various pathways, in terms of potential dose to people, depends on the growing stage of the crop in the affected area and the types of agriculture. As a general principle, however, short-term releases will result in peak levels of activity being reached immediately in fresh crops ready for sale. Intake via animal products may take days to reach their peak, iodine or caesium isotopes in milk being the most important examples, or weeks as was observed for caesium in sheep meat in various parts of Europe after the Chernobyl accident.

5.6.2   Radionuclides in soil

The main source of radioactivity in soils is uranium and its daughter radionuclides, but fallout from weapons testing and accidents have added artificial radionuclides, which have substantially increased background radioactivity in some cases. Once they are on or in soils, uptake of a radionuclide into plants and animals is usually determined by its chemistry, speciation and the nature of the soil, notably its pH and organic content. For a radionuclide to remain in the rooting zone, it has to be held there by ion exchange, with aluminium silicates or organic matter, or as various oxides. The degree to which a radionuclide partitions itself between soil particles and soil solution is described by the distribution coefficient (K_d ml/g), which is the ratio of the quantity of a radionuclide held in unit weight of solid fraction of soil to the quantity dissolved per unit volume of water (Eisenbud and Gesell, 1997). The more tightly a radionuclide is sorbed onto soil particles, the higher the K_d, and the less available for uptake into plants. This does, however, mean that the specific nuclide will remain longer in the upper horizons of the soil. Examples from opposite ends of the K_d spectrum are iodine, which has a K_d of approximately 50 ml/g, and the actinides, which have $K_d > 10^5$ ml/g (NCRP, 1984). K_d values are of particular importance when considering the potential ability of radioactivity stored in underground repositories to reach the diet of human populations many years later.

The transfer of radioactivity from soils to plants is described by the transfer factor (TF) or concentration ratio (CR), which is the ratio of the concentration of a nuclide in wet vegetation to that in dry soil. A wide range of TFs has been obtained for most nuclides—summaries of these data can be obtained from a number of references (IAEA, 1994; NCRP, 1984), but among the most significant radionuclides TFs are highest for strontium and iodine and low for the actinides, with intermediate values for caesium.

An important example of the effect of soil type on soil transfer factors has been seen with radiocaesium deposited after the Chernobyl accident. In rich agricultural soils, radiocaesium is tightly sorbed onto clay particles and is largely unavailable for plant uptake. In contrast, caesium remains available for uptake over very long periods of time in acidic, organic soils with low clay mineral and potassium content and high levels of organic matter. Radiocaesium recycling between plants and organic soils is responsible for the majority of Chernobyl-derived dose in the longer term in northern Europe (Desmet and Sinnaeve, 1992; Absalom et al., 1995; Hird et al., 1996).

For radionuclides with high K_d and low TF (in particular the actinides and uranium), resuspension and rain splash become the most important routes for contamination and thus ingestion, as well as soil ingestion by animals.

5.6.3   Radionuclides in water

Natural sources are the main contributors to radioactivity in water, in particular potassium-40 in sea water and radium and its daughters in fresh water that flows through uraniferous rock. However, in some areas, discharges from the nuclear industry are the dominant source. In the United Kingdom, for example, historical liquid discharges from the Sellafield reprocessing plant located on the Irish Sea coast have been the source of considerable quantities of long-lived artificial radioactivity.

The behaviour of radioactivity once released to a body of water is much more complex and uncertain than for an atmospheric release. Apart from the nature of the receiving body (river, lake, estuary or sea), very many other factors are important. These include currents, tides, types, amounts and movement of sediment, salinity, degree of vertical mixing, and the type and speciation of the radionuclide. The concentration of some radionuclides, including strontium and caesium isotopes, is proportional to salinity and these can be transported long distances from the original source. For example, characteristic radionuclide ratios of two isotopes of caesium, discharged from Sellafield, have long been studied in the Irish Sea and the English Channel (Kershaw *et al.*, 1992; MAFF, 1963 *et seq.*). Recently, this Sellafield 'signal' has been identified near the Arctic (Smith *et al.*, 1997). Fish can concentrate these nuclides from the sea water itself, but also from feeding on plankton that have themselves absorbed the radioactivity from the sea water. Figure 5.2(a) shows the concentration of radiocaesium in plaice caught in the Irish Sea between 1963 and 1991, and the levels correlate closely with the discharges of this nuclide from Sellafield (MAFF, 1994).

Other radionuclides, including plutonium and radioruthenium, are adsorbed onto sediments and particulate matter, which usually sinks rapidly to the sea bed. However, such contaminated sediments may act as sources or sinks of radioactivity, depending on whether there is erosion, movement and redeposition. Filter feeders, such as molluscs, and bottom dwellers, including *Nephrops norvegicus* and other crustacea, are vulnerable to contamination from these radionuclides. Figure 5.2(b) shows the levels of plutonium in winkles caught near the Sellafield site between 1977 and 1991, and again these relate to the changing discharges of this element from the plant (MAFF, 1994).

The behaviour of radionuclides in ground water (aquifer systems) is very topical because of the perceived need for storage of both high-level and intermediate-level nuclear waste in underground repositories. One of the key questions here is the degree to which long-lived radionuclides may migrate out of the repository in ground water and eventually reach human populations.

5.7 Detection methods

The measurement of man-made radioactivity in food is a difficult task because of the generally low activity concentrations involved, and in some cases the presence of conflicting natural radionuclides (for a review of techniques and problems, see DoE, 1989 and IAEA, 1989). However, the use of radioactive materials as tracers is testimony to the sensitivity of detectors. For example, 100 Bq of caesium-137, which is relatively easily measured, represents 3×10^{-11} g of caesium.

Techniques roughly divide into groups depending on the radiations emitted by the radionuclides under assay (Table 5.4). However, other divisions could also be made, for instance by the amount of pretreatment and chemical separation required. Although nearly all radionuclides emit some penetrating γ-radiation as a consequence of each decay, this radiation is not always energetic or characteristic enough to be detected and identified. Thus, quite often, some specific chemical separation has to be carried out with inactive carriers before detection is possible. In some cases, the specific activity (Bq per unit mass of the element) of a radionuclide is so low, for example for uranium and thorium, that the radioactive species is detected with more sensitivity by chemical methods. In fact, chemical and physical detection techniques, such as mass spectrometry (Halverson, 1987) are now becoming more common, particularly for the heavier elements such as the actinides.

Some radionuclides present special problems because of the low penetration of their emitted radiations and low levels of contamination. For example, α emitters, such as plutonium-239, are normally present in small, but significant, amounts and can therefore only be detected by quite complex and time-consuming chemistry (usually by ion exchange or solvent extraction) followed by α spectrometry of an electroplated source. Methods of counting that involve detection of the characteristic or specific energies of emitted radiations are useful because they enable the use of an internal yield tracer. However, counting times of the plutonium separated from food samples contaminated with, say, fallout or routine discharges (see section 5.9) could be several days. For such counting techniques, and many others in radiochemical analysis, the reduction of the background count rate is of paramount importance.

Low-energy β emitters, such as ^{3}H, ^{14}C and sulphur-35 (^{35}S) present somewhat similar problems but are solved by other counting techniques. Suitable concentration followed by liquid scintillation or gas counting may be used. The former involves the intimate mixing of a light-emitting 'phosphor' with the radionuclide in a liquid matrix and counting in a photomultiplier system. Although β emission produces a continuous energy spectrum, liquid scintillation does allow fairly simple mixtures of

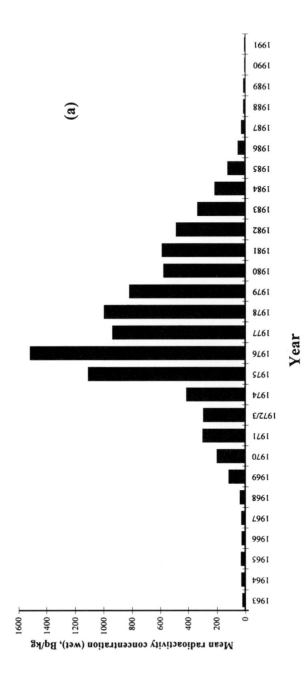

Figure 5.2 Radioactivity in seafood caught in the vicinity of the reprocessing plant at Sellafield, West Cumbria, UK (reproduced, with permission, from MAFF, 1994) (a) caesium-137 in plaice 1963–91 (b) plutonium-239/plutonium-240 in winkles 1977–91.

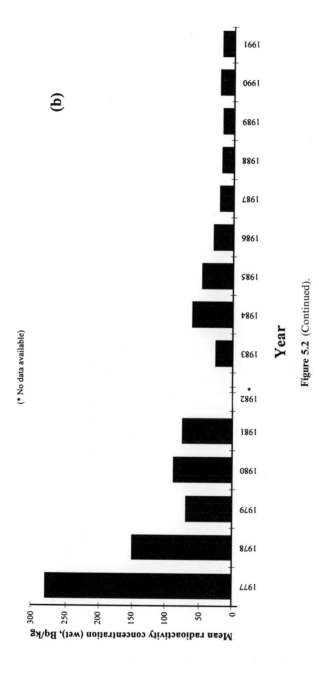

Figure 5.2 (Continued).

Table 5.4 Analytical methods and typical limits of detection in food surveillance programmes

Radionuclide group	Sample preparation and principal detection method[a]	Typical LOD[b] (Bq/kg wet weight) (e.g. for milk, crops, meat)
Low-energy β emitters	Separation and concentration, then liquid scintillation counting	^{3}H, 10; ^{35}S, 1.0; ^{14}C, 15
Low-energy γ emitters	Separation and concentration, then γ counting using thin sodium iodide crystal	^{125}I, 0.05–0.1; ^{129}I, 0.01–0.05
β/γ Emitters	Minimal treatment (homogenisation only), then γ spectrometry, using germanium detector	Between 0.5 and 10 for most nuclides
Pure β emitters and some β/γ emitters	Ashing and chemical separation, then gas proportional counting	^{90}Sr/^{134}Cs/^{137}Cs, 0.2; ^{99}Tc, 0.04–0.1; ^{106}Ru, 0.1; ^{210}Pb, 0.01
α Emitters	Ashing and chemical separation, then α spectrometry. Non-radiometric methods are also used, e.g. colorimetry (Th), fluorimetry (U).	Actinides and Th, 0.0002; U, 0.004; ^{210}Po, 0.01.
	Activation in a nuclear reactor, and delayed neutron counting used for some nuclides	U, 0.004.

[a]For brief descriptions of these methods see section 5.7. Fuller descriptions are in DoE (1989).
[b]Typical limits of detection (LODs) have been taken from MAFF and SEPA (1997).

radionuclides to be resolved although, in general, β spectrometry is not possible. Other methods for β emitters involve chemical separation followed by ingrowth of a daughter radionuclide, which is counted. Strontium-90, for example, is commonly detected by counting its daughter, yttrium-90, which has a particularly energetic β emission. Traditionally, β counting has used end-window Geiger–Muller instruments, but the development of gas flow counters that operate in the proportional region (where the counting pulse is proportional in size to the energy of the incident radiation) with low, stable backgrounds and windowless detectors has revolutionised low-level β counting.

Clearly the simplest and easiest approach to analysis is that which requires the minimum of pretreatment, i.e. no ashing or chemical separation, and is isotope specific. This requirement is met in the assay for γ emitters, using either a solid-state high-purity germanium detector or a sodium iodide crystal. Germanium detectors, which have the drawback of having to be used at the temperature of liquid nitrogen, have high energy resolution and low detection efficiency, whereas the opposite is true for sodium iodide crystals. Gamma spectrometry using modern computer-based multichannel analysers and the appropriate software enables complex mixtures of radioisotopes to be assayed routinely.

If a reactor source of neutrons is available, activation analysis can be undertaken, which has the great advantage of extreme sensitivity, particularly for low specific activity radionuclides (for example, iodine-129 and uranium) and trace metals but is limited in the presence of sodium and phosphorus.

Most measurements of radioactivitiy in food or water are of very low concentrations, and analytical data are therefore often given at a limit of detection (LOD). The parameters that affect an LOD include the inherent background of the counter and the time of count for both sample and background. In the case of γ spectrometry, where sample preparation is minimal, sample size is particularly important. When analysing for caesium-137, for example, with a sample size of 1.0 kg and a count time of 2 hours one could expect an LOD of < 5 Bq/kg. If the counting time is quadrupled this drops to < 2.3 Bq/kg, and if the sample size is increased to 2.5 kg the LOD becomes < 1.6 Bq/kg (MAFF, 1988). If chemical methods are employed prior to counting, the efficiency of recovery of the nuclide of interest is vital in obtaining a low LOD. Where a positive result is obtained, i.e. above the LOD, the uncertainty due to counting statistics is commonly expressed at the 95% confidence interval.

All the methods described above are used for the laboratory analysis of natural and artificial radionuclides in environmental and dietary samples. However, after the Chernobyl accident, when large numbers of free-ranging animals became contaminated with radiocaesium (section 5.9.2), methods of live monitoring were developed using hand-held sodium iodide detectors (Meredith *et al.*, 1988; Brynildsen and Strand, 1994). These were lead-shielded to reduce background radiation, weather-proofed for use in the field, and gated to detect the photons from caesium-137 and caesium-134. Held against an animal's rump, with a count time of 30–60 seconds, the methods allowed a lower limit of detection of around 300–500 Bq/kg.

5.8 Natural sources of radioactivity in food

The earth was formed from radioactive materials. Most have long since decayed to the stable nuclides that form the bulk of the earth's crust. However, some, because they have very long half-lives, are still present today. There are a few primordial isotopes, including potassium-40 and radionuclides in the three decay series headed by uranium-238, uranium-235 and thorium-232. Potassium-40 decays into a stable nuclide (calcium-40), but it is an energetic γ emitter and dominates internal exposure of the body. The isotopic abundance of potassium-40 in potassium is 0.012% by weight and it is not enriched in any source with respect to stable potassium. This, and the fact that potassium is an essential element whose concentration is under homeostatic control by the body, means that the dose from ingested potassium-40 is constant at about 165 µSv per year for adults, and 185 µSv for children (UNSCEAR, 1993).

The other primordial radionuclides are the parents of the three decay chains mentioned above. Each nuclide in the chain decays with its own characteristic half-life and radiation emissions to give rise to a different daughter. Eventually, the chains end with stable isotopes of lead. One of the members of the decay chains is a gas, radon, which seeps up from rocks into buildings and is a significant source of radiation dose (see entry for 'radon and daughters', Table 5.1(a)). Important sources of dose from ingestion of members of these decay chains are radioisotopes of radium, lead and polonium. Their concentrations in the diet are dependent on soil and food type.

There are a few radionuclides that are formed in the upper atmosphere from interactions of cosmic rays with stable nuclides. The most important examples that are incorporated into food chains are ^{14}C and ^{3}H. Both are also formed anthropogenically.

UNSCEAR (1993) has produced reference activity concentrations of natural radionuclides in foodstuffs and water, and the most significant data are reproduced in Table 5.5. Annual average doses to adults from ingestion of these radionuclides are given in Table 5.1(b), and are about 64 µSv. Hence, including the dose from potassium-40, natural intakes are annually responsible for 230 µSv to an average adult (Table 5.1(a)).

Large variations may occur in concentration of natural radionuclides in foods. Where soils contain enhanced quantities of uranium or thorium, crops and grazing animals will also show increased levels of various members of these decay chains. One well-known example of such soils is in the Kerala area of India where a thorium-232-rich mineral called monazite occurs extensively. Levels of radium-226 of up to nearly 5 Bq/kg and of thorium-228 of up to 32 Bq/kg have been reported in crops from this area (UNSCEAR, 1993).

The occurrence of natural radioactivity in drinking water is dependent on the geology of the source. Surface waters contain very little activity, while ground waters may contain significant levels of isotopes of uranium, radium and radon. Table 5.5 gives typical concentrations of these radionuclides in water supplies. Ground waters may be derived

Table 5.5 Examples of concentrations of natural radionuclides in food and water (mBq/kg, wet weight)

	$^{238}U + ^{234}U$	^{226}Ra	^{228}Ra	^{210}Pb	^{210}Po
Milk products[a]	1	5	5	40	60
Meat products[a]	2	15	10	80	60
Cereals[a]	20	80	60	100	100
Leafy vegetables[a]	20	50	40	30	30
Roots/fruits[a]	3	30	20	25	30
Fish products[a]	30	100	–	200	2000
Water supplies[a]	1	0.5	0.5	10	5
Bottled waters					
Germany [b]	< 1–140	< 1–1800		3.3–53	0.4–8.9
France[b]	up to 2000	up to 2700			
UK[c]	17–130	< 2–5			< 2–83
Ground waters					
Finland[b]	up to 74 000	up to 5300		up to 10 200	up to 6300
USA[d]	11–74	11–30	15–37	< 3.7	< 3.7

[a]Reference activity concentrations (UNSCEAR, 1993).
[b]Ranges of elevated activity concentrations in ground waters from various sources (UNSCEAR, 1993).
[c]Concentrations in higher activity samples from a survey of UK mineral waters (MAFF, 1994).
[d]Radon-222 levels between 1800 and 22 000 Bq/l. Average concentrations from various studies (Milvy and Cothern, 1991).

from aquifers in highly uraniferous rock, and some exceptionally high levels of radon and its daughter products (mainly isotopes of lead, bismuth and polonium) have been reported in both bottled waters and in some groundwater supplies (Table 5.5).

Industrial processes that involve ores containing elevated levels of uranium and its decay products may also result in some enhancement of natural radioactivity in foods. Phosphate rocks in particular, which can contain relatively high concentrations of uranium and its daughters, are often concentrated in industrial by-products . A good example of this existed on the coast of Cumbria in the United Kingdom, where, until 1992, a plant manufacturing phosphoric acid from imported phosphate ore discharged phosphogypsum as a liquid into the sea. This resulted in a considerable enhancement in polonium-210 and lead-210 in local shellfish. In 1988, 251 Bq/kg of ^{210}Po was reported in flesh from winkles gathered near the plant (Kershaw et al., 1992). These levels have now fallen to around 25 Bq/kg, reflecting lower discharges in recent years and radioactive decay of material already present in the environment (MAFF and SEPA, 1997). The use of phosphate fertiliser itself does not significantly affect radioactivity in food, although its persistent use on farmland will obviously add to the natural radioactivity of soil over time.

Some plants show an exceptional ability to accumulate certain radionuclides. Brazil nuts, in particular, have been shown to contain between 10 and 260 Bq/kg of radium (Penna Franca *et al.*, 1968). The brazil nut tree has a tendency to concentrate barium, and hence also radium which is a chemical congener of barium. A more extreme example of radionuclide concentration in nature occurs with lichens that are consumed by reindeer. Lichens have a well-established tendency to concentrate trace elements from the air, and this includes polonium-210. In Lapland, very high levels of polonium-210 and lead-210 occur in reindeer meat (around 11 000 mBq/kg and 600 mBq/kg respectively, for comparison see Table 5.5) (UNSCEAR, 1993), which is a significant part of the diet of Lapps.

Carbon-14, although of low radiological significance because it is a low-energy β emitter, is regarded as an important contributor to collective doses from radioactivity as it is continually formed cosmogenically in the upper atmosphere, is present in all biota, and indeed is discharged from the nuclear industry (section 5.5). Pure carbon currently has a specific activity of 252 Bq of ^{14}C per kilogram of carbon. Carbon-14 of natural origin in food varies from around 5 Bq/kg wet weight up to around 150 Bq/kg (MAFF and SEPA, 1997), this variation being due solely to the concentration of carbon in the food. The annual dose from ingestion of carbon-14 of natural origin is around 12 μSv. Other cosmogenic radionuclides contribute around 2% of the dose of carbon-14 (tritium, sodium-22 and beryllium-7) and are unimportant as natural food contaminants (UNSCEAR, 1993).

5.9   Anthropogenic sources of radioactivity in food

There are three main sources of man-made radioactivity in the environment: the explosion of atomic bombs in the atmosphere, accidents involving radioactivity, and discharges from nuclear plant and establishments using radioisotopes. Atmospheric weapons testing has given a radiation dose to just about everybody on Earth, with a bias towards the population of the northern hemisphere (section 5.9.3). Accidents have ranged from the catastrophic, causing death of workers from radiation sickness and wide-scale contamination of land, through to very minor incidents, nonetheless of note because of the public reaction they caused. Routine discharges from nuclear establishments generally give measurable doses only to people living close by. In all these cases the main sources of exposure are fission products produced from the splitting of uranium or plutonium or activation products resulting from neutron capture of stable elements to produce, for example, ^{134}Cs and ^{60}Co. The most important fission products, in terms of the food chain, are ^{89}Sr, ^{90}Sr, ^{137}Cs and ^{131}I, while neutron capture by uranium produces the transuranic elements plutonium, americium and curium, most of which are α emitters.

5.9.1 Routine discharges from nuclear plant and other users of radioactivity

Industrial sources of radioactivity comprise nuclear power plant, nuclear fuel recycling facilities and isotope production units. Hospitals and research establishments are generally only minor sources. Emissions of radioactivity to air, water, or in solid wastes are usually controlled by national legislation based on recommendations of the ICRP (see section 5.2). Most routine surveillance for radioactivity in food in countries with a nuclear industry is based around these sites, and such programmes serve to demonstrate that those most at risk by virtue of where they live, their habits and what they eat (called the critical group) are not receiving doses above certain limits. In the United Kingdom, for example, the basic annual limit is 1 mSv per year but dose constraints of 0.3 mSv per year for new facilities and 0.5 mSv per year for existing facilities were introduced in 1995 (Clarke et al., 1993; Cm 2919, 1995). There may be different critical groups for atmospheric and aqueous discharge routes. For example, the effect of aqueous discharges will be assessed by studying the diets of those who consume large quantities of seafood or who spend long periods of time on contaminated shores. They may receive very little dose from terrestrial food chains affected by atmospheric discharges, so a separate assessment would have to be carried out for prodigious consumers of vegetables or milk for example. There are many sources of data from routine surveillance programmes around nuclear plant. In the United Kingdom, operators and government bodies undertake independent monitoring programmes and dose assessments and publish the data in separate annual reports (see Cotter et al., 1992, for examples).

Most radioactivity discharged from the nuclear industry comes from fuel reprocessing plants where spent fuel rods from reactors are treated to recover uranium and plutonium. These are Cap de la Hague and Marcoule in France, Chelyabinsk in Russia, and Sellafield and Dounreay in the United Kingdom. In western Europe, the Sellafield plant has been infamous for the level of its historic discharges. In the 1970s, its discharges to the Irish Sea were around 150 times what they are today, present discharges now being comparable to those of Cap de la Hague. As many of the radionuclides are long lived, and have dispersed far from the Irish Sea (see also section 5.6.3) the site is responsible for around 90% of the collective dose from industrial sources to the European population

(Mayall *et al.*, 1994). In the 1980s, the local critical group were receiving about 3.5 mSv per year from seafood consumption (Kershaw *et al.*, 1992). Doses from this pathway are now about 0.12 mSv (MAFF and SEPA, 1997). Figure 5.2(a) and (b), referred to in section 5.6.3, reflect the changes in aqueous discharges of radionuclides over previous decades. This reduction has been due to a tightening of radiation protection standards and improvements in abatement technology over the past decade or so. The environmental impact of the discharges from this site has been the subject of much scientific and legal scrutiny. To date, no correlation has been observed of adverse effects on those possibly exposed with levels of discharges (COMARE, 1996).

When well operated, most nuclear plants are insignificant sources of dose to the population. For example, in the United Kingdom, nuclear power stations generally contribute less than 10 µSv to local critical groups from food (MAFF and SEPA, 1997). However, there have been accidents at some installations, which have had profound consequences on the nuclear industry and, in some notable cases, on both local and distant populations.

5.9.2 Accidental and large-scale releases of radioactivity from industrial and military plant

There have been numerous releases of man-made radioactivity over the past 50 years, some from military establishments, some at commercial sites, and others involving satellites and transport of nuclear material. Four in particular, are worth mentioning—one in the United States (Three Mile Island), two in the former Soviet Union (Chelyabinsk and Chernobyl), and one in the United Kingdom (Windscale). The quantities of the major radionuclides released in each case are shown in Table 5.6. Good summaries of these and other more minor accidents can be found in Eisenbud and Gesell (1997) and UNSCEAR (1993).

Chelyabinsk, USSR, September 1957
This is a plutonium production centre, located near Kyshtym in the Ural Mountains. In September 1957, an explosion in a tank of high-level radioactive waste led to the atmospheric release of 500 PBq (500×10^{15} Bq) of material. Around 90% of the radioactivity was deposited locally. Strontium-90 was the principal radionuclide present in the remaining 10%, which was dispersed over the population living within a few hundred kilometres of the site. An area of around 20 000 km² received ^{90}Sr deposition in excess of twice that associated with bomb fallout (see section 5.9.3). Over 10 000 people were evacuated from the most contaminated areas, many 8 months after the event. Ingestion of

Table 5.6 Radionuclides of principal importance to ingestion dose released during weapons tests, accidents and from military installations. Values are PBq (petabecquerel; 1 PBq = 10^{15} Bq).

Nuclide	Nuclear weapons tests (1949–64)	Chelyabinsk (1957)	Windscale (1957)	Three-Mile Island (1979)	Chernobyl (1986)
^{3}H	240 000				
^{14}C	220				
^{89}Sr	91 000		0.005		80
^{90}Sr	650	50^a	0.0001		10
^{131}I	700 000		0.7	0.0005	1760
^{134}Cs			0.001		35
^{137}Cs	900		0.02		70
^{210}Po			0.009		
^{239}Pu/^{240}Pu	10				0.042

From Crick and Lindsley (1984); IAEA (1991); UNSCEAR (1982, 1988, 1993); OECD (1995).
^{a}This is 10% of what was released. The remaining 90% was deposited very close to the site.

^{90}Sr was the dominant source of dose in the long term. Eventually food monitoring took place to prevent ^{90}Sr intakes exceeding 50 kBq/year (which incidentally would result in an annual dose to an adult of around 1.5 mSv). Food bans and the destruction of considerable amounts of agricultural produce were implemented. Even so, total doses up to 0.5 Sv were received by a number of people from both external and internal sources (UNSCEAR, 1993).

Apart from this explosion, about 100 PBq of untreated liquid waste from this plant was dumped into the Techa river between 1949 and 1956 (Trabalka and Auerbach, 1990). This caused high doses, both internal and external, to people living along the river and eventually resulted in 7500 people within about 50 km of the plant being evacuated. An alternative dump for this waste was built in 1952, a reservoir called Lake Karachay, 15 km east of Kyshtym, which is so radioactive that unshielded people cannot approach it. This lake partially dried out in 1967 and highly radioactive wind-blown dust contaminated an area of around 1800 km^2. Epidemiological studies have now started on the exposed populations, and significant increases in leukaemias and other cancers have been observed (Kosenko and Degteva, 1994). None of these events was known about in the West until the era of glasnost in 1986.

Windscale, UK, October 1957

This accident was caused by a fire at one of the plutonium-producing air-cooled reactors at Windscale (now Sellafield) in the United Kingdom. The main off-site effect was on local milk, in which the principal nuclide released, ^{131}I, was detected within 24 hours. A maximum permissible concentration of 3700 Bq/l was set, based on a dose beneath a threshold that it was thought, at the time, might exist for thyroid cancer in children. Milk above this level was discarded (Dunster *et al.*, 1958). At one time, this affected around 500 km^2, but because of the short half-life of ^{131}I this was quickly reduced to a much smaller area. Nevertheless, milk from some farms remained restricted for around one and a half months. The highest concentrations of ^{131}I in milk came from a farm 10 miles from the site, and were around 50 kBq/l. The highest doses to children and adults living up to 24 miles from the site were estimated, from thyroid scans, at 160 mSv and 95 mSv, respectively. This accident highlighted the need for pre-planning on what levels of dose (or radioactivity in the environment) should be used for the implementation of countermeasures.

Three-Mile Island, USA, 1979

This accident was caused by a loss of coolant from one of the pressurised water reactors near Harrisburg in Pennsylvania. In spite of the fact that the main release in terms of quantity was the inert gas krypton-85, which would have given mainly a skin dose, this accident received massive publicity, largely because of the public response in self-evacuation. It also demonstrated that the need for dealing with the public's fear of radioactivity is almost as great as that for the technical issues during a nuclear incident. The highest doses received have been estimated as 0.05 mSv (Kemeny, 1979).

Chernobyl, former Soviet Union, April 1986

Commonly rated as the worst industrial accident ever to occur, this event was caused by an explosion and fire at one of the water-cooled reactors at Chernobyl in the Ukraine in April 1986. It resulted, eventually, in the evacuation of around 135 000 people, and has had severe effects on agriculture both in the short and long terms. Abandoned animals had to be destroyed in large numbers, and nearby forests and soil have been so heavily contaminated that wide-scale remediation has been necessary, including the felling and burial of trees and removal of soil. Furthermore, an area of 30 km radius remains uninhabitable to this day. Significant doses were accumulated by people prior to evacuation. Average thyroid doses of about 1 Sv for children, and 0.7 Sv for adults have been calculated (OECD, 1995), with highest exposures around 4 times these values (IAEA, 1991). Beyond 30 km, radioisotopes of iodine and caesium have been the main problem contaminants and the complex meteorology that occurred after the accident took the radioactive plume to Scandinavia and most other areas of Europe. In the weeks and months after the accident, vast numbers of environmental samples were collected and analysed. Elevated levels of ^{131}I and ^{137}Cs in environmental media were widespread in Europe, especially where the radioactive plume had coincided with rainfall. None the less, average European doses over the next 30 years have been estimated as less than 1.5 mSv. Doses outside Europe are estimated as less than 200 μSv, while the southern hemisphere has been largely unaffected.

Within the former Soviet Union, the areas principally affected were in Belarus, the Ukraine and parts of Russia. Surface contamination in excess of 185 kBq/m^2 occurred over an area of 25 000 km^2, which contained a population of 825 000 and where dairy farming and the raising of animals for meat dominate the agricultural activities (IAEA, 1991). The soils were mostly sandy or peaty, of low natural fertility and poor in minerals, especially potassium. This, and the fact that much animal grazing was on unimproved, natural pastures where the contamination of fodder was up to ten times that on regrassed land, has led to long-term contamination of the animal–human food chain from radiocaesium (see also section 5.6.2). Many farmers were self-sufficient in the very products that were most affected (meat and milk). The persistence of radiocaesium in large numbers of grazing animals has necessitated the development of monitoring of animals while alive (see also section 5.7). Some data on radiocaesium and radiostrontium levels in food products for the affected areas are given in Table 5.7 for the years 1986 to 1995. The gathering of wild produce for home consumption (berries and mushrooms) is a significant source of additional exposure in some areas, as these tend to accumulate radiocaesium in greater amounts than other foods because of high uptake factors from soil and a long effective half-life (Skuterud et al., 1997).

The authorities imposed food bans (as well as prohibiting the picking of wild foods) from very early on for both radioiodine and radiocaesium. However, the lack of alternative foods in some cases meant that people had to continue to eat foods that were above the limits. Various agricultural countermeasures were tried including deep ploughing to move most contaminated soil to a depth below roots, land treatments with potassium clay minerals and fertilizers, restricting grazing of cattle to pastures beneath a certain level of deposition, or feeding them crops that were less contaminated. Alternative forms of agriculture, including poultry or pig farming, were also used, as well as processing of the most contaminated milk into cream and butter (see section 5.10). Where countermeasures were successfully applied, it has been estimated that doses from ingestion were reduced by a factor of up to 10. Over time, the proportion of food above the limits set by the authorities for banning has fallen dramatically. For example, in the Gomel district, which is around

Table 5.7 Levels of radioactivity (Bq/kg, wet weight) in foods close to the Chernobyl plant (1986–1995)

	Bragen District (30 km to north)			
	^{134}Cs + ^{137}Cs		^{137}Cs	^{90}Sr
Food product	1986[a]	1990[a]	1994/5[b]	
Milk	1600	260	10	8
Green vegetables	3700	220	3–14[c]	–
Potatoes			7	7
Meat	1700	300	38	–
Fish	–	220		
Berries and mushrooms	2200	1100		
Berries			6540	166
Mushrooms			215 000	563[d]
Game meat			8–60 000[e]	3–100[f]

[a]From IAEA (1991) (medians).
[b]From Aslanoglou et al. (1996) (averages of two villages).
[c]Tomatoes, beets, cabbage, onions, carrots and cucumbers.
[d]Dry weight.
[e]Range is from wild birds at lower end to wild boar at upper end.
[f]Range is from wild boar at lower end to elk at upper end.

150 km from Chernobyl, nearly 70% of milk exceeded the limit for radiocaesium in 1986, but this had decreased to only 1.2% in 1990 (IAEA, 1991). It is not possible, however, to assess the extent to which this is due to countermeasures being applied, rather than to the gradual non-availability of radiocaesium for plant uptake. Over recent years, dairy cattle in the affected areas have been given a feed additive (Ferrocyn, see later text on Norway) during winter to reduce radiocaesium transfer to milk (N.A. Beresford, 1998, personal communication).

Dose estimates for the people living in the most affected areas have been made both by measuring foodstuffs and estimating intakes and from whole-body counting of the γ-rays from the decay of radiocaesium, these being detected from outside the body using arrays of detectors. Using the former method, doses of several tens up to hundreds of millisieverts per year have been calculated. Results from whole-body counting suggest that these estimates are a factor 5–10 too high. In 1994, a duplicate diet study was undertaken of food consumption of adult males living some 100–400 km to the west and south of Chernobyl. Some individual foodstuffs were also measured. Annual doses were estimated at around 20 μSv from food, and around 76 μSv from milk (Shiraishi et al., 1997). In the study of rural villagers closer to the site (Aslanoglou et al., 1996), food measurements and dietary survey information were combined to produce estimates of daily intakes of around 35 Bq of ^{137}Cs, and 20 Bq of ^{90}Sr from local agricultural products, 85% of which were derived from domestically produced milk, pork and potatoes. These intakes would produce an annual dose of around 370 μSv (which is about 40 times greater than might be expected from other fallout, see Table 5.1(a)). Whole-body measurements were also used in this study, and close correlation of radiocaesium body burdens and frequency of consumption of wild mushrooms was observed (Skuterud *et al.*, 1997), which was not found for other foods. Differences in species of mushroom eaten and in preparation methods are likely to have a large effect on intakes. Strand *et al.* (1996) found that 70% of doses from radiocaesium were derived from mushroom consumption.

Outside of the Soviet republics, the most important routes of exposure in the short term, for the majority of people, were milk and fruit and vegetables contaminated as a result of direct deposition. Figure 5.3 shows the level of ^{137}Cs in the UK milk supply from 1963 to 1991. The effect of the Chernobyl accident is clearly seen in 1986. In the longer term, ^{137}Cs has remained in the environment, but in most agricultural soils it has largely become unavailable for plant uptake (see section 5.6.2). There have been some notable exceptions, which again have been due to the types of affected soils, and, as in the USSR, have led to long-term restrictions on the slaughter of animals grazing semi-natural pastures as

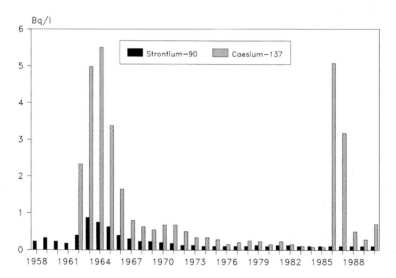

Figure 5.3 Radioactivity in UK milk 1958–91 (mean levels) (reproduced, with permission, from MAFF, 1994).

far away as the United Kingdom (Desmet and Sinneave, 1992; Cremers *et al.*, 1990; Horrill and Howard, 1991). Norway, in particular, was badly affected (Strand, 1994), with reindeer meat reaching around 150 000 Bq/ kg, and lamb around 40 000 Bq/kg (Strand *et al.*, 1990) as a result of these animals grazing on acid, organic soils, as well as eating highly contaminated lichen in the case of the reindeer (see section 5.8). Norway has been notable for pioneering the use of caesium binders in managing the transfer of radiocaesium into free-range animals. In particular, rumen-dwelling boli containing a Prussian Blue compound (ammonium hexacyanoferrate, AFCF) have been developed, which are administered a few months before slaughter; as the AFCF is slowly released into the gut, it combines with radiocaesium in the diet, making it unavailable for gut uptake, and so reducing the amount of activity in the muscle by a factor 2–5 (Hove *et al.*, 1990; Pearce *et al.*, 1989). Studies conducted in the USSR, under the guidance of the Norwegians, have shown the very successful use of such boli in dairy cows, with a 70% reduction in radiocaesium transfer to milk within 1 month of administering two boli (IAEA, 1991). A practical development of this work has been the routine use, in recent years, of a Prussian Blue compound called ferrocyn in the areas around Chernobyl to reduce radiocaesium transfer to milk. For example, in Belarus, owners of private cattle are given 50 kg of a ferrocyn/concentrate mix per winter per cow (Beresford, 1998, personal communication)

Drainage of affected land into freshwater lakes has also resulted in elevated levels of radiocaesium in fish. In Norway, levels reached 60 000 Bq/kg (Strand *et al.*, 1990).

As well as monitoring food, water and live animals, some countries outside of the USSR undertook detailed studies of individuals' diets to assess intakes from the accident via the diet. In the United Kingdom, a week-long duplicate diet study was conducted in areas of high and low deposition in June 1986. The highest annual dose in the year following the study for people living in the high deposition areas was calculated at just under 100 μSv, and from those in the low deposition areas was less than 10 μSv (Mondon and Walters, 1990). In comparison, average doses from ingestion in Norway for the first year after the accident were up to 250 μSv, although the Lapps were calculated to have received 1–3 mSv, 90% of this being due to reindeer meat (Strand *et al.*, 1989).

The collective doses from these accidents are given in section 5.11.

5.9.3 Food contamination from fallout

By far the greatest contribution to global food contamination from anthropogenic sources has been from the 'fallout' from the testing of nuclear weapons in the atmosphere (see section 5.11). Nuclear weapons derive their explosive energy from one or both of two nuclear processes: the fission or splitting of uranium-235 and plutonium-239 in an uncontrolled chain reaction or the fusion of the hydrogen isotopes deuterium and tritium in a thermonuclear process. Fission produces 'fission products' (see section 5.9) whereas fusion, in terms of radio-activity, in theory produces only tritium. However, the conditions necessary for fusion (somewhat similar to those in the sun) usually require a fission primer and therefore, in practice, fission products are also a contaminant of these devices. Neutron (induced) activation products are also produced from the capture of neutrons by stable elements by both types of devices tested. One of the more important of these from the point of view of food contamination is ^{14}C produced from stable nitrogen (^{14}N) in the atmosphere.

A total of 520 atmospheric nuclear tests (8 underwater) have been carried out since 1945. The peak periods for testing were 1952–54, 1957–58 and 1961–62. A limited test ban treaty was signed in 1963 and only sporadic testing in the atmosphere (by the Chinese and the French) has occurred since then—the last test in the atmosphere was in 1980, although tests have continued underground. The total explosive yield of the tested weapons has been estimated to be 545 megatonnes (Mt) (TNT equivalent) with about 90% of this in the northern hemisphere (UNSCEAR, 1993). This location has some significance as there is not a great deal of mixing of atmospheric contamination debris across the equator.

The fallout produced, and in particular its distribution, has depended on the type of device and to some extent its height of detonation. The fallout is generally partitioned between the local area (up to about 100 km from the test site), the troposphere and the stratosphere:

Local fallout consists generally of large particles that have not been a significant food contaminant, except in one or two unusual circum-stances, because tests have mostly been conducted in remote locations. However, recently it has become clear that some of the 87 Russian atmospheric tests carried out at the Semipalatinsk test site, which is located in the northern part of the Kazakhstan Republic, produced considerable local fallout contamination (Takada et al., 1997). There are several villages located downwind from the test site that were not evacuated. For surface tests this local fallout can comprise up to 50% of the fission product inventory but on average has been about 12%.

Tropospheric fallout consists of smaller aerosol particles that are not carried above the tropopause and which are confined to the approximate latitude band of the explosion. These particles follow the prevailing winds and deposit within about 30 days. From the point of view of food contamination, fission products with half-lives of a few days to 2 months will be important and include ^{131}I, ^{140}Ba, ^{89}Sr and ^{90}Sr. This fraction of fallout has comprised about 10% of the inventory.

Stratospheric fallout, which comprises about 78% of the inventory, is made up of small particles that are injected into the stratosphere and then fall out globally over several years. Global fallout from the stratosphere has been equivalent to about 168.5 Mt (UNSCEAR, 1982). The total amounts of the most important fission products and actinides deposited as stratopheric fallout worldwide are shown in Table 5.6. Although relatively small amounts of ^{14}C were released (about 220 PBq compared to an *annual* natural production rate of 1 PBq), by far the greatest collective dose from ingestion will arise from this radionuclide, i.e. 86% of the total, but only 7.7% of this dose will be delivered up to the year 2000.

The legacy of fallout in the late twentieth century is thus from the fission products with the longest half-lives, notably ^{90}Sr and ^{137}Cs, the activation product ^{14}C, and the actinides such as ^{239}Pu. In practice, only ^{90}Sr, ^{137}Cs and ^{14}C are of any global significance in terms of collective dose.

For the purposes of dose or risk estimation from fallout, three main pathways of exposure must be considered: external exposure from ground deposition, inhalation of particulates and radioactive gases, and ingestion of contaminated food (Figure 5.1). Only food ingestion is of relevance here—the inhalation contribution is relatively insignificant from ^{90}Sr and ^{137}Cs and the dose commitment from ingestion of ^{14}C compounds is about 10 000 times that from inhalation of ^{14}C as carbon dioxide. The transfers from deposition to plant uptake, to grazing animal uptake and to dietary intake have already been covered (section 5.6). For most radionuclides including ^{137}Cs and ^{14}C but not ^{90}Sr, dietary intake results mainly from initial retention by crops during or shortly after deposition and much less from delayed root uptake. However, the fact that weapons test fallout deposition occurred throughout the year was also of significance in relation to transfer to the diet. For deposition within a shorter period, e.g after Chernobyl, transfer is more dependent on the agricultural conditions at the time, the stage of the growing season, and any short-term food restrictions that may have been introduced (Muck, 1996). The transfer, over time, of fallout to diet has been assessed by UNSCEAR (1982, 1993) for populations in Argentina, Denmark and the United States, the data being weighted for consumption. These populations were selected because they are assumed to be typical of the global situation. Of the deposited fallout that is transferred to the food chain, 22% takes place from direct deposition, 27% as a lagged transfer (e.g. from stored food and from surface deposition), and the rest from long-term root transfer. The relative contributions of each of these types of transfer for ^{90}Sr and ^{137}Cs in different food groups are shown in

Figure 5.4. (and see Figure 5.3). It can be seen that in the case of ^{90}Sr, lagged transfer and transfer via root uptake are more important routes compared with ^{137}Cs, for which direct deposition is more significant for most food groups. From these data, global doses from food ingestion have been calculated (UNSCEAR, 1993) for a number of the radio-nuclides in weapons test fallout. Globally, ^{14}C is responsible for the greater part of the dose, and therefore of risk. If all the ^{14}C dose is used in the assessment, about 80% comes from the ingestion of food, but of

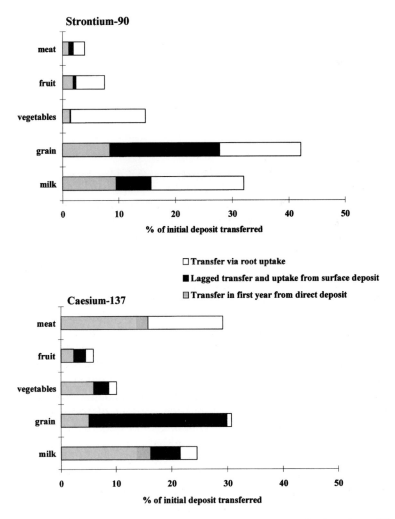

Figure 5.4 Transfer of weapons test fallout to diet: percentage contributions from different foodstuffs over time.

course this will be delivered over a very long time. A large proportion of the dose from ingested ^{90}Sr and ^{137}Cs will have been received in the years during, and immediately after, the period of weapons testing. An example of this food contamination is illustrated in Figure 5.3 which shows milk contamination in the UK since 1958. The increase in ^{137}Cs, due to Chernobyl in 1986 was commented on in section 5.9.2.

Some populations of people living near nuclear test sites have received significantly higher doses as a consequence of ingestion of contaminated food, inhalation of radioactive materials (mostly ^{131}I) and external irradiation. These groups include:

- the 180 000 people who lived downwind from the US Nevada test site in the period 1951–1962;
- the more than 10 000 people who lived near the Semipalatinsk test site in the former USSR in the period 1949 and 1962;
- the inhabitants of the Rongelap and Utirik atolls in the Pacific who were exposed in 1954 to fallout (mainly ^{131}I) from a US test;
- the (relatively few) Australians exposed as a result of the UK tests on the Monte Bello Islands and at Emu and Maralinga.

There is relatively little information on the dose from ingestion of radioactivity among these groups. However, it is thought that the highest-exposed group is the population surrounding Semipalatinsk. Their total collective dose has been assessed at about 4600 man-Sv with about 43% from the ingestion of contaminated food.

5.10   Effects of food processing and decontamination of food

Most studies on the changes in radionuclide concentration following processing and preparation of food have concentrated on radioisotopes of caesium, strontium and iodine, because of their domination in exposures following accidents. A review of this work has been undertaken recently by Green and Wilkins (1995) and the following summary has mostly been extracted from this reference. The commonest definitions used to quantify changes in radionuclide content of foodstuffs are the total recovery of the nuclide (R_t), which is expressed as the total activity in processed food divided by the total activity in raw food, and the fresh mass retention factor (R_w), which is expressed as the activity concentration in fresh processed food divided by that in the raw food. Green and Wilkins (1995) concluded that R_t is the most useful factor for expressing a processing effect, because consumption rates are expressed in terms of fresh weight of food ready for eating, and because radionuclide concentrations refer to fresh raw food. In some situations, such as milk processing, R_w is more appropriate because the food produced resulting from all stages in the process may be consumed. As an example of the difference between R_t and R_w, the R_t for ^{90}Sr in goat-milk cheese is 0.61, i.e. 61% of the activity is retained in the cheese following the processing of the milk. However, 1 kg of milk yields only 120 g of cheese, so the activity concentration of ^{90}Sr in the cheese is a factor of 5 higher than in the original milk, i.e. R_w is around 5. It is therefore important to be aware of the exact meaning of any processing factor.

Potential losses from the food chain start with direct deposition onto crops. For some crops, when unpalatable parts are discarded before preparation, up to 98% of the surface radioactivity can be removed. Washing can remove activity, which for some fruit can be up to 20% of the surface deposition. In the case of the short-lived radionuclide ^{131}I, storage time that allows for physical decay will of course decontaminate affected food. Milk, for example, can be processed into powder and stored. However, once longer-lived radionuclides are incorporated into food, some cooking processes can reduce levels. In the wet cooking of meats and fish, up to half the activity may pass into the cooking liquors. However, the latter are frequently consumed as gravy, etc., and so no net losses from the diet occur. Variable effects have been observed in the wet cooking of vegetables that could be related to nonstandard practices such as cooking times and addition of salt. In industrial processes, where cooking liquors are likely to be discarded, wet cooking may, in some cases, be a useful way of decontaminating foods.

For radionuclides that concentrate in the skeleton, intakes can be reduced by filleting; for example, in fish only 20% of radium, lead and polonium in the whole animal is contained in the fillets. However, in crustaceans, with an exoskeleton, some cooking processes, for instance the preparations of soups and croquettes, use the whole animal and variable amounts of radionuclides initially in the exoskeleton may be present in the edible fraction. A large proportion of radioactivity associated with filter feeders can be removed by depuration in clean water as this removes sediment from the gut. An R_t of 0.17 for radiocaesium has been observed in depurated mussels, which implies for caesium, at least, that 80% of the activity is associated with the gut contents.

Retention factors for cereals are important because edible parts can be processed into several different products with different radionuclide contents. For example, in the production of flour, R_t is 0.1–0.6 for radiocaesium and 0.1–0.4 for actinides. Bran comprises 10–40% of the grain by mass but can be the main contributor to total activity in the grain. Plutonium in bran, for example, is eight times more concentrated than in the grain. Removal of the outer husk of rice, followed by polishing, reduces radionuclide concentrations to 10–25% of the original.

Most studies on the effects of industrial processes have concentrated on milk products, again mainly with radioisotopes of caesium, strontium and iodine. Products from the fatty portions, such as cream and butter, retain only a small proportion of total radioactivity, which mainly goes to the aqueous fraction. However, the latter (as whey) is now used extensively in the food industry and so, unless it is discarded, the radioactivity is simply switched from raw milk to a different product. The way in which cheese is coagulated will also have an effect. For example, if acid coagulation (and not rennet) is used, strontium is retained in the curd used for hard cheese production, rather than mostly passing to the whey. Table 5.8 shows R_w values for some milk products.

In summary, for food prepared in the home from fresh ingredients, the variables involved suggest that apart from obvious methods such as removal of outer leaves of some crops, it is unlikely that major losses of radioactivity will occur. However, there may be some circumstances when judicious use of industrial food processing may significantly reduce levels of some radionuclides in foods. It seems unlikely though, that the public would accept food that had been decontaminated, especially when alternative foodstuffs would be easily available.

Drinking water supplies are very unlikely to become contaminated as a result of an accidental release of radioactivity to the extent that they would require decontamination. None the less, water treatment can be used to substantially decontaminate supplies. For example, sand filters can remove up to nearly 80% of some types of radioactivity (DoE/WO, 1990). In the case of water containing high levels of dissolved radon, for instance in Devon, UK, degassing plant installed in mains water supplies has long been known to be highly effective in reducing the amount of radon in water supplied to homes (Hoather and Rackham, 1963).

Table 5.8 Fresh mass retention factors (R_w)[a] for milk products

	^{137}Cs	^{90}Sr
Whole milk	1	1
Whole milk powder	8	9
Cream	0.6	0.5
Butter	0.2	0.2
Skimmed milk	1	1
Skimmed milk powder	10	
Hard cheese whey	1	0.7
Hard cheese (acid coagulation)	0.5	7
Whey powder	20	

From Green and Wilkins (1995).
[a]R_w is the ratio of activity concentration in processed food to that in raw food, both in terms of fresh mass (see section 5.10).

5.11 Summary and perspectives

Collective dose is a convenient way of expressing the total effects of different sources of radiation on populations (see section 5.5). The collective doses that will be received from all pathways over a 50-year period from a number of significant sources of radioactivity are shown in Table 5.9. As can be seen, by far the greatest contribution comes from naturally occurring radionuclides, which continues at a more or less constant dose rate, with about half this coming from the inhalation of radon. There is of course very little that can be done to avoid this and man has evolved in this radiation field.

Anthropogenic sources, by definition, can be controlled and these are, at least locally, significant. Nevertheless the largest source is still the legacy from the testing of nuclear weapons in the atmosphere. The dose rate is decreasing as there have been no tests since 1980 but the dose from, for example, carbon-14 in food will continue for about the next 10 000 years. For an individual exposed to fallout over the last 50 years, the dose

Table 5.9 Estimates of collective dose over a 50-year period (1000 man-Sv) from various sources

Source of radionuclide	From ingestion	Local/regional total	Global total
Weapons testing fallout			
^{14}C	25 800		
^{137}Cs	677		
^{90}Sr	406		
^{3}H	176		
Total (all radionuclides)			30 000
Accidents			
Chelyabinsk, 1957		2.5	
Windscale, 1957		2.0	
Three Mile Island, 1979		0.04	
Chernobyl, 1986			600
Nuclear power			
Reactor operation		3.7	
Reprocessing:		4.6	
Sellafield		0.31[a]	4.1
Other			
Releases into Techa River (Chelyabinsk)		1.5	
Natural radiation	62 000		650 000[b]

From UNSCEAR (1993).

[a]This is the European collective dose. Local collective dose is 33 man-Sv (i.e. 0.033 thousand man-Sv).

[b]Half of this dose is from inhalation of radon and its daughters.

over that period from food contaminated as a result is equivalent to approximately one fifth of the annual dose from natural radiation. In contrast to these rather low global doses, people living near to test sites have received quite substantial doses from external radiation, inhalation and, occasionally, the intake of contaminated food. This is the case for the intake of iodine-131 by children living near to the US Nevada test site, the Pacific test site and particularly the Semipalatinsk site in the former USSR. Thyroid doses as high as 10 Gy have been estimated for some of these people. Doses of the same magnitude have been also estimated for some of the unfortunate people living near plutonium production facilities such as the Russian reprocessing plant at Mayak in the Urals. The contamination near the latter site is a continuing problem.

Discharges from civil nuclear power programmes have produced relatively small doses; for instance, the global collective dose from effluents released from the nuclear fuel cycle is around 10 000 man-Sv, which is 3000 times less than that from weapons testing. Local doses from the fuel cycle have been of some significance but only near reprocessing plants. The annual individual doses from food contaminated by discharges from the Sellafield plant in the United Kingdom have been as high as 3.5 mSv but are now about 0.15 mSv which is equivalent to a risk of fatal cancer of about 7.5 in 1 million.

Although the impact of routine discharges from the nuclear power industry on food and agriculture is largely under strict surveillance and control, occasional accidental releases have had serious and widespread consequences. By far the worst of these was the accident at Chernobyl in 1986, which has caused extensive disruption of agriculture in the European part of Russia. Nearly a million people have had their food supplies seriously affected and about 15% of these people were evacuated or relocated. The accident caused some disruption of certain foods in nearly every country of western Europe. Even in 1998 there were still restrictions on the movement and slaughter of lamb from certain areas in the United Kingdom. However one of the spin-offs from this disaster has been an increased awareness of the need for emergency preparedness for large-scale accidents, and this has been extended to international arrangements on banning levels for foods. Accident scenarios are rehearsed more frequently than before the Chernobyl accident, and networked radiation early warning systems are now in place. Whether these agreements would prevent national governments from taking unilateral action to protect their own (agricultural) interests, or indeed using more restrictive levels, in the event of another serious accident, is a matter of conjecture.

The overall message from this brief review of radioactivity in food must be that this contamination is monitored and under at least as strict control as other (chemical) pollutants. Considerably more is known about the long-term effects of intake of radioactive materials than of most other pollutants and this is reflected in the controls exercised. The highest doses come from natural sources that are outside our control and the doses from man's activities continue to decline. In normal circumstances, many countries operate expensive, highly technical laboratories to monitor virtually background levels of radioactivity. However, there is a continued need to maintain expertise and vigilance in radiation monitoring so that arrangements are in place for a rapid response to serious environmental contamination.

References

Absalom, I.P., Young, S.D. and Crout, N.M.J. (1995) Radio-caesium fixation dynamics: measurement in six Cumbrian soils. *European Journal of Soil Science*, **46** 461-469.

Aslanoglou, X., Assimakopoulos, P.A., Averin, V., Howard, B.J., Howard, D.C., Karamanis, D.T. and Stamoulis, K. (1996) Impact of the Chernobyl accident on a rural population in Belarus, in *The Radiological Consequences of the Chernobyl accident: Proceedings of the First International Conference, Minsk, Belarus, 18–22 March 1996* (eds A. Karaoglou, G. Desmet, G.N. Kelly and H.G. Menzel), European Commission and the Belarus, Russian and Ukrainian Ministries of Chernobyl affairs, Emergency Situations and Health, EUR 16544 EN.

Attwood, C.A., Titley, J.G., Simmonds, J.R. and Robinson, C.A. (1996) Generalised derived limits for radioisotopes of strontium, ruthenium, iodine, caesium, plutonium, americium and curium. *Document of the NRPB*, **7** (1).

BEIR (1989) Report of the 'Biological Effects of Ionizing Radiation' Committee of the US National Academy of Sciences and National Institutes of Health (BEIR V).

Beresford, N.A., Mayes, R.W., Hansen, H.S., Crout, N.M.J., Hove, K. and Howard, B.J. (1998) Generic relationship between calcium intake and radiostrontium transfer to the milk of dairy ruminants. *Radiation and Environmental Biophysics*, in press.

Brynildsen, I.L. and Strand, P. (1994) A rapid method for the determination of radioactive caesium in live animals, carcasses and its practical application in Norway after the Chernobyl nuclear reactor accident. *Acta. Veterinaria Scandinavica*, **35** 401-408.

Bull, N.L. and Wheeler, E.F. (1986) A study of different dietary survey methods amongst 30 civil servants. *Human Nutrition: Applied Nutrition*, **408** 60-66.

Carter, M.W. (ed.) (1988) *Radionuclides in the Foodchain*. International Life Sciences Institute Monographs, Springer-Verlag, New York.

Clarke, R.H., Fry, F.A., Stather, J.W. and Web, G.A.M. (1993) 1990 recommendations of the ICRP: Recommendations for the practical application of the Board's statement. *Documents of the NRPB*, **4** (1) 9-22.

Cm 2919 (1995) *White Paper on Review of Radioactive Waste Management Policy: Final Conclusions*. HMSO, London.

COMARE (1996) *The Incidence of Cancer and Leukaemia in Young People in the Vicinity of the Sellafield Site, West Cumbria: Further Studies and an Update of the Situation Since the Publication of the Report of the Black Advisory Group in 1984*. Report by the Committee on Medical Aspects of Radiation in the Environment (COMARE), Published by Department of Health, Wetherby, Yorkshire, UK.

Cotter, A.J.R., Myatt, A., Hunt, G.J. and Walters, C.B. (1992) *Monitoring of Radioactivity in the UK Environment: An Annotated Bibliography of Current Programmes*. Fisheries Research Data Report No 72. Ministry of Agriculture, Fisheries and Food, Directorate of Fisheries Research, Lowestoft, UK.

Council Regulation (Euratom) No. 2218/89 of 18/7/89 amending Regulation 3954/87 laying down maximum permitted levels of radioactive contamination of foodstuffs and of feedingstuffs following a nuclear accident or any other case of radiological emergency. *Official Journal* No. L211, 22/7/89.

Council Regulation (Euratom) No. 3954/87 of 22/12/87 laying down maximum permitted levels of radioactive contamination of foodstuffs and of feedingstuffs following a nuclear accident or any other case of radiological emergency. *Official Journal* No. L371, 30/12/87.

Cremers, A., Elsen, A., Valcke, E., Wauters, J., Sandalls, F.J. and Gauder, S.L. (1990) The sensitivity of upland soils to radiocaesium contamination, in *Transfer of Radionuclides in Natural and Semi-natural Environments* (eds G. Desmet, P. Nassimbeni and M. Belli), Elsevier Applied Science, London, pp 238-248.

Crick, M.J. and Lindsley, G.S. (1984) An assessment of the radiological impact of the Windscale reactor fire, October, 1957. *International Journal of Radiation Biology*, **46** 479-506.

Desmet, G. and Sinnaeve, J. (eds) (1992) *Radiation Protection. Evaluation of Data on the Transfer of Radionuclides in the Food Chain*, CEC, Directorate-General, Science, Research and Development.

DoE (1989) *Sampling and Measurement of Radionuclides in the Environment: A Report by the Methodology Sub-group to the Radioactivity Research and Environmental Monitoring Committee*. HMSO, London.

DoE/WO (1990) *The Impact of Accidental Releases of Radioactivity on Water Services*. Department of the Environment Water Inspectorate Report by the Joint Working Group on Water Services, Department of the Environment, London, UK.

Dunster, H.J., Howells, H. and Templeton, W.L. (1958) District surveys following the Windscale incident, October 1957, in *Proceedings of the Second UN Conference on Peaceful Uses of Atomic Energy*. United Nations, New York.

Eisenbud, M. and Gesell, T. (1997) *Environmental Radioactivity from Natural, Industrial and Military Sources*, 4th edn, Academic Press, New York.

Green, N. and Wilkins, B.T. (1995) *Effects of Processing on Radionuclide Contents of Foods: Derivation of Parameter Values for Use in Radiological Assessments*, National Radiological Protection Board, M587, Chilton, UK.

Halverson, J.E. (1987) Application of mass spectrometry to radionuclide metrology. *Nuclear Instruments and Methods in Physics Research* A254.

Hird, A.B., Rimmer, D.L. and Livens, F.R. (1996) Factors affecting the sorption and fixation of caesium in acid organic soil. *European Journal of Soil Science*, **47** 97-104.

Hoather, R.C. and Rackham, R.F. (1963) Some observations on radon in water and its removal by aeration. *Journal of the Institute of Water Engineering Science*, **17** 13-22.

Horrill, A.D. and Howard, D.M. (1991) Chernobyl fallout in three areas of upland pasture in West Cumbria. *Journal of Radiological Protection*, **11** 249-257.

Hove, K., Hansen, H.S. and Strand, P. (1990) Experience with the use of caesium binders to reduce radiocaesium contamination of grazing animals, in *Environmental Contamination Following a Major Nuclear Accident* (eds S. Flitton and E.W. Katz), IAEA-SM-306 2, Vienna, pp 181-189.

HSE (1992) *The Tolerability of Risk from Nuclear Power Stations*, Report published by the Health and Safety Executive, HMSO, London.

IAEA (1989) *Measurement of Radionuclides in Food and the Environment: A Guidebook*, Technical reports series No. 295, International Atomic Energy Agency, Vienna.

IAEA (1991) *The International Chernobyl Project: Assessment of Radiological Consequences and Evaluation of Protective Measures*, Technical Report, International Atomic Energy Agency, Vienna.

IAEA (1994) *Handbook of Transfer Parameter Values for the Prediction of Radionuclide Transfer in Temperate Environments*, (Technical Reports Series No. 364), International Atomic Energy Agency, Vienna.

IAEA (1996) *International Basic Safety Standards for Protection against Ionising Radiation for the Safety of Radiation Sources*, International Atomic Energy Agency Safety Series No. 115, Vienna.

ICRP (1991) *1990 Recommendations of the International Commission on Radiological Protection*, Annals of the International Commission on Radiological Protection **21** (1–3), Pergamon Press, Oxford (ICRP Publication 60).

ICRP (1996) *Age-dependent Doses to Members of the Public from Intake of Radionuclides: Part 5. Compilation of Ingestion and Inhalation Dose Coefficients*, Annals of the International Commission on Radiological Protection **26** (1), Pergamon Press, Oxford (ICRP Publication 72).

Kemeny, J.G. (1979) *President's Commission on the Accident at Three Mile Island. Report of the Task Force on Public Health and Safety*, US Government Printing Office, Washington, DC.

Kershaw, P.J., Pentreath, R.J., Woodhead, D.S. and Hunt, G.J. (1992) *A Review of Radioactivity in the Irish Sea. Aquatic Environment Monitoring Report (32)*, MAFF Directorate of Fisheries Research, The Library, MAFF Fisheries Laboratory, Pakefield Road, Lowestoft, Suffolk, NR33 OHT, UK.

Kosenko, M.M. and Degteva, N.A. (1994) Cancer mortality and radiation risk evaluation in the Techa River population. *Science of the Total Environment*, **142** 73-89.

MAFF (1963 *et seq*) *Radioactivity in Surface and Coastal Waters of the British Isles* (annual report from 1963 to 1994), Ministry of Agriculture, Fisheries and Food, Directorate of Fisheries Research, Aquatic Environmental Monitoring Reports, The Library, MAFF Fisheries Laboratory, Pakefield Road, Lowestoft, Suffolk, NR33 OHT, UK.

MAFF (1988) *Terrestrial Radioactivity Monitoring Report for 1987: Radioactivity in Food and Agricultural Products in England and Wales. TRAMP/2*, Ministry of Agriculture, Fisheries and Food, Ergon House, 17 Smith Square, London SW1P 3HX.

MAFF (1994) *The Forty-third Report of the Steering Group on Chemical Aspects of Food Surveillance: Radionuclides in Foods. Food Surveillance Paper No. 43*, Ministry of Agriculture, Fisheries and Food, HMSO, London.

MAFF and SEPA (1997) *Radioactivity in Food and the Environment, 1996. Ministry of Agriculture, Fisheries and Food and Scottish Environment Protection Agency Annual Report RIFE-2*, Radiological Safety and Nutrition Division of MAFF, 17 Smith Square, London SW1P 3JR.

Mayall, A., Cabianca, T., Morris, T.P., Nightingale, A., Simmonds, J.R. and Cooper, J.R. (1994) NRPB note for COMARE. NRPB-M453. National Radiological Protection Board, Chilton, Didcot, Oxford, UK.

Meredith, R.C.K., Mondon, K.J. and Sherlock, J.C. (1988) A rapid method for the *in vivo* monitoring of radiocaesium in sheep. *Journal of Environmental Radioactivity*, **7** 209-214.

Milvy, P. and Cothern, C.R. (1991) Scientific background for the development of regulations in drinking water, in *Radon, Radium and Uranium in Drinking Water* (eds C.R. Cothern and P.A. Rebers), Lewis Publishers, Chelsea, MI, pp 1-16.

Mondon, K.J. and Walters, B. (1990) Measurement of radiocaesium, radiostrontium and plutonium in whole diets following deposition of radioactivity in the UK originating from the Chernobyl power plant accident. *Food Additives and Contaminants*, **7** (6) 837-848.

Muck, K. (1996) Long-term effective decrease of cesium concentration in foodstuffs after nuclear fallout. *Health Physics*, **72** 659-673.

NCRP (1984) *Radiological Assessment: Predicting the Transport, Bioaccumulation and Uptake by Man of Radionuclides Released to the Environment*, Report 76, National Council on Radiation Protection and Measurements, Bethesda, MD, USA.

OECD (1995) *Chernobyl 10 Years On*. Organisation for Economic Cooperation and Development, Paris.

Pearce, J., Unsworth, E.F., McMurray, C.H., Moss, B.W., Logan, E., Rice, D. and Hove, K. (1989) The effects of prussian blue provided by indwelling boli on the tissue retention of dietary radiocaesium by sheep. *Science of the Total Environment*, **85** 349-356.

Penna Franca, E., Fiszman, M., Lobao, N., Costa Ribeiro, C., Trindade, H., Dos Santos, P.I. and Batista, D. (1968) Radioactivity of Brazil nuts. *Health Physics*, **14** 95-99.

Royal Society (1992) *Risk: Analysis, Perception, Management, Report of Royal Society Study Group*, The Royal Society, London.

Russell, R.S. (1960) *Radioactivity and Human Diet*, Pergamon Press, Oxford.

Shiraishi, K., Tagami, K., Ban-nai, T., Yamamoto, M., Muramatsu, Y., Los, I.P., Phedosenko, G.V., Korzun, V.N., Tsigankov, N.Y. and Segeda, I.I. (1997) Daily intakes of Cs-134, Cs-137, K-40, Th-232 and U-238 in Ukrainian adult males. *Health Physics*, **75** (5) 814-819.

Simmonds, J.R., Lawson, G., Mayall, A. (1995) *Methodology for Assessing the Radiological Consequences of Routine Releases of Radionuclides to the Environment*, European Commission (Radiation Protection), EUR 15760 EN, Luxembourg.

Skuterud, L., Travnikova, I.G., Balonov, M.I., Strand, P. and Howard, B.J. (1997) Contribution of fungi to radiocaesium intake by rural populations in Russia *Science of the Total Environment*, **193** 237-242,

Smith, J.N., Ellis, K.M. and Kilius, L. (1997) Nuclear submarine sampling of the European reprocessing signal in the Artic Ocean. *Proceedings of Radionuclides in the Oceans, RADOC 96–97, 7–11 April 1997, Norwich/Lowestoft, UK*, in press.

Strand, P. (1994) *Radioactive Fallout in Norway from the Chernobyl Accident: Studies on the Behaviour of Radiocaesiums in the Environment and Possible Health Impacts*, Norwegian Radiation Protection Authority, PO 55, N-1345 Osteras, Norway.

Strand, P., Boe, E., Berteig, L., Strand, T., Trygg, K. and Harbitz, O. (1989) Whole body counting and dietary surveys in Norway during the first year after the Chernobyl accident. *Radiation Protection Dosimetry*, **27** 1163-1171.

Strand, P., Brynildsen, L.I., Harbitz, O. and Tveten, U. (1990) Measures introduced in Norway after the Chernobyl accident. A cost-benefit analysis, in *Environmental Contamination Following a Major Nuclear Accident* (eds S. Flitton and E.W. Katz), IAEA-SM-306 2, Vienna, pp 191-202.

Strand, P., Balonov. M., Skuterud, L., Hovem, K., Howard, B.J., Prister, B.S., Travnikova, I. and Ratnikov, A. (1996) Exposures from consumption of agricultural and semi-natural products, in *The Radiological Consequences of the Chernobyl Accident: Proceedings of the First International Conference, Minsk, Belarus, 18–22 March 1996* (eds A. Karaoglou, G. Desmet, G.N. Kelly and H.G. Menzel), European Commission and the Belarus, Russian and Ukrainian Ministries of Chernobyl affairs, Emergency Situations and Health, EUR 16544 EN.

Takada, J., Hoshi, M., Rozenson, R.I., Endo, S., Yamamoto, M., Nagatomo, T., Imanaka, T., Gusev, B.I., Apsalikov, K.N. and Tchaijunusova, N.J. (1997) Environmental radiation dose in Semipalatinsk area near nuclear test site. *Health Physics*, **73** (3) 524-526.

Trabalka, J.R. and Auerbach, S.I. (1990) One western perspective of the 1957 Soviet nuclear accident, in *Proceedings of a Seminar on the Comparative Assessment of the Environmental Impact of Radionuclides Released during Three Major Nuclear Accidents: Kyshtym, Windscale, Chernobyl*, EUR-13574, pp 41-69.

UNSCEAR (1982) *Ionizing Radiation: Sources and Biological Effects*, United Nations Scientific Committee on the Effects of Atomic Radiation, United Nations, New York.

UNSCEAR (1988) *Sources, Effects and Risks of Ionizing Radiation*, United Nations Scientific Committee on the Effects of Atomic Radiation, United Nations, New York.

UNSCEAR (1993) *Sources and Effects of Ionizing Radiation*, United Nations Scientific Committee on the Effects of Atomic Radiation, United Nations, New York.

WHO (1993) *Guidelines for Drinking Water Quality*, World Health Organisation, Geneva.

6   Trace metal contaminants in food

Farid E. Ahmed

6.1   Introduction

Metals are considered to be the oldest toxins, known to humans since the Stone Age; they posed hazards to humans from the time they were fashioned into spears and arrows to present-day exposure to Space-Age metals, alloys or salts. Lead was mined in ancient Egypt before 2000 BC, as a by-product of smelting silver; arsenic, used for the decoration of tombs in Egypt, was recovered while melting copper; and tin was cited by Theophratus of Erebus (370–287 BC). High natural concentrations of metals in food or water could have led to the first human exposures, and metals leaching from eating utensils or metallic cookware increased the risk of contamination. Antimony was used medicinally as far back as 4000 BC. Hippocrates was the first philosopher/physician to describe symptoms of abdominal colic in a man who extracted metals from soil. On the other hand, many of the metals of known toxicity today have only recently been recognized. For example, cadmium was discovered in 1817 in ores containing zinc carbonate, and although about 80 of the 105 elements in the periodic table are metals, fewer than 30 have been reported toxic to humans. The advent of industrialization has associated many occupational diseases with a variety of toxic metals, in addition to their role as environmental pollutants in recent years due to human activities. On the positive side, many metals have therapeutic applications and some of them are essential nutrients (Hammond and Beliles, 1980; Goyer, 1995; NRC, 1989).

Metals are redistributed in the environment by both natural geological and biological cycles (Beijer and Jernelöv, 1986). Rocks and ores are dissolved by rain and transported to streams and rivers (the hydrosphere). Running water strips additional material from adjacent soils and transports them to the ocean for precipitation as sediments, and air takes them up in water vapour to the atmosphere and precipitates them on earth's soil (the lithosphere) with rain. Biological cycles (the biosphere) bioconcentrate elements through both plants and animals, through biomagnification in food cycles. It is important to realize that the elements can move from one phase, or compartment, to another (e.g. air–water, soil–water, or soil–air) by transport, or chemodynamics, within phases (i.e. intraphase) and between phases (interphase) (Mackay, 1991). An example of an ecological chemodynamic cycle is as follows: a metal is released into one environmental compartment, is partitioned among the various environmental compartments, becomes involved in movements and reactions within each compartment, is partitioned between each compartment and the biota of that compartment and, finally, reaches an active site in an organism at a high enough concentration and stays for long enough time to exert an effect. Interaction among various environmental compartments is presented in Figure 6.1.

Although the natural cycles may exceed the anthropogenic ones, as in the case of mercury, human industrial activities inevitably shorten the residence times of metals in ores, may produce new compounds, or may greatly enhance worldwide distribution of elements, as was demonstrated by the 200-fold increase in lead content of the Greenland ice cap from around 800 BC to the 1920s when lead was added to gasoline to serve as an anti-knock agent (Ng and Patterson, 1981).

Some metals are essential for life (e.g. selenium, zinc, copper, iron, manganese, chromium and molybdenum). Some have no known biological functions, but do not pose serious hazards to man, and some have the potential to cause intoxications. Metals that are essential nutrients can also exert toxic action if the homeostatic mechanism maintaining their physiological limit is disrupted (NRC, 1989).

6.2 Factors influencing toxicity of heavy metals

Because of the elemental nature of metals and their diverse affinities for organic ligands in biological structures, their toxicities arise as multiple organ effects more than with any other class of toxicants. In view of this

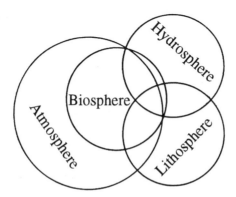

Figure 6.1 Interactions among various environmental compartments. (Adapted from Mackay (1991) and Menzer (1991).)

multiplicity of effects, the concept of critical organ and critical dose have evolved to delineate the most sensitive ones. Thus, the critical organ is the one showing adverse effects at the lowest dose, although other organs and systems may be much more severely affected but only at higher doses. This concept is important with respect to toxicants for which a tolerance greater than zero has been assigned for technical or economic reasons (Nordberg, 1976).

Factors that influence the toxicity of metals at a certain level of exposure are important, particularly in susceptible populations, and include metabolic interactions of essential metals, formation of metal–protein complexes, age and state of development of exposed individuals, lifestyle factors, chemical form or speciation, and immune status of the host (Burns *et al.*, 1995; Fowler, 1991; Oehme, 1978).

The relationship of level of exposure, or dose, to its metabolism in various organs and major excretory pathways is presented in Figure 6.2. Blood, urine and hair are most frequently used to assess exposure to

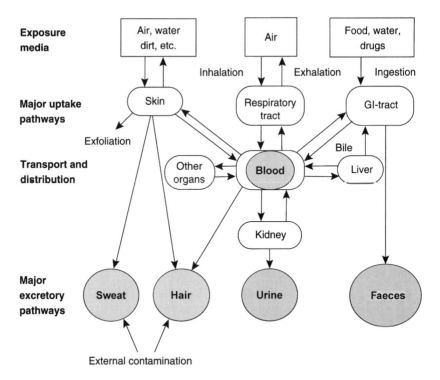

Figure 6.2 Relationship between sources of exposure, transport and distribution to various organs and excretory pathways (tissues useful for biological monitoring are shaded). (From Elinder *et al.* (1994), with WHO permission.)

metals, and are thus referred to as indicator tissues. Blood and urine exposure usually represent recent effects and correlate best with acute toxicity, except in case of urinary excretion of cadmium, which reflects renal damage due to the accumulation of this element in the kidney over a long time (Nordberg, 1972). Quantitation of metals in organs can be carried out by techniques such as atomic absorption, atomic fluorescence, flame emission, electron probe, ion probe, neutron activation analysis, spark source mass spectroscopy, proton-induced x-ray emissions, x-ray fluorescent spectroscopy and scanning x-ray analysis (Jacobs, 1996; Meloan, 1978).

The dose is a function of time and concentration of the metal. A critical determinant of the metabolism of a metal and toxic behaviour is its biological half-life (i.e. the time it takes the body to excrete half of the accumulated metal). The half-life of cadmium and lead is somewhere between 20 and 30 years, whereas it is only a few hours or days for metals such as arsenic, cobalt and chromium. For inorganic mercury, the half life is a few days in blood, and a few months for the whole body (Oehme, 1978).

Toxicity of a metal is determined by the dose at the cellular level and such factors as chemical form (or speciation) and ligand binding. For example, alkyl compounds are lipid-soluble and pass readily across biological membranes causing them to accumulate inside the cell which makes their toxicity differ from the inorganic form (e.g. alkylmercury is primarily neurotoxic, whereas mercuric chloride, $HgCl_2$, is nephrotoxic). The strong attraction between metal ions and organic ligands influences not only their availability for absorption, but the deposition of a metal in the body and its excretory role. Most biologically important metals bind strongly to tissues and therefore are slowly excreted, leading them to accumulate *in vivo*. However, tissue affinities of various metals are not similar; for example, lead has a strong affinity to osseous tissue, but cadmium and mercury localize mainly in the kidney. Partioning of a metal between cells and plasma, and between filterable and non-filterable plasma components provides more accurate information about the presence of a biologically active form of the metal (e.g. ionic calcium in the blood). Speciation of a metal in the urine or plasma may provide a toxicological indication for the metal (e.g. cadmium metallothionein vs total cadmium). Hair might also be a useful indicator of exposure to some metals (e.g. alkyl mercury), but care must be exercised because metal deposits as a result of external contamination may complicate the analysis in spite of washing of the hair (Cherniaeva *et al.*, 1997; Meloan, 1978; Schumacher *et al.*, 1996).

The interaction of a toxic metal with an essential one occurs when they have similar metabolism. Moreover, absorption of a toxic metal from the

gut or lung is often influenced by an essential element if they both share—
or if the toxic metal influences—a homeostatic mechanism, as for lead,
cadmium and iron. An inverse relationship between the protein content
of the diet and cadmium and lead toxicities was observed. Vitamin C was
reported to reduce lead and cadmium absorption, probably because of
increased absorption of Fe^{2+}. Additionally, toxic metals influence the
role of essential metals as cofactors or other metabolic processes;
examples are lead interference with calcium-dependent release of
neurotransmitters; the complex role of lead, calcium and vitamin D in
bone mineralization; a direct role involving lead impairment of the
synthesis of 1,25-dihydroxyvitamin D in the kidney; arsenic alleviating
effects of selenium poisoning in pigs, dogs, chicken and cattle (Ahmed,
1991a; Fullmer, 1997; NRC, 1989; Underwood, 1978).

Many toxic metals were shown to form complexes with proteins
resulting in either their detoxification or protection from toxicities (e.g.
metallothionein complexes with cadmium, zinc, or copper; and hemosi-
derin complexes between iron and proteins) (Fowler et al., 1981; Mistry
et al., 1985; NRC, 1989).

Individuals at either end of their life span (e.g. children or the elderly)
are believed to be more susceptible than adults to toxic effects of metal
exposure, with the major exposure pathway being the diet (Ahmed,
1992a; NRC, 1989). Moreover, life cycle factors such as smoking or
alcohol were shown to exert influence on metal toxicities. Alcohol intake
may influence the diet by interfering with intake, absorption and
availability of essential minerals (Ahmed, 1992b, 1995). Cigarette smoke
contains some toxic metals such as cadmium, and cigarette smoking may
also influence pulmonary functions (Ahmed, 1991b; Goyer, 1995).

Speciation, or the chemical form of a metal, may influence toxicity, not
only for gastrointestinal and pulmonary absorption but also in terms of
distribution through the body. For example, dietary phosphate generally
forms less soluble salts with metals than with their ions (NRC, 1989).

Some metals such as mercury, chromium, nickel, beryllium and
platinum produce variable immune reactions affecting cell-mediated
and humoral responses (e.g. anaphylactic or immediate hypersensitivity,
cytotoxic hypersensitivity, immune complex hypersensitivity, or cell-
mediated hypersensitivity (Burns et al., 1995; Exon et al., 1996).

Carcinogenicity is regarded as a special form of toxicity. Given the long
history of exposure of humans to toxic metals, information on
carcinogenicity has been developing gradually. Table 6.1 summarizes
the carcinogenicity of metals in humans and animals as classified by the
International Agency for Research on Cancer (IARC, 1987, 1994). IARC
has given the following classification of carcinogenicity from both human
and animal studies: sufficient, limited, inadequate and no evidence.

Table 6.1 Evidence associating metals with carcinogenicity

Metal	In humans	In animals
Cadmium	Sufficient	Sufficient
Lead (inorganic)	Insufficient	Sufficient
Mercury (inorganic)	Inadequate	Limited[a]
Arsenic	Sufficient	Limited
Selenium	Insufficient	Insufficient
Zinc	Insufficient	Insufficient
Iron complexes		
Dextran	Insufficient	Sufficient
Dextrin	ND[b]	Limited
Chromium		
Trivalent	Insufficient	Insufficient
Hexavalent	Sufficient	Sufficient

Adapted from IARC (1987, 1994).
[a]For $HgCl_2$.
[b]No data.

Evidence of a specific metal carcinogenicity has been variable. For example, the carcinogenicity of arsenic to human skin was recognized in the nineteenth century, but epidemiological support did not materialize until over 50 years later, and has not been confirmed in animal studies. The reverse is true for lead, which is the only metal shown to be carcinogenic in animals by the oral route of administration, yet evidence from epidemiologic studies is limited to case reports. Animal studies that use routes of administration not usually used by humans (e.g. injection as compared to ingestion or inhalation) have limitations for extrapolation to humans. Moreover, identification of metal carcinogens occupationally is even more complicated because humans are seldom exposed to a single metal. The role of metals such as lead as promoters or cocarcinogens may be important because of their persistence in tissues (IARC, 1987, 1994). Activation of oncogenes and/or inactivation of tumour suppressor genes have been proposed as the overall mechanism of carcinogenesis of some metals such as arsenic, chromium and nickel (Landolph *et al.*, 1996).

Short-term tests (e.g. Ames *Salmonella* test) do not seem responsive for assessing metal mutagenicity or genotoxicity of most metals (De Flora *et al.*, 1994; Pesheva *et al.*, 1997). However, recent studies have shown the capacity of metal carcinogens (e.g. lead, cadmium) to cause oxidative damage to DNA and inhibits its repair; to produce strand breaks and rearrangements *in vivo* and *in vitro* leading to chromosomal aberrations and intragene recombination; and to inhibit cell cycle progression. These are all factors that lead to mutation, or abnormal gene expression, or abnormal cellular differentiation and cell death (Godet *et al.*, 1996; Hartwig, 1994; Ou *et al.*, 1997; Poirier and Littlefield, 1996; Skoczynska, 1997).

6.3   Dietary intake of environmentally important metals

Heavy metals, unlike synthetic chemicals, are not man-made, nor are they destroyed by humans. Environmental pollution is affected by environmental transport and chemodynamics (i.e. human or anthropogenic contribution to air, water, soil and food and interactions among various environmental compartments) and by altering speciation (Beijer and Jernelöv, 1986; Mackay, 1991). The epidemiology of heavy-metal food intoxication has generally been associated with one of three patterns of occurrence: (1) environmental pollution, (2) accidental inclusion during processing, and (3) contamination from containers during processing or storage of food. Most epidemiological investigations of heavy-metal intoxications have been of the retrospective type in which subjects with the effect are compared with intoxication-free control subjects with regard to the suspected causative agent or agents, and little has been done with the prospective type of long-term exposure, or cohort study, in which a group of apparently healthy people is followed for some time for the occurrence of an effect as a result of exposure to low-level metal contaminants in human food (NRC, 1989). A useful epidemiological model for the flow of heavy metals such as cadmium and mercury into the human food is shown in Figure 6.3.

Estimates of dietary intake of trace metal contaminants are carried out periodically in various countries to estimate the dose (intake) of the contaminant, which are then combined with hazard analysis and dose–response information to derive a characterization of the risk for that contaminant. The term acceptable daily intake (ADI) develop̓d in the mid 1950s by the US Food and Drug Administration (FDA), represents a level of daily intake that, if not exceeded, or if exceeded only rarely in the life of an individual, should permit an insignificant risk of 'carcinogenic' health effects (NRC, 1980). On the other hand, the term reference dose (RfD) is specifically applied for toxicity values derived for evaluating 'non-carcinogenic' health effects (EPA, 1991). Both ADI and RfD were established similarly, although the RfD was derived using a stricter method developed by the US Environmental Protection Agency (EPA) (Beck et al., 1993), and both are expressed in milligrams (mg) of contaminant per kilogram (kg) of body weight (bw).

At the international level, the United Nations (UN) Joint Food Agriculture Organization (FAO)/World Health Organization (WHO) Expert Committee on Food Additives (JECFA) expresses the tolerable intake of cumulative environmental contaminants having toxicological effect(s), e.g. cadmium, lead, mercury, when taken in the diet on a weekly basis to allow for variations in intake levels, and hence the term

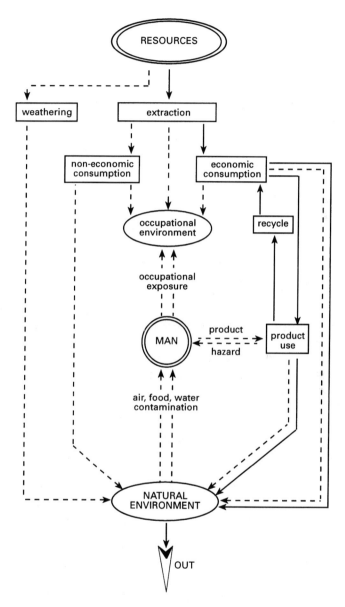

Figure 6.3 A schematic model for flow of trace elements through food (from MacGregor (1975)).

Provisional Tolerable Weekly Intake (PTWI) was developed (UNEP/FAO/WHO, 1988).

An international 'Food Programme' that is part of the UN's systemwide Global Environmental Monitoring System (GEMS) compares certain priority food contaminants in foods of major dietary importance in more than 1200 individual foods (see Chapter 13). This chapter is primarily concerned with the GEMS/Food data on three priority trace metal contaminants (cadmium, lead and mercury) gathered in the 1970s and 1980s, in addition to reports prepared for the Codex Alimentarius Commission (CAC) and its various subsidiary bodies under the Joint FAO/WHO Food Standards Programme, and articles in the worldwide open literature on these three metals. To complete the picture, other metals of toxicological (i.e. industrial, occupational) but not environmental importance as potential food contaminants are also presented below.

6.3.1 Cadmium

Cadmium is unique among toxic metals because knowledge of it is relatively recent; it was only discovered as an element in 1817, and its industrial use was minor until about 50 years ago. Cadmium is a by-product of zinc and lead mining and smelting, which are important sources of environmental pollution. In addition, the use of municipal sludge and compost on agricultural land, incineration of wastes and combustion of fuel constitute other important sources of pollution (Hammond and Beliles, 1980; Nordberg, 1972). Cadmium is used mainly in electroplating and galvanizing because of its noncorrosive properties, as a stabilizer for plastics, as a component in pigments and as a cathode material for nickel–cadmium batteries. It occurs in industrial heavy-metal emissions, phosphate rock used in fertilizers, and the mining and metal industries. Cadmium exposure comes from the ambient air, drinking water, tobacco, the working environment, soil, dust and food, with food being the main source of exposure to cadmium in non-occupational settings (IARC, 1976), and the main source of cadmium entering food is cadmium-containing soil (Sherlock, 1984). Total daily intake from food, water and air in North America and Europe varies considerably, with a range from 10 to 40 µg/day (Goyer, 1995).

Cadmium is commonly found in its metallic form, and as sulphides and sulfates. It is more readily taken up by plants than any other metal. Factors contributing to the presence of cadmium in soils are fallout from air, use of cadmium-containing water for irrigation, and addition of cadmium to fertilizers. Commercial phosphate fertilizers usually contain < 20 mg/kg. Moreover, commercial sludge used to fertilize fields contributes up to 1500 mg of cadmium per kg of dry material (Anderson and Hahlin, 1981). Compared to other heavy metals, cadmium is mobile in the aqueous environment, and thus tends to be widely distributed. It is readily absorbed and accumulated in plants grown in contaminated soils and is bioconcentrated in organisms eating these plants (Menzer, 1991).

6.3.1.1 Dietary intake

In 1988, JECFA established a PTWI of 7 μg/kg bw for adults and infants over an accumulative period of 50 years at an exposure rate equivalent to 1.0 μg/kg bw/day for adults. Intakes above this level may be tolerated provided they are not sustained over long periods (WHO, 1989). No maximum limits have been recommended for cadmium in foods by the Codex CAC. However, several countries have established limits varying from 0.01 mg/kg in milk or eggs, to 2.0 mg/kg in finfish and shellfish. The International Organization for Standards (ISO) adopted limits of release for cadmium from ceramic foodware, varying from 0.17 mg/dm^2 for flatware to 0.25–0.5 mg/l of extraction solution from hollow-ware (ISO, 1981).

Dietary intake of cadmium by adults in 24 countries submitting data to GEMS/Food are shown in Figure 6.4. Very few countries monitored the same food consistently over several years, and specific foods varied considerably from country to country, making it difficult to compare countries, although a trend within a particular country could be established. Typical cadmium levels reported to GEMS/Food are shown in Table 6.2. Lowest cadmium levels were reported in milk, eggs, fruit, and meat muscle; median levels were found in potatoes, cereal and finfish; highest concentrations were present in molluscs and crustacea, and in kidney (the organ most affected by cadmium), in which contamination was found to increase with the age of the animal. In Slovenia, the mean concentrations of cadmium during the years 1989 through 1993 in bovine meat, liver and kidney were 4, 94 and 373 μg/kg wet weight, respectively, while those in the corresponding pig tissue were 10, 88 and 393 μg/kg wet weight (Doganoc, 1996). Invertebrates (both crustacea and bivalves) tend to accumulate metallic cadmium in large amounts by binding to various high molecular mass metallothioneine ligands. In a study undertaken in four geographical regions in Spain, the average total dietary cadmium intake, provided mainly by finfish and shellfish, was about 24–45% of the PTWI (Cuadrado et al., 1995). A recent study in the United States on cadmium content of clams and oysters showed average levels of 90, 50, 510 and 1100 μg/kg wet weight in hardshell clams, softshell clams, Eastern oysters and Pacific oysters, respectively (Capar and Yess, 1996). There is a differential affinity between crustacean muscle and hepato-pancreas, the latter organ containing 10–20 times the concentration of the

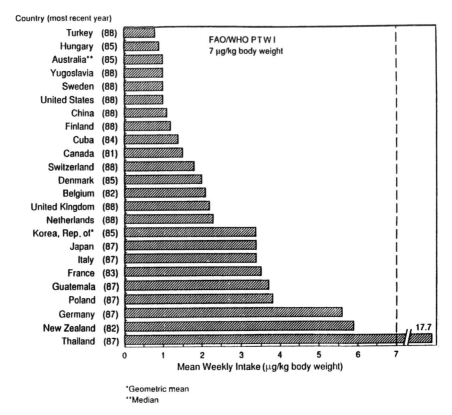

Figure 6.4 Dietary intake of cadmium by adults. (From Jelinek (1992), with permission.)

Table 6.2 Cadmium levels reported in foods

Food	μg/kg
Milk, eggs	5
Fruit	10
Meat muscles	15
Vegetables	25
Cereal/cereal products	30
Fish	35
Molluscs/crustaceans	350
Kidney	500

From UNEP (1992), with permission.

former. Because hepatopancreas may be considered a delicacy, or marketed as 'brown crab meat', the potential for ingesting large amounts of cadmium when eating lobster or crab is increased (Ahmed, 1991b; De Gregori *et al.*, 1994; McKenzie-Parnell *et al.*, 1988).

Figure 6.4 shows that the average intake from Thailand (17.7 µg/kg bw/week) was well above the PTWI, whereas intakes from New Zealand and Germany (Lorenz *et al.*, 1986) were just below the PTWI, and those from Poland, Guatemala, France, Italy, Japan and Korea were around 50% of the PTWI. At the lower end were intakes from Finland, China, United States, Sweden, former Yugoslavia, Australia, Hungary and Turkey (≤1 µg/kg bw/week).

Canada (Dabeka *et al.*, 1987), Denmark, Finland, the Netherlands and the United States identified cereals and their products, followed by potatoes and other vegetables, as the largest contributors to this intake. Although high cadmium levels were found in animal kidneys, molluscs and crustaceans, these foods constitute much less of the average diet (Jelinek, 1992). In Belgium, it was reported that consumption of mussels or kidneys once a week would result in intakes that approximate PTWI (Fouassin and Fondu, 1980). Similarly, it was noted in Denmark that above average consumption of beef kidney, mussels from contaminated water, or wild mushrooms would result in intakes that would exceed the PTWI (Andersen, 1981). Meals that included wild mushrooms and liver led to higher intakes of cadmium in Finland (Louekari *et al.*, 1987). A Canadian study during 1986–88 (Dabeka and McKenzie, 1995) indicated mean and range of cadmium concentrations in retail foods in five Canadian cities at 9.96 (< 0.02–167) ng/g or, by using the standard conversion factor of 0.12 and multiplying by 7 (for days in one week) (see Chapter 13), a mean intake level of 8.37 µg/kg bw. A study of Canadian Inuit living in the Eastern Arctic community of Qikiqtarjuaq on Baffin Island showed a relatively high concentration of cadmium (> 1 mg/kg, or > 84 µg/kg bw) from ringed seal liver, mussels and kelp consumed by adult men and women (> 20 years old) and children (3–12 years old) (Chan *et al.*, 1995). In a study of the general population of the Republic of Croatia, the mean weekly dietary intake of cadmium per person was 121.4 µg, which represents 24.4% of the PTWI (Sapunar-Postruznik *et al.*, 1996). The first Dutch National Food Consumption Survey (1987–88) (*n* = 5898, ages 1 to 85 years) on 226 food products showed that intake of cadmium did not exceed the ADI, and has actually declined between 1976–78 and 1988–89 (Brussard *et al.*, 1996). A study to compare non-occupational background exposure of 202 Chinese in Beijing, Shanghai, Nanning and Tainan, and 72 Japanese women in Tokyo, Kyoto and Sendai during 1993–95 showed geometric mean cadmium daily intakes of 9.9 µg/day for the Chinese and 32.1 µg/day for the Japanese women, of which boiled rice accounted for 31.1% and 32.7% of the total dietary burden for the Chinese and Japanese women, respectively (Zhang *et al.*,

1997). A Korean survey conducted in Seoul, Pusan, Chunan and Haman showed a dietary cadmium geometric mean of 21.2 μg/d for the four cities in combination, of which boiled rice accounted for 23% of the total daily cadmium intake (Moon et al., 1995).

The cadmium content of various species of fruits gathered from regions not directly exposed to air pollution from industrial plants and traffic in Poland during 1989–91 showed highest cadmium levels in raspberries and strawberries (mean 0.02 mg/kg), and lowest in apples and pears (mean 0.001–0.006 mg/kg); levels in fruits and soft drinks from heavily industrialized areas were much higher (Wojciechowska-Mazurek et al., 1995).

Studies in Belgium (Buchet et al., 1983), China (Vahter and Slorach, 1990), Cuba (Beltran Llerandi et al., 1989), Poland, Sweden (Slorach et al., 1983), the United Kingdom and former Yugoslavia (Vahter and Slorach, 1990) indicate that urban surroundings did not lead to elevated cadmium intakes. On the other hand, intakes of cadmium in individuals living in industrial areas in Poland (Gzyl, 1997; Marzec and Bulinski, 1988; Szymczak et al., 1984) and the United States (Gunderson, 1995) were much higher than those living in non-industrial areas. Similarly, consumption of home-grown vegetables or fruits in an area in Denmark close to a lead smelter led to higher cadmium intake (~70% of the PTWI as compared to an average intake of ~35% of PTWI) (Andersen, 1981). In the United Kingdom, a survey conducted in Shipham, after a finding of substantial soil contamination by cadmium due to leakage from a zinc mine operating since the mid-nineteenth century, showed that about 6% of the inhabitants had estimated dietary intakes higher than PTWI when locally grown produce was consumed, whereas none of the households who did not consume Shipham-grown produce had estimated intakes exceeding the PTWI (Morgan et al., 1988).

Dinnerware contaminated with overglaze design, which was made in the United States before 1970, was shown to release toxic metals into food substances when treated with acids. For example, when these dishes were filled with 4% acetic acid for 24 h, cadmium concentrations up to 15 μg/ml were released; 25% of dishes (26 of 98) exceeded the FDA limits of 0.5 μg/ml. Higher concentrations of cadmium were extracted by 1% solutions of citric and lactic acids, as well as by basic solutions of sodium citrate and sodium tripolyphosphate, and by commercial food substances such as sauerkraut, pickle and orange juices, and low-lactose milk (Sheets, 1997).

Too few countries submitted dietary intake data to GEMS/Food for enough years to allow accurate trends to be derived. Of those submitting, it seems that there is a downward trend for Japan, United Kingdom and

the United States (Figure 6.5). A recent study conducted in Japan from
19 sites to compare exposure to cadmium during the 1990s (1991 to 1994)
and the 1980s (1979 to 1983) showed a reduction from 38 to 30 μg/day
(Watanabe *et al.*, 1996).

Ten countries reported to GEMS/Food cadmium intakes by infants
and young children (Figure 6.6). Average intakes for Cuba and Poland
exceeded the PTWI. Intakes from Finland, Germany, Canada, UK and
Niger were around 50% of the PTWI. The lowest average intakes (≤2 μg/
kg bw/week) were reported from the United States, Philippines and
Australia. On a body weight basis, the average daily intakes of infants are
higher than those for adults, but it must be noted that the PTWI was
derived on the basis of cadmium accumulation over a 50-year period.
However, none of these intakes exceeded the PTWI in recent years.

Data on cadmium intake of infants in West Germany show that
average cadmium intakes associated with different levels of cadmium in

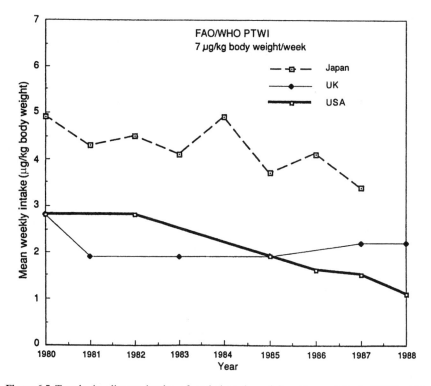

Figure 6.5 Trends in dietary intake of cadmium by adults. (From Jelinek (1992), with
permission.)

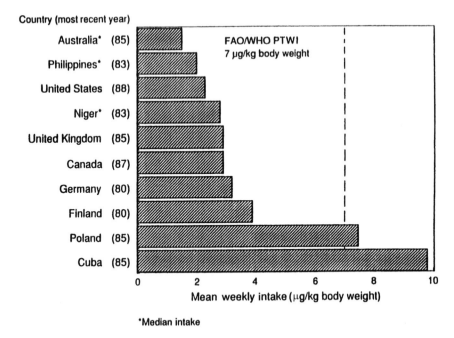

Figure 6.6 Dietary intake of cadmium by infants and children. (From Jelinek (1992), with permission.)

drinking water (1 and 6 μg/l) for several age groups exceeded the PTWI, particularly at the high levels. The WHO guideline for cadmium in drinking water is 5 μg/l (Muller and Schmidt, 1983). However, higher intakes above PTWI may be a health risk only if sustained for long periods, as the PTWI for cadmium is based on exposure for 50 years (UNEP/FAO/WHO, 1988). Moreover, recent data from Duisburg, Germany, show that the medium and maximum dietary intakes for children were 30.2% and 98.7% of the PTWI, respectively (Wilhelm *et al.*, 1995). The 1987–88 Dutch survey showed that ∼1.5% of young people exceeded their PTWI (Brussard *et al.*, 1996). In Canadian studies, cadmium intake by infants consuming soy-based infant formula was greater than those drinking milk-based formula (Dabeka, 1989). A Swedish study on the intake of cadmium exclusively by breast-fed infants had intakes of 0.1 μg/kg bw compared to a much higher adult intake of 1.0 μg/kg bw (Larsson *et al.*, 1981). These results clearly indicate that cadmium does not accumulate in human milk.

A major non-occupational source of respirable cadmium is cigarettes. A cigarette contains 1–2 μg cadmium, and 10% of the cadmium in a cigarette (0.1–0.2 μg) is inhaled. Thus smoking one or more packs of cigarette a day may double the daily absorbed burden of cadmium (Goyer, 1995). In studies in Berlin on the effect of smoking on cadmium levels in human milk, the medium level of cadmium was $0.07\,\mu g/kg$ in the milk of non-smoking mothers and $0.12\,\mu g/kg$ in those who smoked 10 cigarettes per day (Radisch et al., 1987), indicating that smoking increased cadmium content in breast milk, but the level was still less than in infant formula. These levels in human milk are in agreement with the Canadian studies (Dabeka et al., 1986).

In conclusion, exposure to dietary cadmium represents a human heath risk. Thus, diet intake studies should be carried by more countries and appropriate measures should be taken to minimize its occurrence in the diet, especially in animal organs, shellfish, vegetables, fruits and grains from areas of known cadmium contamination. Well-designed dietary studies should be conducted in locations of potential cadmium contamination, such as those near mining and metal extraction operations, phosphate fertilizer plants, high-cadmium bearing strata in the soil, municipal sludge-deposition areas and shellfish growing areas affected by improperly treated industrial or municipal discharges (UNEP, 1992).

6.3.1.2 Toxicity

Acute toxicity may result from the ingestion of relatively high concentrations of cadmium such as may occur in contaminated beverages or food. Nausea, vomiting and abdominal pain were found following consumption cf drinks containing approximately $16\,mg/l$ of cadmium. Recovery was rapid and without apparent long-term effects (Nordberg, 1972). Inhalation of cadmium fumes or other cadmium-containing materials may produce an acute chemical pneumonitis and pulmonary oedema (Friberg et al., 1986).

The principal long-term effects of low-level exposure to cadmium are chronic obstructive pulmonary disease and emphysema, and chronic renal tubular disease. Cadmium has an extremely long half-life in humans. The kidney, particularly the cortex, was identified as the target organ relative to low-levels of exposure to cadmium. The first adverse functional change is usually a low molecular mass proteinurea. There may also be effects on the cardiovascular and/or skeletal systems (Friberg et al., 1986).

Cadmium has been correlated positively with major human poisoning incidents as a contaminant released by the activities of the 'Kamioka Mine' through waste water used for irrigation of rice fields in the Jinzu River Valley in Japan (Nogawa et al., 1983). The osseous illness is known as itai-itai (ouch-ouch) disease because afflicted patients continually moan owing to severe internal pains. It is a chronic osteoporotic and osteomalic condition that primarily affects middle-aged to elderly multiparous females who lived in that area for more than 30 years. Since approximately 1940, an estimated 200 farm women suffered this painful disease; half of them died from it by the end of 1965 (Kobayashi, 1978). Although the highest accumulation of cadmium is found in bone, leading to osteomalachia with attendant spontaneous multiple bone fractures, the liver and the kidney also have a propensity for accumulating the metal, and the kidney is often seriously damaged in chronic occupational exposures. Clinically, patients suffer tubular dysfunction resulting in amino aciduria, proteinuria and glycosuria, and decreased renal tubular absorption of phosphate. The observed proteinuria is principally tubular, consisting of low molecular mass proteins whose tubular reabsorption has been impaired by cadmium injury to proximal tubular lining cells. The predominant protein is a β_2-microglobulin, a low molecular mass protein ($\sim$6500 Da), in addition to other minor proteins such as retinol-binding protein lysozyme, ribonuclease and immunoglobulin light chains, which have been identified in the urine of workers with excessive exposure to cadmium (Lauwerys et al., 1980). The presence of high molecular mass proteins in the urine, such as albumin and transfernin, indicates that some markers may actually have a mixed proteinuria and suggests a glomerular effect as well, but the pathogenesis of glomerular lesions in cadmium nephrology is not presently understood (Goyer, 1995). Persons with renal tubular dysfunction resulting from excessive dietary ingestion of cadmium-polluted rice did not show reversal of nephrology as long as 10 years after reduced exposure, when the β_2-microglobinuria exceeded 1000 µg/g of creatinine. This may reflect the level of body burden, and/or the shifting of cadmium from liver to kidney (Kido et al., 1988).

The half-life of cadmium in human kidneys is uncertain, but it may be as long as 30 years. Consequently, it has been conjectured that the critical concentration in kidney cortex that produces tubular dysfunction in 10% of the population by age 50 ($\sim$200 µg/g), and about 300 µg/g for 50% of the general population, could be used to establish maximum levels of daily exposure (Kjellström et al., 1977). There is a pattern of liver and kidney cadmium levels increasing simultaneously until the average renal cortex cadmium concentration is about 300 µg/g and the average liver level is about 60 µg/g. At higher cadmium liver levels, the level in the renal cortex is disproportionately low, as cadmium is lost from the kidney. Animal studies suggest that cadmium nephrotoxicity follows the slow release and renal excretion of cadmium metallothionein from the liver and other soft tissues.

Cadmium metallothionein is toxic when taken up by the proximal tubular cell complex, whereas cadmium chloride at greater concentrations in proximal tubular cells was not toxic (Dorian *et al.*, 1995). Daily intake of food containing 140–260 µg/d of cadmium for more than 50 years, or workroom air exposure of 50 µg/m^3 for more than 10 years, have produced renal dysfunction (WHO, 1992). An epidemiological study of the dose–response relationship of cadmium intake from eating rice found that the total cadmium intake over a lifetime that produced adverse health effects was 2000 mg for both men and women (Nogawa *et al.*, 1989). Cadmium is of dietary concern because background dietary exposures were estimated to yield kidney concentrations of about one-quarter the hypothesized critical level. The segment of the US population at greatest risk appears to be older adults (ATSDR, 1989a). Studies of maternal–fetal tissue have shown that cadmium crosses the placenta and accumulates in the fetus (Goyer, 1995).

Cadmium has been considered as a Category 1 (human) carcinogen primarily on the basis of its induction of pulmonary tumors (IARC, 1994). Animal studies showed that the inorganic salt produced sarcomas in rats following injection. Cadmium chloride produced a dose-dependent increase in the incidence of lung cancer in rats following inhalation, and low incidence of prostatic cancer after injection into the ventral prostate. While evidence for a relationship between cadmium and cancer of the prostate in humans is debatable, studies in rats demonstrated carcinogenicity in the ventral prostate by oral or parenteral exposures, as well as by direct injection (Waalkes and Rehm, 1994).

6.3.1.3 Treatment

Chelating agent therapy was detrimental to the host as it increases cadmium uptake by the kidney, which in turn leads to nephrotoxicity in spite of increased urinary excretion of cadmium. Since susceptibility to cadmium-induced toxicity is particularly influenced by the ability of the body to provide binding sites on metallothionein, protection may be produced by induction of metallothionein by dietary zinc, cobalt, or selenium; however, the only effective treatment for cadmium-induced toxicity is to eliminate the source of exposure to this element (Nogawa and Kido, 1996).

6.3.2 Lead

Of all the heavy metals, lead has probably the longest history of environmental contamination and toxicity to humans. It is the most ubiquitous toxic metal and is detectable in practically all phases of the inert environment and in all biological systems. There is evidence that lead in the environment has increased during the past 200 years (Shukla and Leland, 1973). For this reason, lead poisoning, or plumbism, has been intensely studied and a large body of information is available on its toxicity. Because lead is toxic to most living things at high exposure, and there is no demonstrated biological need for it, a major issue is to determine the dose at which it becomes toxic. Specific toxicities vary with the age and circumstances of the host, but the major risk is toxicity to the nervous system. The most susceptible populations are children, particularly toddlers, infants in the neonatal period and the fetus (CDC, 1991).

Lead is naturally present in the soil. Environmental lead is a product of storage batteries, ammunition, solder, pigment colours and dyes, galvanizing and plating, pipes, and insecticides. It is also a constituent or a contaminant of houseware materials such as crystal and pewter. Tetraethyllead is an antiknock additive in gasoline, introduced into the environment through exhaust fumes from leaded gasoline used in vehicles; although in recent years this use has been drastically curtailed, the non-point-source nature of the contaminant has resulted in concentrations of lead that have remained for many years in soil and water after its use was abandoned (Menzer, 1991). Lead is easily taken into the body by inhaling lead dust, absorbing lead-based chemicals through the skin, or ingesting lead present in food and water. In the home, lead may be present in drinking water because lead pipes were formerly used in domestic plumbing, and lead-based solder was used with copper pipes. Lead was also added to paints used on toys and furnishings as well as walls, and can be present in enamelled kitchenware and pottery glazes, and in the solders in cans containing food and drink (UNEP, 1992).

6.3.2.1 Dietary intake
Because it is often present as a contaminant, lead is one of the most frequently monitored metals. Exposure of the general population occurs by inhalation and by ingestion of food and water. Data were collected by GEMS/Food from 26 countries covering a wide variety of foods. In 1972, JECFA established a PTWI for lead from all sources of 50 µg/kg bw for adults. Because of increased sensitivity of infants and children, JECFA lowered the PTWI for this group of population to 25 µg/kg bw in 1986 (UNEP/FAO/WHO, 1988). Kitchenware, especially improperly glazed ceramic ware, can be a major source of lead in the diet. ISO adopted limits for release of lead from ceramic foodware of $1.7\,mg/dm^2$ for flatware, and 2.5–5.0 mg/l of extraction solutions for hollow-ware (ISO, 1981).

Figure 6.7 shows that generally shellfish and finfish have higher lead content than milk, fruit, vegetables and meat, and levels in molluscs and crustaceans are higher than in fish. Lead levels in kidney and liver are substantially higher than meat muscle. A recent study conducted in the

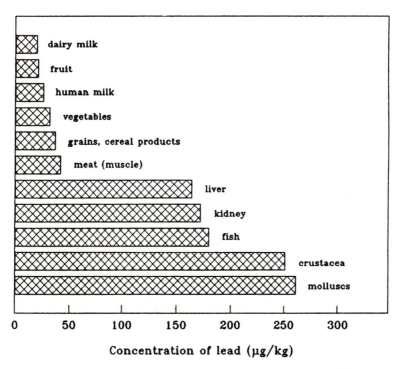

Figure 6.7 Typical lead levels in foods (1980–85). (From UNEP/FAO/WHO (1988), with permission.)

United States on lead content of clams and oysters showed average levels of 240, 300, 110 and 60 µg/kg wet weight in hardshell clams, softshell clams, Eastern oysters and Pacific oysters, respectively (Capar and Yess, 1996). In Slovenia, the mean concentrations of lead during the years 1989 through 1993 in bovine meat, liver and kidneys were 50, 100 and 140 µg/ kg wet weight, respectively, and those in the corresponding pig tissue were < 50, 60 and 60 µg/kg wet weight (Doganoc, 1996). Grains and cereal produce were found to contain low levels of lead, even after processing grain to flour, as did commercially baked products. A study between 1990 and 1995 of rice consumed worldwide by local populations yielded a total of 1528 samples (~ half from Japan) from 17 areas, 10 in Asia and 7 from outside. The highest and lowest geometric means in Asia were in Indonesia (38 ng/g) and Australia (2 ng/g); those outside Asia were in Spain (58 ng/g) and the United States (3 ng/g). There has been no substantial reduction of lead in rice worldwide during the past decade (Zhang *et al.*, 1996).

Published data indicates that vegetables grown in industrial mining areas show high levels of lead (up to 13 700 µg/kg), and levels in leaves were higher than in roots of the plant (Gzyl *et al.*, 1984; Marchwinska *et al.*, 1984). High levels of lead were reported in grains, vegetables and fruits in suburban and city-centre areas of Nairobi (Dickinson *et al.*, 1987). The lead levels in vegetables are somewhat higher than fruits, probably because of lead translocation from soil to the edible portion of vegetables compared with fruits grown on trees and bushes. Of all vegetables studied, spinach, with a large surface area compared to weight, usually contains high lead levels. Levels in potatoes were usually near the limit of detection. In some cases, the 90th centile levels of lead in canned fruit juices and nectars exceeded Codex maximum contaminant levels. The lead content of various species of fruit in Poland gathered during 1989–91 from non-industrialized areas showed highest levels in strawberries, raspberries and carrots (~0.1 mg/kg on average), and lowest levels in apples and pears (mean 0.010–0.089 mg/kg). The content of fruits and soils from industrialized areas was much higher than rural or non-industrialized areas (Wojciechowska-Mazurek *et al.*, 1995). On the other hand, levels of about 20 µg/kg were reported in human and dairy milk (UNEP/FAO/WHO, 1988). In Finland, a nationally representative sample of low-fat milk and cheese collected from dairies and eggs collected from distributors, together with samples of imported cheese and egg products collected in surveillance studies, showed that the mean lead content was 1.7 µg/kg in milk, 17 µg/kg in Finnish cheese, 17–60 µg/kg in imported cheese, 1 µg/kg in eggs and 6–72 µg/kg in imported dry egg products (Tahvonen and Kumpulainen, 1995). In Varanasi, India, a recent study on lead content of cattle milk sampled from three sites along the road with different automobile traffic densities showed that milk collected from an area of heavy traffic contained 4.6–7.2 mg/kg of lead, which is much higher than the FDA permissible level of 0.3 ppm (Bhatia and Choudhri, 1996). In a Canadian study conducted during 1986–88, the mean and average dietary lead concentrations were 23.2 µg/kg (or 19.49 µg/kg bw) and < 0.4–523 µg/kg, respectively. Estimated dietary intake over all ages and sexes was 24 µg/d. The average lead level in canned food decreased from 73.6 to 46 µg/kg between 1986 and 1988 (Dabeka and McKenzie, 1995). A study of Canadian Inuit living in the Eastern Arctic community of Qikiqtarjuaq on Baffin Island showed high lead concentration (> 1 mg/kg, or > 84 µg/kg bw) from ringed seal liver, mussels and kelp consumed by adult men and women (> 20 years old) and children (3–12 years old) (Chan *et al.*, 1995). In Croatia, the mean weekly dietary intake of lead per person was 701 µg, which represent 19.9% of the PTWI (Sapunar-Postruznik *et al.*, 1996). Representative market basket diets from four different geographical regions of Spain (Galicia, Valencia, Andalucia and Madrid) showed that the daily dietary amount of lead was between 37 and 521 µg/day; the Madrid population had the highest average lead intake and exceeded the PTWI because of consumption of contaminated vegetables and cereals (Cuadrado *et al.*, 1995). In the German Democratic Republic, an estimate of 34 and 25 µg/ day of lead was found for adult men and women, respectively (Muller and Anke, 1995). In 1992 a study in China of 603 specimens of food gathered from markets in Beijing, Shanghi, Jiangsu, Sichuan and Guangdong showed that 96.2% of cereals, 84.4% of vegetables, 97.6% of dairy products, 96% of eggs, and 100% of meat and fish were within national standards for allowable lead, and daily dietary intake levels were lower than the PTWI (Yang *et al.*, 1995).

The average weekly lead intakes of adults are presented in a descending order in Figure 6.8. In spite of the large database, very few countries

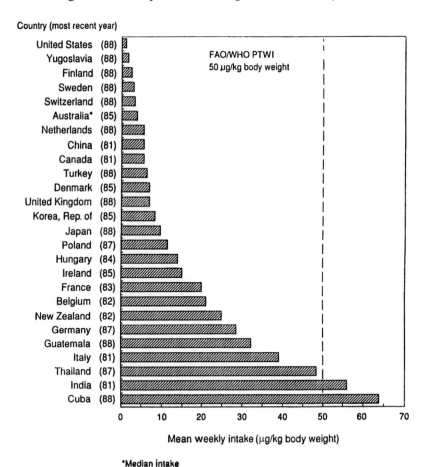

Figure 6.8 Dietary intake of lead by adults. (From Jelinek (1992), with permission.)

reported results on the same food over several years, thus limiting the possibility of identifying time trends within a country. Nevertheless, GEMS/Food provides some indication of the current levels of lead in a broad group of foods. As in the case of cadmium, comparison between various countries could not be made because of different study approaches, years of monitoring, and inadequacies in analytical quality control methods. As a general rule, the 95th centile consumers of food had intakes that were twice the average consumption of the whole population, while the ratio between the average and the 95th centile of a particular food appeared to be roughly three times the mean consumption. Thus, the intake of a contaminant by the average adult population should be viewed with concern if it approaches the PTWI, since a considerable segment of the population may exceed the PTWI because of varying dietary habits (Jelinek, 1992).

The average adult dietary intake of lead in Cuba, India and Thailand was essentially equal to or above the PTWI, and intakes reported from Italy, Guatemala, Germany, New Zealand, Belgium and France were one-half or more of the PTWI. Lower intakes were reported by Sweden, Finland, former Yugoslavia and the United States. The recent Dutch National Food Consumption Survey (1987–88) showed that intake of lead did not exceed the ADI, and had actually declined between the intervals 1976–78 and 1988–89 (Brussard et al., 1996). A study to assess and compare non-occupational background exposure of Chinese and Japanese women in the years 1993 through 1995, showed a geometric mean dietary lead intake of 25.8 µg/day for Chinese compared with 11.6 µg/day for Japanese women, of which boiled rice accounted for 3.6% and 12.1% of the total dietary burden for Chinese and Japanese women, respectively (Zhang et al., 1997). A survey of the Korean cities of Seoul, Pusan, Chunan and Haman for 141 healthy, non-smoking women aged 21–56 years showed the dietary geometric mean intake of lead for the four cities combined to be 20.5 µg/day, with intake from boiled rice representing 12% of total daily lead intake (Moon et al., 1995). In an Indian study, the mean intake of lead by adults was ~60 µg/kg bw; thus exceeding the PTWI of 50 µg/kg bw (Jathar et al., 1981). In a 1980 study reported to GEMS/Food by Denmark, the average weekly adult intake of lead was ~8 µg/kg bw. An appreciable increase in lead intake was noted with consumption of a quarter of a litre of wine per day (24 µg/kg bw), in areas where vegetables were grown near heavy traffic (30 µg/kg bw), and in areas where adults lived near a secondary lead smelter outside Copenhagen (32.6 µg/kg bw). Mean lead levels in wine of about 20–100 µg/kg have also been reported from Belgium (Fouassin and Fondu, 1980), the United Kingdom (Sherlock et al., 1986) and the Netherlands (Edel et al., 1983). The UK study reported levels up to 1890 µg/kg lead in the wine first poured from bottles having lead caps. Reports from France indicated poisoning in a patient who consumed wine containing around 2000 µg/kg lead (Conso *et al.*, 1984).

GEMS/Food data from Poland, in addition to recent published data (Gzyl, 1997; Jakubowski *et al.*, 1996) indicate an increase in dietary intakes of lead, and higher blood lead levels, near industrial areas; the values decrease as the distance from the source of emission/discharge increase. Lower lead intakes were reported in rural areas in Japan and Korea where traffic is low (Ikeda *et al.*, 1989; Watanabe *et al.*, 1987). Recent data from South Africa also show lower atmospheric lead concentrations in residential and rural areas compared with industrial areas (Nriagu *et al.*, 1996). The presence of lead plumbing registered a dramatic increase in dietary lead intake, as seen in a study in Ayr, UK, where the 90th centile intake was 114 µg/kg bw/week (i.e. more than twice the PTWI) as compared to the UK national average of around 7 µg/kg bw (Sherlock *et al.*, 1982).

Of the few countries that reported lead intakes for several years, a downtrend was observed in Finland, the United Kingdom, and especially the United States, where the trend is more pronounced in recent years (Figure 6.9), probably as a result of the introduction of pollution controls and conversion to non-leaded soldered food cans (UNEP, 1992). Similarly, a recent Japanese study comparing lead exposure in the 1990s (1991–94) with the 1980s (1979–83), on samples collected from 19

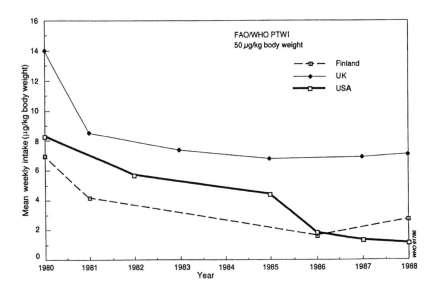

Figure 6.9 Trends in dietary intake of lead by adults. (From Jelinek (1992), with permission.)

sites showed a dramatic reduction of the dietary lead geometrical mean from 32.2 to 7.1 µg/day (Watanabe *et al.*, 1996).

In 1993, WHO proposed a guideline value of 10 µg/l (48.3 nmol/l) for lead in drinking water. A phased approach in Europe may result in a temporary parametric value of 25 µg/l within 5 years, and a 10 µg/l value 15 years later. The current European Community Directive and the French decree stipulate a maximum admissible concentration (MAC) of 50 µg/l (Vilagines and Leroy, 1995). The EPA has suggested that each 1 µg/l of lead in water can lead to an increased blood lead level of approximately 0.2 µg/l for a child (EPA, 1989a). In order to comply with those values, it would be necessary to replace all internal lead plumbing and supply pipes at a cost of 143 billion French francs for France and 347 billion French francs for Europe (Vilagines and Leroy, 1995). A study conducted on 1812 mothers with a live infant born between October 1991 and September 1992 in Loch Katrine water supply area, Glasgow, subjected to maximal water treatment to reduce plumbosolvency, found that 17% of households had tap water lead concentration of 10 µg/l in 1993 compared with 49% in 1981. Tap water lead accounted for 62% and 76% of cases of maternal blood concentrations above 5 and 10 µg/dl (0.24 and 0.48 µmol/l) for 1993 and 1981, respectively. An estimated 13% of infants were exposed via bottle feeds to tap water lead concentrations exceeding WHO guidelines of 10 µg/l (48.3 nmol/l) (Watt *et al.*, 1996). On the other hand, in Ibadan, Nigeria, a city with less industrialization, tap water lead at various city locations was < 5 µg/l (Omokhodion, 1994). The widespread use of lead piping and soldering of water tanks in some countries contributes to incidences of lead poisoning much higher than those attributable to leaded gasoline (Grobler *et al.*, 1996).

Tin-coated copper utensils contribute to lead found in food, as reported in Australia, where these utensils contributed more lead to vegetables, fish and meat than aluminium containers or copper utensils (Reilly, 1985). In Mexico, variable leaching of lead into food stored or cooked in lead-glazed ceramicware, at different pH values, resulted in a range from 8 to > 2000 ppm (regulatory level for rejection is 2 ppm) (Gonzalez and Craigmill, 1996). In the United States, lead concentrations up to 610 µg/ml were released after treatment with 4% acetic acid, and one-half of tested dishes (78 of 149) exceeded the FDA level of allowable concentration of 3.0 µg/ml. Repeated extractions with acetic acid showed that even after 20 consecutive 24 h leachings, many dishes released high lead concentrations exceeding the FDA limits (Sheets, 1997). A survey conducted on 676 domestic and 5222 imported ceramic lots in the United States to determine the extent of lead leaching from dinnerware, performed by the FDA from January to February 1992, found that 17 and 46 lots, representing 2.5% and 0.9% of domestic and imports, respectively, exceeded the 1991 guidelines (Baczynskyj and Yess, 1995).

Lead solder used in cans is a major controllable source of lead in food, and food processors in many countries have switched to non-soldered cans. This is particularly important in the case of infant food. Figure 6.10 shows the levels of lead in a number of foods before and after the introduction of non-soldered cans, which has cut lead concentrations on average to between one-fifth and one-tenth of previous levels. Attempts to decrease lead content by improving processing operations for food packed in lead-soldered cans have achieved only a 50% reduction (UNEP, 1992).

As for infants and children, the PTWI for lead refers to the maximum intake from all sources. Thus, not only the median and mean intakes should be below 25 µg/kg bw, but also the 90th centile intake, because some of the young will ingest substantial amounts of lead from paint dust and soil by hand-to-mouth activities. One of the highest mean weekly intakes (118 µg/kg bw) was reported in an Austrian study after introduction of mixed feeding of adult foods (e.g. vegetables) at the age of around 4 months. This was due to environmental contamination (Haschke and Steffan, 1981). Average dietary lead intakes for infants and children in 14 countries are shown in Figure 6.11. Intakes slightly below the PTWI were reported for Switzerland and Finland, whereas intakes over 50% of the PTWI were reported for Poland, Germany and the United Kingdom. The lowest intakes were from Guatemala, Zaire, United States and Niger.

Higher levels of lead were reported in breast milk from mothers living in urban (10–30 µg/kg) as compared to rural areas (1–2 µg/kg) (Lechner et al., 1980; Dabeka et al., 1986; Sternowsky and Wessolwoski, 1985). In Poland, the weekly intake for children living in an industrial area was 33 µg/kg bw, which is about twice that for children living in a non-industrial area. Dietary lead intake by 47 children (aged 5–8 years) in Duisburg, Germany, showed median and maximum intakes of 22.2% and 72.1% of the PTWI, respectively (Wilhelm et al., 1995). Mean intakes of infants of around 105 µg/kg bw were reported in the German study, with high lead content in drinking water (Muller and Schmidt, 1983). Similarly, mean intakes of around 360 µg/kg were reported in a study in Ayr, UK, where some dwellings had lead plumbing, and water from the pipes was drunk or used to dilute dehydrated infant formulas and infant cereals (Sherlock et al., 1982). Very low intakes (∼2–3 µg/kg bw) were reported for 0–1-month-old infants fed breast milk in studies in Sweden (Larsson et al., 1981) and Canada (Dabeka and McKenzie, 1988). The Canadian investigators noted a correlation between lead levels in breast milk, heavy traffic and occupancy of older houses (Dabeka et al.,

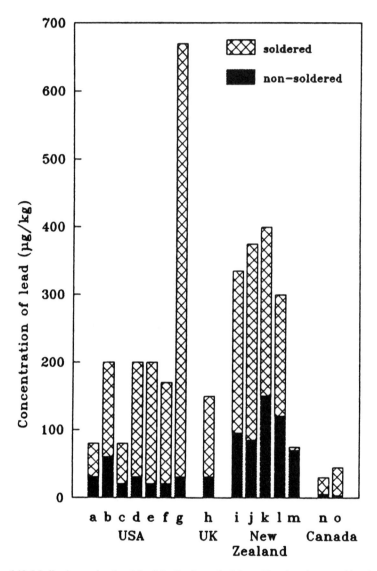

Figure 6.10 Median/mean levels of lead in foods packed in soldered and non-soldered cans: a, apple juice; b, apple sauce; c, orange juice; d, cherries; e, baked beans; f, green beans; g, tuna; h, ravioli; i, vegetables; j, fruit; k, fish; l, meat; m, spaghetti; n, infant formula, concentrate; o, infant formula, ready to use. (From UNEP/FAO/WHO (1988), with permission.)

1986). In Egypt, a recent study on lead levels in mother's milk from 20 different governorates throughout the country showed values much higher than the PTWI in the crowded governorates of Alexandria, Assuit and Cairo, attributable to heavy automobile traffic (using leaded

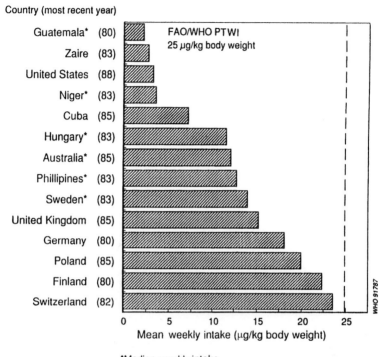

Country (most recent year)

Guatemala* (80)
Zaire (83)
United States (88)
Niger* (83)
Cuba (85)
Hungary* (83)
Australia* (85)
Phillipines* (83)
Sweden* (83)
United Kingdom (85)
Germany (80)
Poland (85)
Finland (80)
Switzerland (82)

FAO/WHO PTWI
25 µg/kg body weight

WHO 91787

Mean weekly intake (µg/kg body weight)

*Median weekly intake

Figure 6.11 Dietary intake of lead by infants and children. (From Jelinek (1992), with permission.)

gasoline) and lead water pipelines (Saleh *et al.*, 1996). In Sweden, lead in milk obtained at 6 weeks after delivery from women living in the vicinity of a lead and copper smelter was higher (0.9 µg/l) than in the control group (0.5 µg/l) (Hallen *et al.*, 1995). A US study in Chicago of 40 reconstituted infant milk formulas reported two samples having lead concentrations of 17 and 70 µg/l, which are above the current EPA action level of 15 µg/l for safe water. These samples were prepared using cold tap water run for 5 and 30 seconds, respectively, from the plumbing of houses > 20 years old where galvanized lead pipes were common (Baum and Shannon, 1997).

Very few countries reported dietary intakes in infants long enough to determine whether a trend was established. The data in Figure 6.12 for Australian and American infants (up to 11 months) and children (1–4 years) indicate a probable downtrend in Australia, and a definite downtrend in the United States from 1980 to 1988, which coincided with the total conversion in 1982 to non-soldered containers for infant

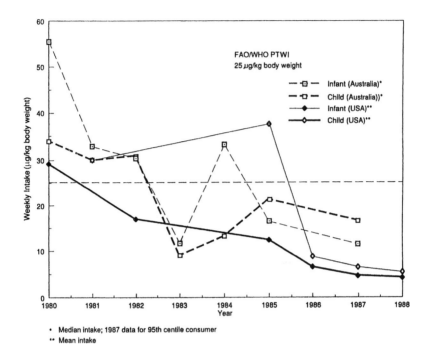

Figure 6.12 Trends in dietary intake of lead by infants and children. (From Jelinek (1992), with permission.)

formula, infant juices, puréed baby food and evaporated milk, and with a drastic reduction in lead content in gasoline (Capar and Rigsby, 1989). FDA has also taken steps to minimize or further reduce sources of lead in the diet from lead glazes or ceramic ware, leaded crystalware, dietary supplements, bottled water, and lead capsules on wine bottles (Bolger *et al.*, 1996).

Lifestyle factors of pregnant women were shown to influence prenatal exposure of newborns to lead. In a study conducted in two hospitals in Quebec, Canada, on 430 mothers and their newborns, a dose–response relationship was found between cigarette smoking and alcohol consumption of mothers and cord blood lead levels. An average increase of about 15% (0.013 μmol/l) in cord blood lead level was estimated for every 10 cigarettes smoked per day. Mean blood lead levels in babies whose mothers did not smoke during pregnancy, but who drank alcohol moderately, was 17% higher than for non-smoking mothers who abstained from alcohol. Both cigarette smoking and alcohol intake made significant and independent contributions to cord blood lead concentration (Rhainds and Levallois, 1997). A West German study during

1987–88 on the effect of lifestyle factors on blood lead levels found that alcohol consumption (especially of wine) accounted for the largest proportion of variability in blood lead levels, followed by both age and smoking. Other significant contributory factors were gender, haematocrit, calcium intake and consumption of milk and milk products. Filterless cigarettes were associated with higher blood lead levels than filter-tipped cigarettes. In addition, smoking cigars, cigarillos, or pipes resulted in higher blood lead levels than smoking only cigarettes. Alcohol consumption and smoking were independent contributors to blood lead levels in both men and women, but the effect of alcohol consumption was stronger in women than in men (Weyermann and Brenner, 1997). In Saudia Arabia, blood lead levels were significantly higher in current male smokers than in non-smokers and previous smokers (al-Saleh, 1995). A study of 567 adults in five Polish towns with no large industrial emitters, and 539 adults living in the vicinity of zinc and copper mills, showed that cigarette smoking resulted in a significant increase in lead blood levels for both males and females in both localities (Jakubowski et al., 1996).

Overall, lead intakes near or above the PTWI were reported for adults and infants and young children from both developed and developing countries. A downtrend was observed during conversion from lead-soldered to non-soldered cans and during the phasing out of lead additives in gasoline. However, higher dietary intakes still occur in industrial areas, in areas with high traffic density, and in areas with high lead concentrations in tap water, and in the case of adults as a result of consumption of wine from bottles with lead caps. In view of these findings, both developed and developing countries should continue their surveys of lead, which should be carried out on a continuing basis. More duplicate diet studies should be carried out in areas with heavy industrial activities, especially mining and metal processing, in areas with high lead levels in the tap water, and in areas of high traffic density.

Fewer surveys were carried out on dietary intakes of lead by infants and children, especially over long periods and/or by a consistent approach. However, a definite decreasing trend was noted in the United States and Japan. The lowest intake was noted among breast-fed infants. A correlation of lead levels in breast milk occurred with proximity to heavy traffic, and with occupancy in older houses. In addition, lead intake far in excess of the PTWI of 25 µg/kg bw occurred in localities where tap water contained elevated lead levels. A study on 98 recently post-partum women from Mexico city showed that in spite of a recent decline in environmental lead exposure, the potential for delayed toxicity from bone lead stores remains a significant public health concern, as mobilization of lead from bone can be markedly enhanced during bone turnover in pregnancy and lactation, resulting in lead exposure to the fetus and breast-fed infant. Moreover, consumption of high-calcium food was found to provide protection against lead accumulation in bone (Hernandez *et al.*, 1996).

In view of these findings, and since lead passes the placental barrier readily, efforts should continue to reduce levels in foods for infants and children, in foods consumed by women of child-bearing age and in tap water. In addition, more duplicate dietary studies for lead should be carried out with pregnant and nursing women (especially those drinking alcohol and smoking), and with infants and young children (Jelinek, 1992; UNEP, 1992).

6.3.2.2 *Toxicity*

The toxic effects of lead form a continuum from clinical or overt effects to subtle or biochemical effects, which involve several organ systems and biochemical activities. The most critical effects in children include the nervous system, where lead interferes with synaptic mechanisms of transmitter release (NRC, 1993). For adults with excess occupational, or even accidental, exposure, the effects are peripheral neuropathy, and/or chronic neuropathy. However, the critical effect for adults in the general population appears to be hypertension (Goyer, 1995). The toxic effects of lead and the minimum blood lead levels at which the effects are most likely observed are shown in Table 6.3.

Nearly all environmental exposure to lead, including dietary, is to the inorganic compounds. Organic lead compounds such as tetraethyl lead may be absorbed in large quantities through the skin, but in these forms their toxicities are primarily a problem in the petroleum industry. All forms of lead toxicity are less frequent in adults; any occurrence is usually acute and occupationally related (EPA, 1989a).

Haematological effects provide biochemical indicators of lead exposure in the absence of chemically detectable effects, but lead-induced anaemia is often associated with other detectable effects or synergistic factors. Other target organs are the gastrointestinal and reproductive systems. There are at least five pools of lead in the body, two of which reside in the skeleton (90%), in cortical and in trabecular bone. Lead in cortical bone is similar in half-life to cadmium ($\sim$20 years). Other body compartments for lead include the kidney, lung and central nervous system. It is not surprising, therefore, that major lesions and clinical signs in humans suffering frank plumbism are referable to blood (anaemia), brain (convulsions, paralysis) and kidney (proteinuria) (NRC, 1993).

Chronic toxicity to young children who ingest lead-based paint chips or lead in soil, housedust from paint, industrial dust and automotive emissions is a familiar occurrence. Oral ingestion of inorganic lead is considered the primary port of entry into young humans. Of the lead

Table 6.3 Lowest observed effect levels (LOEL) from lead-related health effects

Effects	Blood lead concentration (µg/dl)	
	Children	Adults
Neurological		
Encephalopathy (overt)	80–100	100–112
Hearing deficit	20	
IQ deficits	10–15	–
In utero effects	10–15	–
Peripheral neuropathy	40	40
Hematological		
Anaemia	80–100	80–100
U-ALA[a]	40	40
B-Epp[b]	15	15
ALA inhibition	10	10
Py-5-N[c] inhibition	10	–
Renal		
Neuropathy	40	
Vitamin D metabolism	< 30	
Blood pressure (males)	–	30
Reproduction		40

From Goyer (1995), with permission.

[a]Aminolaevulinic acid in urine.

[b]Concentration of erythrocyte protoporphyrin.

[c]Enzyme pyrimidine-5-nucleosidase inhibition results in accumulation of nucleotides in red blood cells altering their energy metabolism, and affecting their membrane stability and survival.

ingested, only 5–15% is absorbed in adults, but considerably more in children. Increased blood lead levels in infancy and early childhood may be manifest in older children and adolescents as decreased attention span, reading disabilities, and failure to graduate from high school. Studies on pregnant women who ingested very low levels of lead suggest that this may result in learning and behaviour disabilities in neonates and pre-school children. There are no specific indicators of neurological effects associated with exposure to lead. Children in the lower socioeconomic groups may begin to manifest language deficit by the second year of life, which may be preventable in the higher socioeconomic groups. In addition, increased risks for antisocial and delinquent behaviour following a developmental course was associated with lead exposure. The most sensitive indicators are psychomotor tests or mental development indices (e.g. the Bayley Scales for infants) and broad measures of IQ, such as full-scale WISC-RIQ scores for older children. Blood lead levels at 2 years old are more predictive of a longer term adverse neurological effects than umbilical cord blood lead concentration (Needleman *et al.*, 1990, 1996).

Lead's toxicological mode of action depends on its molecular configuration; inorganic lead is the form most available environmentally and contaminates foods, producing clinical signs different from those of organic forms (e.g. tetraethyl lead), and is distributed differently in the body. Tetraethyl and tetramethyl lead are rapidly dealkylated by the liver to the trialkyl metabolites, which are responsible for toxicity. These metabolites in turn are slowly converted to inorganic lead (Hammond and Beliles, 1980). Lead acts as an inhibitor of δ-aminolaevulinic acid dehydratase (ALA-D) and haem synthetase, which leads to anaemia. The sensitivity of certain individuals to the effects of lead on haem metabolism may be related to genetic polymorphism of haem among people in the general population. It has been suggested that ALA-D phenotypes might be used as indices of individual susceptibility to lead (Ming *et al.*, 1997; Todd *et al.*, 1996). The metal causes necrosis of neurons, myelin sheath degeneration and, especially, brain vascular damage with increased cerebrospinal fluid pressure. These effects eventually lead to encephalopathy and elevated mental retardation in children (NRC, 1993).

Lead nephropathy is one of the oldest recognized health effects of lead since 1902, mainly due to industrial exposure (Kim *et al.*, 1996), but, with progressive reduction of exposure in the workplace and the introduction of sensitive biological indicators of renal toxicity, it is a vanishing disease. An increase in blood pressure is considered the most critical adverse health effect from lead exposure in adults. Lead affects blood pressure directly by altering sensitivity of vascular smooth muscle to vasoactive stimuli, and indirectly by altering neuroendocrine input to vascular smooth muscle. Clinically apparent lead toxicity has long been associated with sterility and neonatal death in humans. Gametocytic effects have been shown to occur in both male and female animals. A few clinical studies showed increased chromosomal aberrations in workers with blood levels $> 60\,\mu g/dl$. Reduction in sperm counts and abnormal sperm motility were observed in lead battery workers at blood levels of $40\,\mu g/dl$, and decreased testicular endocrine function was reported in smelter workers with mean blood lead levels of $60\,\mu g/dl$. Lead crosses the placental barrier, and there is a good correlation between maternal and fetal blood value of $45\,\mu g$ or greater (Goyer, 1995; NRC, 1993).

Lead is classified as a 2B carcinogen by IARC (IARC, 1987). Lead induction of renal adenocarcinoma in rats and mice is dose related, and has not been reported at levels below those which result in nephrotoxicity (EPA, 1989a).

Lead was also shown to exert ecotoxicological effects. For example, it can cause mortality in birds, or can independently affect populations through effects on the food base, avian behaviour, reproductive success and migratory pattern, and it increases vulnerability to cold stress, hunting and predation (Burger, 1995).

6.3.2.3 Treatment

The most effective way to treat lead toxicity is removal of individuals from the source(s) of exposure. Chelation has a role in the treatment of symptomatic workers or children, or petrol sniffers, and is warranted in adults with blood lead levels above 60 μg/100 ml following assessment of biological and clinical parameters of exposure (Burns and Currie, 1995; Trachtenbarg, 1996). The Centers for Disease Control in Atlanta has established guidelines to assist in evaluating exposure factors for lead toxicity in children (CDC, 1991). For children with severe lead poisoning, chelation is the standard procedure, even though the mortality rate may be 25–35% when ethylenediaminetetraacetic acid (EDTA) or British Anti-Lewisite (BAL) are used individually; when both agents were combined, mortality was reduced. The oral chelating agent *meso*-2.3-dimercatosuc-cinic acid (Succimer; DMS), has been used at lead blood levels of 45 μg or greater. While Succimer was shown to lower blood levels, its effectiveness in removal of lead from the brain, and in reversing cognitive functions and behavioural development, has not been demonstrated (Porru and Alessio, 1996).

6.3.3 Mercury

Mercury was used for many centuries in ancient China, and has been found in Egyptian tombs 3500 years old. There has always been an aura of magic surrrounding this lustrous liquid since ancient times; its name is shared by both a Roman god and a mysterious planet. Since before Christ (and even today) magic properties have been associated with mercury. During the Middle Ages, it occupied a central role among alchemists intent on transmuting base metals into gold, and was carried as amulets to ward off diseases, to treat deadly ones such as syphilis, and to protect from evil. Throughout successive centuries it has been used to treat almost every ailment known to man. Even today it is used in many medicines and incorporated in pesticides because of its therapeutic and protective properties. Nevertheless, its toxic characteristics have not been unnoticed. It has long been widely condemned as a drug with an unreasonable margin of safety, and occupational intoxications attribu-table to mercury have been reported since the Middle Ages (Goldwater, 1972).

The most extensive episode of human mercury poisoning resulted in Iraq from contamination of bread in 1971 and 1972 made erroneously from 'Green Revolution' grains that were intended for use as seeds and not to be consumed as food, and which had been treated for weevil control with the mercurial fungicide Panogen 15 containing 2.2% active ingredient of cyano(methylmercury)guanidine; the mercury equivalent is 1.5%. More than 6000 cases were reported and 500 people died

(Bakir *et al.*, 1973). As a result of this episode, the EPA cancelled and suspended all uses of alkylmercury compounds, as did other countries.

Mercury poisoning still occurs today, as man is a poor learner from history. The death of Dartmouth College Chemistry Professor Karen Wetterhahn in June, 1997 as a result of spills of dimethylmercury on the latex glove she wore months earlier has prompted the US Occupational Safety and Health Administration (OSHA) to tighten safety standards associated with occupational exposure to these highly toxic substances (Lewis, 1997).

The major source of environmental mercury is the natural de-gassing of the earth's crust, including land areas, rivers and oceans, estimated at 2700–6000 tonnes per year. On average, about 10 000 tonnes of mercury are mined every year, but there is considerable variation from year to year. Total man-made release into the atmosphere ranges from 2000 to 3000 tonnes every year. Mining, smelting and industrial discharges have been a major factor of environmental contamination in the past but, today, they have been reduced by 99% for chloralkali plants, which are one of the largest users of mercury. Moreover, use of mercury in the paper pulp industry has been reduced dramatically in many industrial countries, and even banned in Sweden since 1966. Industrial activities not directly employing mercury or its products still give rise to substantial amounts of this metal. For example, fossil fuel may contain as much as 1 ppm of mercury. Nearly 5000 tonnes are estimated to be emitted every year from burning coal and natural gas, and from refining petroleum products. Mercury in the atmosphere is believed to be the major global pathway of transport of this element (Beijer and Jernelöv, 1986), but it is not possible to make a complete inventory of the amounts from human activities and those from natural sources.

Mercury comes in three forms: elemental (metallic, Hg^0); inorganic salts (monovalent or mercurous, Hg^+; and divalent or mercuric, Hg^{2+}); and organic mercury compounds such as the phenylmercuric salts and alkylmercury compounds, which have been used as fungicides and herbicides. Each has its own toxicity and health effects. Regardless of the source, both inorganic and organic mercury undergo transformation in the environment. For example, metallic mercury may be oxidized to the inorganic divalent mercury, particularly in the presence of organic material as in aquatic environments; divalent inorganic mercury contained in discharges of industrial effluents into rivers or seas may, in turn, either be converted to metallic form under appropriate reducing conditions or alkylated to dimethylmercury by anaerobic bacteria. In this organic form it is easily absorbed after ingestion and has a half-life varying from 60 to 120 days in man, but up to 20 years in fish where it is the predominant form (Al-Shahristani and Shihab, 1974).

Today, mercury is still commonly used in thermometers, batteries and fluorescent lights, and in industrial processes such as production of paints and fungicides (Goyer, 1995).

6.3.3.1 Dietary intake

JECFA established PTWI of 300 μg for the general population, of which no more than 200 μg should be present as methylmercury; these amounts are equivalent to 5 and 3 μg/kg bw, respectively. Data were, however, considered insufficient to recommend methylmercury PTWIs for the most sensitive populations, namely pregnant women and nursing mothers (Jelinek, 1992). The Codex CAC has no limits for dietary mercury. However, some countries have established limits for total mercury rather than methylmercury, with a range of 0.5–2.0 mg/kg (DFO, 1988; FDA, 1987; Mackay et al., 1975).

Extensive data on mercury in food from 16 countries that submitted data to GEMS/Food on weekly dietary intake of adults are presented as average intakes in Figure 6.13. The highest average intakes, from Poland and Denmark, were about 60% of the PTWI for methylmercury and about 40% of the PTWI for total mercury. Average intakes for Cuba, Belgium, Germany and Guatemala were about 50% of the PTWI for

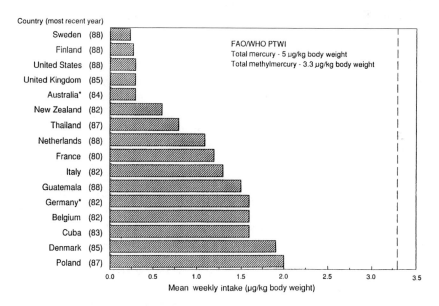

*Median weekly intake

Figure 6.13 Dietary intake of mercury by adults. (From Jelinek (1992), with permission.)

methylmercury. The lowest intakes were for Sweden and Finland, corresponding to 0.23 and 0.27 µg/kg bw, respectively. In Belgium, dietary intake exceeded the PTWI for total mercury in about 2% of study participants (Buchet *et al.*, 1983). The data from Belgium, Germany and The Netherlands indicate that finfish contributed around 20% of total mercury dietary intake; a high of about 85% was found in Finland and the United States; whereas France and the United Kingdom were in between (Jelinek, 1992). Surveys conducted in the United Kingdom showed that the diet of the general population contained 0.3 µg/kg bw, whereas the average dietary mercury level in coastal communities receiving industrial discharges was about 1.45 µg/kg bw. Average fish consumption in these communities was about 50 g/day as compared to 20 g/day for the whole country. Nearly 12% of the surveyed population exceeded the PTWI for methylmercury (Sherlock *et al.*, 1982). In Spain, dietary intake of mercury in four different geographical locations provided mainly by fish, was about 9–17% of the PTWI (Cuadrado *et al.*, 1995). In a study of the Canadian Inuit Eastern Arctic Community of Qikiqtarjuaq on Baffin Island, elevated concentrations of mercury (> 50 µg/100 g) were reported in ringed seal liver, narwhal mattak, beluga meat and beluga mattak. The average daily intakes of total mercury for adult men, women and children were 97, 65 and 38 µg, respectively, which were higher than the Canadian average value of 16 µg, and the average weekly intake of all age groups exceeded the 5 µg/kg/day established by JECFA (Chan *et al.*, 1995).

Monitoring carried out worldwide in the late 1960s and early 1970s in 31 countries showed that mercury-contaminated food, other than seafood, was generally below 30 µg/kg. In contaminated freshwater areas, mercury levels averaged 500–700 µg/kg, or higher, in finfish; in uncontaminated fresh water the level was about 200 µg/kg; in oceans, the average was about 150 µg/kg or less. Large carnivorous species such as shark, swordfish and tuna had mercury levels ranging from 200 to 1500 µg/kg (Galal-Gorchev, 1987).

Because of the extensive mining activities in the Mediterranean, where mining sites account for nearly 50% of the world's production of mercury, the area was surveyed by the UNEP Coordinated Pollution Monitoring and Research Programme in the 1970s and early 1980s. Species like mullet, bonito, tuna, goatfish and mussels were sampled (Table 6.4). Striped mullet and the albacore tuna from coastal mining regions near the Monte Amiato area of the northern Tyrrhenian coast of Italy showed mean mercury levels of 694 and 1050 µg/kg, respectively, and maximum levels of 7900 and 2300 µg/kg, respectively, while other non-mining areas of Italy showed an average of 160 µg/kg. Concentrations in mussels showed mean and maximum mercury concentrations of

Table 6.4 Marine finfish species with high levels of mercury

Species	Mean/median (µg/kg)		90th centile/maximum range (µg/kg)
	Range	Typical value	
Dogfish	160–720	450	500–1170
Garfish	130–685	550	250–1250
Mackerel	890–1540	1200	–
Meka	–	1600	3000
Milkfish	–	640	–
Perch	50–870	500	1700
Purbeagle	300–2100	1800	4200
Shark	100–2500	1000	1600–8350
Snapper	–	380	990
Swordfish	400–1500	1000	1400–3000
Tuna	10–1500	300	50–2000

From UNEP/FAO/WHO (1988), with permission.

232 and 7000 µg/kg, respectively (UNEP/ FAO/WHO, 1983). Other studies showed that fish caught off the coast of Tuscany contain considerably higher mercury levels than those caught from the Strait of Gibraltar (UNEP/ FAO/WHO, 1988). Goat fish sampled from the eastern Mediterranean, also an area of extensive mercury mining activity, showed mean and maximum levels of 426 and 1112 µg/kg, respectively. Moreover, generally mercury levels were higher in finfish and shellfish sampled from the Mediterranean sea than in those from the Atlantic, Pacific, or Indian Oceans. A comprehensive and systematic assessment of mercury in these areas needed is before definitive conclusions can be reached (UNEP/FAO/WHO, 1988; UNEP, 1992).

Assuming no dietary exposure to mercury except from fish, and that mercury in fish is the methyl form (MeHg), then Figure 6.14 can be used to evaluate safe fish consumption for adults. For example, if the average MeHg in fish is 0.5 mg/kg, fish consumption by a 60 kg individual should not exceed 400 g/week (~60 g/day). In countries with higher fish consumption, i.e. 700 g/person/week, the concentration of MeHg in fish should not exceed 0.2 mg/kg (UNEP/FAO/WHO, 1983, 1988).

As for the intake of dietary mercury by infants and children, data in Table 6.5 for six countries participating in GEMS/Food show that all of them are all below the PTWI of 5.0 µg/kg bw. Comparison of the intakes per kg bw with the adult intakes from the same countries (see Figure 6.13) shows that infants and young children have intakes at least equal to or higher than those of adults. In a Swedish survey, breast milk from fishermen's wives who consumed fish from coastal areas of the Baltic Sea, the Gulf of Bothnia or Lake Malar showed average levels of 3.1 µg/kg for

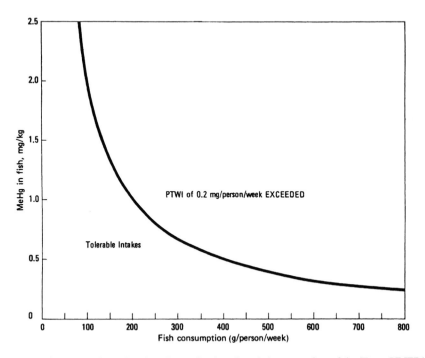

Figure 6.14 A graph for estimating dietary intake of methylmercury from fish. (From UNEP/FAO/WHO (1983), with permission.)

Table 6.5 Dietary intake of total mercury by infants and children[a]

Country	Year	Age group	Weekly intake (µg/kg bw)	
			Median	Mean
Australia	1984	9 months	0.02	
	1984	2 years	0.5	
Finland[b]	1980	3 years		1.5
Poland[c]	1985	1–3 years		3.1
	1985	3–7 years		2.5
Sweden[d]	1988	3 months		2.6
USA	1985	6–11 months		0.3
	1986–88	6–11 months		0.4
	1985	2 years		0.6
	1986–88	2 years		0.5
Zaire	1983	3 months	1.7	

Modified from Jelinek (1992), with permission.
[a]GEMS/Food, unless otherwise referenced.
 FAO/WHO PTWI: 5.0 µg/kg bw.
[b]Mykkanen et al. (1986).
[c]Szymczak et al. (1987).
[d]Skerfving (1988), among high fish consumers.

total mercury and 0.6 µg/kg for methylmercury in the milk (Skerfving, 1988). Using the standard conversion factor of 0.12 and multiplying by 7, these levels would be equivalent to intakes of 2.6 and 0.5 µg/kg bw for total and for methylmercury, respectively, which are well within the PTWIs. In a recent study of lactating women in Sweden, breast milk mercury levels for women who consumed fish contaminated with methylmercury showed a mean value of 0.6 µg/kg, the equivalent of 0.5 µg/kg bw (Oskarsson *et al.*, 1995).

In conclusion, although the average adult intake did not exceed the PTWI, in several countries it amounted to an appreciable portion of the population, especially if the 90th centile intakes were considered instead. Thus, dietary intake studies should be continued and expanded to include additional coastal countries and more developed countries, as their representation has been low. These studies should target consumers of fish and those living near areas of metal pollution. In addition, dietary monitoring of infants for mercury should be increased, as all reports showed their mercury intake equals or exceeds that of adults. Well-designed duplicate portion studies should include other risk groups such as pregnant and nursing women (Jelinek, 1992; UNEP, 1992).

6.3.3.2 Toxicity

Occupational studies on mercury exposure, mainly of elemental or inorganic mercury, showed that effects at relatively low exposure levels were primarily neurological and renal (Ratcliffe *et al.*, 1996). Clinically, low-dose exposure to mercuric salts or elemental mercury vapour was shown to induce an immunological glomerular disease, which is the most common form of mercury-induced nephropathy, characterized by two phases: an early phase of an anti-basement membrane glomerulonephrits, and a later phase of a superimposed immune complex glomerulonephritis (Pelletier and Druet, 1995). Exposed people may develop proteinuria, which is reversible following cessation of exposure to mercury. On the other hand, high doses of mercuric chloride, $HgCl_2$, were shown to be directly toxic to the renal tubular lining (Goyer, 1995).

Organic mercury intoxications, which are the most important in terms of health effects from environmental exposure, lead to c-mitosis and chromosomal aberrations that cause cellular damage, especially in the kidney and the brain (De Flora *et al.*, 1994). Neuronal damage and axonal demyelination result in the clinical symptoms of paraesthesia, incoordination, tremor and epileptic seizures (Bakir *et al.*, 1973).

Clinical manifestations of neurotoxic effects following intake of methylmercury are parathaesia, ataxia, neurathenia, vision and hearing loss, and finally coma and death. The cortex of the cerebellum and cerebrum were histopathologically shown to be selectively involved, with focal necrosis, lysis and phagocytosis of the neurons, and neuronal replacement by supporting glial cells. The overall acute effects result in cerebral oedema; on the other hand, prolonged effects result in destruction of the grey matter and subsequent gliosis leading to cerebral atrophy (Goyer, 1995; Tollefson and Cordle, 1986). It appears that all forms of mercury cross the placenta and reach the fetus (Amin-Zaki *et al.*, 1979), although most of what is known about these transplacental effects has been learned from animal studies (Fowler and Woods, 1977; Goyer, 1995).

6.3.3.3   Treatment

Treatment of mercury poisoning should aim at lowering its concentration in critical organ(s) or site of injury. In severe cases, haemodialysis is the first step, followed by infusion of chelating agents such as cysteine or penicillamine. In less severe cases of inorganic mercury poisoning, chelation with BAL may be effective (Goyer, 1995). However, chelation therapy is not very useful for alkylmercury exposure (Berlin, 1986). Selenium readily complexes with methylmercury and is believed to provide a protective action against its toxic effects (Inskip and Piotrowski, 1985; NRC, 1989).

6.3.4   Arsenic

Arsenic intoxication has been widely associated with occupational, chemical and chemotherapeutic exposures. Arsenic has no known vital functions in humans and is ubiquitous in the biosphere. The main sources of occupational exposure are chemical plants manufacturing pesticides, herbicides, wood preservatives and other agricultural products; the mining industry as a by-product of its activities; smelting operations, primarily among roaster workers in copper, zinc and lead smelters; and glass manufacturing, in which arsenic is added to raw material. Lead–arsenic alloys are also used because they are more rigid than pure lead, and anti-fouling paints and material used for sludge control formation in lubricating oils also contain arsenic (Buck, 1978; Hammond and Beliles, 1980). Arsenic exists as the toxic trivalent form (arsenic trioxide, As_2O_3, sodium arsenate, arsenic trichloride, etc.), as the less toxic pentavalent form (arsenic pentoxide, arsenic acid, lead arsenate, calcium arsenate, etc.), and as numerous organic forms (arsanilic acid, bimethyl arsenate, etc.). Potency depends on the chemical form of the element, with As(III) being the most potent, followed by As(V), then monomethylarsenate and finally dimethylarsenate (ATSDR, 1989b; EPA 1987).

Arsenic is ubiquitous in distribution, mostly present in soil (Hammond and Beliles, 1980). Its presence in air and water is influenced by proximity to point emissions. For example, in areas near copper smelters, its concentration in air may exceed $1 \, \mu g/m^3$. Drinking water usually contains a few micrograms of arsenic per litre or less, and most US drinking water supplies contain $< 5 \, \mu g/l$. It has been estimated that about 350 000 people in the United States might drink water contaminated with $> 50 \, \mu g/l$ of arsenic. Levels exceeding $50 \, \mu g/l$ were also reported in Nova Scotia, Canada, due to the content of arsenic in bedrock and in Japan, Cordoba (Argentina) and Taiwan from artesian well water (Hopenhayn-Rich et al., 1996).

When ingested, inorganic arsenic may cause acute or chronic toxicity. Arsenic is also of concern as a carcinogen, causing skin cancer, particularly dermal basal cell and squamous cell carcinoma, pulmonary carcinoma, and haemangiosarcomas. Its toxicity is dependent upon the oxidation state and route of exposure (ATSDR, 1989b; IARC, 1987). However, the predominant form of arsenic that exists in the edible parts of animals is the organic form, either arsenobetaine or arsenocholine, which have been shown to have no toxic effects in animals or humans (EPA, 1987; Tam et al., 1982). Although arsenobetaine constitutes the bulk of arsenic in fish, available studies are inconclusive regarding the presence of the more toxic inorganic forms or of organic forms that may be metabolized to inorganic forms in all fish (Buck, 1978; ATSDR, 1989b).

Most foods (meat and vegetables) in the United States contain minute levels of arsenic, averaging 0.2 mg/kg, although seafood may contain higher concentrations; an aquatic food mean of 2.6 mg/kg, with an average of 0.1 to 6.0 mg/kg was reported for catfish and shrimp, respectively, and up to 51 mg/kg was reported in some benthic marine animals. Dietary surveys indicate that beef contains $< 0.5 \, mg/kg$ in the muscle, liver and kidneys in the United States (Russell, 1978). Dairy products, which account for $\sim 31\%$ of the dietary intake, average $3.3 \, \mu g/$ kg (Mahaffey et al., 1981). Levels of arsenic in climbing beans of harvest years 1913 to 1986 kept in home-sterilized glass jars in Germany showed a decreasing trend; from 1913 to 1955 all levels were $> 5 \, \mu g/kg$, whereas after 1955 only levels below $5 \, \mu g/kg$ were found, presumably because of discontinued use of arsenic-containing pesticides (Boppel, 1995). The FAO/WHO PTWI is $182 \, \mu g/day$ and the FDA ADI is $45 \, \mu g/day$ (FAO, 1989; FDA, 1987).

In its chronic manifestation, arsenic has been associated with gastroenteritis, nephritis, hepatomegaly, peripheral symmetrical and central neuropathy, and a number of skin lesions including plantar and palmar hyperkeratosis and generalized melanosis. Some of these lesions seem related to destruction of capillary endothelium with consequent oedema and circulatory failure. Liver injury is characteristic of chronic exposure, manifesting itself initially as jaundice, which may progress to cirrhosis and ascites. Toxic effects on hepatic parenchyma cells result in elevation of blood liver enzymes in humans, and in alteration of the ultrastructure of mitochondria and loss of glycogen in animals. In addition, peripheral vascular diseases have been observed associated with chronic exposure in Taiwan and Chile, manifested by acrocyanosis and Raynaud's phenomenon, which is suspected to progress to endarteritis obliterans and gangrene of the lower extremities, leading eventually to black foot disease (EPA, 1987; Engel and Smith, 1994).

Arsenic compounds are inducers of metallothionein *in vivo*. At the molecular level, the metal is known to uncouple oxidative phosphorylation; it reacts with sulphydryl groups, leading to disruption of cellular metabolism, causes direct damage to DNA and inhibits its repair (ATSDR, 1989b). Moreover, the element causes teratogenicity in animals, especially hamsters (Earl and Vish, 1978), and crosses the placenta in humans (Kagey *et al.*, 1977).

Epidemiological evidence supports the conclusion that arsenic is a human carcinogen associated mainly with cancer of the skin by ingestion, with lung cancer by inhalation and with bladder cancer by drinking water (EPA, 1987; Hopenhayn-Rich *et al.*, 1996; IARC, 1987).

British Anti-Lewisite has been used for treatment of acute dermatitis and pulmonary symptoms associated with excessive arsenic exposure, and also for treatment of chronic arsenic poisoning, but its therapeutic usefulness has been questioned (Goyer, 1995).

Thus, although dietary arsenic does not seem at the moment to pose a public health risk, potential problems may occur in the future with continued increased use of arsenical herbicides and defoliants, especially in foods associated with cottonseed by-products; in areas where coal is used as a fuel source, and the resultant fly-ash contaminates the soil and water; with arsenic emissions into run-off waters from geothermal power plants contaminating rivers that receive these discharges; and near gold mines. Monitoring of these sources of environmental pollution should be increased as a precaution against the potential toxic effect of this element.

6.3.5   Selenium

Selenium at a low concentration is an essential element, and is an anti-oxidant present in essential enzymes such as glutathione peroxidase and haem oxidase and with the co-factor ubiquinone; but at high doses it causes toxicity to humans and animals, including embryotoxicity, teratogenicity and mutagenicity (Harr, 1978; NRC, 1989; Schamberger, 1985).

Selenium exists in a number of chemical forms: elemental selenium (Se^0), selenite $Se(IV)$, selenate $Se(VI)$ and selenide (Se^{2-}); these forms may bond either with other metals, or with organic substances such as amino acids. The selenates are the most soluble compounds and can easily enter biological systems, whereas the selenites and elemental selenium are relatively insoluble, which may affect their absorption and distribution within the body. Selenium has been reported in urban air, presumably as a product of fossil fuel combustion (fly-ash) and manufacture of paints, alloys, photoelectric batteries and rectifiers, and in water from seleniferous areas and foods, particularly seafood, meat, milk products and grains (Ewan, 1978; NRC, 1989).

Toxicity to animals arises in seleniferous (alkaline, oxidizing) soils when certain plants known to be accumulators of selenium (e.g. *Astragalus*, *Oonopsis*, *Stanleya* and *Zylorhiza*) concentrate the element up to 10 g/kg in the edible tissues. When these plants are grazed by cattle, sheep, horses and swine, animals develop a condition known as 'alkali disease' and 'blind stagger' after ingesting < 50 and > 100 mg/kg, respectively. Signs of poisoning include anorexia, tooth and hair loss, watery diarrhea, lassitude, progressive paralysis and eventually death. The areas of the world where human toxicities have been reported include locations in China, Venezuela, and Nebraska and South Dakota in the United States (Harr,1978; NRC, 1989).

FAO/WHO has no set PTWI for selenium, whereas the US National Academy of Sciences (NAS) recommended a range of 50–200 µg/day of selenium for adults (NRC, 1980). Metabolic studies in North America showed that daily dietary selenium intakes of 70 and 50 µg for men and women, respectively, are required to maintain balance. The average dietary selenium intake for US men between 1974 and 1982 was 108 µg/day (NRC, 1989).

Data from China indicate that a daily intake of < 20 µg selenium leads to Keshan disease, an endemic cardiomyopathy, first reported in Keshan County, China in 1935, due to consumption of maize and rice grown in selenium-deficient soils. The disease occurs most frequently in children under 15 years and women of child-bearing age (Yang *et al.*, 1987). The role of selenium deficiency in the aetiology of atherosclerotic cardiovascular diseases in the United States, however, remains uncertain (NRC, 1989).

The role of selenium deficiency as a cancer risk factor is unsettled because the evidence is inconsistent and insufficient, and there is a lack of enough dietary studies to reach a firm conclusion, although respiratory and gastrointestinal tumours have been implicated most often than other forms of cancer (IARC, 1987, 1994; Moore *et al.*, 1996; NRC, 1989).

Experimental and epidemiological evidence showing that selenium possesses anti-carcinogenic effects, presumably due to its inhibition of malonaldehyde (a carcinogenic product of pre-oxidative tissue damage), or gluathione, is condradictory and inconclusive at present (NRC, 1989).

Selenium interacts biologically with some vitamins, nutrients and other trace metals, and this interaction will impact on its toxicity, and/or deficiency. For example, if intake of vitamin E (i.e. α-tocopherol) is low, susceptibility to selenium toxicity will increase in experimental animals, whereas resistance to its toxicity will increase if vitamin E intakes are increased. Dietary sulphate, linseed meal and methionine were reported to confer partial protection against selenium poisoning. Moreover, selenium forms insoluble complexes with silver, copper, calcium, cobalt, mercury, thallium, zinc and silver. In addition, arsenite increases the biliary excretion of selenium, and thus enhances selenium excretion in urine. The exact mechanisms of these interactions are only partially understood, although it is recognized that their occurrence affects the general population (Ewan, 1978; Goyer, 1995).

6.3.6 Other trace metal contaminants

In addition to the above environmentally important dietary, trace metal contaminants, other elements, which may pose toxicological hazard to humans under certain conditions, are present in the diet in trace amounts (e.g. copper, iron, tin, zinc, antimony, aluminium, magnesium, manganese, molybdenum, chromium). These contaminants/essential elements are discussed briefly below.

6.3.6.1 Copper

This element is widely distributed in nature in the form of native copper, as well as several oxide, carbonate and sulphide ores, and is an essential element that is widely present in food and water. Copper is a component of several metabolic enzymes such as tyrosinase, catalase, peroxidase, cytochrome oxidase, superoxide dismutase, amine oxidase and uricase. Copper is also essential for utilization of iron, and a complex interrelationship among copper, iron and lead exists in the developing erythrocyte. Although copper is not an effective inducer of metallothionein as cadmium or zinc, copper–metallothionein binding is a normal storage form of copper, particularly in infants, and its accumulation in lysozymes facilitates the excretion of copper (Winge and Mehra, 1990).

The National Research Council (NRC) of NAS has set an estimated safe and adequate daily dietary intake (ESADDI) level of copper at 0.5–1.0 mg/day for infants; 1.0–2.5 mg/day for children; and 2.0–3.0 mg/day for those over 11 years old (NRC, 1980). The EPA maximum contaminant level for copper in drinking water is 1.3 mg/l (EPA, 1985).

Outbreaks of poisoning due to copper have involved acid foods prepared or stored in copper vessels (Russell, 1978).

The healthy adult body contains about 80 mg of copper. Foods with the greatest natural amount of copper are oysters (containing up to 137 mg/kg) and beef liver (11 mg/kg), whereas most other foods of animal origin contain between 2 and 4 mg/kg (Russell, 1978). Pennington *et al.* (1986) estimated that the mean intakes of 25- to 30-year-old women and men in the United States were 0.93 and 1.24 mg/day, respectively, whereas intakes of older women and men were lower (0.86 and 1.17 mg/day, respectively). The Continuing Survey of Food Intakes of Individuals (CSFII) conducted by the US Department of Agriculture (USDA), indicated that copper intake based on four non-consecutive days averaged 0.8 mg/day for women 19 to 50 years old, whereas for men 19 to 50 years old it was 1.6 mg/day based on one day's intake (USDA, 1987). Schrauzer *et al.* (1977) reported copper intake in 28 countries to range from 1.6 to 3.3 mg/day. Concentrations of copper in human milk in Croatia ranged from 0.27 to 1.35 mg/l (Mandic *et al.*, 1997). Dietary intake of healthy 5- to 9-year-old children living in the Duisburg area of North Rhine, Westphalia, showed a median daily copper intake of 1.0 mg, which was in the lower range recommended in Germany (Laryea *et al.*, 1995).

Individual tissues differ generally in their susceptibility to variations in dietary copper intake. Subnormal levels of dietary copper are reflected in subnormal blood copper concentrations. Elements such as zinc, cadmium and iron depress copper absorption and can reduce plasma copper concentrations when ingested at high dietary levels. Copper also plays a role in iron absorption and mobilization (Underwood, 1978). The metabolism of copper is negatively influenced by zinc, and the observed associations between copper and cancer, and other chronic diseases, may reflect the opposite effect of zinc. Elevated plasma copper levels are characteristic of acute and chronic infections (NRC, 1989).

Acute poisoning results from ingestion of excessive amounts of copper salts, e.g. copper sulphate, and may lead to death, with histopathological findings of centrilobular hepatic necrosis. Haematological risks such as haemolytic anaemia due to copper poisoning have also been reported. There are two genetically inherited inborn errors of copper metabolism. (1) Wilson's disease, sometimes referred to as hepatolenticular degeneration, is characterized by excessive accumulation of copper in liver, brain, kidneys and cornea, leading to clinical abnormalities in these organs, although the biochemical defect is not known. Chelation with either pencillamine or Trien [triethylenetetramine (2HCl)] has been effective. (2) Menke's disease is characterized by peculiar kinky hair, neurological impairment, severe mental retardation and death before the age of 3 years. There is extensive degeneration of the cerebral cortex and the white matter, although the basic biochemical defect is not known (Goyer, 1995).

6.3.6.2   Iron

Iron is an essential element present in all body cells, as a component of haemoglobin and myoglobin and also as a constituent of certain oxidative enzymes such as cytochrome and xanthine oxidase. The principal ores of iron are oxides, and its industrial uses are many, mainly in the fabrication of steel. Urban air and water may contain high amounts of iron from individual or natural sources. Iron has been used therapeutically since 1500 BC in treatment of maladies such as acne, alopecia, haemorrhoids, gout, pulmonary diseases, excessive lacrimation, weakness, oedema and fever. Its main therapeutic use today is in the treatment of iron deficiency anaemia (Goyer, 1995).

The availability of iron in the food supply has increased since 1909 chiefly because of the enrichment of flour and cereal products. Sixty-seven per cent of the total iron is contained in blood haemoglobin, and 24% is stored in the liver as ferritin; in case of excess intake, it is stored as haemosiderin, also in the liver. The normal body burden is about 4 g (Hammond and Beliles, 1980; NRC, 1989).

Excessive dietary intake of iron due to use of iron pots in food preparation and beer brewing in South Africa leads to a disease termed 'Bantu siderosis', more frequently in men than women. Histopathologically there is marked accumulation of iron in Kupffer cells in the liver and in the reticuloendothelial cells of the spleen and bone marrow. High dietary iron intake from other sources (e.g. wine), or frequent blood transfusion were reported to play a role in the aetiology of haematochromatosis. Dietary intakes for 15-year-old boys and girls from two different regions of Sweden ranged from 7 to 35 and from 6 to 27 mg/day, with daily median intakes of 18.7 and 14.2 mg, respectively (Samuelson et al., 1996). Higher dietary iron intake among 299 urban pre-school children in Maryland (ages 9 months to 15 years) was associated with lower lead blood levels (Hammad et al., 1996).

Occupational long-term exposure to iron oxide, mainly Fe_2O_3 has resulted in mottling of lungs, siderosis, which is a benign pneumoconiosis without significant physiological impairment. However, haematite miners in certain areas reported from 50% to 70% higher death rates due to lung cancer, as well as tuberculosis and interstitial fibrosis, than the normal population. The increased incidence of cancer may be attributable to other factors such as smoking, radioactivity and/or exposure to silica and other minerals in the mines (Boyd et al., 1970). Dose levels of iron among iron workers developing pneumoconiosis have been reported to exceed 10 mg Fe/m^3 (Goyer, 1995).

Treatment of acute iron poisoning is directed towards removal of ingested iron from the gut by inducing vomiting or gastric lavage, and providing therapy for systemic effects such as acidosis and shock. Chelation with desferrioxamine is used for treatment of iron overload in acute exposures, and removal of tissue iron in haemosiderosis. Repeated phlebotomy was reported to remove as much as 20 g of iron per year. Ascorbic acid appears to increase iron excretion by as much as twice normal levels (Brown, 1983).

6.3.6.3 Tin

Cassiterite (SnO_2), the only important ore of tin, is used for manufacture of tinplate, food packaging and solders, bronze and brass. Stannous and stannic chlorides are used in dyeing textiles. Organic compounds have been used in fungicides, bactericides and slimicides, as well as plastics and stabilizers. Chronic inhalation of tin in the form of the element or fumes leads to benign pneumoconiosis. Tin hydride (SnH_4) is more toxic to mice and guinea-pigs than is arsine; however, its effects appear to be limited to the central nervous system. Inorganic tin salts given by injection produce diarrhoea, muscle paralysis and twitching. Acute burns or subacute dermal irritation have been attributed to tributyltin among workers. Triphenyltin has been shown to be a potent immunosuppressor. Excessive industrial exposure to triethyltin has been reported to produce headaches, visual effects and electroencephalograph (EEG) changes that were very slowly reversed. Experimentally, triethyltin produced depression and cerebral oedema, and hyperglycaemia which may be related to centrally mediated depletion of catecholamines from the adrenal gland. Orally, tin or its organic compounds require intake of relatively large doses (500 mg/kg for 14 months), and the use of tin in food processing did not show appreciable hazards. The average US daily intake, mostly from food as a result of processing, is estimated at 17 mg (WHO, 1980).

6.3.6.4 Zinc

Zinc is ubiquitous in the environment and is present in most foodstuffs, water and air. It is an essential element and a constituent of more than 200 enzymes, and plays an important role in nucleic acid synthesis and metabolism, cell replication and tissue repair and growth through its function in nucleic acid polymerases. Zinc also displays biologically important interactions with hormones. Zinc does not accumulate with continued exposure, but body contents are modulated by homeostatic mechanisms that act primarily on its absorption in the liver (Walsh et al., 1994).

Atmospheric zinc levels are higher in industrial areas. The average daily intake of zinc in the United States is approximately 12–15 mg, mostly from food. The content in substances in contact with galvanized copper or plastic pipes may be increased. The richest sources of zinc are shellfish, especially oysters, beef and other red meat. Poultry, milk, certain dairy products like yogurt, hard cheese, eggs, legumes, nuts and whole-grain cereals are good sources of zinc, while vegetables are lower, although they take up zinc from soil. Many dietary factors, including other minerals, phytates and dietary fibre may adversely affect zinc absorption. Zinc from animal sources appears to be better absorbed than that from plant sources (Hammond and Beliles, 1980).

The recommended dietary allowance (RDA) for zinc is 15 mg/day for people aged 11 years or older (NRC, 1980), but zinc available in the US food supply is only ~12.3 mg per capita (NRC, 1989). Zinc concentrations in human milk in Croatia ranged from 0.62 to 15.0 mg/l (Mandic et al., 1997). Dietary intake of healthy 5- to 9-year-old children living in Duisburg area of North Rhine, Westphalia, showed median daily zinc intakes of 5.3 mg, which were below the recommended levels in Germany (Laryea et al., 1995). Studies in Andhra Pradesh, India, showed that rice grown in zinc-deficient soil had a lower zinc content; however, serum zinc levels of adults and grown children seemed to be influenced by protein content and source of protein in the diet, in addition to zinc levels in the diet (Sunanda et al., 1995). Dietary zinc does not appear to contribute to chronic degenerative diseases, including cancer, in humans (IARC, 1994; NRC, 1989). Excessive intake of zinc was found to interfere with absorption and utilization of iron and copper. Zinc deficiency in newborn humans may be manifest by dermatitis, loss of hair, impaired healing, susceptibility to infection and neuropsychological abnormalities. Dietary inadequacies due to alcohol consumption may be associated with dermatitis, night blindness, testicular atrophy, impotence and poor wound healing (Ahmed, 1995). Other degenerative disorders such as ulcerative colitis, malabsorption syndrome, chronic renal disease and haemolytic anaemia are also prone to zinc deficiency. Many drugs alter zinc homeostasis, particularly metal-chelating agents and some antibodies such as penicillin and isoniazid (Goyer, 1995).

6.3.6.5 Antimony

The primary ore of antimony is stibnite (Sb_2S_3). Important uses are in lead alloys, pewter, bearing alloys, rubber, matches, arsenic, enamels, paints, lacquers, storage battery grids, textiles and insecticides (e.g. antimony potassium tartarate). Antimony may contaminate food as a result of use of rubber or soldered tinfoil for packaging, and leaching from cheap enamelled vessels. Both trivalent and pentavalent organic compounds have been administered parenterally as parasiticides, with the former being more effective. Moreover, the trivalent compounds have been used as emetics and expectorants, but this use has largely been abandoned because of toxicity. The trivalent forms, after being absorbed, concentrate in red blood cells, while the pentavalent compounds concentrate in plasma. Trivalent forms accumulate in the liver and are excreted slowly, particularly in the faeces. The pentavalent forms tend to concentrate in the liver and spleen, and are excreted in the urine. Acute poisoning has resulted from accidental or intentional ingestion of antimonids, giving rise to symptoms similar to arsenic poisoning (e.g. vomiting, watery diarrhoea, collapse, irregular respiration and lowered temperature, with death occurring within a few hours. Vomiting increases the chance of recovery. Cardiac fibrillation due to a direct effect on the heart leads to death. Liver toxicity, characterized by jaundice and fatty degeneration, pulmonary congestion and oedema have been reported (Winship, 1987). Antimony miners have shown disablement, albeit a form of silicosis (Hammond and Beliles, 1980).

6.3.6.6 Aluminium

Aluminium is the second most plentiful element in earth's crust, the principal ore being bauxite. Its concentration in sea water and fresh water is usually quite low, but is increasing as a result of acid rain. Aluminium is widely used as a building material and in other uses where light weight and corrosion resistance are important. Aluminium oxide is used industrially as an abrasive and a catalyst. Medicinally, various soluble salts of aluminium have been used as astringents, septics and antiseptics. The insoluble salts are used as antacids and as antidiarrhoeal agents. Inhalation of aluminium hydroxide is used as a preventive and curative agent for silicosis (Hammond and Beliles, 1980).

Intake of aluminium in food ranges from 3 to 5 mg/day. Dietary sources include aluminium used as a filler in pickles and cheese, or as a major component of some types of baking powder. It also leaches from cooking utensils during preparation of food (Rao and Rao, 1995). Dietary intake of aluminium in children aged 5–8 years in Duisburg, Germany, showed the medium and maximum values to be 2.9% and 11.2%, respectively, of the FAO/WHO PTWI (Wilhelm et al., 1995). Little ingested aluminium is absorbed, so the total body content is extremely low and does not accumulate with age. Massive oral doses of aluminium lead to gastrointestinal irritation and rickets due to interference with phosphate absorption (NRC, 1989). There is no evidence that aluminium is carcinogenic to humans (IARC, 1987, 1994; NRC, 1989).

The incidence of Alzheimer's disease was correlated with aluminium concentration in County district water supplies in England and Wales (Martyn et al., 1989). The primary lesion in Alzheimer's disease is postulated to be an impaired permeability of the blood–brain barrier that allows neurotoxins such as aluminium to reach the central nervous system (NRC, 1989). In patients with the disease, high levels of intracellular aluminium have been found in neurofibrillary tangles of the hippocampal region. Similar increases in peroral aluminium have been found among the Chamorro population of Guam, parts of South Japan and New Guinea afflicted with amyotrophic lateral sclerosis or Parkinsonian dementia. In these areas, the soil and water are rich in aluminium and poor in calcium and manganese, suggesting that aluminium may be a common aetiologic factor in several related neurological conditions (Miller *et al.*, 1984; Perl, 1985). However, it may be possible that accumulation of aluminium in brain tissue is secondary to the degeneration in the affected neurons (NRC, 1989).

6.3.6.7   Magnesium
The primary ores of magnesium are magnesite and dolomite. It can also be obtained from brine wells and salt deposits. Magnesium is used in alloys, particularly those requiring light weight, in electrical conducting material such as wires and ribbon for radios, as well as for incendiary material such as flares. Magnesium is an essential nutrient and a co-factor for many enzymes, particularly metalloenzymes associated with phosphate. Deficiency in humans has been associated with magnesium-losing renal disease, alcoholism and low-magnesium-induced ischaemic heart disease, and in animals as a result of ingestion of grasses grown in manganese-poor soil. Deficiency causes neuromuscular irritability, calcification, and cardiac and renal damage, all of which can be prevented by supplementation. Nuts, cereal, seafood and meats are high dietary sources of magnesium. The average city water contains about 65 µg/l, but varies considerably with the hardness of water (Rubenowitz *et al.*, 1996). The magnesium concentration of human milk is about 30 mg/l, while infant formulas provide about 40–80 mg/l, and cow's milk approximately 130 mg/l; about 55–75% of magnesium is absorbed in pre-term and term infants. Thus, infant diets are likely to supply adequate amounts of magnesium for growth (Lonnerdal, 1995). Magnesium has many medicinal uses including use as antacids or cathartics, for anti-inflamation and anti-depressant action and as an antidote against poisoning. Magnesium toxicity is similar to that of zinc (Hammond and Beliles, 1980; Goyer, 1995).

6.3.6.8   Manganese
The principal ore of manganese is pyrolusite (MnO_2). Manganese and its compounds are used for making steel alloys, dry cell batteries, electrical coils, ceramics, matches, glass, dyes, fertilizers and welding rods and as

animal feed additives. Primary medicinal uses are as antiseptics and germicides, as potassium permanganate ($KMnO_4$), applied dermally because of its astringent action. Manganese is an essential element and a co-factor in a number of enzymatic reactions, particularly those related to phosphorylation and cholesterol and fatty acid synthesis. It is also involved in the metabolism of biogenic amines and participates in the regulation of carbohydrate metabolism (Hurley and Keen, 1987).

Although manganese is present in urban air and most water supplies, the principal intake is derived from food. The best sources of manganese are plant foods, especially vegetables; the germinal portions of grains, fruits, nuts, tea and some spices are rich in manganese. Cereal was estimated to contain between 10 and 100 mg/kg (NRC, 1989). Daily manganese intake ranges from 2 to 9 mg; gastrointestinal absorption is < 5%. Manganese is transported in plasma bound to a β_1-globulin, believed to be a transferrin, and concentrates in mitochondria in organs such as the pancreas, liver, kidney and intestines. The biological half-life in the body is 37 days. Manganese readily crosses the blood–brain barrier and its half-life in the brain is longer than in the rest of the body. Estimates of the human requirement for manganese are based on balance studies, with approximately 35 to 70 µg/kg bw/day resulting in a positive balance. The daily dietary intake in the United States has been estimated to range from 2 to 4 mg. The NRC ESADDI is 1.0–3.0 mg/day for children up to 10 years old, and 2.5–5.0 mg/day for adolescents and adults (NRC, 1980). Manganese toxicity has been observed by inhalation in occupational settings, leading to pneumonitis and central nervous system disorders. Dietary manganese appears to be non-toxic (NRC, 1989).

6.3.6.9   Molybdenum

The most important mineral source of molybdenum is molybdenate (MoS_2). Industrial uses of this metal include the manufacture of high-temperature steel alloys, use in gas turbine and jet aircraft engines, and production of catalysts, lubricants and dyes. The United States is the world's major producer of molybdenum. Molybdenum is widely distributed in nature, being an essential element. It is a co-factor for the enzymes xanthine oxidase and aldehyde oxidase. In plants, it plays an important role in bacterial fixing of atmospheric nitrogen at the start of protein synthesis, and is added in trace amounts in fertilizers to facilitate plant growth. Because of these functions, it is ubiquitous in food. Since plankton tend to concentrate molybdenum to 25 times the levels in seawater, shellfish also tend to accumulate this element. Average daily human intake in food is about 350 µg. Its concentration in air averages <0.005 µg/m^3. It is present in more than one-third of freshwater supplies

and in certain areas at concentrations approaching 1 µg/l. The requirement for molybdenum can only be estimated from balance studies in
humans. The NRC ESADDI is 0.15–0.5 mg for adults (NRC, 1980).
Intakes above 0.5 mg appeared to compromise copper balance. Intakes of
10–15 mg/day have been associated with gout in Russia (NRC, 1989).
There are no data documenting molybdenum toxicity in man due to
industrial exposure (Goyer, 1995; Hammond and Beliles, 1980).

6.3.6.10   Chromium

Chromium is a generally abundant element in the earth's crust. Chromite
($FeCr_2O_4$) is the most important chromium ore. Oxidation states range
from Cr(II) to Cr(VI), but the trivalent and hexavalent forms are only of
biological significance. Sodium chromate and dichromate are the
principal substances for the production of all chromium chemicals.
Metallurgic-grade chromite is usually concentrated into ores of several
types of ferrochromium or other chromium alloys containing cobalt or
nickel. Chromium plating is one of the major uses for this metal in steel
fabrication (ferrochrome). Sodium dichromate ores are used for the
production of paint and pigment, and as anticorrosives in cookware,
boilers and oil drilling rods. Chrome salt used in leather tanning is
another major use is another of chromium. Medicinal use of chromium is
limited to external application of chromium trioxide as a caustic, and
intravenous sodium radiochromate to evaluate the life span of red blood
cells. Trivalent chromium is the most commonly found form in nature.
There is no evidence that Cr(III) is converted into Cr(VI) in biological
systems. However, Cr(VI) readily crosses cell membranes and is reduced
intracellularly to trivalent chromium (Katz and Salem, 1993). Cr(III) is
considered as an essential trace nutrient, as a component of the glucose
tolerance factor for insulin action forming a ternary complex with the
insulin receptor, and thus facilitating attachment of insulin to the binding
sites. Human chromium deficiency may occur in infants suffering from
protein–calorie malnutrition and in elderly people with impaired glucose
tolerance (Goyer, 1995).

Chromium in ambient air originates from industrial sources, particularly ferrochrome production, cement production, ore refining, chemical
and refractory processing and combustion of fossil fuels. In rural areas,
chromium in the air is $< 0.1 \, \mu g/m^3$; in industrial cities it ranges from
0.015 to 0.33 µg/m³. Particulates from coal-fired power plants may
contain from 2.3 to 31 mg/kg, but this is reduced to 0.19–6.6 µg/kg by fly-
ash collection. Chromium precipitates and fallouts eventually are carried
to water by run-off, where they are deposited as sediments. The
chromium content in food is low, and most intake of chromium is from
food, estimated to be under 100 µg/day with trivial quantities from most

water supplies and ambient air (Goyer, 1995; Hammond and Beliles, 1980).

Tissue concentrations of chromium in the general population vary considerably from region to region. The highest content of $7\,\mu g/kg$ occurs in lungs of persons in New York or Chicago, with lower concentrations in liver and kidney. In persons with excess exposure, blood chromium concentration is between 20 and $30\,\mu g/l$ and is evenly distributed between erythrocytes and plasma. Urinary excretion is independent of the oxidation state of chromium administered to animals, and is $< 10\,\mu g/$ day for humans in the absence of excessive exposure (Goyer, 1995). Systemic toxicity to chromium compounds occurs largely from accidental exposures, occasional attempts to use chromium as an agent of suicide, and from therapeutic uses, with major effects being acute tubular glomerular damage. Hexavalent chromium is corrosive and causes chronic ulceration and perforation of the nasal septum and other skin surfaces. Additional allergic skin reactions may also occur and are independent of the dose. Trivalent chromium compounds are considered less toxic than the hexavalent compounds, and are neither irritating nor corrosive. Cr(VI) was shown to be carcinogenic to lungs in humans and animals (Langard and Norseth, 1986).

There is no firm evidence to show that the above industrial and occupational metal contaminants constitute a public health risk at the minuscule levels in which they are present in the diet (Goyer, 1995; Menzer, 1991; NRC, 1989; Russell, 1978).

6.4 Steps to reduce contamination

In industrialized countries, as illustrated by the United States, improper industrial discharges still remain a serious environmental problem. For example, it is estimated that 36 million tonnes of hazardous wastes are generated annually by industry, practically all of which are categorized as toxic. Ninety per cent of this waste is believed to be disposed of in an unsound manner: 48% in surface impoundments, 30% in inadequate land fills, 10% by improper burial and 2% by other non-specific means (Neely et al., 1981). Environmental problems such as these prompted the US Congress to pass the Comprehensive Environmental Response, Compensation and Liability Act (CERCLA, Superfund) in 1980, and in 1986 it passed the Superfund Amendments and Reauthorization Act (SARA) (USC 9604 (i) (2)) to increase funding levels for clean-up and strengthened clean-up strategies (Fay and Mumtaz, 1996). The most frequently found combinations of metals in soil were chromium and lead (20.5%), and chromium, cadmium and lead (12.0%). Since 1980, approximately 40 000 uncontrolled waste sites have been identified in the United States, and approximately 1300 of these sites constitute the current national Superfund priority list sites for remediation (Johnson, 1995). The programme, however, is plagued with problems, and it is uncertain whether legislative measures will effectively redeem contaminated sites within affordable cost any time in the near future (GAO, 1987). In Mexico, it is estimated that about 90% of the hazardous waste is not treated adequately and is disposed of improperly, leading to widespread pollution of the environment (Diaz-Barriga, 1996).

A primary source of the majority of inorganic contaminants in water is thought to be the emission and release of solid/liquid waste (fly-ash) resulting from the burning of coal and other fossil fuels (Menkes and Fawcett, 1997; Moore, 1990), and the evidence shows that contaminants move from one environmental compartment into another (Mackay, 1991).

In developing countries, on the other hand, the situation is grimmer. The widespread and frequent contamination of foods indicates that this problem will continue to persist (Gzyl, 1997; Hernandez-Avila et al., 1996; Jung and Thornton, 1997; Menkes and Fawcett, 1997; Nriagu et al., 1996; Romieu et al., 1997; Shen et al., 1996) unless a serious effort on the part of the governments in these countries is taken to establish regulatory actions, which may be tailored on those of one or another developed country, and to monitor compliance with these legislative actions with sufficient resources; all measures that will eventually reduce these severe public health hazards (Diaz-Barriga, 1996; Romieu et al., 1997).

Results on the extent of food contamination due to trace metals worldwide indicate that measures taken in the 1970s and 1980s by scientists, legislators, regulators, industry, and consumers, together with their advocates in developed countries, have lowered contamination of food during the 1990s (Bavazzano and Cotti, 1994; Dabeka and McKenzie, 1995; MMWR, 1997; Oskarsson et al., 1995; Schumacher et al., 1996; Stromberg et al., 1995). However, in spite of these public health advances, the situation is not foolproof, as accidental contamination still happens (Karita et al., 1997; Powell et al., 1995).

Food contamination, which may be either environmental or natural, can occur during growth, processing, transport or storage (Table 6.6). It is much more difficult to control contamination once it has happened. Thus, an enlightened strategy is to aim at preventing it in the first place because, if contaminated food reaches consumers, it may not be possible in many instances to trace or retrieve the product, even if monitoring methods are available. Moreover, there will be additional cost involved in surveying, interpreting toxicities and monitoring susceptible segments of the population (e.g. children, immunosuppressed individuals, and

Table 6.6 Common sources of metal contaminants in food and steps to reduce contamination

Sources of contamination	Steps to reduce contamination
Vehicle emissions (lead)	Use catalytic converters and lead-free gasoline
Land fills (lead)	Reduce industrial discharge
Agriculture pesticides (cadmium)	Restrict disposal of hazardous waste
Industrial discharges and effluents (cadmium, lead, mercury)	Institute hygienic conditions for livestock rearing and seafood aquaculture
	Monitor food crops and seafood
Processing-related (cadmium, lead)	Introduce non-lead-soldered tins and hygienic food processing and preparations
	Improve transportation and storage facilities
	Monitor processed foods
Cooking of food (lead, cadmium)	Use non-toxic material for cooking utensils
	Decontaminate waters and sediments
	Remove toxicants from food
	Control exposed animals
	Improve consumer awareness
	Issue advisories

Adapted from Ahmed (1991b; 1992c,d) and UNEP (1992).

pregnant and lactating women). Thus, in the long run, preventive measures are much cheaper and will require fewer resources (GAO, 1987; Tengs *et al.*, 1995; UNEP, 1992; Vilagines and Leroy, 1995).

Lead levels in rivers and lakes can often be reduced by regulating releases from mining, smelting and refining operations and emissions from battery, lead alkyl and pigment plants. In addition, atmospheric releases of lead can be further lowered by decreasing emissions from vehicle exhausts through use of both unleaded gasoline and catalytic converters, especially in urban settings, given their high vehicular density (Flegal and Smith, 1995; Jung and Thornton, 1997). Lead can be reduced in drinking waters by replacing lead pipes or soldered copper ones with non-lead plumbing (Grobler *et al.*, 1996; Watt *et al.*, 1996). Atmospheric cadmium from mining and smelting operations and other industrial processes such as electroplating, is a major source of water and nearby crop contamination. Emissions and effluents from these sources need to be restricted and reduced to protect the environment and eventually the food supply (Gzyl, 1997). Moreover, industrial discharges into water bodies, as in the case of mercury, must be curtailed. For example, measures taken in the United States during the 1970s have led to a reduction in the amount of mercury released into water by 60-fold (from 540 tonnes to <9 tonnes per year) within a relatively short time (EPA, 1989b).

The processing of food is now an industrial activity that requires control points at various critical stages of the process to ensure the safety and quality of food through good manufacturing and control practices in food processing plants (Ahmed, 1992c). Technological advances have reduced the chances of metal contamination. For example, the development of acid-resistant ceramic wares has decreased the probability of lead or cadmium entering food accidentally through contact with cooking utensils, and the introduction of non-lead-soldered tins has drastically reduced lead levels in food (Baczynskyj and Yess, 1995; Bolger *et al.*, 1996). Regulatory bodies should ensure that these and other good manufacturing practices are followed during processing, particularly during canning (UNEP, 1992). Additionally, inspection of food plants by inspectors from pertinent government agencies and sampling and analysis should be carried out regularly using up-to-date scientific methods. Good science must also be employed by regulatory agencies to ensure that the regulations are effective and credible, that resources are not wasted on meaningless and/or outdated testing, and that food costs are reasonably maintained, otherwise consumers will lose confidence in the integrity of the regulatory system (Ahmed, 1992d) and the poor will suffer the most if costs escalate (Ames and Gold, 1997).

It is important that regulations apply to industrial operations in order to prevent pollution of the environment as a result of toxic discharges; to emissions and effluents, which will eventually contaminate food; to agricultural operations to reduce excessive contamination of both food and agriculture workers; and to processors to ensure that wholesome food will be produced that is not be contaminated during any steps of the processing or packaging operations. In addition, a national monitoring programme will ensure the effectiveness of these regulations and supply governments with information on the extent of contamination if and when it occurs, so that immediate action can be taken (Ahmed, 1991b; Merrill, 1995; UNEP/FAO/WHO, 1988).

Information should also be made available to consumers about the food they buy. This can be done through proper labelling of foods. In addition, local authorities should issue advice or warnings, when necessary, to alert populations about the dangers of environmental pollution and its relation to food contamination, as for instance when localized water pollution affects fishery products (Ahmed, 1991b; NOAA, 1989).

Reduction of non-tariff barriers in local and international food trade should be strengthened by promoting cooperation and information exchange on food so as to prevent dumping of substandard foods on developing countries. These steps will ensure confidence in international food supplies, and will bring greater economic benefits to both exporting and importing countries (Ahmed, 1992e; UNEP, 1992).

References

Ahmed, F.E. (1991a) Effect of diet on progression of chronic renal disease. *Journal of the American Dietetic Association*, **91** 1266-1272.

Ahmed, F.E. (ed.) (1991b) *Seafood Safety*, Institute of Medicine, US National Academy of Sciences, National Academy Press, Washington, DC.

Ahmed, F.E. (1992a) Effect of nutrition on the health of the elderly. *Journal of the American Dietetic Association*, **92** 1102-1108.

Ahmed, F.E. (1992b) Effect of diet on cancer and its development in humans. *Environmental Carcinogenesis Ecotoxicology Reviews*, **C10** (2) 141-180.

Ahmed, F.E. (1992c) Hazard analysis critical control point (HACCP) as a public health system for improving food safety. *Journal Occupational Medicine Toxicology*, **1** (4) 349-359.

Ahmed, F.E. (1992d) Assessment of cattle inspection in the United States from a public health perspective. *Journal of Occupational Medicine Toxicology*, **1** (2) 209-222.

Ahmed, F.E. (1992e) Programs of safety surveillance and control of fishery products. *Regulatory Toxicology Pharmacology*, **15** 14-31.

Ahmed, F.E. (1995) Toxicological effects of alcohol on human health. *CRC Critical Reviews in Toxicology*, **25** (4) 347-367.

al-Saleh, I.A. (1995) Lead exposure in Saudia Arabia and its relationship to smoking. *Biometals*, **8** (3) 243-245.

Al-Sharistani, H. and Shihab, K.M. (1974) Variation of biological half-life of methylmercury in man. *Archives Environmental Health*, **28** 342-344.

Ames, B.N. and Gold, L.S. (1997) Environmental pollution, pesticides and the prevention of cancer: misconceptions. *FASEB Journal* (July 21) 1-14.

Amin-Zaki, L., Majeed, M.A., Elhassani, S.B., Clarkson, T.W., Greenwood, M.R. and Doherty, R.A. (1979) Prenatal methylmercury poisoning, clinical observations over five years. *American Journal of Diseases of Children*, **133** 172-177.

Andersen, A. (1981) *Lead, Cadmium, Copper and Zinc in the Danish Diet*, Publication No. 52, Miljøministeriet, Statens Levnesmiddelinstitut, Søborg, Denmark.

Anderson, A. and Hahlin, M. (1981) Cadmium effects from phosphorus fertilization in field experiments. *Swedish Journal Agriculture Research*, **11** 2.

ATSDR (Agency for Toxic Substances and Disease Registry) (1989a) *Toxicological Profile for Cadmium*, prepared by Life System, Inc for ATSDR, US Public Health Service in collaboration with US Environmental Protection Agency, ATSDR/TP-88/08, Washington, DC.

ATSDR (Agency for Toxic Substances and Disease Registry) (1989b) *Toxicological Profile for Arsenic*, prepared by Life Systems, Inc for ATSDR, US Public Health Service in collaboration with US Environmental Protection Agency, ATSDR/TP-88/02, Washington, DC.

Baczynskyj, M.W. and Yess, N.J. (1995) US Food and Drug Administration monitoring of lead in domestic and imported ceramic dinnerware. *Journal of the Association of Official Analytical Chemists*, **78** (3) 610-614.

Bakir, F., Damluji, S.F., Amin-Zaki, L., Murtadha, M., Khalidi, A., al-Rawi, N.Y., Tikkriti, S., Dahair, H.I., Clarkson, T.W., Smith, J.C. and Doherty, R.A. (1973) Methylmercury poisoning in Iraq. *Science*, **181** 230-241.

Baum, C.R. and Shannon, M.W. (1997) The lead concentration of reconstituted infant formula. *Journal of Toxicology: Clinical Toxicology*, **35** (4) 371-375.

Bavazzano, P. and Cotti, G. (1994) Biological monitoring of lead in the study of urban pollution due to automobile traffic. *Epidemiologia e Prevenzione*, **18** (58) 27-34.

Beck, B.D., Conolly, R.B., Durson, M.L., Guth, D., Hattis, D., Kimmel, C. and Lewis, S.C. (1993) Improvements in quantitative noncancer risk assessment. *Fundamental Applied Toxicology*, **20** 1-14.

Beijer, K.J. and Jernelöv, A. (1986) Sources, transport and transformation of metals in the environment, in *Handbook on the Toxicology of Metals, General Aspects*, vol. 1, 2nd edn, Elsevier, Amsterdam, pp 68-74.

Beltran Llerandi, G., Grillo, M., Jones, O. and Garcia Roche, M.O. (1989) Estimation of the daily intake of cadmium which may be consumed by students 12–17 years old in secondary schools in the city of Havana. *Nahrung*, **33** (4) 315-318.

Berlin, M. (1986) Mercury, in *Handbook on the Toxicology of Metals*, 2nd edn (eds L. Friberg, G.F. Nordberg and C. Nordman), Elsevier, Amsterdam, pp 386-445.

Bhatia, I. and Choudhri, G.N. (1996) Lead poisoning of milk—the basic need for the foundation of human civilization. *Indian Journal of Public Health*, **40** (1) 24-26.

Bolger, P.M., Yess, N.J., Gunderson, E.L., Troxell, T.C. and Carrington, C.D. (1996) Identification and reduction of sources of dietary lead in the United States. *Food Additives and Contaminants*, **13** (1) 53-60.

Boppel, B. (1995) Arsenic, lead and cadmium in home-made preserves of fruit and vegetables from previous years. 2. Climbing beans since the harvest year, 1913. *Zeitschrift für Lebensmittel-Untersuchung und -Forschung*, **201** (1) 12-13.

Boyd, J.T., Dolls, R., Faulds, J.S. and Leiper, J. (1970) Cancer of the lung in iron ore (hematite) miners. *British Journal of Industrial Medicine*, **27** 97-105.

Brown, E.B. (1983) Therapy for disorders of iron excess, in *Biological Aspects of Metal-Related Diseases* (ed. B. Sakar), Raven Press, New York, pp 263-278.

Brussard, J.H., Van Dokkum, V., Van der Paauw, C.G., De Vos, R.H., De Kort, W.L. and Lowik, M.R. (1996) Dietary intake of food contaminants in The Netherlands (Dutch Nutrition Surveillance System). *Food Additives and Contaminants*, **13** (5) 561-573.

Buchet, J.P., Lauwerys, R., Vandevoorde, A. and Pycke, J.M. (1983) Oral daily intake of cadmium, lead, manganese, copper, chromium, mercury, cadmium, zinc and arsenic in Belgium: a duplicate metal study. *Food and Chemical Toxicology*, **21** 19-24.

Buck, W.B. (1978) Toxicity of inorganic aliphatic organ arsenicals, in *Toxicity of Heavy Metals in the Environment*, part 1 (ed. F.W. Oehme), Marcel Dekker, New York, pp 357-374.

Burger, J. (1995) A risk assessment for lead in birds. *Journal of Toxicology and Environmental Health*, **45** (4) 369-396.

Burns, C.B. and Currie, B. (1995) The efficacy of chelation therapy and factors influencing mortality in lead intoxicated petrol sniffers. *Australian and New Zealand Journal Medicine*, **25** (3) 197-203.

Burns, L.A., Meade, B.J. and Munson, A.E. (1995) Toxic responses of the immune system, in *Casarett & Doull's Toxicology: The Basic Science of Poisons*, 5th edn (eds C.D. Klassen, M.O. Amdur and J. Doull), McGraw-Hill, New York, pp 355-402.

Capar, S.G. and Rigsby, E.J. (1989) Surveys of lead in canned evaporated milk. *Journal of the Association of Official Analytical Chemists*, **72** 416-417.

Capar, S.G. and Yess, N.J. (1996) US Food and Drug Administration survey of cadmium, lead and other elements in clams and oyesters. *Food Additives and Contaminants*, **13** (5) 553-560.

CDC (Centers for Disease Control) (1991) *Preventing Lead Poisoning in Young Children—A Statement by the Centers for Disease Control*, US Department of Health and Human Services No. 99-2230, US Government Printing Office, Washington, DC, pp 51-64.

Chan, H.M., Kim, C., Khoday, K., Receveur, O. and Kuhnlein, H.V. (1995) Assessment of dietary exposure to trace metals in Baffin Inuit food. *Environmental Health Perspectives*, **103** (7–8) 740-746.

Cherniaeva, T.K., Matveeva, N.A., Kuzmichev, IuG. and Gracheva, M.P. (1997) Heavy metal content of the hair of children in industrial cities. *Gigienia i Sanitarria*, **3** 26-28.

Conso, F., Tulliez, M., Fauvarques, M.O. and Nenna, A.D. (1984) Sideroblastic anemia, lead poisoning and wine from Bordeaux. *Annales Medicine Interne*, **135** (1) 43-45.

Cuadrado, C., Kumpulainen, J. and Moreivas, O. (1995) Lead, cadmium and mercury contents in average Spanish market basket diets from Galicia, Valencia, Andulucia and Madrid. *Food Additives and Contaminants*, **12** (1) 107-118.

Dabeka, R.W. (1989) Survey of lead, cadmium, cobalt and nikel in infant formula and evaporated milks and estimation of dietary intakes of the elements by infants, 0–12 months old. *Science of the Total Environment*, **89** 279-289.

Dabeka, R.W. and McKenzie, A.D. (1988) Lead and cadmium levels in contaminated infant foods and dietary intake by infants 0–1 year old. *Food Additives and Contaminants*, **5** 333-342.

Dabeka, R.W. and McKenzie, A.D. (1995) Survey of lead, cadmium, fluoride, nickel, and cobalt in food composites and estimation of dietary intakes of these elements by Canadians in 1986–1988. *Journal of the Association of Official Analytical Chemists*, **78** (4) 897-909.

Dabeka, R.W., Kapinski, K.F., McKenzie, A.D. and Bajdik, C.D.C. (1986) Survey of lead, cadmium and floride in human milk and correlation with environmental and food factors. *Food and Chemical Toxicology*, **24** 913-921.

Dabeka, R.W., McKenzie, A.D. and Lacroix, M.A. (1987) Dietary intakes of lead, cadmium, arsenic and fluoride by Canadian adults: a 24-hour duplicate diet study. *Food Additives and Contaminants*, **4** 89-102.

De Flora, S., Bennicelli, C. and Bagnasco, M. (1994) Genotoxity of mercury compounds. *Mutation Research*, **317** (1) 57-79.

De Gregori, I., Delgado, D., Pinochet, H., Gras, N., Munoz, L., Bruhn, C. and Navarrete, G. (1994) Cadmium, lead, copper and mercury levels in fresh and canned bivalve mussels *Tagelus dombeii* (Navajuela) and *Semelle solide* (Almeja) from the Chilean Coast. *Science of the Total Environment*, **148** (1) 1-10.

DFO (Canadian Department of Fisheries and Oceans) (1988) *Fish Inspection Regulations, Amendment List*, June 13, Government of Canada, Ottawa.

Diaz-Barriga, F. (1996) Hazardous residues in Mexico. Evaluation of health risks. *Salud Publica de Mexico*, **38** (4) 280-291.

Dickinson, N.M., Lepp, N.W. and Surtan, G.T.K. (1987) Lead and potential health risks from subsistence food crops in urban Kenya. *Environmental Geochemistry and Health*, **9** 37-42.

Doganoc, D.Z. (1996) Lead and cadmium concentration in meat, liver and kidney of Slovenian cattle and pigs from 1989 to 1993. *Food Additives and Contaminants*, **13** (2) 237-241.

Dorian, C., Gattone, V.H. and Klassen, C.D. (1995) Discrepancy between the nephrotoxic potencies of cadmium-metallothionein and cadmium chloride and renal concentration in the proximal convuluted tube. *Toxicology and Applied Pharmacology*, **130** 161-168.

Earl, F.L. and Vish, J.J. (1978) Teratogenicity of heavy metals, in *Toxicity of Heavy Metals in the Environment*, part 2 (ed. F.W. Oehme), Marcel Dekker, New York, 617-640.

Edel, W., Kremers, G.J., Pieters, J.J.L., Schudeboom, L.J. and Staarink, T. (1983) *Surveillance Programme "Man and Nutrition", Results up to and Including 1980*, Ministry of Welfare and Cultural Affair, Government Prints Office, The Hague.

Elinder, C.-G., Friberg, L., Kjellstrom, T., Nordberg, G.F. and Vouk, V. (1994) *Biological Monitoring of Metals*, World Health Organization, Geneva.

Engel, R.R. and Smith, A.H. (1994) Arsenic in drinking water and mortality from vascular disease: an ecologic analysis in 30 counties in the United States. *Archives of Environmental Health*, **49** (5) 418-427.

EPA (US Environmental Protection Agency) (1985) National drinking water regulations, 40 CFR Part 141. *Federal Register*, **50** 46967.

EPA (US Environmental Protection Agency) (1987) *Special Report on Ingested Inorganic Arsenic: Skin Cancer and Nutritional Essentiality, Risk Assessment Form*, EPA, Washington, DC.

EPA (US Environmental Protection Agency) (1989a) *Air Quality Criteria for Lead*, Addendum, EPA/600/8-89/049, vol. 1, Office of Health and Environmental Assessment, EPA, Washington, DC, pp A1-A67.

EPA (US Environmental Protection Agency) (1989b) *Methyl Mercury (CASRN 22967-926 (02/01/89)*, Integrated Risk Information System, DIALCOM, Washington, DC.

EPA (US Environmental Protection Agency) (1991) *Risk Assessment Guidance for Superfund, Volume I: Human Evaluation Manual; Supplemental Guidance*, "Standard Default Exposure Factors", Interim final, OSWER Directive: 9285.6-03, 25 March, Office of Emergency and Remedial Responses, Toxic Integration Branch, EPA, Washington, DC.

Ewan, R.C. (1978) Toxicology and adverse effects of mineral imbalance with emphasis on selenium and other minerals, in *Toxicity of Heavy Metals in the Environment*, part 1 (ed. F.W. Oehme), Marcel Dekker, New York, pp 445-489.

Exon, J.H., South, E.H. and Hendrix, K. (1996) Effect of metals on the humoral immune response, in *Toxicology of Metals* (eds L.W. Chang, L. Magos and T. Suzuki), CRC Lewis Publishers, Boca Raton, FL, pp 797-810.

FAO (Food and Agriculture Organization) (1989) *Seafood Safety Regulations Applied for Fish by Major Importing Countries*, FAO Fisheries Circular 25, FAO, Rome.

FAO (Food and Agriculture Organization) (1997) *Fish Inspector*, **38** 1-6 (FAO, Rome).

FAO/WHO (Food and Agriculture Organization/World Health Organization) (1972) *Evaluation of Certain Food Additives and of Contaminants Mercury, Lead, and Cadmium*, Joint FAO/WHO Expert Committee on Food Additives, FAO Nutrition Meetings Report Series No. 51, Rome.

Fay, R.M. and Mumtaz, M.M. (1996) Development of a priority list of chemical mixtures occuring at 1188 hazardous waste sites, using the HazDat database. *Food and Chemical Toxicology*, **34** (11–12) 1163-1165.

FDA (US Food and Drug Administration) (1987) *Compliance Policy Guide, Chapter 41—Pesticides, 7141.01*, April 1, FDA, Washington, DC.

Flegal, A.R. and Smith, D.R. (1995) Measurements of environmental lead contamination and human exposure. *Reviews of Environmental Contaminants and Toxicology*, **143** 1-45.

Fouassin, A. and Fondu, M. (1980) Evaluation de la teneur moyenne en plomb et en cadmium de la ration alimentaire en Belgique. *Archives Belges de Medecine Scociete Hygiene Medecine du Travail et Medecine Legale*, **38** 453-467.

Fowler, B.A. (1991) Risk with inorganic contaminants is seafood, in *Proceedings of Symposium on Seafood Safety issues* (cd. F.E. Ahmcd), Institute of Medicine, US National Academy of Sciences,Washington, DC, pp 149-159.

Fowler, B.A. and Woods, J.S. (1977) The transplacental toxicity of methylmercury to fetal rat liver mitochondria: morphometric and biochemical studies. *Laboratory Investigation*, **36** 122-130.

Fowler, B.A., Carmichael, N.G., Squibb, K.S. and Engel, D.W. (1981) Factors affecting trace metal uptake and toxicity to estuarine organisms. II. Cellular mechanisms, in *Biological Monitoring of Marine Pollutants* (eds F.J. Vernberg, E. Calabrese, F.P. Thurberg and W.B. Vernberg), Academic Press, New York, pp 145-163.

Friberg, L., Elinder, C.-G., Kjellström, T. and Nordberg, G.F. (1986) *Cadmium and Health: A Toxicological and Epidemiological Appraisal, General Aspects, Effects and Response*, vols 1 and 2, CRC Press, Boca Raton, FL.

Fullmer, C.S. (1997) Lead-calcium interactions: involvement of 1,25-dihydroxyvitamin D. *Environmental Research*, **72** (1) 45-55.

Galal-Gorchev, H. (1987) *Mercury in Fish Products*, Codex Committee on Food Additives, CXFA 87/18, Addendum 2, FAO/WHO, Geneva.

GAO (US General Accounting Office) (1987) *Superfund: Extent of Nation's Potential Hazardous Waste Problem Still Unknown*, GAO/RCED-88-44, Washington, DC.

Godet, F., Babut, M., Burnel, D., Verber, A.M. and Vasseur, P. (1996) The genotoxicity of iron and chromium in electroplating effluents. *Mutatrion Research*, **370** (1) 19-28.

Goldwater, L.J. (1972) *Mercury, A History of Quicksilver*, York Press, Baltimore, MD.

Gonzalez de Mejia, E. and Craigmill, A.L. (1996) Transfer of lead from lead-glazed ceramics to food. *Archives of Environmental Contamination and Toxicology*, **31** (4) 581-584.

Goyer, R.A. (1995) Toxic effects of metals, in *Casarett & Doull's Toxicology: The Basic Science of Poisons*, 5th edn (eds C.D. Klassen, M.O. Amdur and J. Doull), McGraw-Hill, New York, pp 691-736.

Grobler, S.R., Theunissen, F.S. and Maresky, L.S. (1996) Evidence of undue lead exposure in Cape Town before the advent of leaded petrol. *South African Medical Journal*, **86** (2) 169-171.

Gunderson, F.L. (1995) FDA Total Diet Study, July 1986–April 1991. Dietary intakes of pesticides, selected elements and other chemicals. *Journal of the Association of Official Analytical Chemists*, **78** 1353-1363.

Gzyl, J. (1997) Assessment of Polish population exposed to lead and cadmium with special emphasis to the Katowice Province on the basis of metal concentrations in environmental compartments. *Central European Journal of Public Health*, **5** (2) 93-96.

Gzyl, J., Kucharski, R. and Morchwinska, E. (1984) Evaluation of lead exposure in inhabitants of certain areas of the Legnicka-Glogow copper mining district by indicator method. *Roczniki Panstwowego Zakladu Higieny*, **35** 399-403.

Hallen, I.P., Jorhem, L., Lagerkvist, B.J. and Oskarsson, A. (1995) Lead and cadmium levels in human milk and blood. *Science of the Total Environment*, **166** 149-155.

Hammad, T.A., Sexton, M. and Langenberg, P. (1996) Relationship between blood lead and dietary iron intake in preschool children. A cross sectional study. *Annals of Epidemiology*, **6** (1) 30-33.

Hammond, P.B. and Beliles, R.B. (1980) Metals, in *Casarett and Doull's Toxicology: The Basic Science of Poisons* (eds J. Doull, C.D. Klassen and M.O. Amdur), 2nd edn, Macmillan, New York, pp 409-467.

Harr, J.R. (1978) Biological effects of selenium, in *Toxicity of Heavy Metals in the Environment*, part 1 (ed F.W. Oehme), Marcel Dekker, New York, pp 393-443.

Hartwig, A. (1994) Role of DNA repair inhibition in lead- and cadmium-induced genotoxicity: a review. *Environmental Health Perspectives*, **102** (Suppl. 3) 45-50.

Haschke, F. and Steffan, I. (1981) Lead intake with food of young infants in the years 1980/1981. *Wiener Klinische Wachenschrift*, **93** 613-616.

Hernandez-Avila, M., Gonzalez-Cossio, T., Palazuelos, E., Romieu, I., Aro, A., Fishbein, E., Peterson, K.E. and Hu, H. (1996) Dietary and environmental determinants of blood and bone lead levels in lactating postpartum women living in Mexico city. *Environmental Health Perspectives*, **104** (10) 1076-1082.

Hopenhayn-Rich, C., Biggs, M.L., Fuchs, A., Bergoglio, R., Tello, E.E., Nicolli, H. and Smith, A.H. (1996) Bladder cancer mortality associated with arsenic in drinking water in Argentina. *Epidemiology*, **7** (2) 117-124.

Hurley, L.S. and Keen, C.L. (1987) Manganese, in *Trace Elements in Human and Animal Nutrition*, vol. I (ed. W. Mertz), Academic Press, New York, pp 185-223.

IARC (1976) *Monograph on the Evaluation of Carcinogenic Risks of Chemicals to Man, Vol 11, Cadmium, Nickel, some Epoxides, Miscellaneous Industrial Chemicals and General Consideration of Volatile Anaesthetics*, International Agency for Research on Cancer, Lyon.

IARC (1987) *Monograph on the Evaluation of Carcinogenic Risks of Chemicals to Humans, suppl 7, Overall Evaluation of Carcinogenicity: An Updating of IARC Monographs*, vols 1 to 42, International Agency for Research on Cancer, Lyon.

IARC (1994) *Monograph on the Evaluation of Risks to Humans. Cadmium, Mercury, Beryllium and the Glass Industry*, vol. 58, International Agency for Research on Cancer, Lyon.

Ikeda, M., Watanabe, T., Koizumi, A., Fujita, H., Nakatsuka, H. and Kasahara, M. (1989) Dietary intake of lead among Japanese farmers. *Archives of Environmental Health*, **44** 23-29.

Inskip, M.J. and Piotrowski, J.K. (1985) Review of health effects of methylmercury. *Journal Applied Toxicology*, **5** 113-133.

ISO (International Organization for Standardization) (1981) *Ceramic Ware in Contact with Food—Release of Lead and Cadmium—Part 2: Permissible Limits*, International Standard 6486/2, ISO, Geneva.

Jacobs, R.M. (1996) Techniques employed for the assessment of metals in biological systems, in *Toxicology of Metals* (eds L.W. Chang, L. Magos and T. Suzuki), CRC Lewis Publishers, Boca Raton, FL, pp 69-79.

Jakubowski, M., Trzcinka-Ochocka, M., Razniewska, G., Christensen, J.M. and Starek, A. (1996) Blood lead in the general population in Poland. *International Archives of Occupational and Environmental Health*, **68** (3) 193-198.

Jathar, V.S., Pendarkar, P.B., Raut, S.J. and Panday, V.K. (1981) Intake of lead through food in India. *Journal of Food Science and Technology, India*, **18** 240-242.

Jelinek, C.F. (1992) *Assessment of Dietary Intake of Chemical Contaminants*, United Nations Environmental Programme, Nairobi.

Johnson, B.L. (1995) Nature, extent and impact of Superfund hazardous waste sites. *Chemosphere*, **31** (1) 2415-2428.

Jung, M.C. and Thornton, I. (1997) Environmental contamination and seasonal variation of metals in soils, plants and waters in the paddy fields around a Pb–Zn mine in Korea. *Science of the Total Environment*, **198** (2) 105-121.

Kagey, B.T., Bumgarner, J.E. and Greason, J.P. (1977) Arsenic levels in maternal-fetal tissue sets, in *Trace Substances in Environmental Health XI* (ed. C.D. Hemphill), University of Missouri Press, Columbia, MO, pp 252-256.

Karita, K., Shinozaki, T., Tomita, K. and Yaro, E. (1997) Possible oral lead intake via contaminated facial skin. *Science of the Total Environment*, **199** (1–2) 125-131.

Katz, S.A. and Salem, H. (1993) The toxicology of chromium with respect to its chemical speciation: a review. *Journal of Applied Toxicology*, **13** (2) 217-224.

Kido, J., Honda, R., Tsuritani, I., Yamaya, H., Ishizaki, M., Yamaa, Y. and Nogawa, K. (1988) Progress of renal dysfunction in inhabitants environmentally exposed to cadmium. *Archives of Environmental Health*, **43** 213-217.

Kim, R., Rotnitsky, A., Sparrow, D., Weiss, S., Wager, C. and Hu, H. (1996) A longitudinal study of low-level lead exposure and impairment of renal function: The Normative Aging Study. *Journal of the American Medical Association*, **275** (15) 1177-1181.

Kjellström, T., Ervin, P.-E. and Rahnster, B. (1977) Dose–response relationship of cadmium-induced tubular proteinuria. *Environmental Research*, **13** 303-317.

Kobayashi, J. (1978) Pollution by cadmium and the itai-itai disease in Japan, in *Toxicity of Heavy Metals in the Environment*, part 1 (ed. F.W. Oehme), Marcel Dekker, New York, pp. 190-260.

Landolph, J.R., Dews, M.P., Ozbun, L. and Evans, D.P. (1996) Metal-induced gene expression and neoplastic transformation, in *Toxicology of Metals* (eds L.W. Chang, L. Magos and T. Suzuki), CRC Lewis Publishers, Boca Raton, FL, pp 321-355.

Langard, S. and Norseth, T. (1986) Chromium, in *Handbook on Toxicology of Metals* (eds L. Friberg, G.F. Nordberg and V.B. Vouk), Elsevier, Amsterdam, pp 185-210.

Larsson, B., Slorach, S.A., Hagman, V. and Hofrander, Y. (1981) WHO collaborative breast feeding study II. Levels of lead and cadmium in Swedish human milk. *Acta Pediatrica Scandinavica*, **70** 281-284.

Laryea, M.D., Schnittert, B., Kersting, M., Wilhelm, M. and Lombeck, I. (1995) Macro-nutrietnt, copper and zinc intakes of young German children as determined by duplicate food samples and diet records. *Annals of Nutrition and Metabolism*, **39** (5) 271-278.

Lauwerys, R.R., Roles, H.A., Bernard, A. and Bucket, J.P. (1980) Renal response to cadmium in a population living in an ferrous smelter area in Belgium. *International Archives of Environmental Health Perspectives*, **45** 271-274.

Lechner, W., Battista, H.J. and Dienstel, F. (1980) The lead content of human milk in high and low traffic areas of Tirol. *Gynakologische Rdschau*, **20** (Suppl. 2) 268-270.

Lewis, R. (1997) Researchers' death inspires actions to improve safety. *The Scientist*, **11** (21) 1,4.

Lonnerdal, B. (1995) Magnesium nutrition of infants. *Magnesium Research*, **8** (1) 99-105.

Lorenz, H., Ocker, H.D., Bruggemann, J., Weigert, P. and Sonneborn, M. (1986) Cadmium contents in cereal samples in the past—a comparison with the present. *Zeitschrift für Lebensmittel-Untersuchung und -Forschung*, **183** (6) 402-405.

Louekari, K., Jolkkonen, L. and Varo, P. (1987) Exposure to cadmium from foods, estimated by analysis and calculation, comparison of methods. *Food Additives and Contaminants*, **5** 111-117.

MacGregor, A. (1975) Analysis of control methods: mercury and cadmium pollution. *Environmental Health Perspectives*, **12** 137-148.

Mackay, D. (1991) *Multimedia Environmental Models: The Fugacity Approach*, Lewis Publishing, Chelsea, MI.

Mackay, N.J., Williams, R.J., Kacprzac, J.L., Kazacos, M.N., Collins, A.J. and Auty, E.H. (1975) Heavy metals in cultivated oysters (*Crassostrea commercialis* and *Saccostrea cucullata*) from the estuaries of New South Wales. *Australian Journal of Marine and Freshwater Research*, **26** 31-46.

Mahaffey, K.R., Capar, S.G., Gladen, B.C. and Fowler, B.A. (1981) Concurrent exposure to lead, cadmium and arsenic: effects on toxicity and tissue metal concentration in the rat. *Journal Laboratory and Clinical Medicine*, **98** 463-487.

Mandic, Z., Mandic, M.L., Grgic, Z., Klape, T., Primorac, L. and Hasenay, D. (1997) Copper and zinc content in human milk in Croatia. *European Journal of Epidemiology*, **13** (2) 185-188.

Marchwinska, E., Kucharski, R. and Gzyl, J. (1984) Cadmium content of precipitating dust, soil and vegetables grown in the upper Silesian industrial region. *Roczniki Panstwowego Zakladu Higieny*, **35** 23-28.

Martyn, C.N., Barker, D.J.P., Osmond, C. and Harris, E.C. (1989) Geographical relation between Alzheimer's disease and aluminum in drinking water. *Lancet*, **1** 59-62.

Marzec, Z. and Bulinski, R. (1988) Evaluation of cadmium, mercury and lead intake with reproduced diet. *Roczniki Panstwowego Zakladu Higieny*, **39** 344-348.

McKenzie-Parnell, J.M. (1984) Bioavailability of trace elements in foodstuffs and beverages, in *Changing Metal Cycles and Human Health* (ed. J.O. Nriagu), Springer-Verlag, New York, pp 187-198.

McKenzie-Parnell, J.M., Kjellström, T.E., Sharma, R.P.-E. and Robinson, M.F. (1988) Unusually high intake and fecal output of cadmium, and fecal output of other trace elements in New Zealand adults consuming dredge oysters. *Environmental Research*, **46** 1-14.

Meloan, C.E. (1978) Quantitative analysis of environmental and biological contamination of heavy metals, in *Toxicity of Heavy Metals in the Environment*, part 2 (ed. F.W. Oehme), Marcel Dekker, New York, pp 797-944.

Menkes, D.B. and Fawcett, J.P. (1997) Too easily lead? Health effects of gasoline additives. *Environmental Health Perspectives*, **105** (3) 270-273.

Menzer, R.E. (1991) Water and soil pollution, in *Casarett & Doull's Toxicology: The Basic Science of Poisons*, 4th edn (eds C.D. Klassen, M.O. Amdur and J. Doull), McGraw-Hill, New York, pp 872-902.

Merrill, R.A. (1995) Regulatory toxicology, in *Casarett & Doull's Toxicology: The Basic Science of Poisons*, 5th edn (eds C.D. Klassen, M.O. Amdur and J. Doull), McGraw-Hill, New York, pp 1011-1023.

Miller, R.G., Kopfler, F.C., Keltry, K.C., Stober, J.A. and Ulmer, N.S. (1984) The occurrence of aluminum in drinking water. *Journal of the American Water Works Association*, **77** 84-91.

Ming, L.O., Surif, S. and Abdullah, A. (1997) Lead exposure among Malaysian School children using delta-aminolevulinic acid as an indicator. *Science of the Total Environment*, **193** (3) 207-213.

Mistry, P., Lucier, G.W. and Fowler, B.A. (1985) High affinity lead-binding proteins from rat kidney cytosol: mediate cell-free nuclear translocation of lead. *Journal of Pharmacology and Experimental Therapy*, **232** 462-469.

MMWR (Mortality & Morbidity Weekly Report) (1997) Update: blood lead levels—United States, 1991–1994. *MMWR*, **46** (7) 141-146.

Moon, C.S., Zhang, Z.W., Shimbo, S., Watanabe, T., Moon, D.H., Lee, C.U., Lee, B.K., Ahn, K.D., Lee, S.H. and Ikeda, M. (1995) Dietary intake of cadmium and lead among the general population in Korea. *Environmental Research*, **71** (1) 46-54.

Moore, F.R., Urda, G.A., Krishna, G. and Theiss, J.C. (1996) Genotoxicity evaluation of selenium sulfide in in vivo and in vivo/in vitro micronucleus and chromosome aberration assays. *Mutation Research*, **367** (1) 33-41.

Moore, J.W. (1990) *Inorganic Contaminants of Surface Water*, Springer-Verlag, New York.

Morgan, H., Smart, G.A. and Sherlock, J.C. (1988) The Shipham report. An investigation into cadmium contamination and its relationship for human health. Intakes of metal. *Science of the Total Environment*, **75** (1) 71-100.

Muller, J. and Schmidt, E.H.F. (1983) Heavy metals in the infant diet, in *Health Evaluation of Heavy Metals in Infant Formula and Junior Food* (eds E.H.F. Schmidt and A.G. Hildbrandt), Springer-Verlag, New York, pp 1-12.

Muller, M. and Anke, M. (1995) Investigation into the oral lead exposure of adults in the former German Democratic Republic. *Zeitschrift für Lebensmittel-Untersuchung und -Forschung*, **200** (1) 38-43.

Mykkanen, H., Rasanen, L., Ahola, M. and Kimppa, S. (1986) Dietary intakes of mercury, lead, cadmium and arsenic by Finnish children. *Human Nutrition: Applied Nutrition*, **40A** 32-39.

Needleman, H.L., Schell, A., Bellinger, D., Leviton, A. and Allred, E. (1990) Long-term effects of childhood exposure to lead at low dose: an eleven-year follow-up report. *New England Journal of Medicine*, **322** 83-88.

Needleman, H.L., Riess, J.A., Tobin, M.J., Biesecker, G.E. and Greenhouse, J.B. (1996) Bone lead levels and delinquent behavior. *Journal of the American Medical Association*, **275** (5) 363-369.

Neely, N., Gilles Pie, D., Schauf, F. and Walsh, J. (1981) *Remedial Actions of Hazardous Waste Sites, Survey and Case Studies*, Report 43019-81-05, Environmental Protection Agency, Washington, DC.

Ng, A. and Patterson, C. (1981) Native concentrations of lead in anscient Arctic and Antarctic ice. *Geochimica et Cosmochim Acta*, **45** 2109-2121.

NOAA (National Oceanic and Atmospheric Administration) (1989) *State-Issued Fish Consumption Advisories: A National Perspective* (ed. D. Zeitlin), National Ocean Pollution Program Office, NOAA, Washington, DC.

Nogawa, K. and Kido, T. (1996) Itai-Itai disease and health effects of cadmium, in *Toxicology of Metals* (eds. L.W. Chang, L. Magos and T. Suzuki), CRC Lewis Publishers, Boca Raton, FL, pp 363-369.

Nogawa, K., Kawano, S., Kato, T. and Sakamoto, M. (1983) The prevalence of itai-itai disease and the mean cadmium concentration in rice produced by individual villages. *Nippon Eiseigaku Zaschi*, **37** 843-847.

Nogawa, K., Honda, R., Kido, T., Tsuritani, I., Yamada, Y., Ishizaki, M. and Yamaha, H. (1989) A dose–response analysis of cadmium in the general environment with special reference to total cadmium intake limit. *Environmental Research*, **48** 7-16.

Nordberg, G.F. (1972) Cadmium metabolism and toxicity. *Environmental Physiology and Biochemistry*, **2** 7-36.

Nordberg, G.F. (ed.) (1976) *Effects and Dose–Response Relationship of Toxic Metals*, Elsevier Scientific, Amsterdam.

NRC (National Research Council) (1980) *Recommended Dietary Allowance*, 9th edn, US National Academy of Sciences, National Academy Press, Washington, DC.

NRC (National Research Council) (1989) *Diet and Health: Implications for Reducing Chronic Disease Risk*, National Academy Press, Washington, DC.

NRC (National Research Council) (1993) *Measuring Lead Exposure in Infants, Children and Other Sensitive Populations*, National Academy Press, Washington, DC.

Nriagu, J., Jinabhai, C., Naidoo, R. and Coutsoudis, A. (1996) Atmospheric lead pollution in Kwa Zula/Natal, South Africa. *Science of the Total Environment*, **19** (1–2) 69-76.

Oehme, F.W. (1978) Mechanisms of heavy metals inorganic toxicities. *Toxicity of Heavy Metals in the Environment*, Part 1, Marcel Dekker, New York, pp 69-85.

Omokhodion, F.O. (1994) Blood lead and tapwater levels in Ibadan, Nigeria. *Science of the Total Environment*, **151** (3) 187-190.

Oskarsson, A., Palminger-Hallen, I. and Sundberg, I. (1995) Exposure to toxic elements via breast milk. *Analyst*, **120** (3) 765-770.

Ou, Y.C., Thompson, S.A., Kirchner, S.C., Kavanagh, T.J. and Faustman, E.A. (1997) Induction of growth arrest and DNA damage inducible genes Gadd45 and Gadd153 in primary rodent embryonic cells following exposure to methylmercury. *Toxicology and Applied Pharmacology*, **147** 31-38.

Pelletier, L. and Druet, P. (1995) Immunotoxicology of metals, in *Handbook of Experimental Pharmacology*, vol. 115 (eds R.A. Goyer and M.G. Cherian), Springer-Verlag, Heidelberg, pp 77-92.

Pennington, J.A., Young, D.E., Wilson, D.B., Johnson, R.D. and Vanderveen, J.E. (1986) Mineral content of foods and total diets in the selected minerals in Food Survey, 1982–1984. *Journal of the American Dietetic Association*, **86** 876-891.

Perl, D.P. (1985) Relationship of aluminum to Alzheimer's disease. *Environmental Health Perspectives*, **63** 149-153.

Pesheva, M.G., Chankova, S.G., Avramova, Ts V., Malanov, D.V. and Genova, G.K. (1997) Genotoxic effects of cadmium chloride in various test-systems, *Genetika*, **32** (2) 183-188.

Poirier, L.A. and Littlefield, N.A. (1996) Metal interactions in chemical carcinogens, in *Toxicology of Metals* (eds L.W. Chang, L. Magos and T. Suzuki), CRC Lewis Publishers, Boca Raton, FL, pp 289-298.

Porru, S. and Alessio, L. (1996) The use of chelating agents in occupational lead poisoning. *Occupational Medicine*, **46** (1) 41-48.

Powell, J.J., Greenfield, S.M., Thompson, R.P., Cargrello, J.A., Kendall, M.D., Landsberg, J.P., Watt, F., Delves, H.T. and House, I. (1995) Assessment of toxic metal exposure following the Camelford water pollution incident: evidence of acute mobilization of lead into drinking water. *Analyst*, **120** (3) 793-798.

Radisch, B., Luck, W. and Nau, H. (1987) Cadmium concentration in milk and blood of smoking mothers. *Toxicology Letters*, **36** 147-1520.

Rao, K.S. and Rao, G.V. (1995) Aluminum leaching from utensils—a kinetic study. *International Journal of Food Science and Nutrition* **46** (1) 31-38.

Ratcliffe, H.E., Swanson, G.M. and Fischer, L.J.C. (1996) Human exposure to mercury: a critical assessment of the evidence of adverse health effects. *Journal of Toxicology and Environmental Health*, **49** (3) 221-270.

Reilly, C. (1985) The dietary significance of adventitious iron, zinc, copper and lead in domestically prepared food. *Food Additives and Contaminants*, **2** 209-215.

Rhainds, M. and Levallois, P. (1997) Effect of maternal cigarette smoking and alcohol consumption on blood lead levels of newborns. *American Journal of Epidemiology*, **145** (3) 250-257.

Romieu, I., Lacasana, M. and McConnell, R. (1997) Lead exposure in Latin America and the Carribean. Lead Research Group of the Pan-American Health Organization. *Environmental Health Perspectives*, **105** (4) 398-405.

Rubenowitz, E., Axelsson, G. and Rylander, R. (1996) Magnesium in drinking water and death from acute myocardial infarction. *American Journal of Epidemiology*, **143** (5) 456-462.

Russell, L.H., Jr. (1978) Heavy metals in foods of animal origin, in *Toxicity of Heavy Metals in the Environment*, part 1 (ed. F.W. Oehme), Marcel Dekker, New York, pp 3-23.

Saleh, M.A., Ragab, A.A., Kamel, A., Jones, J. and el-Sebae, A.K. (1996) Regional distribution of lead in human milk from Egypt. *Chemosphere*, **32** (9) 1859-1867.

Samuelson, G., Bratteby, L.E., Berggren, K., Elverby, J.E. and Kempe, B. (1996) Dietary iron intake and iron status in adolescents. *Acta Paediatrica*, **85** (9) 1033-1038.

Sapunar-Postruznik, J., Bazulik, D., Kubala, H. and Bulint, L. (1996) Estimation of dietary intake of lead and cadmium in the general population of the Republic of Croatia. *Science of the Total Environment*, **177** (1–3) 31-35.

Schamberger, R.J. (1985) The genotoxicity of selenium. *Mutation Research*, **154** 29-48.

Schrauzer, G.N., White, D.A. and Schneider, C.J. (1977) Cancer mortality correlation studies. IV. Association with dietary intakes and blood levels of certain trace elements, notably Se antagonists. *Bioinorganic Chemistry*, **7** 35-56.

Schumacher, M., Belles, M., Rico, A., Domingo, J.L. and Corbellaj (1996) Impact of reduction of lead in gasoline on the blood and hair lead levels in the population of Tarragona Province, Spain, 1990–1995. *Science of the Total Environment*, **184** (3) 203-209.

Sheets, R.W. (1997) Extraction of lead, cadmium and zinc from overglaze decoration on ceramic dinnerware by acidic and basic food substances. *Science of the Total Environment*, **197** (1–3) 167-175.

Shen, X., Rosen, J.F., Guo, D. and Wu, S. (1996) Childhood lead poisoning in China. *Science of the Total Environment*, **181** (2) 101-109.

Sherlock, J.S. (1984) Cadmium in the foods and the diet. *Experientia*, **40** 152-156.

Sherlock, J.C., Smart, G., Forbes, G.I., Moore, M.R., Patterson, W.J., Richards, W.N. and Wilson, T.S. (1982) Assessment of lead intakes and dose-response for population in Ayr exposed to a plumbosolvent water supply. *Human Toxicology*, **1** 115-122.

Sherlock, J.C., Pickford, C.J. and White, G.F. (1986) Lead in alcoholic beverages. *Food Additives and Contaminants*, **3** 347-354.

Shukla, S.S. and Leland, H.V. (1973) Heavy metals: a review of lead. *Journal of the Water Pollution Control Federation*, **45** 1319-1331.

Skerfving, S. (1988) Mercury in women exposed to methylmercury through fish consumption, and in their newborn babies and breast milk. *Bulletin of Environmental Contamination and Toxicology*, **41** 475-482.

Skoczynska, A. (1997) Lipid peroxidation as a toxic mode of action for lead and cadmium. *Medycyna Pracy*, **48** (2) 197-203.

Slorach, S., Gustafson, T.-B., Jorhem, L. and Mattson, M. (1983) Intakes of lead, cadmium and certain other metals via a typical Sweedish weekly diet. *Var Foeda* (Suppl.) **1** 3-16.

Sternowsky, H.J. and Wessolowski, R. (1985) Lead and cadmium in breast milk. Higher levels in urban vs rural mothers during the first 3 months of lactation. *Archives of Toxicology*, **57** 41-45.

Stromberg, V., Schutz, A. and Skerfving, S. (1995) Substantial decrease of blood lead in Swedish Children, 1978–1994, associated with petrol lead. *Occupational and Environmental Medicine*, **52** (11) 764-769.

Sunanda, L., Sumathi, S. and Venkatasubbaiah, V. (1995) Relationship between soil zinc, dietary zinc and zinc nutritional status of humans. *Plant Foods for Human Nutrition*, **48** (3) 201-207.

Szymczak, J., Regulska, B., Ilow, R. and Biernat, J. (1984) Cadmium, lead and mercury levels in meals from canteens for young people. *Roczniki Panstwowego Zakladu Higieny*, **35** 328-332.

Szymczak, J., Regulska, B., Ilow, R. and Zechalko, A. (1987) Cadmium, lead and mercury levels in daily food rations of children aged 1–7 years. *Roczniki Panstwowego Zakladu Higieny*, **35** 328-332.

Tahoven, R. and Kumpulainen, J. (1995) Lead and cadmium contents in milk, cheese and eggs on the Finnish market. *Food Additives and Contaminants*, **12** (6) 789-798.

Tam, G.K.H., Charbonneau, S.M., Byrce, F. and Sandi, E. (1982) Excretion of a single oral dose of fish-arsenic in man. *Bulletin of Environmental Contamination and Toxicology*, **28** 669-673.

Tengs, T.O., Adams, M.E., Pliskin, J.S., Safran, D.G., Siegel, J.E., Weinstein, M.C. and Graham, J.D. (1995) Five-hundred life-saving interventions and their cost-effectiveness. *Risk Analysis*, **15** (3) 369-389.

Todd, A.C., Wetmur, J.G., Moline, J.M., Godbold, J.H., Levin, S.M. and Landrigan, P.J. (1996) Unraveling the chronic toxicity of lead: an essential priority for environmental health. *Environmental Health Perspectives*, **104** (Suppl. 1) 141-146.

Tollefson, L. and Cordle, F. (1986) Methylmercury in fish: review of residue levels, fish consumption and regulatory action in the United States. *Environmental Health Prespectives*, **68** 203-208.

Trachtenbarg, D.E. (1996) Getting the lead out: when is treatment necessary? *Postgraduate Medicine*, **99** (3) 201-202, 207-218.

Underwood, E.J. (1978) Interactions of trace elements, in *Toxicity of Heavy Metals in the Environment*, part 2 (ed. F.W. Oehme), Marcel Dekker, New York, pp 641-668.

UNEP (United Nations Environment Programme) (1992) *The Contaminants of Food*, Nairobi.

UNEP/FAO/WHO (1983) *Assessment of the Present State of Pollution by Mercury in the Mediterranean Sea, and Proposed Control Measures*, UNEP/WG9119, Athens.

UNEP/FAO/WHO (1988) *Assessment of Chemical Contaminants in Food*, Report of the Results of UNEP, Food and FAO and WHO Programme on health-related environmental monitoring, Geneva.

USDA (US Department of Agriculture) (1987) Nationwide Food Consumption Surveys, Continuing Survey of Food Intakes of Individuals, Women 19–50 Years and Their Children, 1986, Report No. 86-1, Nutrition Monitoring Division, Human Nutrition Information Service, Hyattsville, MD.

Vahter, M. and Slorach, S. (1990) *Exposure Monitoring of Lead and Cadmium: An International Pilot Study within the WHO/UNEP Human Exposure Assessment Location (HEAL) Programme*, WHO and UNEP, Nairobi.

Vilagines, R. and Leroy, P. (1995) Lead in drinking water, determination of its concentration and effects of new recommendations of the World Health Organization (WHO) on public and private network management. *Bulletin de l'Academie Nationale de Médecine*, **179** (7) 1393-1408.

Waalkes, M.P. and Rehm, S. (1994) Cadmium and prostate cancer. *Journal of Toxicology and Environmental Health*, **43** 251-269.

Walsh, C.T., Sandstead, H.H., Prasad, A.S., Newberne, P.M. and Fraker, P.J. (1994) Zinc: Health effects and research priorities in the 1990's. *Environmental Health Perspectives*, **102** (Suppl. 2) 5-46.

Wahlström, B. (1989) Sunset for dangerous chemicals. *Nature*, **341** 276.

Watanabe, T., Cha, C.W., Song, D.B. and Ikeda, M. (1987) Pb and Cd levels among Korean populations. *Bulletin of Environmental Contamination and Toxicology*, **38** 189-195.

Watanabe, T., Nakatsuka, H., Shimbo, S., Iwami, O., Imai, Y., Moon, C.S., Zhang, Z.W., Iguchi, H. and Ikeda, M. (1996) Reduced cadmium lead burden in Japan in the past 10 years. *International Archives of Occupational and Environmental Health*, **68** (5) 305-314.

Watt, G.C., Britton, A., Gilmour, W.H., Moore, M.R., Murray, G.D., Robertson, S.J. and Womersley, J. (1996) Is lead in tap water still a public health problem? An observational study in Glasgow. *British Medical Journal*, **313** (7063) 979-981.

Weyermann, M. and Brenner, H. (1997) Alcohol consumption and smoking habits as determinants of blood lead levels in a national population sample from Germany. *Archives of Environmental Health*, **52** (3) 233-239.

WHO (World Health Organization) (1980) Tin and organotin compounds: a preliminary review. *IPCS Environmental Health Criteria*, **15**, WHO, Geneva.

WHO (World Health Organization) (1989) Toxicological evaluation of certain food additives and contaminants. *WHO Additive Series* **24**, Cambridge University Press, Cambridge.

WHO (World Health Organization) (1992) Cadmium. *IPCS Environmental Health Criteria*, **134**, WHO, Geneva.

Wilhelm, M., Lombeck, I., Kouros, B., Wuthe, J. and Ohnesorge, F.K. (1995) Duplicate study on the dietary intake of some metals/metaloids by children in Germany. Part II. Aluminum, cadmium and lead. *Zentralblatt für Hygiene und Umweltmedizin*, **197** (5) 357-369.

Winge, D.R. and Mehra, R.K. (1990) Host defences against copper toxicity. *International Review of Experimental Pathology*, **31** 47-83.

Winship, K.A. (1987) Toxicology of antimony and its compounds. *Advances in Drug Reactions Acute Poisoning Review*, **2** 67-90.

Wojciechowska-Mazurek, M., Zawadzka, T., Karlowski, K., Starska, K., Cwiek-Ludwicka, K. and Brulinska-Ostrowska, E. (1995) Content of lead, cadmium, mercury, zinc and copper in fruit from various regions in Poland. *Roczniki Panstwowego Zakladu Higieny*, **46** (3) 223-238.

Yang, G.Q., Qian, P.C., Zhu, L.Z., Huang, J.H., Liu, S.J., Lu, M.D. and Gu, L.Z. (1987) Human selenium requirement in China, in *Selenium in Biology and Medicine* (eds G.F. Combs, Jr, O.A. Levander, J.E. Spallholz and J.E. Oldfield), Van Nostrand Reinhold, New York, pp 589-607.

Yang, H., Zou, Z. and Jin, C. (1995) Monitoring of lead content in foodstuff in China. *Chinese Journal of Preventive Medicine*, **29** (2) 96-98.

Zhang, Z.W., Moon, C.S., Watanabe, T., Shimbo, S. and Ikeda, M. (1996) Lead content of rice collected from various areas in the world. *Science of the Total Environment*, **191** (1–2) 169-175.

Zhang, Z.W., Moon, C.S., Watanabe, T., Shimbo, S., He, F.S., Wu, Y.Q., Zhou, S.F., Su, D.M., Qu, J.B. and Ikeda, M.C. (1997) Background exposure of urban populations to lead and cadmium: comparison between China and Japan. *International Archives of Occupational and Environmental Health*, **69** (4) 273-281.

7 Pesticides

Jana Hajšlová

7.1 Introduction

7.1.1 What pesticides are and why they are used

Plants, the world's main source of food, are affected by different pests and by competition from weeds. Devastation of crops may be caused by up to 100 000 viruses, bacteria, mycoplasma-like organisms, fungi, algae and parasitic higher plants. From the 800 000 known species of insects, about 10 000 cause extensive economic losses, and approximately 30% of the 3000 known nematodes commonly attack crop plants. Almost 1800 of 30 000 weeds around the world pose serious concerns in crop production (Holland, 1996). The degree of damage varies greatly depending on climate and agricultural region; substantial problems result from the introduction of new plant species and cultivars in plantation and cash-crop farming as well as from new monocultures. It has been estimated that ~10–30% of the world's food crops are destroyed owing to pest and plant diseases; losses are much higher in emerging countries (Ware, 1989).

As a result of scientific advance, 'pesticides', chemical agents that may be classified according to the target pest to be damaged (e.g. insecticides, fungicides, molluscicides, bactericides, herbicides), have been developed as extremely important aids to world agricultural production over the last five decades. The internationally adopted definition of the Food and Agriculture Organization of the United Nations (FAO) (Edwards, 1986) is as follows: 'Pesticide means any substance or mixture of substances intended for preventing, destroying, attracting, repelling or controlling any pest including unwanted species of plants or animals during the production, storage, transport, distribution and processing of food, agricultural commodities, or animal feeds or which may be administered to animals for the control of ectoparasites. The term includes substances intended for use as a plant growth regulator, defoliant, desiccant, fruit thinning agent, or sprouting inhibitor and substances applied to crops either before or after transport. The term normally excludes fertilizers, plant and animals nutrients, food additives and animal drugs.' A glossary of terms related to pesticides has been published by the International Union of Pure and Applied Chemistry (IUPAC) (FAO, 1995).

Chemical control of pests and weeds has become an important and, in some circumstances, indispensable measure leading not only to increased yields of crops but also to improved qualitative characteristics of food and feed. Apart from use in agriculture, horticulture, forestry and livestock production, public health programmes are the most important other area for pesticide use; for example, vector-borne diseases can be controlled by insecticides and molluscicides.

7.1.2   History of pesticides

Since the dawn of civilization, man has been opposed by ravages wrought by pests and crop diseases in his efforts to produce adequate supplies of food. The idea of combating these by the use of chemicals is not new (Cremlyn, 1978). The fumigant value of burning sulphur was widely appreciated in ancient Greece and Rome. Similarly, the use of soda and olive oil for the treatment of legume seeds and the application of arsenic against field pests were also recorded at that period. In medieval times, the salts of arsenic were employed by the Chinese. Nicotine contained in extracts of tobacco, pyrethrum and soap also represent examples of historical pesticides.

The first systematic scientific studies into the use of chemicals for crop protection were initiated in the middle of the nineteenth century. Investigation of the use of new asrsenic compounds led to the introduction (1867) of Paris green (impure copper arsenite) as an insecticide. This agent was later used successfully in the United States to check the spread of the Colorado beetle. At that time (1896), another valuable chemical treatment for the control of pathogenic fungi such as potato blight or vine mildew, Bordeaux mixture (copper sulphate, lime and water), was discovered in France. It was soon noticed (1896) that it caused the leaves of yellow charlock to turn black in the vicinity of the application. This observation is probably the origin of the idea of selective herbicides. Several other inorganic substances had been shown to kill weeds selectively without damaging the crop. Among other landmarks, the introduction of the first organomercury seed dressings in the second decade of the twentieth century (1913) should be noted.

The 1930s can be considered as the real beginning of the modern era of synthetic organic pesticides (Holland, 1996). Important examples of compounds introduced are: dinitro-*ortho*-cresol (1932), controlling weeds in cereals; thiram (1934), the first of the dithiocarbamate fungicides; pentachlorophenol (1936), a wood preservative against fungi and termites; and TEPP, the first organophosphorus insecticide. In 1939, the powerful insecticidal properties of DDT were discovered and this compound became the most widely used single insecticide in the world;

following the 'success' of DDT, several other organochlorine compounds were found to be potent contact insecticides. Next was the discovery of the first hormone herbicides, phenoxyacetic acids MCPA and 2,4-D. In 1945 soil-acting carbamate herbicides were discovered, and the invention of insecticidal carbamates followed shortly afterwards. In the period 1950–55, many groups of pesticides that are still in use became available. Herbicidal urea derivatives were developed, and important fungicides such as captan and glyodin appeared on the market. Research on organophosphorus compounds resulted in the introduction of the broad-spectrum insecticide malathion which, in contrast to previous compounds of this group, had a relatively low mammalian toxicity. Between 1955 and 1960, the newcomers included herbicidal triazines and quaternary ammonium herbicides. In the following period, the most important second group of systemic fungicides appeared, represented by benomyl and, soon afterwards, the soil-acting herbicide glyphosate was discovered. The new insecticides tetramethrin, resmethrin and bioresmethrin, synthetic pyrethroids with greater activity than natural pyrethrins, became available. The introduction of a new generation of pesticide groups (herbicidal sulphonylureas, and new systemic fungicides such as metalaxyl and triadimefon) continued during the next decades.

Analytical work on pesticide residues in the diet began in the 1960s, but, based on the alarming new findings on environmental hazards resulting from its use, DDT and some other persistent organochlorine insecticides (aldrin, dieldrin) were banned from use all over the world at the beginning of the 1970s. The discovery of some subtle long-term effects of some other pesticides in the environment on non-target organisms, including human beings, resulted in cancellation of registration of several other pesticides in many countries (e.g. toxaphene, dinoseb and others).

Improved knowledge of host–pest interactions is reflected by new strategies employed in the design of modern pesticides and their formulations. At the same time, new methods of pesticide application are being developed by many research institutions around the world, aiming to reduce the risk of pesticide poisoning.

7.2   Main classes of pesticides

Several criteria, mainly the target pest and the chemical structure, can be used for classification of pesticides (WHO, 1990a; Hassall, 1990). The most common approach is to combine functional and chemical classification. Table 7.1 shows the main groups of pesticides, listing representatives of particular groups.

Table 7.1 Classification of pesticides—overview of main groups

Group according to **Target pest**, *application and action* and chemical class	Common name of pesticides representing particular group[a]
Insecticides	
Organochlorines	Aldrin, dieldrin, DDT, dicofol, chlordane, endrin, HCH, heptachlor, methoxychlor, toxaphene
Non-systemic	
Organophosphorus compounds	Azinphos methyl, diazinon, dichlorvos, fenitrothion, malathion, mevinphos, parathion ethyl, parathion methyl, pirimiphos-methyl, tetrachlorvinphos
Carbamates	Carbaryl, methomyl, propoxur
Systemic	
Organophosphorus compounds	Dimethoate, disulfoton, phorate, phosphamidon, trichlorfon, vamidothion
Carbamates	Aldicarb, carbofuran, pirimicarb, methomyl, oxamyl
Synthetic pyrethroids	Allethrin, bifenthrin, bioresmethrin, cyfluthrin cyhalothrin, cypermethrin, deltamethrin, fluvalinate
Specific acaricides	
Non-fungicidal	
Organochlorine	Chlorobenzilate, tetradifon
Organotin	Cyhexatin
Fungicidal	
Dinitro compounds	Binapacryl, dinocap
Molluscicides	
Aquatic	
Botanical	Endod
Other chemical	Niclosamide, sodium pentachlorophenate, triphenmorphe, tributyltin, trifenyltin
Terrestrial	
Carbamates	Aminocarb, mexacarbate, methiocarb
Others	Mataldehyde
Protectant fungicides	
Non-systemic	
Dithiocarbamates	Mancozeb, maneb, metiram, propineb, thiram, zineb
Phthalimides	Captan, folpet, captafol, dichlofluanid
Dinitro compounds	Binapacryl
Organomercurials	Phenylmercury
Organotin compounds	Fentin (acetate and hydroxide)
Chlorine-substituted aromatics	Chlorothalonil, dichlone, dicloran, quintozene (PCNB), tecnazene (TCNB)
Cationic detergents	Dodine acetate, glyodin acetate
Others	Iprodione, procymidone
Eradicant fungicides	
Systemic	
Antibiotics	Blasticidin, cyclohexamide, kasugamycin, streptomycin
Benzimidazoles	Benomyl, thiophanate-methyl, thiabendazole
Morpholines	Dodemorph, tridemorph
Pyrimidines	Ethirimol, bupirimate

Table 7.1 (Continued)

Group according to **Target pest**, *application and action* and chemical class	Common name of pesticides representing particular group[a]
Eradicant fungicides	
Systemic	
Piperazines	Triforine
Others	Benomyl, carboxin, thiabendazole, thiophanate-methyl
Herbicides	
Foliar applicaiton, systemic or translocated	
Phosphonoamino acids	Glyphosate, glufosinate
Benzoic acids	Chlorfenprop-methyl, dicamba, 2,3,6-TBA
Chlorinated aliphatic acids	Dalapon, TCA
Oxyphenoxy acid esters	Cycloxydim, diclofop-methyl, fenoxaprop, fluazifop-butyl, haloxyfop-methyl, quizalofop-ethyl
Phenoxyalkanoic acids	2,4-D, 2,4-DB, dichlorprop, mecoprop, MCPA, MCPB, Silvex, 2,4,5-T
Quaternary ammonium compounds (bipyridiliums)	Diquat, paraquat
Foliar application, contact	
Benzonitriles	Bromoxynil, dichlobenil, ioxynil
Benzothiadiazoles	Bentazon
Carbanilates	Phenmedipham
Cyclohexenones	Clethodim, cycloxydim, sethoxydim
Dinitrophenols	Dinoseb
Diphenyl ethers	Acifluorfen, lactofen, nitrofen, oxyfluorfen
Soil application	
Acetanilides	Alachlor, butachlor, metolachlor, propachlor
Amides and anilides	Benzoylprop-ethyl, diphenamid, naptalam, pronamide, propanil
Carbanilates and carbamates soil	Asulam, barban, bendiocarb, carbetamide, chlorpropham, propham, triallate
Dinitroanilines	Benefin, phendimethalin, trifluralin
Pyridazinones and pyridinones soil	Amitrole, dimethazone, fluridone, norflurazon, oxadiazon, pyrazon
Pyridinoxy and picolinic acids	Clopyralid, fluroxypyr, picloram, triclopyr
Phenylureas or substituted ureas	Diuron, fenuron, fluometuron, linuron, monolinuron, siduron
Sulphonylureas	Chlorimuron-ethyl, chlorsulfuron, metsulfuron-methyl, sulfometuron-methyl, thiameturon-methyl
Thiocarbamates	Butylate, cycloate, EPTC, molinate, pebulate, thiobencarb, triallate
Triazines	Ametryn, atrazine, cyanazine, desmetryne, hexazinone, methoprotryne, metribuzin, prometon, propazine, terbutrizine, terbutryne
Uracils or substituted uracils	Bromacil, lenacil, terbacil
Dessicants, Defoliants	
Organophosphates	Merphos, DEF
Phenol derivatives	Dinoseb, PCP
Quaternary ammonium compounds (bipyridiliums)	Diquat, paraquat

Table 7.1 (Continued)

Group according to **Target pest**, *application and action* and chemical class	Common name of pesticides representing particular group[a]
Plant growth regulators	
Growth promoters	
Auxins	2,4-D, MCPB, NAD
Cytokinins	BA, PBA, adenine, kinetin
Gibberellins	Gibane, GA3
Ethylene generators	Ethephon
Inhibitors and retardants	
Quaternary ammonium compounds	Chlormequat, mepiquat
Hydrazides	Maleic hydrazide, daminozide
Triazoles	Paclobutrazol, uniconazole
Rodenticides	
Fumigants	Aluminium phosphide, calcium cyanide, chloropicrin, methyl bromide
Anticoagulants	
Hydroxycoumarins	Brodifacoum, difenacoum, coumafuryl, caumatetralyl, warfarin
Indandiones	Chlorophacinone, diphacinone, pindone
Non-coagulants	
Arsenicals	Arsenious oxide, sodium arsenite
Benzenamine	Bromethalin
Botanicals	Red squill, strychnine
Thioureas	Antus promurit
Others	Fluoroacetamide, fluoroacetate, norbormide sodium, zinc phosphide

[a]Most chemical pesticides have common names agreed by the International Organization for Standardization through its Technical Committee 81 (ISO/TC 81). The principles for coining these common names are explained in ISO 257:1988.

7.2.1   Pesticide formulations

With the exception of substances used as fumigants, pesticides are rarely used or applied in pure form. Technical grade compounds representing the active ingredients of commercial products must be subjected to various grinding, mixing or dissolving processes with the aim of making a usable form for direct application or for dilution and subsequent application. In general terms, formulation is the processing of a pesticide compound by any method that will improve its properties of storage, handling, application, effectiveness or safety. The formulation may be sold under the formulator's trade name or it may be custom-formulated for another company.

Table 7.2 is a summary of common pesticide formulations (Holland, 1996; Hassall, 1990). It should be noted that the effective state of an

Table 7.2 Common formulations of pesticides

Formulation	Type
Sprays (insecticides, herbicides, fungicides)	Emulsifiable concentrates; water-miscible liquids; wettable powders; water-soluble powders; oil solutions; soluble pellets for water-hose attachments; flowable or sprayable suspensions; flowable microencapsulated suspensions; ultralow-volume (ULV) concentrates
Dusts (insecticides, fungicides)	Undiluted toxic agent; toxic agent with active diluent; toxic agents with inert diluent; aerosol dust
Aerosols (insecticides, repellents, disinfectants)	Pushbutton; total release
Granules (insecticides, herbicides, algicides)	Inert carrier impregnated with pesticide soluble granules, e.g. dry flowable herbicides
Fumigants (insecticides, nematicides, herbicides)	Stored products and space treatment, e.g. liquids, gases, moth crystals; soil treatment liquids that vaporize
Impregnates (insecticides, fungicides, herbicides)	Polymeric materials containing a volatile insecticide; polymeric materials containing non-volatile insecticides; shelf papers containing a contact insecticide; mothproofing agents for woollens
Slow-release insecticides	Microencapsulated; adhesive tapes for pest control operators and homeowners; resin strips containing volatile organophosphate fumigant
Fertilizer combinations with herbicides, insecticides, or fungicides	
Baits (insecticides, molluscicides, rodenticides, and avicides)	

active ingredient at the time of application has little to do with whether it is solid or liquid in the pure state. It should be also observed that the state in which an active ingredient is applied may be different from its state at the moment of uptake by the target organism.

7.3   Physicochemical properties of pesticides related to their post-application fate

Many groups of chemical compounds are employed for crop protection (Table 7.1). A comprehensive data set of basic characteristics (nomenclature, physical chemistry, commercialization, applications, analysis, mammalian toxicology, ecotoxicology and environmental fate) for most modern pesticides (and related agents in temporary use) in agriculture can be found in *The Pesticide Manual* (Tomlin, 1997) which has been issued for 30 years by the British Crop Protection Council. These compounds possess a wide range of physicochemical properties due to their diverse chemical structures. This is well illustrated in Table 7.3, which characterizes 'modern' pesticides with a high incidence in food crops (for more details see Table 7.7).

Several other monographs and databases (Wauchope *et al.*, 1992; Augustijn-Beckers *et al.*, 1994; Heller and Herner, 1990; Hornsby *et al.*, 1996; Howard, 1991) on properties and environmental fate of pesticides, including 'classic' persistent organochlorine compounds, are available. Distinct differences between the properties of compounds of this group (which were banned in almost all countries) and those that are still in use for crop protection are well documented (Verschueren, 1983). High values of bioconcentration factor (BCF) reflect their low degradability and highly hydrophobic nature (Table 7.4).

Relationships between some characteristics of pesticides and their behaviour both in the environmental compartments and during food processing are summarized below (Hartley and Graham-Bryce, 1980).

Water solubility
Solubility values provide considerable insight into the fate of pesticides in the environment as well as into losses during processing of residue-containing crops. Highly soluble pesticides are likely to be easily removed by water from the surface of the plant and in soil they do not tend to partition. Such (polar) compounds are more likely to biodegrade, their rate of hydrolysis and oxidation typically being higher. Washout from the atmosphere by rain and fog is also influenced by solubility. On the other hand, substances possessing good solubility are less likely to volatilize from water (the Henry's law constant should be considered for prediction of the rate of pesticide evaporation from water). Thus, such losses from aqueous solutions are not significant.

Vapour pressure
Volatilization from water is also influenced by the vapour pressure of the pesticide. Considering the form in which the pesticide will be transported in the atmosphere, those substances with vapour pressures less than $\sim 110^{-7}$ mPa will be associated mostly with particulate matter. In food processing (drying, milling, baking, roasting, etc.), substances with high vapour pressures will be lost to some extent due to volatilization. The extent of these losses depends both on the character of the matrix and the form of the residue. Volatilization of ionized compounds does not occur.

Octanol–water partition coefficient (K_{ow})
The octanol–water partition coefficient has been shown (Kenaga, 1980) to correlate well with adsorption of pesticides to soil and sediments. Unless other factors operate (such as rapid biodegradation), compounds with

high K_{ow} values, such as organochlorine compounds listed in Table 7.4, tend to bioaccumulate in the fat portion of organisms. K_{ow} (reported usually as log P_{ow}) was chosen (FAO, 1997) to represent the solubility of a pesticide in fat. A compound is designated as fat-soluble when log P_{ow} exceeds 4 (with several exceptions) and is not so designated when P_{ow} is less than 3. Interpretation varies between values of 3 and 4. High K_{ow} is known to be typical for persistent compounds that are largely (bio)accumulated in the fat portion of organisms. The ability of a compound to enter the cuticule of plants (the distribution of residues in plants) is also indicated by the K_{ow} value.

Dissociation constant (K_a)
The acid constant, as the negative log (pK_a), is given for pesticides that are likely to dissociate under environmental pH conditions (values between 5 and 8). Almost a third of active ingredients in current use are capable of ionization (see phenols, carboxylic acids and amines in Table 7.1). The degree of ionization affects such processes as photolysis (the absorption spectrum of the ionized form may be different from that of the non-ionized one), solubilization, evaporation from water (the ionic form of residue is more water soluble and does not evaporate from water), etc. Soil or sediment adsorption and bioconcentration are also influenced by the degree of ionization.

Soil sorption coefficient (K_{oc})
The soil organic carbon sorption coefficient K_{oc} is often reported (McCall *et al.*, 1980) to document soil sorption, which is the main process affecting pesticide pollution potential. It is calculated by measuring the ratio K_d of the sorbed to soluble concentrations of the pesticide after equilibration of the pesticide in a water–soil slurry and then dividing by the weight fraction of organic carbon in soil, F_{oc}: $K_{oc}=K_d/F_{oc}$. Pesticides that are strongly sorbed (adsorbed or absorbed) by soil particles are likely to be more persistent because they are protected from degradation and volatilization by the binding. Sorbed pesticides will not easily leach to ground water; they will wash off the field only under erosive conditions (attached to moving soil particles).

Bioconcentration factor (BCF)
BCF is reported for hydrophobic chemicals (Table 7.4) having a tendency to partition from the water column and bioconcentrate in aquatic organisms. BCF is given by the concentration of the chemical in the organism at equilibrium divided by the concentration of the chemical in water. The correlation with physical properties such as K_{ow} was mentioned earlier.

Table 7.3 Some important physicochemical characteristics of common pesticides

Pesticide	Chemical Abstracts Reference No.	IUPAC name	Molecular mass	Water solubility (mg/l)(°C)	Vapour pressure (mPa)(°C)	log K_{ow}[a]
Acephate	30560-19-1	O,S-Dimethyl acetylphosphoramidothioate	183.2	790 000 (20)	0.226 (24)	−0.89
Benomyl	17804-35-2	Methyl 1-(butylcarbamoyl) benzimidazol-2-yl carbamate	290.3	4 (25)	< 0.0049 (25)	1.37
Bioresmethrin	28434-01-7	5-Benzyl-3-furylmethyl (1R,3R)-2,2-dimethyl-3-(2-methylprop-1-enyl) cyclopropanecarboxylate	338.4	<0.3 (25)	18.6 (25)	>4.70
Captan	133-06-2	N-(Trichloromethylthio) cyclohex-4-ene-1,2-dicarboximide	300.6	3.3 (25)	<1.3 (25)	2.79
Carbaryl	63-25-2	1-Naphthyl methylcarbamate	201.2	120 (20)	0.041 (23.5)	1.59
Carbendazim	10605-21-7	Methyl benzimidazol-2-ylcarbamate	191.2	29 (pH 4) (24)	0.09 (20)	1.38 (pH 5)
Chlorothalonil	1897-45-6	Tetrachloroisophthalo nitrile	265.9	0.81 (25)	0.076 (25)	2.89
Chlorpyrifos	2921-88-2	O,O,-Diethyl-O-3,5,6-trichloro-2-pyridyl phosphorothioate	350.6	1.4 (25)	2.7 (25)	4.70
Chlorpyrifos-methyl	5598-13-0	O,O-Dimethyl-O-3,5,6-trichloro-2-pyridyl phosphorothioate	322.5	2.6 (20)	5.6 (25)	4.24
Cyhalothrin	68085-85-8	(RS)-α-Cyano-3-phenoxybenzyl-(Z)-(1RS, 3RS)-(2-chloro-3,3,3-trifluoropropenyl) 2,2-dimethylcyclopropane carboxylate	449.9	0.000004 (20)	0.001 (20)	6.8
Cypermethrin	52315-07-8	(RS)-α-Cyano-3-phenoxybenzyl-(1RS,3RS;1RS,3SR)-3-(2,2-dichloro-vinyl-2,-2-dimethylcyclopropane carboxylate	416.3	0.004	0.00023 (20)	6.6

Name	CAS number	Chemical name				
Deltamethrin	52918-63-5	(S)-α-Cyano-3-phenoxybenzyl (1R,3R)-3-(2,2-dibromovinyl)-2,2-dimethylcyclopropane carboxylate	505.2	<0.0002 (25)	0.0133 (25)	4.6
Diazinon	333-41-5	O,O-Diethyl-O-2-isopropyl-6-methylpyrimidin-4-yl phosphorothioate	304.3	60 (20)	12 (25)	3.3
Dichlorvos	62-73-7	2,2-Dichlorovinyl dimethyl phosphate	221.0	8000 (25)	2100 (25)	1.9
Dicofol	115-32-2	2,2,2-Trichloro-1,1-bis(4-chlorophenyl)ethanol	370.5	0.8 (25)	0.053	4.28
Dimethoate	60-51-5	O,O-Dimethyl-S-methylcarbamoylmethyl phosphorodithioate	229.2	23 300 (pH 5) (20)	1.1 (25)	0.70
Endosulfan	115-29-7	(1,4,5,6,7,7-Hexachloro-8,9,10-trinorborn-5-en-2,3-ylenebismethylene) sulphite	406.9	0.32 (22)	0.83 (20)	4.74
Fenitrothion	122-14-5	O,O-Dimethyl-O-4-nitro-m-tolyl phosphorothioate	277.2	21 (20)	18 (20)	3.43
Fenvalerate	51630-58-1	(RS)-α-Cyano-3-phenoxybenzyl (RS)-2-(4-chlorophenyl)-3-methylbutyrate	419.9	<0.01 (25)	0.0192 (20)	5.01
Folpet	133-07-3	N-(Trichloromethylthio) phthalimide	296.6	1 (20)	1.3 (25)	3.11
Malathion	121-75-5	Diethyl (dimethoxythio phosphorylthio) succinate	330.3	145 (25)	5.3 (30)	2.75
Parathion	56-38-2	O,O-Diethyl-O-4-nitrophenyl phosphorothioate	291.3	11 (20)	0.89 (20)	3.83

Table 7.3 (Continued)

Pesticide	Chemical Abstracts Reference No.	IUPAC name	Molecular mass	Water solubility (mg/l)(°C)	Vapour pressure (mPa)(°C)	log K_{ow} [a]
Permethrin	52645-53-1	3-Phenoxybenzyl-(1RS,3RS;1RS, 3SR)-3-(2,2-dichlorovinyl)-2, 2-dimethylcyclopropane carboxylate	391.3	0.2 (20)	0.07 (20)	6.10
Pirimiphos-methyl	29232-93-7	O,O-Dimethyl-O-2-diethylamino- 6-methylpyrimidin-4-yl phosphorothioate	305.3	9.9 (pH 5.2) (30)	2 (20)	4.20
Vinclozolin	50471-44-8	(RS)-3-(3,5-Dichlorophenyl)- 5-methyl-5-vinyl-1,3-oxazolidine-2, 4-dione	286.1	2.6 (20)	0.016 (20)	3.00

[a] n-Octanol–water partition coefficient.

Table 7.4 Persistent organochlorine hydrocarbons and other priority pollutants

Pesticide	Chemical Abstracts Reference No.	Water solubility (mg/l) (20°C)	Vapour pressure (mPa) (20°C)	K_{ow}	K_{oc}	Half-life in soil (days)	Half-life in air (days)	log BCF (fish)
pp-DDT[a]	50-29-3	1.70×10^{-3}	1.50×10^{-4}	1.00×10^{6}	2.40×10^{5}	700–5000	3	4.468
pp-DDE[a]	72-55-9	1.30×10^{-3}	1.80×10^{-5}	5.80×10^{5}	1.00×10^{6}	NA	NA	NA
HCB[a]	118-74-1	5.00×10^{-3}	2.50×10^{-3}	2.00×10^{5}	1.40×10^{4}	1300	627	4.342
Toxaphene[a]	8001-35-2	5.50×10^{-1}	8.90×10^{-4}	6.60×10^{4}	1.00×10^{5}	300–5000	NA	4.522
Aldrin[a]	309-00-2	2.70×10^{-2}	1.00×10^{-5}	3.20×10^{6}	6.70×10^{3}	53	0.16	4.029
Dieldrin[a]	60-57-1	1.90×10^{-1}	4.00×10^{-4}	2.10×10^{4}	NA	2550	2.2	4.100
Lindane	58-89-9	7.40	3.00×10^{-3}	4.00×10^{3}	1.00×10^{3}	400	7.9	2.505
Pentachlorophenol	87-86-5	1.90×10^{-2}	7.00×10^{-3}	1.30×10^{5}	NA	NA	23	2.892
Atrazine	1912-24-9	33	4.00×10^{-5}	5.60×10^{2}	41	60	0.1	1.000

[a]Not registered for use in agriculture.
K_{ow}, n-octanol-water partition coefficient.
K_{oc}, soil sorption coefficient.
BCF, bioconcentration factor.
NA, not available.

Other information about pesticides, e.g. hydrolytic stability of pesticides in aqueous solutions (expressed as half-lives), has been reported (Tomlin, 1989). However, the experimental conditions (pH, temperature) under which measurements were carried out are often very different and sometimes these data are not comparable. It is important to note that relatively large differences occasionally exist in reported physicochemical constants, depending on the source of information. Moreover, only limited information is available on the post-application fate of compounds that have been introduced recently to the market.

7.4 Input of pesticides into the environment: targeted application

In contrast with other environmental contaminants, the input of pesticides into the environment occurs under controlled conditions that should conform to 'Good Agricultural Practice' (GAP). This term is applied (Holland, 1996) to the use of pesticides under nationally approved conditions that are necessary for effective and reliable pest control. In general, GAP may encompass a range of levels of pesticide applications up to the highest authorized use, applied in a manner that leaves the smallest amount of residue practicable.

To hit the insects or fungi living on or in crop plants, the cuticle of leaves is treated with appropriate pesticide preparations. In principle (Hassall, 1990), the insect cuticle and/or fungal hyphal walls and spore walls represent a direct target, an intermediate barrier that must be penetrated before the pesticide reaches the particular site of action. Similarly, herbicides have to overcome this barrier to reach the respective objective in the plant cell (Kerney and Kaufman, 1988; Ahrens, 1994).

As already mentioned, pesticides of different kinds are applied to foliage for different reasons; consequently the degree of penetration that is desirable also varies. While pesticides with systemic action penetrate, from deposited gases or particulate matter, right through the cuticle and are transported around the plant, quasi-systemic substances (insecticides, fungicides) exhibit only local internal movement in cuticular wax, exhibiting persistent contact action. Some substances remain after treatment as superficial deposits possessing entirely local contact action.

Application of pesticides may of course be directed to soil (Hassall, 1990; Hance, 1984). The physicochemical properties of this intermediate target are clearly very different from those of waxy cuticular surfaces. Systemic compounds are taken predominantly into the roots of plants and are translocated to the above-ground parts, where they exert toxic effects. They often tend to be transported towards regions of growth, giving protection to the vital and vulnerable tissue near meristems.

Plants may also absorb pesticides via roots from the soil. The uptake is followed by a subsequent translocation of such systemic pesticides throughout the leaves and fruits. The amount of pesticide taken from the soil is limited by the amount of bioavailable compound, which for non-polar substances is negatively related to the K_{oc} value compared with the positive correlation between plant uptake into roots and K_{ow}. It is estimated (Briggs, 1982) that the optimum for plant uptake from soil is c. log $K_{ow} \sim 0.5$ and for uptake from aqueous solutions is log $K_{ow} \sim 1.8$.

7.5   The fate of pesticides in the environment after application

Once applied at a site for pest control, the pesticidal compound is exposed to many agents capable of transforming it into various other forms. After entering both target and non-target biota, pesticides are subjected to attack by detoxification enzymes the effects of which will be discussed later. However, the major proportion of the applied pesticide does not immediately enter any organism, remaining in soil, water and air where it is subjected to further transformation and transport to different locations as well as uptake by organisms at that site (Fuhr, 1982). Figure 7.1 is a simplified scheme illustrating the various processes to which pesticides applied for plant protection are subjected (direct application to soil and/or to plant surfaces is the main route of pesticide input to the environment). Basic aspects of the fate of these contaminants, in particular environmental compartments and organisms, is discussed briefly below with respect to food chain contamination.

7.5.1   Agents involved in pesticide degradation

In general, factors responsible for pesticide degradation can be classified as (i) chemical, (ii) physical and (iii) biological. Under field conditions the breakdown of pesticides is influenced by a combination of these factors, and their relative importance depends to a great extent on the use pattern of the chemical, its physical properties and its chemical structure (section 7.5.2 and 7.5.3).

Light and heat are the physical agents of primary importance. Photolysis of residues on the vegetation, on the soil surface or in water contribute significantly to pesticide dissipation (Zepp, 1991; Crosby, 1985). Direct photoreactions account for only a part of sunlight-induced reactions. Other reactions involving photochemically produced reactive transients such as hydroxyl, hydroperoxyl/superoxide, organoperoxyl and other radicals as well as singlet molecular oxygen may influence the

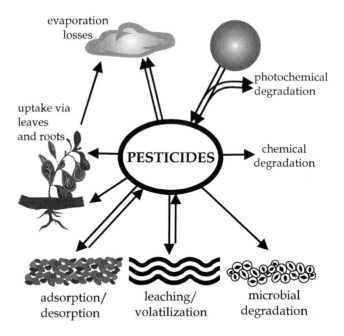

Figure 7.1 The fate of pesticides after application.

fate of agrochemicals in the environment. Thermal decomposition can also be induced by solar radiation.

Chemical reactions are caused by various agents (Gäb *et al.*, 1985), mainly by water. Breakdown of most pesticides occurs in aqueous solution, especially in conjunction with extremes of pH. Oxidation products are often generated via reaction with oxygen and its reactive forms such as ozone, superoxides and peroxides. Chemical oxidations as well as reductions can progress in the presence of inorganic (mostly metallic) redox reagents.

Microorganisms such as bacteria, fungi and actinomycetes represent the most important group of pesticide degraders in soil and water. The main degradation strategies are as follows (Racke and Coats, 1990):

- *Co-metabolism*: biotransformation of pesticide proceeds simultaneously with normal metabolic pathways of microbial cell.
- *Catabolism*: pesticide is utilized as a nutrient or energy source by microorganisms, mainly bacteria that have adapted (following repeated exposure) to utilize the pesticide molecule as a source of carbon or nitrogen.

Plants, invertebrates and vertebrates are further degradation agents. The latter group possesses the most sophisticated enzymatic system capable of biodegrading xenobiotics. These systems are most effective in birds and mammals; the spectrum of transformation reactions is very broad and the rates of detoxification and elimination are typically high (section 7.5.1.1).

7.5.1.1 Transformation of pesticides in biota

All living organisms possess defence mechanisms protecting them from small quantities of diverse kinds of foreign compounds including pesticides (Aizawa, 1982). Accumulation, which results in attainment of the toxic concentration at the site of action, occurs when the respective substance enters an organism more quickly than it can be eliminated (with or without metabolism). Numerous anatomical, physiological and biochemical factors interact to determine the rate of penetration and distribution of a particular pesticide in the living organism, the rate and route of metabolism and the mechanism of excretion of the foreign substance.

Most substances used as pesticides are poorly soluble in water and their oxidation and/or hydrolysis results in insertion or uncovering of polar groups (Khan, 1982, 1996). Following such biotransformation within Phase I (for which hydrolases and oxygenases are commonly responsible), primary metabolites often, but not always, undergo secondary changes in a Phase II whereby they are conjugated with small polar endogenous molecules to yield more excretable products. The nature of the conjugating substance is characteristic of the respective organism. For instance, glucuronic acid is the most typical endogenous conjugant in mammals, birds and some fish (glycine is an alternative in some species), while glucose is common for higher plants and some invertebrates. In higher animals, transportation of pesticide metabolites in the bloodstream takes place, which in turns leads to their eventual excretion. In spite of some differences, the metabolism of pesticides in invertebrates is analogous to that in plants. In the latter case, it has been suggested (Lawrence and McLean, 1991) that lignin, in addition to being a support material, serves as an excretory system for storage of toxic or unwanted foreign compounds in plants.

There are many possible ways in which a pesticide can be degraded in the biotic environment. As an example (Hassall, 1990), Figure 7.2 shows three major routes by which the widely used herbicide atrazine may be metabolized. It is noticeable that in some plants a special 'catalyst', benzoxazinone (2,4-dihydroxy-7-methoxy-1,4-oxazine-3-one glucoside) is present that facilitates non-enzymic hydrolysis of chlorotriazines yielding hydroxytriazines having no herbicidal activity (in contrast to mono-dealkylated products, which retain some activity). As a second example,

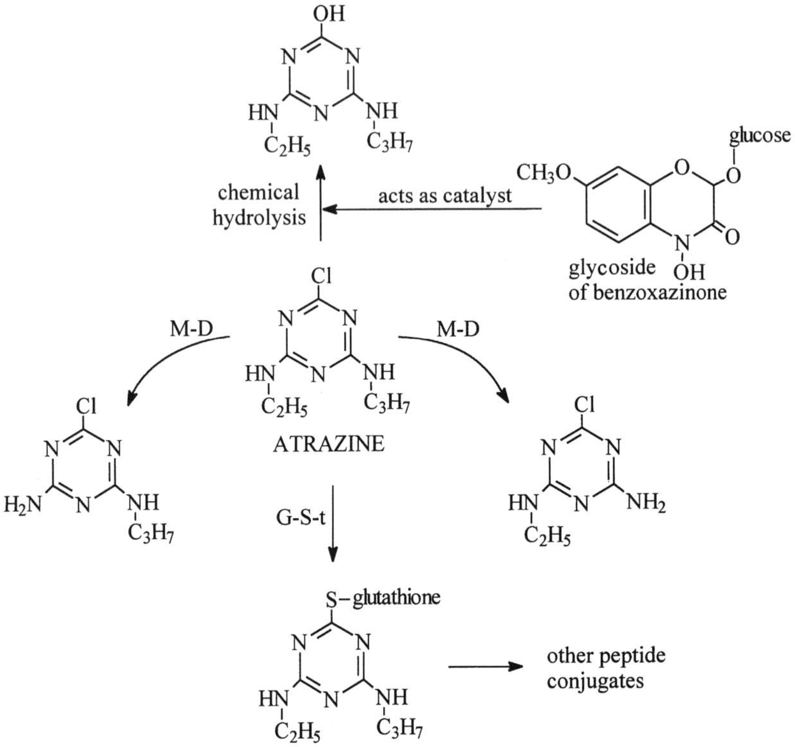

Figure 7.2 Metabolism of atrazine (Hassall, 1990).

Figure 7.3 shows the principal routes of initial metabolism of carbaryl . Oxidation and hydrolytic mechanisms yield various polar primary metabolites.

It has to be emphasized that metabolism has a fundamental role in relation to selectivity of pesticide action and thus plays an important part in determining the safety of these substances to man and other non-target organisms.

7.5.1.2 *Pesticide transformation products, character of residues*

In general, pesticide transformation is a detoxification process resulting in inert substances. Nevertheless, some products may also be toxic to the target pest. The following represent examples of parent pesticides and pesticides formed by their degradation: DDT–dicofol; acephate–methamidophos; benomyl–carbendazim; aldicarb–aldoxycarb.

Of special importance are 'activation' reactions yielding products of greater significance than that of the parent pesticide (Coats, 1991). The most typical examples are organophosphorus insecticides, which are

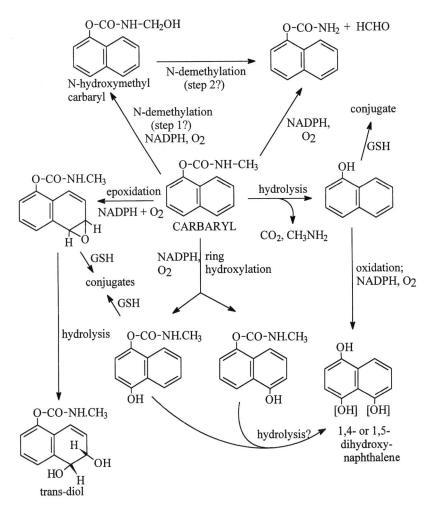

Figure 7.3 Metabolism of carbaryl (Hassall, 1990). The '?' at various reaction steps indicates that detailed experimental data are not available. The course of the reaction is deduced on the basis of theoretical assumptions.

activated *in vivo* to potent acetylcholinesterase inhibitors, e.g. the conversion of parathion to paraoxon requiring loss of sulphur ($P=S$ to $P=O$) (reaction 1).

$$C_2H_5O-\overset{\overset{S}{\|}}{\underset{\underset{C_2H_5O}{|}}{P}}-O-\!\!\!\left\langle\right\rangle\!\!\!-NO_2 \xrightarrow[O_2]{NADPH} C_2H_5O-\overset{\overset{O}{\|}}{\underset{\underset{C_2H_5O}{|}}{P}}-O-\!\!\!\left\langle\right\rangle\!\!\!-NO_2 \qquad (1)$$

parathion   paraoxon

Compounds that are designed for activation as a result of a degradative reaction are called proinsecticides. For instance, carbofuran derivatives were designed to be hydrolysed to carbofuran in insects, but not in mammalian systems providing a good example of selectivity (reaction 2).

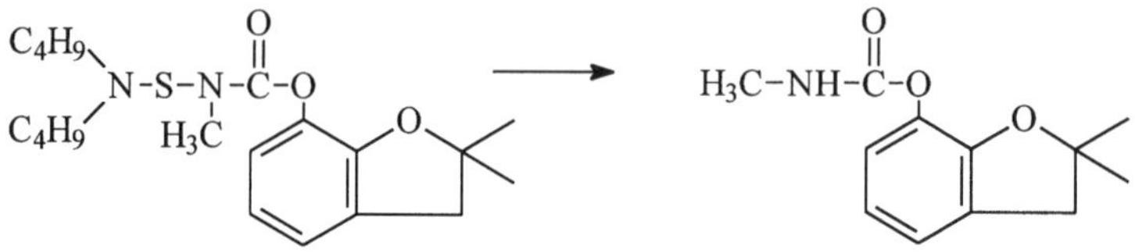

$$\text{carbosulfan} \longrightarrow \text{carbofuran} \tag{2}$$

Under certain conditions, including cooking/processing (section 7.6), some parent compounds may be transformed into mutagens or carcinogens. The best-known example is formation of ethylene thiourea (ETU) from ethylene-bisdithiocarbamates (EBDCs), which are widely used against pathogenic fungi in many crops (reaction 3).

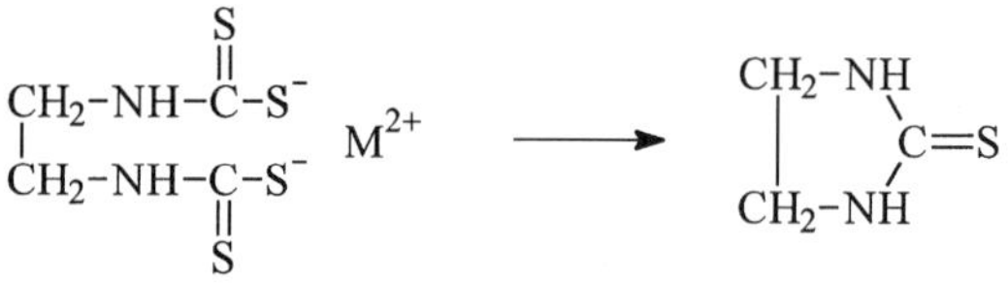

$$\text{EBDC} \longrightarrow \text{ETU} \tag{3}$$

The discovery that unsymmetrical dimethylhydrazine (UDMH) is formed from daminozide, a plant growth regulator/ripening agent (active ingredient of Alar), resulted in it being banned for agricultural use (reaction 4).

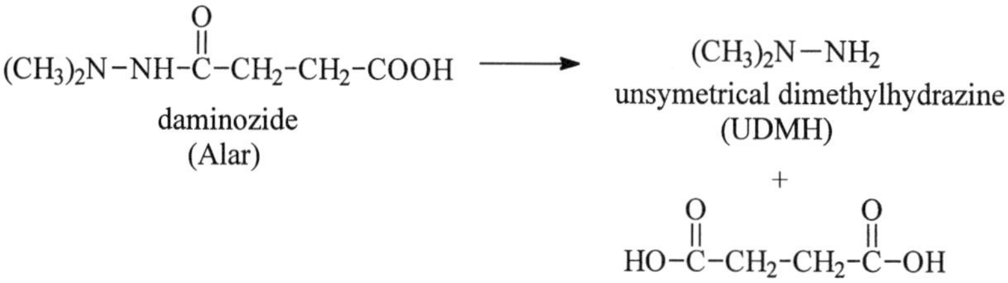

$$(CH_3)_2N-NH-\overset{\overset{\displaystyle O}{\|}}{C}-CH_2-CH_2-COOH \longrightarrow (CH_3)_2N-NH_2 \tag{4}$$

daminozide (Alar)

unsymetrical dimethylhydrazine (UDMH)

$+$

$$HO-\overset{\overset{\displaystyle O}{\|}}{C}-CH_2-CH_2-\overset{\overset{\displaystyle O}{\|}}{C}-OH$$

Several pesticide degradation products are relatively persistent and may be of concern as contaminants of environmental resources (Somasundaram and Coats, 1991). Primary metabolites of herbicidal triazines, namely dealkylated species (Figure 7.2), chlorinated phenols originating during (both biotic and abiotic) transformation of lindane, pentachlorphenol, and phenoxyalkanoic acids are just a few examples of degradation products often occurring in ground water. Major degradation products of DDT such as DDE and DDD are typical examples of contaminants that may bioaccumulate in the human food chain.

To assess environmental significance chiefly in terms of the bioavailability of pesticide residues in soil and plants, two groups of residues should be distinguished (Khan, 1996; Stratton and Wheeler, 1986): (i) those extractable with solvents and (ii) those not extractable with solvents. The first group involves both freely extractable residues and conjugates bound to natural components in plants; the latter group corresponds to residues (parent compounds or their degradation products) that are either firmly retained by the soil organic fraction or bound in the pesticide–plant complex. The true nature of bound residues in soil and plants is still poorly understood. Utilization of radiolabelled tracer compounds for studies of pesticide fate led to the detection and quantitation of residues undetectable by any conventional analytical technique. According to these types of study (Khan, 1991) the possible soil/plant burden of total residues has been rather underestimated. Recent studies indicate that absorption of bound residues from soils by plants may occur. It has been suggested that the ratio of uptake of residues from freshly treated soils to that from soil containing bound residues is about 5:1.

7.5.2 Behaviour of pesticides in soil

Pesticides are designed to be applied mainly on terrestrial areas. Thus, they enter the soil either by direct treatment or by being washed off plant surfaces during rainfall. Apart from volatilization from plant surfaces (Willis and McDowel, 1987), this is the main transport route for the major part of the applied formulations. Depending on the phenotype and the density of the plant culture, it is estimated that on average 35–50% of the plant protection material is deposited on soil immediately after spraying (Lake, 1977).

The ecochemical behaviour of pesticides in soil has already been mentioned. Basically, it involves residence, movement and metabolism (Somasundaram and Coats, 1991; Hill and Arnold, 1978; Ballschmitter, 1992). The formation of residues in soil is mainly dependent on the extent of processes such as (i) microbial metabolism; (ii) metabolism by other organochemical systems such as higher soil organisms and plants, including formation of soluble conjugates; (iii) degradation by photolysis; and (iv) other abiotic processes such as hydrolysis and oxidation. In general terms, the persistence of residues in soil is promoted by their low water solubility and the high binding capacity of both organic and inorganic constituents. Soil residues taken up by resistant plants (Wagenet and Hutson, 1991) represent one of the potential sources of food chain contamination. Contamination of ground water due to leaching (Gustafson, 1989) of some mobile pesticides (e.g. s-triazines) and/or their degradation products, and thus the potential risk of occurrence of pesticides in drinking water, also has to be considered.

7.5.3   Transfer of pesticides to the atmosphere

The widespread application of pesticides in particulate sprays and dusts ensures that appreciable contamination of the air is a consequence. A great deal of what is applied may never reach the target; a portion of material will drift as droplets/particles or in gaseous form. It has been estimated roughly that 10–20% of pesticide material could be subjected to either short-range or long-range drift. The type of formulation, spraying technology and physicochemical parameters of pesticides, as well as climatic factors and the geometry of the target culture play important roles in loss to the atmosphere (Scheunert and Klein, 1985).

Post-application volatilization (Spencer et al., 1973) from treated fields represents further significant pesticide input into the lower atmosphere, but over a much longer period. Physicochemical properties of pesticides, particular soils and other environmental variables play an important role in volatilization from soil (Jury et al., 1983). Similarly, transfer from water may occur (Mackay et al., 1975). In any case, long-range transport of some pesticides occurs via the atmosphere (in the form of vapours and/ or sorbed on solid particles): e.g. persistent chlorinated pesticides and related compounds such as HCB, toxaphene, DDT, dieldrin, penta-chlorophenol, aldrin, endosulfan and lindane are typical. Thus, contamination of food chains at remote places may occur due to emissions elsewhere.

7.6   Effects of processing/cooking procedures on residues

A comprehensive review of pesticide changes during various processing treatments of plant crops has been published by the IUPAC Commission on Agrochemicals (Holland et al., 1994). Changes of pesticide residues in commercially processed fruits and vegetables were summarized by Elkins (1989). In general, the procedures used in industrial food processing and domestic cooking have often been shown to have dramatic effects on residue levels. Large decreases in residues normally occur, as shown in a recent study (Schattenberg et al., 1996) to determine the effects of normal household preparation on pesticides in many food commodities that was initiated by the findings of a committee of the US National Academy of

Sciences (NAS) indicating that Environmental Protection Agency (EPA) tolerances may be too high for infants and small children. Earlier, the NRC (1993) had initiated a screening study targeting 24 common pesticides in 17 types of raw fruit and vegetables. Of 243 samples, 97 (40%) contained residues above the detection limit. After normal processing, this number dropped to 47 (19%). Residue levels were reduced significantly by 29–98%.

However, residues may occasionally concentrate in some food fractions. The ratio of the residue in the processed commodity to the residue in the raw commodity is called the concentration (reduction) factor. If this ratio is greater than 1, the residue is said to concentrate upon processing/cooking. For values equal to or less than 1, no concentration occurs. In addition to these quantitative changes, processing/cooking can also lead to the formation of breakdown products that may be of toxicological concern (section 7.7).

Both commercial processes and typical domestic cooking practices are characterized in greater detail below since, for the process of risk assessment, pesticide levels in food 'on the plate' are essential (section 7.9.2.1).

7.6.1 Drying

Moisture losses during drying may lead to apparently elevated residues in a commodity. Whether this process is considered to result in 'concentration' of residues depends on its nature. It should be noted that Codex Maximum Residue Limits (MRLs, section 7.9) have been assigned for some secondary food commodities of plant origin such as artificially dried fruits, herbs and vegetables as well as for secondary food commodities of animal origin such as artificially or naturally dried meat and fish products. Generally, higher temperatures during drying promote volatilization of surface residues. In addition, solar radiation can cause photolysis/transformation of sensitive pesticides. Mild conditions typical of freeze drying or vacuum drying do not result in extensive reduction of residues.

7.6.2 Peeling, hulling and trimming

Many processing/cooking procedures start with removal of the outer surfaces of the commodity. These operations often result in substantial reduction of residues (frequently by one order of magnitude), especially when pesticides were applied (pre- or post-harvest) directly on the crop; the majority of insecticides and fungicides undergo only very limited movement or penetration through the cuticle. Thus, residues of non-systemic pesticides on citrus fruits, bananas, melons, pineapples, kiwi fruit and similar crops are almost completely eliminated by peeling. However, some major crops such as apples or tomatoes may be consumed whole so that there is no reduction of residues. Removal of the outer leaves (e.g. of lettuce, cabbage) or thick parts of the stem (e.g. of broccoli, asparagus) from a crop food before cooking will frequently result in residues of non-systemic pesticides considerably lower than those that occur in the commodity as it moves in commerce. Anderegg and Madisen (1983) demonstrated that removal of seven outer leaves from cabbage treated with several cyhalothrin sprays resulted in 99% reduction of residues. In peas and some beans, the edible part is protected by the pod, which is usually discarded, thus the risk of occurrence of residues is lower.

Peelings and trimmings obtained as by-products from commercial operations may contain significantly higher residues compared with the original whole commodity. This fact should be taken into account in terms of possible bioaccumulation of lipophilic residues in the food chain whenever these materials are used for production of other food products or animal feeds or, in the case of citrus peels, for obtaining essential oils. Similarly, the hulls of cereals contain the majority of pesticide residues. For instance, 70% and 90% reductions of pirimiphos-methyl residues in rice were achieved by husking and polishing respectively (Desmarchelier, 1980).

Systemic pesticides cannot be removed so effectively by peeling, husking or trimming since they may enter the flesh of crops. Thus, for potatoes grown in soil in which phorate was incorporated, the residues were reduced by only 35% by peeling (Kleinschmidt, 1971).

7.6.3 Washing, blanching

Unit operations in processing/cooking typically include washing the raw product with fairly large amounts of water, frequently using high-pressure sprays, and often incorporating surfactants or other washing aids. Washing has been shown to reduce levels of residues that can be dissolved or physically dislodged from the raw product. The extent of reduction depends upon many factors.

- *Location of residue*: While surface residues are amenable to removal, the levels of systemic residues are little affected.
- *The age of residue*: The proportion of residue that can be removed by washing declines in many cases with time, probably owing to movement into the cuticular waxes or deeper layers. Although there are abundant data in the literature illustrating the effect of

washing procedures on residue levels, the age of the residues is often not specified. In experiments in which the pesticide is applied as a post-harvest dip, the high effectiveness of washing often achieved does not reflect real conditions and therefore these data are less valuable.

- *Water solubility*: Polar, water-soluble pesticides are more easily removed than lipophilic compounds. This is due not only to better solubility in water but also to reduced propensity to move into waxy layers.
- *Temperature and type of wash*: Hot washing and blanching are more effective in removal of residues than is cold washing. Detergents are significantly beneficial in increasing the effectiveness of washing. Domestic rinsing is often less thorough compared to industrial processes. Hot caustic washes used in some commercial peeling operations commonly result in significant reduction of residues, especially of hydrolysable pesticides. Based on the available data, the relatively low efficiency of washing was occasionally documented for more lipophilic organophosphorous pesticides such as parathion, fenitrothion and diazinon (Farrow *et al.*, 1969; Kubacki and Lipowska, 1980; Archer, 1975).

7.6.4 Cooking

Processes and conditions used in cooking food are very variable. Many factors, but mainly time, temperature, pH value and moisture changes, influence the quantitative effect on residue levels. Volatilization and hydrolysis are the main reasons for losses recorded for many pesticides, e.g. for malathion in boiled spinach 100% (Geisman, 1975), in rice 92% (Ishikura *et al.*, 1984), for carbaryl in boiled tomatoes 69% (Farrow *et al.*, 1968). The combined effect of pH and cooking temperature on the partial reduction of organophosphates in beef muscle was found to be more efficient than the individual factors alone (Coulibaly and Smith, 1994).

However, heat treatment does not necessarily result in a reduction in pesticide residues. No reduction of thiabendazole was reported in baked potatoes (Friar and Reynolds, 1991). No reduction or a relatively very small reduction of synthetic pyrethroids in boiled, baked and/or fried plant products was reported in a number of studies (Joia *et al.*, 1985; Mestres and Mestres, 1992; JMPR, 1982). Pesticides of the DDT group also represent compounds possessing low volatility and relatively high stability to hydrolysis. For instance, no decrease of DDT level occurred in boiled potatoes (Lamb *et al.*, 1968). On the other hand, relatively high reductions (up to 40%) of the DDT complex, dieldrin, hexachlorobenzene, the chlordane complex, toxaphene and heptachlorepoxide were found in cooked (baked, chair-broiled) Chinook salmon and carp (Zabik *et al.*, 1995). Probably it was the reduction of fat content (in which these compounds are solubilized) in the processed samples rather than thermal decomposition of these compounds that was responsible for these losses.

7.6.5   Juicing

Residue levels in fruit juices are lower when the primary commodity is peeled or cored prior to juicing rather than pressed whole. However, commercial operations are based generally on processing the whole fruit. Partitioning of residue between the fruit skin/pulp and juice depends on its properties. Moderately to highly lipophilic pesticides (e.g. parathion, folpet, captan, synthetic pyrethroids) are poorly transferred into juices and their residues are further reduced by clarification operations such as centrifugation or (ultra)filtration (Cabras *et al.*, 1987). Clarification by addition of gelatin or tannin to precipitate colloidal materials may also be effective in residue reduction. Compared to thick juices, clear products can be assumed to have lower residue levels. The by-products (pulp, pomace, skin) often retain a substantial proportion of lipophilic residues such as parathion, folpet, captan or synthetic pyrethroids. It should be noted that relatively high concentration factors indicating further 'increase' of residues are due to the simple loss of moisture on drying the by-products. For fruit juices MRL applies to the whole commodity (not concentrated) or the commodity reconstituted to the original juice concentration. However, in processing studies the distribution of residue in individual products should always be recorded on the dry-weight basis.

7.6.6   Canning

Large-scale canning is a commercial food manufacturing process comprising several operations such as washing, peeling, cooking and/or concentration, and each of them may contribute to changes in concentration (mostly reduction) of pesticide residues. For instance, only 13–14% of parathion was carried through the process of juice and ketchup production from contaminated tomatoes (Muhammad and Kawara, 1985). Similarly, malathion and other organophosphorus pesticides were shown to be unstable during canning of food crops (Farrow *et al.*, 1969; Elkins *et al.*, 1972). Comprehensive study (Lentza-Rizos, 1995) on the fate of iprodione during canning of treated peaches showed a typical pattern of dissipation of residues during this multistep process.

7.6.7 Milling

The seeds are protected from pesticides to varying degrees during the growing season by husks that are finally removed before processing and/ or consumption. However, residues of insecticides from post-harvest application of pesticides, which is commonly used for reduction of losses caused by storage pests, are often detected in this commodity. Most residues are contained in the outer portion of grains; consequently, residues in bran are consistently higher than in whole grain. Bran/grain concentration factors in the range from 2 to 6 were recorded in many studies (JMPR, 1988, 1989, 1991, 1992), for organophosphates (e.g. fenitrothion, pirimiphos-methyl, chlorpyrifos-methyl, malathion) and pyrethroids (e.g. bioresmethrin, deltamethrin). On the other hand, a distinct decrease of residues occurs in flour, this reduction being higher for white flour than for wholemeal flour. Similarly, the milling of rice substantially removes residues (commonly more than 90%) that are mostly attached to husk (Kleinschmidt, 1971). It is evident that the early milling stages, especially the cleaning process, may result in considerable reduction of many surface residues. However, more polar pesticides such as glyphosate can enter the grain by translocation, leaving some residues in flour.

Cooking of cereals may further reduce residue levels. However, it is difficult to assess quantitatively on the basis of the available information the extent to which dilution of residues actually occurs because moisture increases in baked bread or boiled rice compared to the raw material. To avoid misinterpretation of the analytical data, a moisture-free basis should be used for all calculations.

7.6.8 Processing of plant oils and fats

Pre-harvest treatment of most oil/fat-bearing crops results mostly in very low or undetectable residues in recovered oil because most seeds and nuts are protected by the husk or shell, which are removed prior to further processing. If present in seeds, polar pesticides are not transferred into oil, they remain in the meal after pressing or extraction. Seeds in pods are protected from most pesticides applied during the growth season except for systemic pesticides.

Post-harvest treatments used to prevent losses due to pest damage during storage and transportation may result in levels of residues in oilseeds, nuts and pulses higher than those from pre-harvest application. Lipophilic residues such as synthetic pyrethroids (e.g. cyhalothrin, cypermethrin, fenvalerate, permethrin) and organophosphates (e.g. diazinon, dichlorvos, etrimfos, malathion, parathion, pirimiphos-methyl, chlorpyrifos) can concentrate in extracted or expressed oil (Bartsch, 1974; Miyahara and Saito, 1993). Accumulation in crude ('virgin') oil was shown to occur when the whole, contaminated commodity (e.g. olives) was processed. High concentration ratios can be expected in crops and raw materials such as maize or citrus peel giving low yields of oil.

Common refining processes largely remove most pesticide residues. Alkali-refining and especially deodorization are the most efficient steps (Kubacki and Lipowska, 1980). However, some chemically stable, low vapour pressure pesticides such as synthetic pyrethroids may be less effectively removed during this treatment.

7.6.9   Production of alcoholic beverages

Residues of fungicides applied in viticulture may be transferred into the final product from treated grapes, depending on their physicochemical properties, namely water solubility and rate of hydrolysis. While practically all the residues of benomyl were transferred to wine (Gnaegi and Dufour, 1972), significant reductions of folpet, captafol, propiconazole, triadimefon/triadimenol and vinclozoline residues were observed during the wine-making process (Lemperle et al., 1970; JMPR, 1987). Adsorption of residues on the solids produced during fermentation contributes largely to residue reduction. Degradation of residues can also occur during storage of bottled wine (Kawar et al., 1979).

Contrary to wine production, production of beer results typically (with the exception of some highly soluble pesticides such as glyphosate) in undetectable residues. This is due not only to degradation during malting (Fahey et al., 1969; Gajduskova, 1974) and significant dilution of the residues that may be contained in barley and hops used as raw material, but is also due to the losses occurring during filtering and fining.

7.6.10   Toxic degradation products originating during processing/cooking

In general, decomposition (hydrolysis, oxidation, etc.) caused by physicochemical factors involved in processing and cooking procedures is desirable, since practically no harmful products compared to the parent compound arise in this way. However, there are some chemicals that will create undesirable degradation products that are more toxic than the parent compound and which in some cases are even carcinogenic. One of the best-known examples of such a highly hazardous compound is daminozide (active ingredient of the plant growth regulator Alar) from which unsymmetrical 1,1-dimethylhydrazine (UDMH) may be formed (section 7.5.1.2) during cooking. An intensive media campaign was focused on the 'Alar crisis' prior to the banning of this agrochemical, which may serve as a good illustration of the problems related to risk communication (Ames and Gold, 1989; Moore, 1989).

Another group of pesticides posing some health risk is a widely used class of fungicides, ethylene bisdithiocarbamates (EBDCs), represented mainly by mancozeb, maneb and zineb. Toxicological concern (WHO, 1988) on the carcinogenic potential of EBDCs is related to the metabolite or breakdown product ETU (similarly, propylene thiourea, PTU, can originate from propylene bisdithiocarbamates). Although the levels of ETU in field crops are relatively low (Lentza-Rizos, 1990), currently not exceeding 0.05 mg/kg, levels may be significantly higher in heat-treated products, especially in those with higher pH values (e.g. tomatoes). High conversion rates of parent EBDCs (up to 50%) were recorded during some processes involving cooking and canning (Newsome, 1976; Phillips et al., 1977; Marshall and Jarvis, 1979). Although virtually water insoluble, EBDCs are removed to a great extent from the crops by washing. They are not very lipophilic and thus not solubilized by surface waxes. Various water bath additives, including sodium hypochlorite followed by detergents, and/or sodium sulphite, may increase the efficiency of removal of ETU precursors (Marshall and Jarvis, 1979; Marshall, 1982). ETU, which is moderately water soluble, can also be reduced by washing. The persistence of ETU residues in canned products depends on many factors, including the presence of ascorbic acid (Hajšlová et al., 1986). This antioxidant lowered the yield of ETU from parent EBDCs while simultaneously stabilizing its residues by preventing oxidative decomposition. Great concern was paid to the formation of ETU in the brewing process from hops containing EBDCs (Nitz et al., 1984). Attempts to reduce its levels were not successful. In contrast, in a study of the wine making process, no ETU residues were found in wine, although low levels of this chemical were detected in grape juice and must. Both ETU and parent zineb were reported to be completely adsorbed onto grape and wine suspended solids (Ripley et al., 1978).

7.7 Toxicity of pesticides to humans

Although designed to control pests, pesticides can also be toxic to non-target organisms, including humans, since species from insects to man share the same basic enzyme, hormone, and other biochemical systems. Exposure to pesticides may be classified in several ways, e.g. as acute and chronic, occupational and non-occupational, intentional and non-intentional, accidental or incidental (Al-Saleh, 1987). Within each type of exposure, pesticides may enter the body dermally, by inhalation or orally. In this section, special attention is paid to long-term exposure of humans to pesticides via the diet.

Generally, the severity of adverse effects resulting from dietary exposure to pesticides depends, as in the case of other contaminants, on the dose as well as on the mechanisms of absorption, distribution, metabolism and excretion (WHO, 1978, 1987; Hopper and Oehme, 1989). The toxic effects also depend on the health status of the individual. Dehydration and malnutrition are likely to increase sensitivity to pesticides. Standards for classification of pesticides according to toxicity hazard were issued by the World Health Organization (WHO, 1996)[1]. As already mentioned, toxicity is usually divided into two types, acute or chronic, based on the number of exposures to a pesticide and the time it takes for toxic symptoms to develop. Principles for the toxicological assessment of pesticide residues in food were established (WHO, 1990b) for both types of toxicity by WHO within the International Programme on Chemical Safety (IPCS). Test methodologies for conduct and evaluation of carcinogenity, reproductive effects, neurotoxicity, geno-toxicity and immunotoxicity are continually being up-dated. The relationships between chemical pesticides and occurrence of these adverse effects have been reviewed (Merlo *et al.*, 1994; Weisenburger, 1993; Dich *et al.*, 1997; Sawada, 1995).

In spite of numerous studies conducted on pesticide toxicology (Kuiper, 1996), the mechanism of toxicity for mammals has been well characterized for only a few groups of compounds. For example, the mechanism for organophosphorus and carbamate insecticides results in inhibition of cholinesterase (among the commonly used compounds of this group, parathion and aldicarb are especially toxic) and nitrophenols and higher chlorinated phenols are inhibitors (uncouplers) of oxidative phosphorylation (Hassall, 1990). Fat-soluble substances (organochlorines such as DDT, HCH and other persistent substances) are accumulated in the body and, when stored in fatty tissues, they are not readily metabolized (WHO, 1990a). However, in times of poor nutrition or relative starvation, the deposits of these compounds are mobilized and they are released into the bloodstream, with the possibility of toxic effects if concentrations reach a high enough level.

In general, within the body the pesticide may be metabolized, stored in fat, or excreted unchanged (section 7.5.1.1). As discussed earlier, metabolism will probably make the pesticide more soluble and thus

---

[1] Based on the oral or dermal LD_{50} for the rat, the following classes of acute toxicity are recognized: Extremely hazardous, Ia; Highly hazardous, Ib; Moderately hazardous, II; Slightly hazardous, III.

more easily excretable (e.g. fat-soluble pyrethroid insecticides are hydrolysed to water-soluble compounds). An increase in toxicity may occur *in vivo*, e.g. oxidation of thiophosphorus insecticides to their oxygen analogues may yield more potent cholinesterase inhibitors (section 7.5.1.2). Similar effects result from hydrolysis of carbofuran, yielding the more toxic furathiocarb.

Although none of the tested synthetic pesticides in common use were classified by the International Agency for Research on Cancer (IARC) as carcinogenic in humans, some of them, such as amitrol, chlorophenoxy herbicides, DDT and toxaphene, were classified as 'possibly carcinogenic to humans' (Group 2B). The best-known degradation products that were shown to be potent carcinogens in animal studies were mentioned previously (section 7.6.10).

One of the recent concerns about the risk posed by pesticides to humans is their ability to alter or disrupt the body's hormone or endocrine system (Olea, 1996; Stevens *et al.*, 1997). Exposure to some notoriously persistent, bioaccumulative, organochlorine pesticides (such as DDT, and metabolites, toxaphene, Mirex, dieldrin, heptachlor and heptachlorepoxide), implicated in disruption of hormones, occurs primarily through foods such as meat, dairy products and fish. However, it is not only persistent/bioaccumulative and chronically toxic pesticides that exhibit endocrine disruption potential. Many 'modern' pesticides are under active review because of ecological or health concerns, and such reviews should include hormone-disrupting potential. Among compounds undergoing such testing are the widely used herbicides such as 2,4-D, alachlor and atrazine. Fungicides possessing endocrine-disrupting potential include benomyl and EBDCs and, in the case of insecticides, include carbaryl, dicofol, permethrin and other synthetic pyrethroids, parathion and lindane.

In summary, it should be emphasized that very few epidemiological data are available for the evaluation of the health effects of pesticides on humans. Although only a small part of the population is likely to receive a pesticide dose high enough to cause acute effects, many more may be at risk of developing chronic effects, depending on the type of pesticide to which they are exposed.

7.8   Occurrence of pesticides in the human food supply

A consumer's dietary intake of pesticide residues comes from four identifiable sources: (i) on-farm use, (ii) post-harvest use, (iii) use on imported foods and (iv) banned pesticides in the environment. Sources (i) to (iii) represent contamination mainly due to residues of 'modern' pesticides and can be characterized as 'from intentional use', whereas (iv) represents residues from banned pesticides that can be characterized as 'from unintentional use'. To get information on residues in the food supply, a number of countries have established national monitoring programmes for a wide range of pesticides. Information on food contamination is obtainable through two different but complementary approaches: (i) regulatory or commodity monitoring, which commonly focuses on raw agricultural commodities and measures levels of residues in individual lots either of domestic or imported foods for determining compliance with established tolerances or guidelines; (ii) total diet studies in which dietary intakes of pesticides are determined by analysing food as consumed. In both cases, time-trends in contamination may be obtained from data collected over a sufficiently long period. An indication of the effectiveness of measures introduced to reduce food residues is obtained in this way. The overall objectives of food contamination monitoring (van der Kolk, 1991; Galal-Gorchev, 1993) are to safeguard health, to improve the management of food and agricultural resources, and to prevent economic losses. Monitoring data obtained within surveillance programmes (random sampling, no prior knowledge or evidence of illegal pesticide residues) are valuable for identification of particular pesticide–commodity combinations that occur frequently. A summary of residue monitoring data sorted in order of incidence has been published by Hamilton *et al.* (1997) based on data available from Australia, Brazil, Denmark, New Zealand, Sweden, and the United States. As an example, pesticides that were found in more than 10% of a particular commodity sampled are shown in Table 7.5.

Grain protectants used on cereals show the highest incidence of residues. For fruits and vegetables the only generalization that can be stated is that those types with high surface areas and 'fuzzy and waxy' coatings and some root foods tend to have higher frequency of positive reports. Appraisal of the whole data set consisting of 208 items (combinations) shows that residues of acephate, benomyl/carbendazim, carbaryl, chlorothalonil, chlorpyrifos, cypermethrin, deltamethrin, dicofol, dimethoate, dithiocarbamates, endosulfan, fenitrothion, fenvalerate, malathion, parathion, permethrin and vinclozolin were identified to account for most of the positive findings. This is not surprising since all of these pesticides are widely used all over the world.

Considering a whole group of commodities representing fruit and vegetables, pesticides with an incidence rate above 10% are much less likely as seen in Table 7.6, assembled from the recent results of the Swedish monitoring programme of the National Food Administration (NFA, 1995, 1996, 1997). In contrast with the list shown in Table 7.5 in which they are not included, thiabendazole, imazalil, captan and

Table 7.5 Pesticide/commodity combinations characterized by high frequency of incidence from the combined data of several national monitoring programmes (Hamilton *et al.*, 1997)

Pesticide	Commodity	No.of samples tested	Percentage of samples with residues[a]
Carbaryl	Sorghum	156	96.8
Fenitrothion	Cereal grains	12759	74.5
Chlorothalonil	Celery	375	65.6
Vinclozolin	Strawberry	509	48.1
Vinclozolin	Kiwi fruit	126	43.7
Carbaryl	Oats	130	33.9
Endosulfan	Lettuce, leaf	483	32.5
Dithiocarbamates	Tomato	866	27.8
Chlorpyrifos	Tomato	3613	27.8
Chlorpyrifos	Peppers	1428	27.7
Endosulfan	Melons	925	26.9
Endosulfan	Spinach	183	21.9
Chlorpyrifos	Kiwi fruit	127	20.5
Dithiocarbamates	Pome fruit	135	19.9
Endosulfan	Cucumber	1659	19.8
Permethrin	Lettuce, head	2451	18.3
Dithiocarbamates	Grapes	285	17.9
Carbaryl	Barley	188	17.6
Endosulfan	Lettuce, head	2169	16.8
Permethrin	Celery	422	15.4
Dicofol	Citrus fruit	1022	14.5
Permethrin	Spinach	183	12.6
Dithiocarbamates	Cucumber	613	11.9
Endosulfan	Common bean	574	11.7
Dimethoate	Peas	837	11.5

[a]Only examples with an incidence greater than 10% are listed.

Table 7.6 Pesticides most commonly found in fresh fruit and vegetables analysed within the Swedish surveillance programme (NFA, 1995; 1996; 1997)

	Incidence Rate (1994) %		Incidence Rate (1995) %		Incidence Rate (1996) %
Thiabendazole	12.6	Thiabendazole	8.8	Thiabendazole	6.8
Imazalil	9.2	Imazalil	7.7	Imazalil	5.7
Captan	6.0	Captan	5.2	Endosulfan	5.2
Methamidophos	5.5	Endosulfan	5.2	Methidathion	4.4
Endosulfan	5.3	Methidathion	5.1	Captan	4.3
Methidathion	5.0	Methamidophos	4.2	Methamidophos	3.6
Vinclozoline	4.5	Dimethoate	3.6	Dimethoate	3.3
Dithio-carbamates	3.7	Azinphos-methyl	3.5	Dithio-carbamates	3.2
Dimethoate	3.5	Chlorothalonil	3.1	Chlorothalonil	3.1
Procymidon	3.4	Dithio-carbamates	2.7	Azinphos-methyl	2.4

methamidophos were among the most common pesticides in each of the three annual reports.

To illustrate the typical incidence of pesticide residues in the food supply, recent data obtained within several surveillance programmes (NFA, 1995, 1996, 1997; FDA, 1995; WPPR, 1995, 1996, 1997) for two categories of food staples are presented in Tables 7.7 and 7.8. In spite of the different designs of the respective monitoring programmes, the incidence of samples above the Maximum Residue Limits (MRLs) is relatively very low, not exceeding 5% of the total samples collected.

However, it should be borne in mind, that although residue data collected under this approach are probably the most extensive in the world, they are not necessarily statistically representative of the overall residue situation for a particular pesticide, commodity or place of origin, since surveillance programmes are often directed at areas where residues are expected to be found. Thus the samples are not collected in a truly random manner. Poor comparability of data among different countries is also partly due to the variable total number and distribution of samples of a given commodity analysed for a particular residue and partly due to differences in the lists of pesticides targeted in particular monitoring programmes. Another important consideration in determining the usefulness of monitoring data is the analytical methodology applied. Three outcomes are possible when testing for residues: a residue may (i) not be detected; (ii) be detected, but in an amount insufficient to measure accurately; or (iii) be detected and measured. The detection rates and,

Table 7.7 Summary of monitoring results for cereals (surveillance sampling in Sweden, UK and USA) (NFA, 1995, 1996, 1997; FDA, 1995a,b; WPPR, 1995, 1996, 1997)

Monitoring programme	Total number of samples	Number of pesticides analysed	Percentage of residues detected	
			Below or at MRL[a]	Above MRL[a]
Sweden, 1994	518		26.1	2.0
Sweden, 1995	432	up to 253	11.3	0.9
Sweden, 1996	317	up to 253	11.7	1.3
UK, 1994	327	24	42	1
UK, 1995	301	15	21	0
UK, 1996	162	21	31	4
USA, 1994		329		
Domestic	529		37.1	1.5
Import	266		33.4	1.9
USA, 1995		331		
Domestic	389		39.8	0.8
Import	238		19.4	0.8

[a]MRL, Maximum Residue Limit.

Table 7.8 Summary of monitoring results for fresh fruit and vegetables (surveillance sampling in Sweden, UK and USA) (NFA, 1995, 1996, 1997; FDA, 1995a,b; WPPR, 1995, 1996, 1997)

Monitoring programme	Total number of samples	Number of pesticides analysed	Percentage of residues detected	
			Below or at MRL[a]	Above MRL[a]
Sweden, 1994	4454		41.2	3.3
Sweden, 1995	3573	up to 253	44.2	4.6
Sweden, 1996	3861	up to 253	33.8	3.6
UK, 1994	898	up to 102	57	2
UK, 1995	1021	up to 103	44	2
UK, 1996	528	up to 93	41	2
USA, 1994				
Domestic	3209		42	1
Import	4400		31	4
USA, 1995				
Domestic	3022		46	2
Import	4292		33	4

[a]MRL, Maximum Residue Limit.

consequently, the estimated average residue levels depend on the limit of detection (LOD). To get valuable information, this should be sufficiently low to allow unambiguous measurement of residues even below the existing MRLs. Strict adherence to quality assurance/quality control (QA/QC) principles, both in the process of sampling and laboratory analysis, is the requirement for generation of valid data (OECD, 1992; Garfield, 1991; EEC, 1994; Hill, 1997).

Market basket studies or total diet studies are conducted in number of countries with the aim of assessing typical dietary intakes of various groups of potentially toxic chemicals including pesticides. The methods employed (WHO, 1987b; Pennington, 1992) for this purpose vary from country to country depending on the specific objectives and available resources. Most total diet studies involve laboratory analyses of individual foods, composite foods or composite diets. To obtain estimates of dietary intake or exposure, food consumption data are needed in the case of analysis of individual or composite foods. Food consumption methodologies are commonly based on 'food balance sheets', household food consumption surveys and individual consumption surveys (FAO/WHO, 1997). Vulnerable subgroups of populations such as children are sometimes considered separately. Possible health risks may be detected in this way.

At the international level, governments, Codex Alimentarius Commission and other relevant institutions, as well as the public, are informed on the levels and trends of priority contaminants in food, the contribution to human exposure and their significance for public health and trade via the reports and documents issued by the joint UNEP/FAO/WHO 'Global Environment Monitoring System (GEMS) Food Contamination Assessment Programme' (GEMS, 1988, 1995; UNEP, 1992), which was established in 1976[2]. Persistent organochlorine pesticides (aldrin, dieldrin and the DDT complex in milk and butter, endosulfan and endrin in animal fats/oils and fish, HCB and HCH in human milk and heptachlor in total diet), and widely used organophosphates (diazinon in cereals, fenitrothion in vegetables and fruit, malathion in drinking water and parathion plus methylparathion in total diet), are monitored within this programme. Increased attention is being paid to the quality and validity of the information submitted in order to improve the comparability of data generated from different sources.

7.9 Legislative and regulatory issues

The rapidly increased use of pesticides in agriculture after the Second World War led many governments to enforce regulations on the sale and use of pesticides in order to protect users of pesticides, consumers of treated foodstuffs, domestic animals and, at a later stage, the environment. Simultaneously with the need to solve general environmental problems, the economics of global trade created an incentive for harmonious international regulatory standards for these chemicals and their residues in traded commodities. The historical background as well as the current activities of the international community in this and related fields is outlined briefly below. (The characterization of specific features of national pesticide registration processes and regulations is beyond the scope of this chapter) (WHO, 1978, 1987).

7.9.1 Institutions and bodies involved in development of international standards

Since its inception in 1949, food safety issues have represented one of the basic parts of the overall health mandate of the World Health Organization (WHO). Concern about the toxic hazards posed by pesticides, noted in documents published in 1953, initiated respective programmes within WHO. The need to solve growing international problems resulting from pesticide application urged the Food and

---

[2] GEMS/Food-EURO programme was established in 1992 to reflect specific priorities and needs of countries located in WHO region of Europe.

Agriculture Organization (FAO) to convene a Panel of Experts on the use of pesticides in agriculture in 1959. Among the many issues discussed, the need to undertake joint studies by FAO and WHO on hazards arising from pesticide residues in and on food and feedstuffs was emphasized (FAO, 1959). The importance of establishing principles to govern the setting up of pesticide tolerances and the feasibility of preparing an international code for toxicological data was also emphasized. In 1961, a joint meeting of this FAO Panel and WHO Expert Committee on Pesticide Residues was held to implement these recommendations. The concept of 'permissible level' calculated from Acceptable Daily Intake (ADI)[3], the food factor and the average weight of a consumer were accepted. Nevertheless, no conclusions were reached on a strategy to estimate internationally acceptable tolerances at that time. Consequently, significant differences in residue tolerances existed even among countries of the same region, although it was accepted that the range of the residues actually remaining when the food is first offered for consumption (i.e. following GAP—for definition see section 7.4) should be taken into account. The FAO Conference on Pesticides in Agriculture held in 1962 strongly urged the need to investigate the reasons for these differences and, if possible, to find a way to harmonize them. In response to this situation, the Codex Alimentarius Commission (CAC) was established in 1962 under the joint sponsorship of FAO and WHO to act as the body representing the international consensus on the adequacy of requirements to protect human health from foodborne hazards. The first Joint Meeting of the FAO Committee on Pesticide Residues in Agriculture and the WHO Expert Committee on Pesticide Residues (referred to as JMPR) in 1963 studied the toxicological properties of a number of pesticides. In the same year, the FAO Working Party on Pesticide Residues identified the approaches essential for arriving at unified tolerances. Since 1966, JMPR has been establishing MRLs[4] for pesticides in food commodities. The forum for achieving agreement among Member countries on international MRLs is provided by the Codex Committee on Pesticide Residues (CCPR), established in 1968. Mutual cooperation between the organizations described above is discussed elsewhere (WHO, 1990a).

---

[3] ADI of a chemical is the estimate of the amount of a substance in food or in drinking water, expressed on a body weight basis, that can be ingested daily over the lifetime without appreciable health risk to the consumer on the basis of all the known facts at the time of evaluation. ADI is expressed in milligrams of the chemical per kilogram of body weight.

[4] MRL is the maximum concentration of pesticide residue (expressed as mg/kg) recommended by CAC to be legally permitted on or in food commodities and animals feeds. MRLs are based on GAP data, and foods derived from commodities that comply with respective MRLs are intended to be toxicologically acceptable.

7.9.2 Process of international standards development

As stated above, international agreement on food standards aimed at protection of public health and ensuring fair practices in international trade is realized through CAC, which comprises at present 150 national governments. It is obvious that the use of Codex MRLs is an important requirement for avoiding non-tariff barriers to international food trade.

To begin the Codex eight step procedure, a pesticide must be placed (on the basis of proposal by national government) on CCPR's Priority List for MRL development (Moy and Wessel, 1998), provided certain criteria are met: (i) the pesticide is available for use as commercial product; (ii) its use gives rise to residues in or on a food commodity moving in international trade; (iii) the presence of a residue may be a matter of public concern; (iv) the residues have potential to create problems in international trade. In addition to these criteria, availability of all relevant data must be ensured. Pesticides on the CCPR Priority List are then evaluated by the independent advisory body, JMPR. Data on pesticide toxicity, agricultural and pest control practices, chemistry and metabolism, supervised field trials (including those conducted at the highest recommended, authorized or registered uses), fate of pesticide residue in the field and during storage and processing of food, residue analytical methodology and other relevant information are considered at the JMPR session by two separate groups of scientists in the FAO panel and WHO groups. At this point, hazard characterization provided by WHO Core Assessment Group may result in establishment or up-dating of an acceptable daily intake (ADI) for particular pesticides or estimation of acute reference dose (section 7.9.3). At the same time, assessment of available information on GAP may result in a proposal for maximum levels for residues of the pesticide on various commodities by FAO Panel experts. These values are used by the Global Environment Monitoring System/Food Contamination Monitoring and Assessment Programme (GEMS/Food) of WHO to calculate estimates of long-term dietary intake or, if necessary, short-term intakes. Reports summarizing information on JMPR evaluations are issued in two parts: Part I—Residues is published by FAO under its FAO Plant Production and Protection Series; Part II—Toxicology is published by WHO under the auspices of the International Programme on Chemical Safety (IPCS).

7.9.3 Assessment of consumer safety

Risk assessment posed by dietary pesticide residues is widely discussed at international level (Winter, 1992; Bates and Gorbach, 1987) and respective recommendations for its analysis are issued (FAO/WHO,

1995). In this process, the importance of the role of exposure assessment has been recognized by the relevant international organizations such as CCPR. This information is necessary for reaching a conclusion on the acceptability of proposed MRLs and the underlying GAP, from a public health point of view. The best assurance that long-term exposure to residues is within safety limits (ADI must not be exceeded) is obtained from dietary intake studies. However, these may not be easily feasible or the pesticide may only recently have been approved for use. Under these circumstances, residue intake must be predicted from the available data. Exposure predictions related to long-term hazards are outlined in greater detail below.

Acute toxic effects may sometimes be encountered following consumption of food containing residues of certain pesticides. The acute reference dose (RfD) was developed (using principles equivalent to those used to derive ADI) to assess acute hazard. For the purpose of risk assessment of acute hazards related to the maximum residue expected in a food as consumed when the pesticide is used according to approved label conditions, MRLs are used for the calculations, especially for food that is consumed raw. The significant variation in residue levels within individual commodity units should also be considered (MAFF, 1995). In practice, it is unlikely that an individual will consume large amounts of different commodities at the same time that contain the same pesticide at its highest permitted concentration (MRL).

7.9.3.1 *Predicting dietary intake at the international level (long-term hazards)*

An international method has been developed by the Food Safety Unit, Division of Food and Nutrition, WHO, for assessing potential long-term (chronic) hazards resulting from ingestion of low levels of pesticide residues in the diet over a lifetime. According to recently revised guidelines (WHO, 1997), in the first phase a rough estimate, 'Theoretical Maximum Daily Intake' (TMDI), is calculated using Codex MRLs and the average daily food consumption of the respective commodity. The products are summed:

$$TMDI_i = \sum MRL_i \times F_i$$

where MRL_i = Maximum Residue Limit for a given commodity and F_i = per capita GEMS/Food regional/cultural food consumption of that commodity.

The $TMDI_i$ is next compared to the ADI of the respective pesticide calculated for a 60 kg person. The $TMDI_i$ value is commonly an overestimate of a true pesticide intake since: (i) only a portion of specific crop is treated with a pesticide; (ii) most treated crops contain residues below the MRL at harvest time; (iii) reduction of residues occurs during storage/processing/cooking of commodity; and (iv) there is a low probability that every food for which an MRL is proposed will have been treated with the pesticide over the consumer's lifetime. In spite of these 'worst case' assumptions, TMDI calculations for most pesticides are less than their ADI. Even if TMDI exceeds ADI, it should not be concluded that the proposed MRL is unacceptable—a more realistic estimate should be made.

According to WHO, the 'International Estimated Daily Intake' (IEDI) incorporating correction factors at the international level should be calculated to refine the intake estimate on the basis of information on median residue levels obtained from supervised field trial data that are submitted to JMPR:

$$\text{IEDI}_i = \sum \text{STMR}_i \times \text{E}_i \times \text{P}_i \times \text{F}_i$$

where STMR_i = Supervised Trial Median Residue level for a given food commodity; E_i = Edible Portion Factor for that food commodity; P_i = Processing Factor for that food commodity; and F_i = GEMS/Food regional consumption of that food commodity.

If ADI is exceeded by the IEDI after all the factors are applied, other additional relevant data are required from governments, industry and other sources for further consideration.

7.9.3.2 Predicting dietary intake at the national level
(long-term hazards)

A similar approach to that described for the international estimate may be adopted at national level. For the calculation of National Theoretical Maximum Daily Intake, the following formula is used:

$$\text{NTMDI}_i = \sum \text{MRL}_i \times \text{F}_i$$

where MRL_i = Maximum Residue Limit (or national maximum limit) for a given food commodity and F_i = national consumption of that food commodity

National Estimated Daily Intake (NEDI) represents a refinement of the IEDI for the calculation of which further factors available on a national level are used: (i) proportion of crop or commodity treated; (ii) proportion of crop or commodity produced domestically; (iii) monitoring and surveillance data; (iv) total diet (market basket) studies; and (v) food consumption data including data for the sub-groups of population.

7.10 Analytical problems

A new burden has been placed on residue chemistry in recent years requiring more effort on conventional lines and/or development of alternate technologies. Some of the main reasons are listed below.

- 'Modern' pesticide classes do not always conform to the analytical prerequisites (e.g. N-methylcarbamates degrade under conditions of gas chromatography (GC)).
- The need for metabolite analysis has increased; however, many of these compounds have very different properties from those of the parent compounds (e.g. aldicarb produces a cluster of metabolites of varying physicochemical and toxicological properties).
- The rising number of analyses required places new demands on analytical throughput and cost.
- Tolerance levels have become lower in many cases and the number of registered pesticides has increased.

Choice of the appropriate analytical procedure for a particular sample always requires considerable prethought, because various residues (often multiple) may be present that have a wide range of physicochemical properties (Table 7.2). The first question to be answered is the purpose of the analysis. Obviously, different analytical strategies will be applied to fulfill the following tasks: (i) to study the fate of particular compound(s) in respective environmental compartments; (ii) to conduct broad monitoring surveys; (iii) to perform rapid field pre-screening; and (iv) to execute regulatory enforcement activities. The nature of toxic residue(s) (metabolites/degradation products may also be involved) puts special demands on the detailed procedures to be applied. Also important is a knowledge of the 'history' of contamination and pesticide usage. In any case, the performance characteristics of the chosen method such as sensitivity, selectivity, detection limits and linear range, as well as sample throughput, overall robustness and economic factors, should comply with the purpose or task undertaken.

7.10.1   Multiresidue methods

In practice, information on the field treatment of crops and/or post-harvest application of protectants is hardly available and, similarly, the character of emissions in the locality of the origin of the particular commodity is commonly unknown. Multiresidue methods (MRMs) that are capable of simultaneously determining more than one residue in a single analysis are therefore commonly used by laboratories responsible for surveillance/compliance activities. The accuracy of such methods should be sufficient to meet general regulatory requirements and scientific needs. Apart from identification and subsequent quantitation of specified residues, the presence of unidentified compounds may be recorded or other residues may be detected (although sometimes not reliably measured) by most MRMs.

As can be seen from the general literature (Thier and Frehse, 1986; Thier and Zeumer, 1987; Zweig and Sherma 1984, 1985, 1986) on the pesticide residue analysis, the character of the matrix determines the methodology to be followed. Classification of commodities, and selection of the appropriate extraction method is mainly based on the fat content. For instance, the Netherlands working group on 'Development and Improvement of Residue-Analytical Methods' set the limit between 'fatty' and 'non-fatty' matrices arbitrarily at 5% (GIHP, 1996). The FDA Pesticide Analytical Manual (McMahon and Wagner, 1996) classified as 'fatty' all products with fat above 2%; for further sorting of 'non-fatty' products, water content is considered, the limit between the two sub-groups being 75%.

MRMs have been developed over 30 years and the most notable changes are in instrumental determinative techniques. GC applications have been significantly enhanced by the introduction of high-performance capillary columns, improved detectors and new types such as coupled mass spectrometric detectors. Introduction of advanced HPLC operating modes has extended multiresidue capability to pesticides not amenable to GC determination. A comprehensive, recent review of official multi-residue methods applied to food and environmental matrices documents these trends (Motohashi et al., 1996). In spite of these advances, the basic steps of an analytical flow-chart, i.e. isolation of residues from a representative sample and subsequent separation of co-extracted components from the matrix, continue to be an essential part of any MRM. Among the array of analytes, two groups of MRMs may be recognized: (i) 'multiclass MRMs' provide coverage of residues representing often widely different classes of pesticides, e.g. chlorinated hydrocarbons, organophosphates and carbamates, and the primary extract is examined by several chromatographic determinative steps; (ii) 'selective MRMs' feature in methods that determine multiple residues of chemically related pesticides. Typical of the selective method are N-methylcarbamates, carboxylic acids and phenols, substituted phenylureas and aromatic amines. Derivatization aimed at either enhancement of detection sensitivity (e.g. post-column fluorescence labelling of N-methylcarba-mates with o-phthaldialdehyde) or volatilization of ionic analytes prior to GC (e.g. esterification of phenoxyacetic herbicides), is typically involved in the above procedures (Newsome, 1995). A special procedure is currently applied for determination of dithiocarbamates, which are measured either spectrophotometrically or by headspace GC after conversion to carbon disulphide (Keppel, 1971; Ahmad *et al.*, 1996).

To illustrate the diversity of existing multiresidue methods, several procedures covering a wide range of non-ionic analytes in an array of high-moisture, low-fat matrices are shown in Table 7.9.

7.10.1.1 *Isolation of residues*

For the extraction ('blending type' techniques) of residues from high-moisture commodities, water-miscible solvents, e.g. methanol, acetonitrile and/or acetone, are commonly used. Alternatively, a less polar solvent such as ethyl acetate (desiccant added prior to extraction if necessary) is employed. Liquid–liquid partitioning of residues from the initial extractant to the non-aqueous solvent follows the use of water-miscible solvents. Variation in the polarity of the solvents may affect both the degree to which a particular residue is extracted/re-extracted and the number of other components co-isolated from the matrix (McMahon and Wagner, 1996). Ethyl acetate and acetone (less toxic compared to other polar solvents and with lower boiling points) are involved as primary extraction solvents in CEN Draft European Standard 12393-2 of MRM, which is at present being validated within an international intercomparison study (dichloromethane re-extraction follows the use of acetone).

As an alternative to conventional solvent-based techniques, super-critical fluid extraction (SFE) offers promise as a rapid way of isolating selected pesticides from respective matrices (Hopper and King, 1991; Khan, 1995; Snyder *et al.*, 1992; Lehotay and Eller, 1995). Using the automated instrumental techniques, SFE reduces the amount of manual labour and laboratory space needed. A novel sample preparation approach has been developed for SFE of pesticide residues utilizing the unique characteristics of this technique (small sample size, need to control water content to avoid changes in extraction efficiency, vulnerability of analytes in solid sample without solvent) and still maintaining relevant accuracy. In principle, SFE allows for a higher degree of selectivity in pesticide extraction compared with solvent-based procedures. Combination with solid-sorbent traps for SFE collection thus affords a single-step extraction and clean-up. This analytical strategy was tested for 46 pesticides commonly occurring in grapes, carrots, potatoes and broccoli and applied to a multiresidue method developed by Lehotay *et al.* (1995). Recoveries > 80% were obtained for 39 analytes; values > 50% were recorded for the remaining compounds except two polar compounds, methamidophos and omethoate. Satisfactory comparability of the generated data with results obtained using traditional approaches was reported (Lehotay *et al.*, 1995).

Table 7.9 Examples of multiresidue methods (MRM) applied widely for surveillance/compliance activities of high-moisture low-fat commodities

Analytical step	Characterization of respective MRM cited by the reference[a]					
	[1]	[2]	[3]	[4]	[5]	[6]
Extraction solvent	Ethyl acetate	Acetone	Acetonitrile	Acetone–water, 2:1	Acetonitrile	Acetone
Partition	–	Petroleum ether, dichloromethane	Aqueous NaCl	Dichloromethane (Sodium sulphate added)	SPE-C18[c], phosphate buffer	EtOAc[b] + NaCl
Clean-up	GPC, Bio Beads SX-3, EtOAc[b]–, cyclohexane, 1:1	None, sample dissolved in acetone	None, sample dissolved in toluene	GPC, Bio Beads SX-3, EtOAc[b]–cyclohexane, 1:1	SPE-NH$_2$ silica	OC compounds + pyrethroids —Florisil OP compounds —silica
Additional clean-up	–	–	–	Silica (fractionation)	–	–
Identification[d] quantitation	GC/TSD, GC/FPD, GC/ECD	GC/FPD, GC/HECD (X,S- and N- mode)	GC/MSD, scan mode, (m/z: 100–400)	GC/ECD, GC/NPD, GC/FPD GC/MSD (SIM)	GC/FPD, GC/ELCD, (X-mode)	OP compounds GC/NPD OC compounds + pyrethroids— GC/ECD

[a][1] Anderson and Palsheden (1991); [2] Luke et al. (1981); [3] Liao et al. (1991); [4] Specht et al. (1995); [5] Lee et al. (1991); [6] Nakamura et al. (1994).
[b]EtOAc, ethyl acetate.
[c]SPE, solid-phase extraction.
[d]Abbreviations for the GC detectors are defined in section 7.10.1.3 (see also Chapter 2).

Recently introduced extraction technologies include liquid–solid extraction performed at elevated temperatures, accelerated solvent extraction (ASE) and microwave solvent extraction (Lopez-Avila *et al.*, 1996). Passive diffusion of analytes through polymeric membranes is another promising principle for their isolation (Ahmad *et al.*, 1995). Reduction of requirements for solvents and time may be achieved by application of a column extraction method consisting in addition of an appropriate sorbent to the sample (a flowing-powder technique) and elution of analytes by respective eluants (Kadenczki *et al.*, 1992)

7.10.1.2   Clean-up of extract

Whichever extraction technique is applied, various components with higher molecular size such as lipids, pigments and resins are unavoidably present in extracts from biotic matrices. Although some MRMs attempt to determine as many residues as possible by examining uncleaned extracts using selective detectors (Hsu *et al.*, 1991), in most methods, clean-up of the crude extract is included. This step is designed to permit more definitive identification of residues at lower limits of quantitation and to minimize adverse effects on the instrumentation applied in the determinative step. However, extensive clean-up of extracts may result in the partial loss of some compounds due to their vaporization, irreversible sorption and chemical degradation; moreover, increased labour and cost demands are encountered.

Many MRMs involve fractionation of extracted components based on polarity, adsorption chromatography on a Florisil, alumina or silica column or on solid-phase extraction (SPE); an overview of chromatographic methods applied to the clean-up of plant extracts for pesticide residue analysis was published recently (Tekel and Hatrik, 1996). In any case, the choice of sorbent dictates what chemicals can be recovered (McMahon and Wagner, 1996). Florisil, one of the most popular materials in residue analysis, is particularly suited for fatty foods. In this case, the most effective clean-up is obtained for less polar residues, which can be eluted by solvent mixtures of low polarity, leaving more polar co-extractives on the column. Increased polarity of eluent permitting recovery of polar analytes results in decreased efficiency of crude extract clean-up. Moreover, very polar residues usually cannot be eluted from Florisil, irrespective of the polarity of the eluent. Instead, non-specific hydrophobic sorbents such as charcoal or, more recently, graphitized carbon black (GCB) should be employed. Many other materials have been applied to remove interfering substances from crude extracts: e.g. natural sulphur-containing compounds from *Cruciferaceae* vegetables may be eliminated by silver-loaded alumina and modified silicas, especially C8 and C18 sorbents, are widely used.

Gel permeation chromatography (GPC) is extensively used in MRMs (Holstege et al., 1991; Specht and Tillkes, 1985; Sannino et al., 1995; Steinwandter, 1988). The principle of this technique is separation by molecular size. Thus, structurally dissimilar components with broadly varying polarity may be analysed together, assuming that subsequent GC or HPLC chromatographic resolution is sufficient. Although soft gel Bio Beads SX-3 is still the most widely used material for GPC, rigid styrene–divinylbenzene copolymers (e.g. Pl gel, Envirogel) enable faster analysis and provide better reproducibility of elution volumes over the long-term, and are used increasingly (Johnson et al., 1997). While the adsorption columns/cartridges described above are dedicated for single use only, GPC columns are suitable for multiple use and are amenable to automation.

7.10.1.3 Determinative step

Gas chromatography is the most frequently used technique for the determinative step. High-performance capillaries containing non-polar or semi-polar (mostly polysiloxane-based) low-bleed stationary phases are preferred over traditional packed columns in most MRMs. GC interfaced with mass spectrometry (GC-MS) or with tandem MS (GC/MS-MS) is often employed nowadays (Scharchterle and Feigel, 1996; Cairns, 1992) as a common, secondary, confirmation technique supporting MRMs relying on established element/group-selective detectors. These are represented typically by the electron capture detector (ECD), nitrogen–phosphorus detector (NPD), flame-photometric detector (FPD) and electrolytic conductivity detector (ELCD). In addition to its identification role (electron-impact ionization data are the most useful for this purpose) GC has also become a primary screening tool (Cairn and Baldwin, 1995; Stison, 1995) in which ion trap technology (GC-ITMS) was shown to have tremendous potential for MRM using a single detector for all residues (Lehotay and Eller, 1995). It should be emphasized that acceptable quantitation for regulatory purposes based on any MS technique is closely related to efficient sample clean-up as well as good resolution of chromatographic separation.

High-performance liquid chromatography is used preferably for analysis of non-volatile, polar and thermolabile pesticides and their metabolites (GIHP, 1996; McMahon and Wagner, 1996; Fillion et al., 1995). The reversed-phase mode of HPLC is mostly used with chemically modified silicas (C8, C18, cyano, amino and others) as stationary phases. For specific purposes, ion exchange, ion-pair and other modes are used. To increase the speed of the analysis, narrow or microbore columns may be used. The multiresidue capabilities of HPLC were significantly expanded by coupling it with MS, and application of thermospray, electrospray and particle beam techniques (Liu *et al.*, 1991; Vollmer *et al.*, 1996). Selectivity of traditional LC detectors such as ultraviolet/visible, or more recently diode array (DAD) is rather limited for determining a broad spectrum of analytes.

7.10.1.4 *Matrix effects in GC analysis of pesticides*

Practically no clean-up procedure completely removes all the matrix components from a crude extract. As mentioned in section 7.10.1.2, these substances may exhibit adverse effects on the quality of the data such as (i) masking of residue peaks due to co-elution; (ii) occurrence of false positives; and (iii) inaccurate quantitation. Problems caused by impurities may be encountered both at the detector and the injector site. In the latter case, increased transfer of sample components to the gas-chromatographic column may occur owing to the blocking of active sites within the injector by the sample matrix, thus preventing thermal degradation/ adsorption of analytes (Erney *et al.*, 1993). These phenomena, known as 'matrix-induced chromatographic response enhancement', allow for explanation of recoveries exceeding 100% that are reported for some pesticides in studies utilizing calibration standards dissolved in pure (matrix-free) solvent. Based on a literature survey (Hajšlová *et al.*, 1998), pesticides giving particularly high recoveries can be identified; acephate, methamidophos, omethoate, iprodione, dimethoate, malathíon and methidathion are typical examples. The extent of these matrix effects is related to both the chemical structure and the type of matrix. General guidelines for using matrix-standard calibration solutions in pesticide residue analysis were published recently (Erney *et al.*, 1997). Improved accuracy of results can be achieved for a range of pesticides by the use of matrix matched standards.

7.10.2 *Immunochemical methods*

Biotechnology-based methods represented mainly by immunoassays, bioassays and biosensors have enormous potential in pesticide analysis (Sherma, 1997). Immunoassays (Conaway, 1991; Kaufman and Clower, 1995) are now available in the form of commercial kits. These techniques are based on binding of target analyte (or its derivative) to an antibody. Concentration of analyte is commonly indicated by a colorimetric change produced by enzymatic reaction. Almost direct determination (or after sample dilution) of various classes of pesticides including organophosphates, carbamates, chlorophenoxycarboxylic acids, triazines and benzoyl phenylureas is possible in water, soft drinks or other aqueous food samples. However, some sample preparation (clean-up) has to be carried out when foodstuffs are analysed (particularly fatty foodstuffs) to avoid matrix effects and nonspecific interactions between structurally similar compounds (i.e. cross-reactivity).

Although not covering such a wide range of analytes as the MRMs, the speed, simplicity and low cost of immunoassays make them an ideal tool particularly for sample screening. Alternatively, this technique may be employed as a confirmation method. In addition to these applications, immunoassay-based sample extraction and/or clean-up may offer distinct advantages over 'classical' techniques owing to their selectivity. Integration of immunochemical methods with other techniques for pesticide residue determination has been reviewed recently (Lucas *et al.*, 1995).

7.11 Strategies for the reduction of occurrence of pesticides in the human diet

During the past decades, pesticide residues have become one of the primary concerns of consumers about the safety of the food supply. The increasing demand for 'organic food' reflects the public perception that these products are more 'healthy' compared to conventional foods. In other words, a wholesome diet is often equated with 'pesticide free' (Weddie, 1991). As a consequence of these concerns about the negative impacts of pesticides on wildlife and health, Integrated Pest Management Systems (IPM) are being implemented as an alternative to unilateral pesticide use and serve as a 'surrogate' to those chemical use patterns that have resulted in actual environmental and health problems. IPM consist of the application of all suitable techniques and methods in as compatible a manner as possible to maintain pest population at levels below those causing economic injury (WHO, 1990a). A number of alternative tools have been used to reduce synthetic pesticide input, including introduction of sterile male insects, application of biological pesticides and breeding of pest-resistant plants. In spite of greater use of alternative pest control, the use of pesticides cannot be fully substituted by application of alternative control measures. Nevertheless, as noted at the beginning of this chapter, when used under registered recommended conditions, small, practically undetectable residues are often left. Moreover, development of less hazardous pesticides by the use of less toxic active ingredients continues.

In addition to the introduction of new and improved agricultural practices, there are many other ways of enabling a reduction of dietary risk related to pesticide residues (Moore, 1991; Kuchler *et al.*, 1996). Effective implementation of protective measures based on the complex analysis of monitoring and total diet studies as well as improved legislation represent efficient tools. Education, training and information systems aimed at both the users of pesticides (farmers) and the consumers of treated foods also play an important role in this field. An international code of conduct on the distribution and use of pesticides has been issued by FAO (1990).

References

Ahmad, N., Guo, L., Mandarakas, P., Appleby, S. and Bugueno, G. (1995) Passive diffusion through polymeric membranes: a novel cleanup procedure for azinphos-methyl and azinphos-ethyl residues in fruits and vegetables. *Journal of the Association of Official Analytical Chemists International*, **78** (6) 1450-1454.

Ahmad, N., Mandarakas, P., Farah, V., Appleby, S. and Girson, T. (1996) Headspace gas-liquid chromatographic determination of dithiocarbamate residues in fruits and vegetables with confirmation. *Journal of the Association of Official Analytical Chemists International*, **79** (6) 1417-1422.

Ahrens, W.A. (1994) *Herbicide Handbook*, 5th edn, Weed Science Society of America, Champaign.

Aizawa, H. (ed.) (1982) *Metabolic Maps of Pesticides*, Academic Press, New York.

Al-Saleh, I.A. (1987) Pesticides: a review article. *Journal of Environmental Pathology, Toxicology and Oncology*, **13** (3) 151-161.

Ames, B.N. and Gold, L.S. (1989) Pesticides, Risk and Applesauce. *Science*, **244** 755-757.

Anderegg, B.M. and Madisen, L.G. (1983) Effect of insecticide distribution and storage time on the degradation of ^{14}C malathion in stored wheat. *Journal of Economy and Entomology*, **76** 733-736.

Andersson, A. and Palsheden, H. (1991) Comparison of the efficiency of different GLC multi-residue methods on crops containing pesticide residues. *Fresenius Journal of Analytical Chemistry*, **339** 365-367.

Archer, T.E. (1975) Chemical and thermal effects on parathion residues on spinach. *Journal of Food Science*, **40** 677-687.

Augustijn-Beckers, P.W.M., Hornsby, A.G. and Wauchope, R.D. (1994) The SCS/ARS/CES pesticide properties database for environmental decision-making, II. additional compounds. *Reviews of Environmental Toxicology*, **125** 1-82.

Ballschmitter, K. (1992) Transport and fate of organic compounds in the global environment. *Angewandte Chemie*, **31** (5) 487-664.

Bartsch, E. (1974) Diazinon. II. Residues in plants, soil, and water. *Residue Reviews*, **51** 37-68.

Bates, J.A.R. and Gorbach, S. (1987) Recommended approach to the appraisal of risks to consumers from pesticide residues in crops and food commodities. *Pure and Applied Chemistry*, **59** 611-624.

Briggs, G.G. (1982) Relationship between lipophilicity and root uptake and translocation of nonionized chemicals by barley. *Pesticide Science*, **13** 495-504.

Cabras, P., Meloni, M. and Pirisi, F.M. (1987) Pesticide fate from vine to wine. *Reviews of Environmental Contamination and Toxicology*, **99** 83-117.

Cairn, T. and Baldwin, R.A. (1995) Pesticide analysis in food by MS. *Analytical Chemistry*, **67** 552-557.

Cairns, T. (1992) Regulatory Mass spectrometry for pesticide analysis: past, present and future. *Journal of the Association of Official Analytical Chemists International*, **75** (4) 591-593.

Coats, J.R. (1991) Pesticide degradation mechanisms and environmental activation, in *Pesticide Transformation Products: Fate and Significance in the Environment* (eds L. Somasundaram and J.R. Coats), ACS Symposium Series 459, American Chemical Society, Washington, DC, pp 10-31.

Conaway, J.E. (1991) New trends in analytical technology and methods for pesticide residue analysis. *Journal of the Association of Official Analytical Chemists International*, **74** (5) 715-717.

Coulibaly, K. and Smith, J.S. (1994) Effect of pH and cooking temperature on the stability of organophosphate pesticides in beef muscle. *Journal of Agricultural and Food Chemistry*, **42** 2035-2039.

Cremlyn, R. (1978) *Pesticides and Mode of Action*, Wiley, Chichester.

Crosby, D.G. (1985) Atmospheric reaction of pesticides, in *Pesticide Chemistry: Human Welfare and Environment* (eds J. Miyamoto and P.C. Kerney), vol. 3, Mode of Action, Metabolism and Toxicology, Pergamon Press, Oxford, pp 327-332.

Desmarchelier, J.M., Golding, M. and Horan, R. (1980) Predicted and observed residues of bioresmethrin, carbaryl, fenitrothion, d-fenothrin, methacrifos and pirimiphos-methyl on rice and barley after storage and losses of these insecticides during processing. *Journal of Pesticide Science*, **5** 539-545.

Dich, J., Zahm, S.H., Hanberg, A. and Adami, H.O. (1997) Pesticides and cancer. *Cancer Causes and Control*, **8** (3) 420-443.

Edwards, C.A. (1986) *Environmental Pollution by Pesticides*, Plenum, New York.

EEC (1994) Annex VI to Council Directive 94/43/EC. Placing of plant protection products on the market. *Official Journal of European Communities L 227/31*.

Elkins, E.R. (1989) Effect of commercial processing on pesticide residues in selected fruits and vegetables. *Journal of the Association of Official Analytical Chemists*, **72** 533-535.

Elkins, E.R., Farrow, R.P. and Kim, E.S. (1972) The effect of heat processing and storage on pesticide residues in spinach and apricots. *Journal of Agricultural and Food Chemistry*, **20** (2) 286-291.

Erney, D.R., Gillespie, A.M. and Gilvydis, D.M. (1993) Explanation of the matrix-induced chromatographic response enhancement of organophosphorus pesticides during open tubular column chromatography with splitless or hot on-column injection and flame photometric detection. *Journal of Chromatography*, **638** 57-73.

Erney, D.R., Pawlowski, T.M. and Poole, C.F. (1997) Matrix induced peak enhancement of pesticides in gas chromatography: is there a solution? *Journal of High Resolution Chromatography*, **20** 375-378.

Fahey, J.E., Gould, G.E. and Nelson, P.E. (1969) Removal of gardona and azodrin from vegetable crops by commercial preparative methods. *Journal of Agricultural and Food Chemistry*, **17** 1204-1206.

FAO (1959) *Report on the Meeting of the FAO Panel of Experts on the Use of Pesticides in Agriculture*, Food and Agriculture Organization of United Nations (Meeting Report No. 1959/3), Rome.

FAO (1990) International code of conduct on the distribution and use of pesticides, Prior informed consent in Article 9 as adopted by 25th session of the FAO, Food and Agriculture Organization of the United Nations, Rome.

FAO (1995) *Manual on the Development and Use of FAO Specifications for Plant Protection Products*, 4th edn, FAO Plant Production and Protection Paper No. 128. FAO, Rome.

FAO (1997) *Manual on Submission and Evaluation of Pesticide Residues Data for Estimation of Maximum Residue Levels in Food and Feed*, Food and Agriculture Organization of the United Nations, Rome, p 36.

FAO/WHO (1995) *Application of Risk Analysis to Food Standard Issues*, Report of a Joint FAO/WHO Expert Cosultation (WHO/FNU/FOS/95.3), World Health Organization, Geneva.

FAO/WHO (1997) Food consumption and exposure assessment to chemicals in food, *Report of Joint FAO/WHO Consultation, WHO/FSF/FOS*, World Health Organization, Geneva.

Farrow, R.P., Elkins, E.R., Rose, W.W., Lamb, F.C., Rall, J.W. and Mercher, W.A. (1969) Canning operations that reduce insecticide levels in prepared foods and in solid food wastes. *Residue Reviews*, **29** 73-87.

Farrow, R.P., Lamb, F.C., Cook, R.W., Kimball, J.R. and Elkins, E.R. (1968) Removal of DDT, malathion and carbaryl from green beans by commercial and home preparative procedures. *Journal of Agricultural and Food Chemistry*, **16** 962-966.

FDA (1995) FDA monitoring program (1995) *Journal of the Association of Official Analytical Chemists International*, **78** (5) 111-144.

Fillion, J., Hindle, R., Lacroix, M. and Selwyn, J. (1995) Multiresidue determination of pesticides in fruit and vegetables by gas chromatography-mass-selective detection and liquid chromatography with fluorescence detection. *Journal of the Asssociation of Official Analytical Chemists International*, **78** (5) 1252-1266.

Friar, P.M.M. and Reynolds, S.L. (1991) The effects of microwave-baking and oven-baking on thiabendazole residues in potatoes. *Food Additives and Contaminants*, **8** (5) 617-626.

Fuhr, F. (1982) Fate of herbicide chemicals in the agricultural environment with particular emphasis on the application of nuclear techniques, in *Agrochemicals: Fate in Food and the Environment*, International Atomic Energy Agency, Vienna, pp 63-82.

Gäb, S., Korte, F., Merz, W. and Neu, H.-J. (1985) Assessment of abiotic transformation, in *Pesticide Chemistry: Human Welfare and Environment* (eds J. Miyamoto and P.C. Kerney), vol. 3, Mode of Action, Metabolism and Toxicology, Pergamon Press, Oxford, pp 333-338.

Gajduskova, V. (1974) Dynamics of organophosphorus pesticide residues in butter during manufacture and storage. *Milchwissenschaft*, **29** 278-280.

Galal-Gorchev, H. (1993) Key elements of food contamination monitoring program. *Food Additives and Contaminants*, **10** (1) 1-4.

Garfield, F.M. (1991) *Quality Assurance Principles for the Analytical Laboratory*, AOAC International, Arlington.

Geisman, J.R. (1975) Reduction of pesticide residues in food crops by processing. *Residue Reviews*, **54** 43-54.

GEMS (1988) *Assessment of Chemical Contaminants in Food*, GEMS/Food, World Health Organization, Geneva.

GEMS (1995) *Reliable Evaluation of Low-Level Contamination of Food*, Report of Workshop in the Frame of GEMS/Food EURO, (ICP/EAHAZ.94.12/WS04), WHO Regional Office for Europe, Copenhagen.

GIHP (1996) General Inspectorate for Health Protection. *Analytical Methods for Pesticide Residues*, 6th edn, The Netherlands.

Gnaegi, F. and Dufour, A. (1972) Residues of anti-botrytis fungicides in wines. *Revue Suisse de Viticulture d'Arborioculture d'Horticulture*, **3** 101-106.

Gustafson, D.I. (1989) Groundwater ubiquity score: a simple method for assessing pesticide leachability. *Environmental Toxicology and Chemistry*, **8** 339-357.

Hajšlová, J., Holadová, K., Kocourek, V., Poustka, V. and Godula, M. (1998) Matrix-induced effects: critical point in GC analysis of pesticide residues. *Journal of Chromatography A*, **800** 283-295.

Hajšlová, J., Kocourek, V., Jehlickova, Z. and Davidek, J. (1986) The fate of ethylenebis (dithiocarbamate) fungicides during processing of contaminated apples. *Zeitschrift für Lebensmittel Untersuchung und Forschung*, **183** 348-351.

Hamilton, D.J., Holland, P.T., Ohlin, B., Murray, W.J., Amrus, A., Baptista, G.C. and Kovacicova, J. (1997) Optimum use of available residue data in the estimation of dietary intake of pesticides. *Pure and Applied Chemistry*, **69** (6) 1373-1410.

Hance, R.J. (1984) Herbicide residues in the soil: some aspects of their behaviour and agricultural significance. *Australian Weeds*, **3** (1) 26-35.

Hartley, G.S. and Graham-Bryce, I.J. (1980) *Physical Principles of Pesticide Behaviour*, vol. 2, Academic Press, New York.

Hassall, K.A. (1990) *The Biochemistry of Pesticides: Structure, Metabolism, Mode of Action and Uses in Crop Protection*, 2nd edn, Verlag Chemie, Weinheim.

Heller, S.R. and Herner, A.E. (1990) *ARS Pesticide Properties Database*, USDA-ARS, Systems Research Laboratory, Beltsville.

Hill, A. (1997) Quality control procedures for pesticide residues analysis, *Guidelines for Residue Monitoring in the European Union, Document* 7826/VI/97, Oeiras.

Hill, I.R. and Arnold, D.J. (1978) Transformation of Pesticides in the Environment—the Experimental Approach, in *Pesticide Microbiology. Microbiological Aspects of Pesticide Behaviour in the Environment* (eds I.R. Hill and S.J.L. Wright), Academic Press, London, pp 203-245.

Holland, P.T. (1996) Glossary of terms related to pesticides. *Pure and Applied Chemistry*, **68** (5) 1167-1193.

Holland, P.T., Hamilton, D., Ohlin, B. and Skidmore, M.W. (1994) Effects of storage and processing on pesticide residues in plant products. *Pure and Applied Chemistry*, **66** 335-356.

Holstege, D.M., Scharberg, D.L., Richardson, E.R. and Möller, G. (1991) Multiresidue screen for organophosphorus insecticides using gel permeation chromatography-silica gel clean up. *Journal of Association of Official Analytical Chemists*, **74** (2) 394-399.

Hopper, L.D. and Oehme, F.D. (1989) Chemical risk assessment: a review. *Veterinary and Human Toxicology*, **31** 543-554.

Hopper, M.L. and King, J.W. (1991) Enhanced supercritical fluid carbon dioxide extraction of pesticides from foods using pelletized diatomaceous earth. *Journal of the Association of Official Analytical Chemists*, **74** (4) 661-666.

Hornsby, A.G., Wauchope, R.D. and Herner, A.E. (1996) *Pesticide Properties in the Environment*, Springer-Verlag, New York.

Howard, P.H. (ed.) (1991) *Handbook of Environmental Fate and Exposure Data—Volume III. Pesticides*, Lewis Publishers, Chelsea.

Hsu, J.P., Schattenberg, H.J. and Garza, M.M. (1991) Fast turnaround multiresidue screen for pesticides in produce. *Journal of the Association of Official Analytical Chemists*, **74** (5) 886-892.

Ishikura, S., Onodera, S., Sumiyashiki, S., Kasahara, T., Nakayama, M. and Watanabe, S. (1984) Evaporation and thermal decomposition of organophosphorus pesticides during cooking of rice. *Journal of Food Hygienic Society Japan*, **25** 203-208.

JMPR (1982) *Pesticide Residues in Food*, FAO Plant Production and Protection Paper 46, Rome.

JMPR (1987) *Pesticide Residues in Food*, FAO Plant Production and Protection Paper 86/1, Rome.

JMPR (1988) *Pesticide Residues in Food*, FAO Pland Production and Protection Paper 93/2, Rome.

JMPR (1989) *Pesticide Residues in Food*, FAO Pland Production and Protection Paper 99, Rome.

JMPR (1991) *Pesticide Residues in Food*, FAO Pland Production and Protection Paper 113/1, Rome.

JMPR (1992) *Pesticide Residues in Food*, FAO Pland Production and Protection Paper 116, Rome.

Johnson, P.D., Rimmer, D.A. and Brown, R.H. (1997) Adaption and application of multi-residual method for the determination of a range of pesticides, including phenoxy acid herbicides in vegetation, based on high-resolution gel permeation chromatographic clean-up and gas chromatographic analysis with mass-selective detection. *Journal of Chromatography A*, **765** 3-11.

Joia, B.S., Webster, G.R.B. and Loshiavo, S.R. (1985) Cypermethrin and fenvalerate residues in stored wheat and millet fractions. *Journal of Agricultural and Food Chemistry*, **33** 618-622.

Jury, W.A., Spencer, W.F. and Farmer, W.J. (1983) Use of models for assessing volatility, mobility and persistence of pesticides and other trace organics in soil systems, in *Hazard Assessment of Chemicals*, vol. 2, Academic Press, London, pp 1-43.

Kadenczki, L., Arpad, Z., Gardi, I., Ambrush, A., Gyorfi, L., Reese, G. and Ebing, W. (1992) Column extraction of residues of several pesticides from fruits and vegetables: a simple multiresidue analysis method. *Journal of the Association of Official Analytical Chemists International*, **75** (1) 53-61.

Kaufman, B.M. and Clower, M.C. (1995) Immunoassay of pesticides: an update. *Journal of the Association of Official Analytical Chemists International*, **78** (4) 1079-1090.

Kawar, N.S., Iwata, Y., Disch, M.E. and Gunther, F.A. (1979) Behaviour of dialifor, dimethoate, and methidathion in artificially fortified grape juice processed into wine. *Journal of Environmental Science and Health*, **B20** 505-513.

Kenaga, E.E. (1980) Correlation of bioconcentration factors of chemicals in aquatic and terrestrial organisms with their physicochemical properties. *Environmental Science and Technology*, **14** (5) 553-556.

Keppel, G.E. (1971) Collaborative study of the determination of dithiocarbamate residues by a modified carbon disulphide evolution method. *Journal of the Association of Official Analytical Chemists*, **54** 528-532.

Kerney, P.C. and Kaufman, D.D. (eds) (1988) *Herbicides—Chemistry, Degradation and Mode of Action*, vol. 3, Marcel Dekker, New York.

Khan, S.U. (1982) Bound pesticide residues in soil and plants. *Residue Reviews*, **84** 1-23.

Khan, S.U. (1991) Bound (nonextractable) pesticide degradation products in soils : bioavailability to plants, in *Pesticide Transformation Products: Fate and Significance in the Environment* (eds L. Somasundaram and J.R. Coats), ACS Symposium Series 459, American Chemical Society, Washington, DC, pp 108-121.

Khan, S.U. (1995) Supercritical fluid extraction of bound pesticide residues from soil and food commodities. *Journal of Agricultural and Food Chemistry*, **43** 1718-1723.

Khan, S.U. (1996) *Bound Pesticide Residues*, CRC Press, Boca Raton, FL.

Kleinschmidt, M.G. (1971) Fate of di-syston(o,o-diethyl-S-[2-(ethylthio)ethyl]phosphorodithioate) in potatoes during processing. *Journal of Agricultural and Food Chemistry*, **19** 1196-1197.

Kubacki, S.J. and Lipowska, T. (1980) The role of food processing in decreasing pesticide contamination of foods, in *Food and Health, Science and Technology* (eds G.G. Birch and K.J. Parker), Applied Science, London, pp 215-226.

Kuchler, F., Ralston, K., Unnevehr, L. and Chandran, R. (1996) Pesticide residues—reducing dietary risks. *Agricultural Economic Report No. 728*, United States Department of Agriculture, Washington, DC.

Kuiper, H.A. (1996) The role of toxicology in the evaluation of new agrochemicals. *Journal of Environmental Science and Health Part B Pesticides, Food Contaminants, and Agricultural Wastes*, **31** (3) 353-363.

Lake, J.R. (1977) The effect of drop size and velocity on the performance of agricultural sprays. *Pesticide Science*, **8** 515-520.

Lamb, F.C., Farrow, R.P., Elkins, E.R., Cook, R.W. and Kimball, J.R. (1968) Removal of DDT, parathion, and carbaryl from spinach by commercial and home preparative methods, *Journal of Agricultural and Food Chemistry*, **16** 967-973.

Lawrence, L.J. and McLean, M.R. (1991) Toxicological significance of bound residues in livestock and crops, in *Pesticide Transformation Products: Fate and Significance in the Environment* (eds L. Somasundaram and J.R. Coats), ACS Symposium Series 459, American Chemical Society, Washington, DC, pp 242-253.

Lee, S.M., Papathakis, M.L., Feng, H.C., Hunter, G.F. and Carr, J.E. (1991) Multipesticide residue method for fruits and vegetables: California Department of Food and Agriculture. *Fresenius Journal of Analytical Chemistry*, **339** 376-383.

Lehotay, S.J. and Eller, K.I. (1995) Development of a method of analysis for 46 pesticides in fruits and vegetables by supercritical fluid extraction and gas chromatography/ion trap mass spectrometry. *Journal of the Association of Official Analytical Chemists International,* **78** (3) 821-830.

Lehotay, S.L., Aharonson, N., Pfeil, E. and Ibrahim, M.A. (1995) Development of sample preparation technique for supercritical fluid extraction for multiresidue analysis of pesticides in produce. *Journal of the Association of Official Analytical Chemists International,* **78** (3) 831-840.

Lemperle, E., Kerner, E. and Strecker, H. (1970) Active residues and fermentation after application of anti-botrytis agents in viniculture. *Wein-Wissenschaft,* **25** 313-328.

Lentza-Rizos, C. (1990) Ethylenthiourea in relation to use of ethylenbisdithiocarbamate fungicides. *Reviews of Environmental Contamination and Toxicology,* **115** 1-37.

Lentza-Rizos, C. (1995) Residues of iprodione in fresh and canned peaches after pre- and postharvest treatment. *Journal of Agricultural and Food Chemistry,* **43** 1357-1360.

Liao, W., Joe, T. and Cusick, W.G. (1991) Multiresidue screening method for fresh fruits and vegetables with gas chromatographic/mass spectrometric detection. *Journal of the Association of Official Analytical Chemists,* **74** (3) 554-565.

Liu, Ch., Mattern, G.C., Yu, X., Rosen, R.T. and Rosen, J.D. (1991) Multiresidue determination of nonvolatile and thermally labile pesticides in fruits and vegetables by thermospray MS. *Journal of Agricultural and Food Chemistry,* **39** 718-723.

Lopez-Avila, V., Young, R. and Teplitsky, L. (1996) Microwave assisted extraction as an alternative to Soxhlet, sonication and supercritical fluid extraction. *Journal of the Association of Official Analytical Chemists International,* **79** (1) 142-156.

Lucas, A.D., Gee, S.J., Hammock, B.D. and Seiber, J.N. (1995) Integration of immunochemical methods with other analytical techniques for pesticide residue determination. *Journal of the Association of Official Analytical Chemists International,* **78** (3) 585-591.

Luke, M.A., Froberg, J.E., Doose, G.M. and Masamuto, H.T. (1981) Improved multiresidue gas chromatographic determination of organophosphorus, organonitrogen, and organohalogen pesticides in produce using flame photometric and electrolytic conductivity detectors. *Journal of the Association of Official Analytical Chemists,* **64** (5) 1187-1195.

Mackay, D., Spencer, W.F.D. and Leinonen, P.J. (1975) Rate of evaporation of low/solubility contaminants from water bodies to atmosphere. *Environmental Science and Technology,* **9** 1178-1180.

MAFF (1995) *Consumer Risk Assessment of Insecticide Residues in Carrots,* Pesticide Safety Directorate, Ministry of Agriculture, Fisheries and Food, London.

Marshall, W.D. (1982) Preprocessing oxidative washes with alkaline hypochloride to remove ethylenbis (dithiocarbamate) fungicide residues from tomatoes and green beans. *Journal of Agricultural and Food Chemistry,* **30** 649-652.

Marshall, W.D. and Jarvis, W.R. (1979) Procedures for the removal of field residues of ethylenebis (dithiocarbamate) (EBDC) fungicide and ethylenethiourea (ETU) from tomatoes prior to processing into juice. *Journal of Agricultural and Food Chemistry,* **27** 766-769.

McCall, P.J., Laskowski, D.A., Swann, R.L. and Dishburger (1980) Measurement of sorption coefficient for chemicals and their use in the environmental rate and movement of toxicant, in *Test Protocols for Environmental Fate and Movement of Toxicants* (eds G. Zweig and M. Beroza), Association of Official Analytical Chemistst, Arlington, pp 89-109.

McMahon, B.M. and Wagner, R.F. (eds) (1996) *Pesticide Analytical Manual,* US Food and Drug Administration.

Merlo, F., Bolognesi, C. and Reggiardo, G. (1994) Carcinogenic risk of pesticides. *Journal of Experimental and Clinical Cancer Research,* **13** (1) 5-20.

Mestres, R. and Mestres, G. (1992) Deltamethrin: uses and environmental safety. *Reviews of Environmental Contamination and Toxicology,* **124** 1-25.

Miyahara, M. and Saito, Y. (1993) Pesticide removal efficiencies of soybean oil refining processes. *Journal of Agricultural and Food Chemistry*, **41** 731-734.

Moore, J.A. (1989) Speaking on data: The Alar controversy. *Environmental Protection Agency Journal*, May/June.

Moore, J.A. (1991) The need for common goals in pesticide management that reflect the consumers general interest, in *Pesticide Residues and Food Safety: a Harvest of Viewpoints* (eds B.G. Tweedy, H.J. Dishburger, L.G. Ballantine, J. McCarthy and J. Murphy), ACS Symposium Series 446, American Chemical Society, Washington, DC, pp 262-266.

Motohashi, N., Nagashima, H., Parkanyi, C., Subrahmanyanam, B. and Zhang, G. (1996) Official multiresidue methods of pesticide analysis in vegetables, fruits and soil. *Journal of Chromatography A*, **754** 333-346.

Moy, G. and Wessel, J. (1998) Codex Standards for pesticide residues, in *Development of International Standards* (eds N. Rees and D. Watson), Chapman and Hall, London, in press.

Muhammad, M.A. and Kawara, N.S. (1985) Behaviour of parathion in tomatoes processed into juice and ketchup. *Journal of Environmental Science and Health*, **B20** 499-510.

Nakamura, Y., Tonogai, Y., Sekiguchi, Y., Tsumura, Y., Nishida, N., Takakura, K., Isechi, M., Yuasa, K., Nakamura, M., Kifune, N., Yamamoto, K., Tarasawa, S., Oshima, T., Miyata, M., Kamakura, K. and Ito, Y. (1994) Multiresidue analysis of 48 pesticides in agricultural products by capillary gas chromatography. *Journal of Agricultural and Food Chemistry*, **42** 2508-2518.

Newsome, W.H. (1976) Residues of four ethylenebis (dithiocarbamates) and their decomposition products on field-sprayed tomatoes. *Journal of Agricultural and Food Chemistry*, **20** 967-969.

Newsome, W.H. (1995) An evolution of pesticide residue methodology. *Journal of the Association of Official Analytical Chemists International*, **78** (1) 4-8.

NFA (1995) *Pesticide Residues in Food of Plant Origin 1994*, Rapport 20/95, National Food Administration, Uppsala.

NFA (1996) *Pesticide Residues in Food of Plant Origin 1995*, Rapport 10/95, National Food Administration, Uppsala.

NFA (1997) *Pesticide Residues in Food of Plant Origin 1996*, Rapport 25/95, National Food Administration, Uppsala.

Nitz, S., Moza, P.N., Kokabi, J., Freitag, D., Behechti, A. and Korte, F. (1984) Fate of ethylenebis (dithiocarbamates) and their metabolites during the brew process. *Journal of Agricultural and Food Chemistry*, **32** 600-603.

NRC (1993) *Pesticides in the Diets of infants and Children*, National Academy Press, Washington, DC.

OECD (1992) Organization for Economic Cooperation and Development GLP Guidelines No. 1. The OECD Principles of Good Laboratory Practice. *Environment Monographs No. 45*, Paris.

Olea, M.F. (1996) Xenobioticos estrogenicos: Un objektivo la toxicologia funcional. *Ars Pharmaceutica*, **37** (2) 183-196.

Pennington, J.A.T. (1992) Total diet studies: the identification of core foods in the United States food supply. *Food Additives and Contaminants*, **9** (3) 253-264.

Phillips, W.F., Grady, M.D. and Freudenthal, R. (1977) Effects of food processing on residues of ethylenbisdithiocarbamate fungicides and ethylenthiourea. *Environmental Health Effects Research Services Report*, EPA-600/1-77-021, US Environmental Protection Agency.

Racke, K.D. and Coats, J.R. (eds) (1990) *Enhanced Biodegradation of Pesticides in the Environment*, ACS Symposium Series 426, American Chemical Society, Washington, DC.

Ripley, B.D., Cox, D.F., Wiebe, J. and Frank, R. (1978) Residues of Dikar and ethylenethiourea in treated grapes and commercial grape products. *Journal of Agricultural and Food Chemistry*, **26** 134-136.

Sannino, A., Mambriani, P., Bandini, M. and Bolzoni, L. (1995) Multiresidue method for determination of organophosphorus insecticide residues in fatty processed foods by gel permeation chromatography. *Journal of the Association of Official Analytical Chemists International*, **78** (6) 1502-1512.

Sawada, J. (1995) Immunotoxicity of chemicals. *Bulletin of the National Institute of Hygienic Sciences*, **113** 1-18.

Scharchterle, S. and Feigel, C. (1996) Pesticide residue analysis by gas chromatography - tandem mass spectrometry. *Journal of Chromatography A*, **754** 411-422.

Schattenberg, H.J., Geno, P.W., Hsu, J.P., Fry, W.G. and Parker, R.P. (1996) Effect of household preparation on levels of pesticide residues in produce. *Journal of the Association of Official Analytical Chemists International*, **79** 1447-1453.

Scheunert, I. and Klein, W. (1985) Chemicals between environmental compartments (predicting movement of air-water-soil-biota), in *SCOPE 25: Apprasial of Tests to Predict Environmental Behaviour of Chemicals* (ed. P. Sheeman), Wiley, New York, pp 307-332.

Sherma, J. (1997) Current status of pesticide residue analysis. *Journal of the Association of Official Analytical Chemists International*, **80** (2) 283-287.

Snyder, J.L., Grob, R.L., McNally, M.E. and Oostdyk, T.S. (1992) Comparison of supercritical fluid extraction with classical sonication and Soxhlet extraction for selected pesticides. *Analytical Chemistry*, **64** 1940-1946.

Somasundaram, L. and Coats, J.R. (1991) Pesticide transformation products in the environment, in *Pesticide Transformation Products: Fate and Significance in the Environment* (eds L. Somasundaram and J.R. Coats), ACS Symposium Series 459, American Chemical Society, Washington, DC, pp 2-9.

Specht, W. and Tillkes, M. (1985) Gas-chromatographische bestimmung von rückständen an pflanzenbehandlungsmitteln nach clean-up über gel-chromatographie und mini-kieselgel-säulen-chromatographie. *Fresenius Zeitschrift für Analytische Chemie*, **322** 443-455.

Specht, W., Pelz, S. and Gilsbach, W. (1995) Gas-chromatographic determination of pesticide residues after clean-up by gel-permeation chromatography and mini-silica gel-column chromatography. *Fresenius Journal of Analytical Chemistry*, **353** 183-190.

Spencer, W.F., Farmer, W.J. and Cliath, M.M. (1973) Pesticide volatilization. *Residue Reviews*, **49** 1-47.

Steinwandter, H. (1988) Contributions to the application of gel-chromatography in residue analysis. *Fresenius Zeitschrift für Analytische Chemie*, **331** 499-502.

Stevens, J.T., Tobias, A., Lamb IV, J.C., Tellone, C. and O'Neal, F. (1997) Fifra subdivision F testing quidelines: Are these tests adequate to detect potential hormonal activity for crop protection chemicals? *Journal of Toxicology and Environmental Health*, **50** (5) 415-431.

Stison, S. (1995) Mass spectral method distinguishes pesticides. *Chemical Engineering News*, (Nov. 13) 95.

Stratton, G.D. and Wheeler, W.B. (1986) Characterization of bound residues in plants, in *Quantification, Nature and Bioavailability of Bound ^{14}C- Pesticide Residues in Soil, Plants and Food*, International Atomic Energy Agency, Vienna, pp 71-82.

Tekel, J. and Hatrík, Š. (1996) Pesticide residue analysis in plant material by chromatographic methods: clean-up procedures and selective detectors. *Journal of Chromatography A*, **754** 397-410.

Thier, H.-P. and Frehse, H. (eds) (1986) *Rückstandsanalytik von Pflanzenshutzmitteln*, Georg Thieme Verlag, Berlin.

Thier, H.-P. and Zeumer, H. (eds) (1987) *Manual of Pesticide Residue Analysis*, DFG Pesticide Commission, VCH, Weinheim.

Tomlin, C. (1989) *The Agrochemicals Handbook*, 2nd edn, Royal Society of Chemistry, Nottingham.

Tomlin, C.D.S. (ed.) (1997) *The Pesticide Manual*, 11th edn, British Crop Protection Council, Surrey.

UNEP (1992) *The Contamination of Food*, UNEP/GEMS Environmental Library No. 5, UNEP, Nairobi.

van der Kolk, J. (1991) The role of monitoring in decision making, in *Pesticide Chemistry: Advances in International Research, Development and Legislation* (ed. H. Frehse), VCH, Weinheim, pp 485-491.

Verschueren, K. (1983) *Handbook on Environmental Data on Organic Chemicals*, 2nd edn, Van Nostrand Reinhold, New York.

Vollmer, D.A., Vollmer, D.L. and Wilkes, J.G. (1996) Multiresidue analysis of pesticides by electrospray LC/MS and LC/MS/MS. *LC-GC International*, **14** 216-224.

Wagenet, R.J. and Hutson, J.L. (1991) Modelling pesticide transport and transformation in the soil-plant system, in *Pesticide Chemistry: Advances in International Research, Development and Legislation* (ed. H. Frehse), VCH, Weinheim.

Ware, G.W. (1989) *The Pesticide Book*, 3rd edn, Thomson Publications, Fresno.

Wauchope, R.D., Buttler, T.M., Hornsby, A.G., Augustijn-Beckers, P.W.M. and Burt, J.P. (1992) The SCS/ARS/CES pesticide properties database for environmental decision-making. *Reviews of Environmental Toxicology*, **123** 1-162.

Weddie, P.W. (1991) Pesticide-free tree fruit crops: can we meet consumer demands? in *Pesticide Residues and Food Safety: a Harvest of Viewpoints* (eds B.G. Tweedy, H.J. Dishburger, L.G. Ballantine, J. McCarthy and J. Murphy), ACS Symposium Series 446, American Chemical Society, Washington, DC, pp 63-68.

Weisenburger, D.D. (1993) Human health effects of agrichemical use. *Human Pathology*, **24** (6) 571-576.

WHO (1978) Principles and methods for evaluating the toxicity of chemicals. *Environmental Health Criteria, No. 6*, World Health Organization, Geneva.

WHO (1987a) Principles for the safety assessment of food additives and contaminants in food. *Environmental Health Criteria, No. 70*, World Health Organization, Geneva.

WHO (1987b) Guidelines for the study of dietary intake of pesticides, *WHO Publication No. 87*, World Health Organization, Geneva.

WHO (1988) *Environmental Health Criteria No. 78*, Dithiocarbamate pesticides, ethylenethiourea and propylene thiourea: A general introduction. World Health Organization, Geneva.

WHO (1990a) *Public Health Impact of Pesticides Used in Agriculture*, World Health Organization, Geneva.

WHO (1990b) Principles for toxicological assessment of pesticide residues in food. *Environmental Health Criteria, No. 104*, World Health Organization, Geneva.

WHO (1996) *The WHO Recommended Classification of Pesticides by Hazard Guidelines to Classification 1996-1997*, WHO/PCS/96.3, World Health Organization, Geneva.

WHO (1997) *Guidelines for Predicting Dietary Intake of Pesticide Residues (revised)*, Report GEMS/Food in collaboration with CCPR (WHO/FSF/FOS/97.7), World Health Organization, Geneva.

Willis, G.H. and McDowel, L.L. (1987) Pesticide persistence on foliage, in *Reviews of Environmental Contamination and Toxicology*, vol. 100 (ed. G.W. Ware), Springer Verlag, New York.

Winter, C.K. (1992) Dietary pesticide risk assessment. *Reviews of Environmental Contamination and Toxicology*, **127** 23-67.

WPPR (1995) *Annual Report of the Working Party on Pesticide Residues: 1994*, Ministry of Agriculture, Fisheries and Food Publications, London.

WPPR (1996) *Annual Report of the Working Party on Pesticide Residues: 1995*, Ministry of Agriculture, Fisheries and Food Publications, London.

WPPR (1997) *Annual Report of the Working Party on Pesticide Residues: 1996*, Ministry of Agriculture, Fisheries and Food Publications, London.

Zabik, M.E., Zabik, M.J., Booren, A.M., Nettles, M., Song, J.H., Welch, R. and Humphrey, H. (1995) Pesticides and total polychlorinated biphenyls in Chinook salmon and carp harvested from the Great Lakes: effects of skin-on and skin-off processing and selected cooking methods. *Journal of Agricultural and Food Chemistry*, **43** 993-1001.

Zepp, R.G. (1991) Photochemical fate of agrochemicals in natural waters, in *Pesticide Chemistry: Advances in International Research, Development and Legislation* (ed. H. Frehse), VCH, Weinheim, pp 329-346.

Zweig, G. and Sherma, J. (eds) (1984) *Analytical Methods for Pesticides and Plant Growth Regulators*, vol. XIII, Academic Press, Orlando, FL.

Zweig, G. and Sherma, J. (eds) (1985) *Analytical Methods for Pesticides and Plant Growth Regulators*, vol. XIV, Academic Press, Orlando, FL.

Zweig, G. and Sherma, J. (eds) (1986) *Analytical Methods for Pesticides and Plant Growth Regulators*, vol. XV, Academic Press, Orlando, FL.

8 Veterinary drugs including antibiotics

George Shearer

8.1 Introduction

Veterinary drugs were first used in animals after the Second World War, during which it was found that the use of penicillin and sulphonamides helped to control sepsis in wounded soldiers. Penicillin had accidentally been found by Sir Alexander Fleming in 1929 when he noticed that a microbiological plate that had been left lying around in his laboratory had areas where microorganisms failed to grow if *Penicillium* cultures had grown. A further 10 years or so elapsed before the importance of this find was realised. Florey and Chain worked on isolation of the inhibiting substance in the *Penicillium* cultures and, around 1941, were able to produce sufficient penicillin for it to be used medically (Fleming, 1946).

The penicillin discovery had been overtaken in the 1930s by a substance produced synthetically that was called prontosil, a dyestuff that when used *in vitro* had no antibacterial activity but did inhibit bacterial growth when given to animals. It was discovered that in the animal the dyestuff was split into a substance called sulphanilamide, which did show inhibitory properties.

Thus was started a major industry that still exists as the search for antibacterial substances continues. Antibacterial substances produced naturally by organisms are called antibiotics and substances synthesised in the laboratory are called chemotherapeutics or antibacterials to differentiate their origin. Routine testing of isolates from bacterial sources as well as compounds synthesised in laboratories around the world are routinely tested to find new drugs. There is a wide range of such substances available either as variations on a theme, e.g. different side-chains on a basic structure such as the sulphonamides, or with totally different chemical structures.

The stimulus of need for such compounds in times of war was the spur, but the potential of these 'wonder drugs' for use in peace time medicine soon became recognised. Naturally such drugs were initially used in human medicine, but testing of them undoubtedly took place in animals to ensure their safety, and it was not long before they were being used in farm animals, initially for therapeutic purposes to cure disease but soon for other purposes.

Post-war it was recognised how vulnerable the United Kingdom was because it was not self-sufficient in food, which had to be shipped in at enormous cost. Now we are self-sufficient for all the staple foods that can be produced in this country as a result of a tremendous agrarian revolution that has taken place over the last 40–50 years.

Animals can be fed with home-grown cereals and production units now intensively rear animals, particularly pigs and poultry, where these animals are kept in limited space. Such intensive units were prone to outbreaks of disease that could decimate them, but it was found useful to routinely feed low levels of antibiotics to animals to prevent the initial outbreak. This is known as prophylactic treatment. It was also found that treatment of animals with drugs at low concentrations often improved the growth rates of these animals by quite significant amounts, and so this practice grew as did the use of hormones that are the natural growth promoters.

Thus, animals are treated with drugs for therapeutic, prophylactic and growth promotional purposes.

Different drugs were produced by the pharmaceutical industry to treat different diseases, infections or infestations. Animals in the field sometimes pick up worms that adversely affect their growth, and substances known as anthelmintics have been produced that specifically kill worms. In poultry, a protozoan gut infection known as coccidiosis can be controlled by drugs known as coccidiostats. There is now a wide range of drugs available to treat the ailments of man and animals.

Selected drugs are listed in Table 8.1 to give some picture of the numbers and nature of drugs used. These include presently banned substances as well as permitted drugs.

8.2 Legislation

Usage of drugs and particularly of antibacterials had grown sufficiently by the 1960s that concern was being expressed on the grounds of the development of bacterial resistance, the potential for allergic reactions in humans, and the occurrence in the meat supply of residues of undesirable substances. The Swann Committee was set up to consider these matters and, in 1968, reported that there was no real cause for concern but recommended that drugs used in animal medicine should be different from those used in human medicine. This has been observed since, although the older drugs penicillin, sulphonamides, and tetracyclines are still used in both.

8.2.1 Licensing of drugs

The Medicines Act was promulgated in 1968 and one of its aims was to license all drugs. The Veterinary Products Committee (VPC), an

Table 8.1 Selected list of drugs used in or tested for in animals

Stilbenes
Diethylstilboestrol
Hexoestrol
Dienoestrol
Thyrostat
Methylthiouracil
Phenylthiouracil
Thiouracil
Propyluracil
1-Methyl 2-mercaptimidazole
Steroids
Trenbolone
Nortestosterone
Methyltestosterone
Oestradiol
Oestrone, etc.
Resorcylic lactones
Zeranol
β-Agonists
Clenbuterol
Salbutamol
Mabuterol
Cimaterol
Terbutaline, etc.
Annex IV
Nitrofurans
Chloramphenicol
Dimetridazole
Ronidazole, etc.

ANTIMICROBIALS
β-Lactams
Benzylpenicillin
Amoxicillin
Ampicillin
Cloxacillin
Cephalosporins
Tetracyclines
Tetracycline
Oxytetracycline
Chlortetracycline
Doxycycline
Aminoglycosides
Streptomycin
Neomycin

Gentamicin
Lincomycin
Bambermycin, etc.
Macrolides
Tylosin
Erythromycin
Spiramycin, etc.
Sulphonamides
Sulphadimidine
Sulphadiazine
Sulphapyridine
Sulphathiazole
Sulphaquinoxaline, etc.
Quinolones
Oxolinic acid
Nalidixic acid
Fluoroquinolones
Ciprofloxacin
Enrofloxacin
Danofloxacin
Sarofloxacin

NSAIDs[a]
e.g. Aspirin, Paracetamol

TRACE ELEMENTS
e.g. Arsenic, Cadmium, Lead
Organochlorines/PCBs
DDT
DDE
Dieldrin
PCBs, etc.
Organophosphorus
Dichlorvos
Malathion
Parathion
Coumaphos, etc.
Carbamates/pyrethroids
Cypermethrin
Permethrin
Deltamethrin, etc.
Dyes
Malachite Green
Leuco malachite green
Gentian Violet
Methylene Blue

[a]NSAIDs, non-steroidal anti-inflammatory drugs.

independent group of experts, was set up to consider data presented by the pharmaceutical companies in support of new drugs. This data must show the efficacy of the drug, safety in production and handling, safety to the animals treated, and the depletion pattern and how this was determined. The VPC, if satisfied, recommends the licensing of the drug to the Ministers of both Health and Agriculture, including specification of a withdrawal period.

Although still functioning, the VPC has now been overtaken by a European Committee known as the Committee of Veterinary Medicinal Products (CVMP), which functioned for a number of years prior to being taken under the auspices of the European Medicines Evaluation Agency (EMEA) formed in 1995. The CVMP is responsible for the approval of veterinary drugs Europe-wide, based on similar data presented by the companies.

A more recent task, undertaken initially by the VPC and again taken over by a Working Group of the CVMP, is the recommendation of Maximum Residue Limits (MRLs) for drugs in meat and meat products based on toxicological grounds.

A similar task is undertaken by a joint World Health Organisation and Food and Agriculture Organisation (WHO/FAO) Committee known as the Joint Expert Committee for Food Additives (JECFA). These MRLs are intended to be the World Standards and are approved by the Committee for Residues for Veterinary Drugs in Food (CCRVDF) of the Codex Alimentarius.

The United Kingdom first published some MRLs for a limited list of drugs in their Animals Meat and Meat Products (Examination for Residues and Limits) Regulations in 1991. The CVMP intends to produce MRLs for all veterinary drugs in use under EC Regulation 90/2377 and routinely issues updates. The CCRVDF similarly adds to its list.

Table 8.2 shows typical values for a limited list of drugs issued by all three bodies. It will be noted that different tolerance limits (MRLs) for each drug, tissue and sometimes species are specified. It is also consoling to note that the various groups of experts have arrived at broadly similar conclusions.

8.2.2 Derivation of Maximum Residue Limit (MRL)

The first value derived for a drug is the No Effect Level (NOEL). This is normally based on animal experiments. This is translated into the likely NOEL value for an adult human of 60 kg. Next, the nature of any effect above the NOEL is considered. This may be minor, such as drowsiness or rash; more serious, such as nausea or elevation of blood pressure; or extremely serious, such as teratogenic or carcinogenic effects. The NOEL is divided by a factor such as 10, 100, 1000 or greater depending on the

Table 8.2 Examples of MRLs produced by UK, EU and JECFA

Drug	Species[a]	Tissue[b]	UK[c]	EU[d] concentration - µg/kg	JECFA[e]
Benzylpenicillin	All	M,L,K,F	50	50	50
		Mi	–	4	4
Tetracycline	All	K	600	600	600
(TC,OTC,CTC)[f]		L	300	300	300
		M	100	100	100
		Mi	100	200(TC)	200
				100(OTC, CTC)	–
		E	100	200	200
Trimethoprim	All	M,L,K,F,Mi	50	50	–
Tylosin	C,P,S	M,L,K	–	100	–
Ceftiofur	C	L	–	2000	2000
		K	–	2000	4000
		F	–	600	600
		Mi	–	100	100
Streptomycin	C,P,S,	K	1000	1000	1000
	Po	M,L,F	1000	500	500
		Mi	1000	200	200
Febantel	All	L	1000	1000	500
Fenbendazole		M,K,F	10	10	100
Oxfendazole		Mi	10	10	100
Sulphonamides	All	M,L,K,F	100	100	100
		Mi	100	100	25
Ivermectin	C	L	15	100	100
		F	20	40	40
	P,S,H	L	15	15	
		F	20	20	20
Levamisole	C,P,S,Po	M,K,F,	10	10	10(M,F)
		L	10	100	100(L,K)
Dimetridazole	C,P,S,Po	M,K,L,F		10	10 or 0

[a]C, cattle; P, pig; S, sheep; Po, poultry; H, horse.
[b]M, muscle; L, liver; K, kidney; F, fat; Mi, milk; E, eggs.
[c]UK Regulations 1991.
[d]EU updated (May 1997) Commission Regulations 90/2377.
[e]CCRVDF Working Paper 1996 (personal communication).
[f]TC, tetracycline; OTC, oxytetracycline; CTC, chlortetracycline.

seriousness of the effect to produce an Acceptable Daily Intake (ADI) for an adult. The potential source of this ADI is next considered and extreme intakes such as 300 g meat, 100 g liver, 50 g kidney, 1500 g milk or 100 g egg on a daily basis are used to calculate the MRLs. The values are very safe because of these in-built safety factors.

As can be seen, MRLs are normally in the µg/kg (part per billion, ppb) range ensuring consumers' safety, particularly when it is considered that, when taken for medical purposes, grams of drug are ingested.

Withdrawal periods that were defined at time of licensing and that originally were based on Good Veterinary Practice will now be based on achievement of the MRL.

8.2.3 Drug resistance

Concern about drug resistance still persists despite the continuing efficacy of the oldest antibacterial drugs. Some bacteria are resistant to some drugs but succumb to others, and there is little evidence of truly broad-based resistance. Public health authorities from time to time identify strains with such resistance and a current example is *Salmonella typhimurium* DT104 which is resistant to ampicillin, chloramphenicol, streptomycin, sulphonamides and tetracyclines. It is also suggested that evidence is growing of further resistance of this strain to trimethoprim and fluoroquinolones. In addition, there is concern about emergence of methicillin-resistant strains of staphylococci (MRSE), which is of particular concern when dealing with immunodepressed individuals.

It is clearly prudent to keep a watching brief on such matters.

8.3 Surveillance programmes

8.3.1 Statutory programme

In 1986 the EU promulgated Directive 86/469, which required each Member State to produce a surveillance programme to investigate the occurrence of veterinary drug residues in the national meat supply. This programme was intended only to include the so called Red Meat animals —cattle, pigs, sheep, goats and horses—although it did give room for extension to other meat species.

The Annex to this Directive, shown as Table 8.3, indicated the drugs that were to be tested for and the numbers of samples to be included. It required that some 0.15% of the animals slaughtered in the Member State should be tested for some of the required drugs. Of the samples 0.1% should be taken at the abattoir and 0.05% on farm.

In the United Kingdom, where the slaughter was of the order of 3 million cattle, 14 million pigs and 18 million sheep, this required a total programme of some 50 000 samples, with 34 000 taken at the abattoir and 17 000 on farm. Clearly, over the years, there have been variations in the species numbers slaughtered, but the overall sample numbers have remained remarkably similar. Thus, at present, because of BSE the slaughter of cattle has fallen below 2 million, compensated for by increases in sheep and pig slaughter.

As can be seen, the drugs and drug classes named in Table 8.3 cover a wide range of substances. The Group I substances include a range of compounds that, although permitted for use at one time, have been subsequently banned. They include natural and synthetic hormonal substances used for growth promotion and thyrostats that operate by slowing down the function of the thyroid, thereby allowing weight gain.

In the EU stilbenes, which had been used widely for caponisation of chickens as well as growth promotion of cattle and pigs, and thyrostats were banned by Directive 81/602 and its Regulations 82/626, The Medicines (Stilbenes and Thyrostats Substances) Regulation. Stilbenes had been shown to have carcinogenic properties, which led to their ban, and similar action was adopted fairly well on a worldwide basis around this period.

In 1985, the EU produced Directive 85/649 with Regulation 88/705, The Medicines (Hormone Growth Promoters) (Prohibition of Use) Regulation, and 88/849, The Animals and Fresh Meat (Hormonal Substances) Regulations, which banned the use of all substances having hormone growth promotion action. This included natural (testosterone, oestradiol, etc.) and synthetic (trenbolone, zeranol) hormonal substances against which no scientific reason on grounds of health endangering

Table 8.3 Directive 86/469 defining residues to be analysed for

---

A Groups common to all Member States

Group I
(a) Stilbenes, stilbene derivatives, their salts and esters.
(b) Thyrostatic substances
(c) Other substances with oestrogenic, androgenic or gestagenic action,
    with the exception of substances in Group II

Group II
(a) Substances authorised in accordance with Article 4 of Directive 81/602/EEC
    and Article 2 of Directive 85/649/EEC

Group III
(a) Inhibitors
    Antibiotics, sulphonamides and similar antimicrobial substances
(b) Chloramphenicol

B Specific groups

Group I: Other medicines
(a) Endo- and ectoparasitic substances
(b) Tranquillisers and β-blockers
(c) Other veterinary medicines

Group II: Other residues
(a) Contaminants present in feedstuffs
(b) Contaminants present in the environment
(c) Other substances

---

properties had or has since been levelled. The EU were reasonably alone in taking this action and many countries including the United States continue to use these drugs for growth promotion. This has caused friction between these countries and the EU over the intervening 10 years, with threatened trade wars from time to time. The present situation is that countries permitting the use of hormones for growth-promoting purposes but wishing to export to the EU must issue certificates assuring that the meat being supplied comes from farms or feed lots not using hormones.

On-farm sampling under Directive 86/469 is mainly directed towards identifying the use of these banned substances, although some samples are also taken at the abattoir for evidence of use of hormones or thyrostats. The on-farm samples consist of urine, faeces, blood, water and feedstuffs.

Testing for residues of the wide range of antibiotic or antimicrobial substances permitted for use in animal husbandry is required under Group A III. The term 'antibiotic' in its strictest use indicates that the substance has been isolated from a microorganism and has antibacterial properties. Penicillins, tetracyclines and many other substances fall into this category. Antimicrobial substances or chemotherapeutics are synthetic substances having antibacterial action. Among these are the sulphonamides, nitrofurans and many others, as pharmaceutical companies routinely check all synthesised compounds for antibacterial activity.

'Sulphonamides and similar antimicrobial substances' and 'chloramphenicol' are separated out in this group because they are not readily detected by the usual antimicrobial inhibition tests used to screen for these substances. This will be discussed later in section 8.6.

Group I of Section B of the Annex includes a range of drugs used more specifically in animal husbandry. Animals pick up pests in the course of their life that affect them both internally and externally. Worms and other parasites are ingested and reside within the various organs of an animal, and substances known as anthelmintics are used, particularly in young animals, to clear these from the animals' system. Externally, animals can suffer from infestations of flies and ticks, etc., which are treated with pesticides. An animal licking treated areas could readily ingest such substances, giving rise to residues in the meat. Tranquillisers and β-blockers, particularly carazolol, are used to calm some species of animals and particularly pigs during transport to reduce stress. This is especially the case during transport to slaughter, when stress can cause changes in pH of meat causing it to became tougher than might be the case otherwise. 'Other veterinary medicines' is a catch-all if required; a case in point is the β-agonist class of drugs. These have grown in importance since 1986. It has been shown that their use causes repartition between fat and meat. This term means that animals, instead of laying down energy surpluses as fat, do so as protein or meat. Clenbuterol, sometimes known as 'angel dust', is particularly effective, although other drugs of this class are reported to be in use. These are included in the list shown in Table 8.1.

The Group B II residues required to be analysed for are feedstuff contaminants, such as organochlorine or organophosphorus pesticides, or environmental contaminants picked up by the animals, such as polychlorinated biphenyls, dioxins or toxic trace elements, e.g. lead (Pb), cadmium (Cd) or arsenic (As). Again, a catch-all term such as 'other substances' is included, lest some unusual contaminant be identified at any time.

Surveillance has been on-going in the EU for the past 10 years. Each Member State must submit its sampling and analytical plan annually to Brussels for approval in advance of its commencement and, after completion, the results are submitted. Unfortunately, Brussels has failed to publish the overall results obtained to date. This is partly because the various Member States submit their results in a variety of formats, making correlation difficult, as well as often failing to meet the deadlines for submission.

In the United Kingdom, results of the National Plan or Statutory Surveillance Programme are published in the quarterly publication of the Veterinary Medicines Directorate, called MAVIS. The results show no evidence of abuse of the Group A/Group I hormonal substances in the United Kingdom nor of β-agonists. The antibacterials show a low (less than 1%) incidence of positives in the screening test, usually due to residues of the commoner antibiotics such as penicillins, tetracyclines and aminoglycosides. The evidence for abuse of sulphonamides, particularly in pig kidney, has reduced very significantly over the years from a high of 10–12% initially to the current level, of approximately 1%. Chloramphenicol residues have never been found in the United Kingdom, to my knowledge, and the Group B substances produce only the occasional positive such as an anthelmintic, pesticide or, in the case of horses for export to the continent for human feeding, a consistently high cadmium concentration. This accumulation is peculiar to horses and is something that apparently cannot be avoided.

Directive 86/469 has been superseded by Directive 96/23, which incorporates many of the lessons learnt over the intervening period. As can be seen in Table 8.4, the general coverage of drugs is similar, although laid out more specifically, and there has been a switch of emphasis from abattoir to on-farm sampling. The other feature of 96/23 is that it encompasses many other meat-producing species including poultry, farmed fish, milk, eggs, farmed game, rabbits and honey. The Directive

Table 8.4 Directive 96/23 on measures to monitor substances and residues thereof in live animals and animal products

Group A Substances having anabolic effect and unauthorised substances
1. Stilbenes, stilbene derivatives, their salts and esters
2. Antithyroid agents
3. Steroids
4. Resorcylic acid lactones including zeranol
5. β-Agonists
6. Compounds included in Annex IV to Council Regulation (EEC) No 90/2377
Group B Veterinary drugs and contaminants
1. Antibacterial substances including sulphonamides, quinolones
2. Other veterinary drugs
(a) Anthelmintics
(b) Anticoccidials, including nitroimidazoles
(c) Carbamates and pyrethroids
(d) Sedatives
(e) Non-steroidal anti-inflammatory drugs (NSAIDs)
(f) Other pharmaceutically active substances
3. Other substances and environmental contaminants
(a) Organochlorine compounds including PCBs
(b) Organophosphorus compounds
(c) Chemical elements
(d) Mycotoxins
(e) Dyes
(f) Others

indicates the drug classes that should be tested for in each of these species. This is shown in Table 8.5. It is intended that surveillance of the first two named species, viz. poultry and farmed fish, should be commenced in January 1998 and the others should follow in January 1999. Red meat surveillance will continue under this Directive. Again, the Member States are required to submit their sampling plan for approval by Brussels prior to commencement and to submit the results obtained shortly after the completion of the annual analytical programme.

Each of the industries has to bear the cost of these surveillance programmes, which includes the sample collection, the analytical work and the administration costs. The red meat programme under 86/469 is reported to cost the UK industry about £2.8 million, collected as a headage levy on animals through the abattoirs. While this appears a high cost, it must be placed against the total value of the red meat industry, which is thought to be of the order of £9 billion. In other words, the cost of the surveillance programme represents a tiny 0.03% of the industry's output.

Table 8.5 Residue or substance group to be detected by type of animal, their feedstuffs including drinking water and primary animal products

Type of animal feedstuffs or animal products substance group	Bovine, ovine, caprine, porcine, equine animals	Poultry	Aquaculture animals	Milk	Eggs	Rabbit meat and the meat of wild game[a] and farmed game	Honey
A 1[b]	×	×	×			×	
2	×	×				×	
3	×	×	×			×	
4	×	×				×	
5	×	×				×	
6	×	×	×	×	×	×	
B 1	×	×	×	×	×	×	×
2a	×	×	×	×		×	
b	×	×			×	×	
c	×	×				×	×
d	×						
e	×	×		×		×	
f							
3a	×	×	×	×	×	×	×
b	×			×			×
c	×	×	×	×		×	×
d	×	×	×	×			
e			×				
f							

[a]Only chemical elements are relevant where wild game is concerned.
[b]For drug class involved, see categories in Table 8.4.

8.3.2 Non-statutory surveillance

The United Kingdom has through the years run a parallel Non-Statutory Programme, which has been paid for centrally. In fact, this programme has dealt with many of the species being incorporated in the programmes being included now in Directive 96/23. Its intent was to investigate the possible occurrence of drug residues in other sectors of the meat supply away from home-produced red meat. The programme also targeted imported meats of all types and often the samples were taken at retail outlets where the consumer acquires meat. The programme also endeavoured to look at a wider range of drugs because of the different commodities being tested. Thus, coccidiostats that are used in poultry and rabbit production were looked for, drugs used for specific disease control such as imidocarb for red water disease in cattle (Tarbin and

Shearer, 1992), and some suggested non-licensed use of drugs such as ivermectin in farmed fish for control of sea lice have been included.

This programme similarly showed that the incidence of drug residues was low, although it did highlight some problems. For example, when the United Kingdom introduced an MRL of $10\,\mu g/kg$ for levamisole, it became apparent that imported lamb liver sometimes failed to reach this value. This was drawn to the attention of authorities and the situation was corrected. Another example was the occasional incidence of clenbuterol in imported cattle liver. The country involved indicated it was testing cattle urine and finding no evidence of abuse of this drug, but it switched to testing liver as the drug remains much longer in that organ than in urine. Further advance in detecting residues of this drug has followed because it is now possible to detect residues for months after use in the pigments of the eye. This organ is often being tested routinely where possible (Blanchflower et al., 1993). A third and final example of an identified problem is current with the detection of the coccidiostats nicarbazin and lasalocid in eggs (MAVIS, 1997). This was unexpected because these drugs should not be used in laying birds. However, it has been shown that contamination of non-medicated feeds following production of medicated feeds through the feed mills is the cause of this problem (Kennedy et al., 1996). The low residual drug content of the contaminated feed targets the egg, resulting in the residues. The pharmaceutical and feedstuff industries are actively investigating this problem by way of new formulations of the drug or greater care in cleaning and programming of feeds through the feed mills. Toxicologists have been consulted about these residues and they are content that at the concentrations found they pose no problem to human health, although the egg producers are concerned because nicarbazin is known to have adverse effects on shell quality.

8.3.3   Milk monitoring

The milk industry has imposed the discipline of monitoring for antibacterials in milk for a number of years. Mastitis is a very common disease in milking cows, requiring treatment by these drugs. The presence of antibiotics can have detrimental effects on starter cultures used in yoghurt and cheese making. Antibiotic testing as well as other milk quality tests are performed routinely and frequently on all producers' milk, with positive financial penalties where they default. Despite the huge numbers of producers, their milk is tested randomly at about weekly intervals.

8.4   Causes of residues in the food supply

If drugs are properly used in compliance with requirements of the licence, there should be no, or at most very low and acceptable (sub-MRL), concentrations of residues. Nevertheless, residues do occur and it is necessary to 'police' their use as evidenced by some examples already given.

The most obvious reason for unacceptable residues is failure to observe the withdrawal period. This may be deliberate or accidental. Disease in animals close to slaughter can be financially undesirable and a farmer may decide to send animals to market without observing some or any of the prescribed withdrawal period. Looking at the odds of only 50 000 samples from 35 000 000, it could be reasonable for a farmer to take a chance. Not being caught once, he could repeat the exercise but, eventually, he is liable to be caught as the sampling is country-wide and random. Another reason, again of a financial nature, would occur where there is demand for meat and the price is elevated; the farmer may again take a chance. This is likely to be much more of a one-off event. Yet again, mistakes can be made with the feed close to slaughter, when animals should be on non-medicated feed. On-farm, where both medicated and non-medicated feeds will be present, it is feasible that the wrong feed could be given. Where single animals are being treated, adequate identification is essential otherwise mistakes can be made as a result of forgetfulness or lack of knowledge, particularly where more than one individual is involved. Poor treatment records can readily cause this problem and, even where records are kept, it is essential that all staff are allowed regular access to the records. Such accidents are more common with milk from cows than with animals going to slaughter. Another source of drug residues can arise among pigs if they are transferred to pens or are transported in vehicles in which medicated animals have been kept. The new pigs can pick up residues from feed, layerage or faeces of the earlier pigs. Contamination of the feedstuff is yet another source. The contamination of laying poultry feeds with feeds medicated with the coccidiostats, nicarbazin and lasalocid has been mentioned above. A similar problem was identified and cleared up several years ago when pig feeds were being contaminated with sulphonamide from earlier medicated feeds.

Contamination of feeds arises in the field or store where treatment with pesticides occurs. Residues, particularly organochlorine pesticides which tend to accumulate in body fat, may be detected in carcase meat. It is for this reason that a range of pesticides is required to be tested for by both Directives 86/469 and 96/23. Mycotoxins can be produced by moulds in feeds that have been harvested damp or have not been adequately dried or are improperly stored. Animals can become infected by these substances through the feedstuff, hence again the testing for these highly toxic substances specifically requested by Directive 96/23. Ochratoxins and aflatoxins B1 and B2 are possible residues in animal tissue, while aflatoxins M1 and M2 may be found in milk. A mycotoxin metabolite, zearalenone, has given rise to an interesting problem. It has been shown that this metabolite can be reduced back to zeranol, which is a synthetic oestrogenic compound banned for growth promotion (Erasmuson *et al.*, 1994). The finding of zeranol as a residue in liver poses the authorities in countries where it is banned with the problem of determining whether the zeranol was administered illegally for growth promotion or has been consumed accidentally by the animal in mouldy feed.

Long-acting drugs can also cause trouble. Benzathine penicillin, for example, is injected as a salt that is relatively insoluble and releases penicillin slowly into the animal. A long withdrawal period is recommended, but it has been shown that this is inadequate around the injection site; if this area is sampled, an offence against the MRL is possible (Korsrud *et al.*, 1994). Misuse of drugs is yet another possible source of unacceptable residues. Overdosing of animals in the belief that this might help produce a more rapid cure can easily give rise to the need for extended withdrawal periods because the MRL would be exceeded at the normal time period.

Finally, where illegal drugs are used there are no defined withdrawal periods and therefore it is very much a case of trial and error. Too long a withdrawal may lose the benefits of the illegal use, so there is possibly the temptation to shorten this as much as possible. Hormones and β-agonists require protracted periods of treatment to exert their effect and, equally, withdrawal times may be protracted. Hormones when permitted were supplied as implants to allow slow release. These implants used to be placed behind the ear, which is discarded, but now it is reported that implants may be found in different points on the animal where, it is hoped, they will not be looked for.

8.5 Residence times, depuration

Being substances foreign to the animal's body, veterinary drugs are dealt with as such. Mammals have the ability to detoxify such substances and dispose of them in their urine or faeces. Where substances are chemically changed they are said to be metabolised and the products produced called metabolites. Drugs have many different chemical structures which give them different polarity. This affects their solubility and distribution in the animal. There is also a wide variety of sites where chemical reactions such as hydrolysis, acetylation, oxidation, reduction, etc. can take place in the animal's body.

Administration of drugs to animals can be achieved by a variety of routes; for example they can be given orally via feed or water, they can be injected subcutaneously or intramuscularly, or they can be implanted. The route of choice is dictated by the desired effect of administration, the route of absorption of the drug and the time that it takes the animal to dispose of the drug either as parent drug or as metabolite. This time is known as the residence time and the process of disposal of the drug as depuration. The complete study of this process is known as 'pharmaco-kinetics' but is not discussed here (Riviere *et al.*, 1991; Craigmill *et al.*, 1994). Such data has to be supplied to allow the withdrawal period to be determined. Withdrawal periods can be very variable, with some drugs effectively requiring no time whatsoever while others can have extremely long times. The drugs with low withdrawal times are often those designed to counter intestinal or gut infections. They have very limited potential to be absorbed across the gut wall and may be described as 'drugs of passage'. Coccidiostatic drugs used in poultry production often fall into this category. On the other hand, there are some 'dry cow therapy' drugs used against mastitis that must be administered 50 or more days prior to the cow's milk being taken for human consumption. Between these extremes, drugs such as sulphonamides may be given withdrawal periods from 5 to 14 days depending on their nature and use, while others such as long-acting penicillins, e.g. benzathine penicillin, may require 28 days. This period is often applied generally as a safety period where drugs are being used 'off licence', either in terms of quantity given or for treatment of non-specified species.

Drugs also distribute in different ways around the animal's body. Generally, the animal's bloodstream is utilised to carry the drug around the body and drug concentrations may be found in muscle, the main organs (kidney and liver), fat, in body fluids such as bile, blood and urine, as well as in faeces. The residue analyst may be asked to analyse any of these materials for evidence of residues depending on circumstances. From the living animal, blood and urine would be the normal tissues analysed for determination of the concentration of a drug, and for evidence of illegal drugs, such as hormones, faeces may be targeted. From the dead animal, the organs or muscle are more normally used and, in some cases, bile is used for analysis, again for evidence of the presence of illegal hormone. As a general rule, concentrations of antibiotics are usually highest in kidney, intermediate in liver and lowest in muscle, whereas growth-promoting hormones and β-agonists appear at highest concentrations in liver. Anthelmintics such as levamisole and the benzimidazole class also tend to concentrate in the liver, and the organochlorine pesticides in fat. The tissue in which the highest concentrations occur is termed the target tissue.

In residue surveillance work it is normally the target tissue that is chosen for analysis. Contravention of the MRL in a target tissue may result in removal of the tissue or the whole carcase from the food chain depending on the concentration and the enforcing agency. Where the concentration is well above the MRL, it is most likely that the whole carcase would be condemned on the grounds that the withdrawal period had clearly not been observed but, where the contravention is only marginally high and the likelihood is that the non-target tissues will be below the MRL, the enforcers may condemn the target tissue but let the rest of the carcase into the chain. In the case of illegal drug usage, the authorities will target the most likely tissue to determine the misuse. Thus, the example quoted earlier for β-agonists serves as the best example. These drugs are normally detectable in urine for a few days after administration, for several weeks in liver, and for several months in the coloured pigmentation of the eyeball. At slaughter the latter two tissues are available. Investigations are presently under way to determine whether the β-agonists accumulate in the pigmented (brown or black) hair in cattle to determine whether sampling of this could be used for on-farm determination of misuse of this drug.

Drugs may be metabolised during the retention time. The process is designed to reduce the toxicity of the substance, rendering it more polar so as to aid its excretion. As examples of this, sulphonamides may be acetylated, or form a glycosyl derivative at the free amine, or may even be deaminated (Giera et al., 1982). The tetracyclines undergo the process known as epimerisation, which involves a change of molecular conformation at their C-4 asymmetric centre (Oka, 1995). Hormones and some β-agonists (the salbutamol type) often undergo a process known as glucuronidation or sulphation (Montrade et al., 1993). Chloramphenicol also undergoes this process in kidney. None of these changes is complete; an equilibrium develops so that there is a concentration of both parent and metabolites present. The benzimidazole group of anthelmintics can undergo a number of chemical interchanges dependent on the redox potential present in the tissue. Thus, fenbendazole may be oxidised to the sulphoxide, oxfendazole, which in turn may be oxidised to oxfendazole sulphone. An equilibrium can exist of all three forms at different concentrations in different tissues or indeed at different points in the same tissue. Similar changes are likely to occur with the other compounds of the benzimidazole class where other substitution groups away from the oxidisable sulphur atom differentiate them. Thus, the drug albendazole is always analysed for as the metabolite albendazole sulphone. The nitrofuran group drugs undergo much more dramatic degradation very rapidly after administration (McCalla, 1978). It is not certain whether the parent drug or a metabolite is the effective antimicrobial agent. Other drugs falling into this type are carbadox and olaquindox, both of which are rapidly and irreversibly deoxygenated at one or both ring nitrogens (MacIntosh and Neville, 1984; Aerts, 1988; Rose and Shearer, 1992).

These metabolic changes can or should significantly alter the analytical approach taken when quantifying residues. For most drugs the MRL simply requires the measurement of the parent drug concentration. However, with the sulphonamides it specifically states 'all substances belonging to the sulphonamide group'. This should be taken to mean mixed parent drugs and/or metabolites. For each of the tetracyclines the regulations stipulate 'sum of parent drug and its 4-epimer'. For the benzimidazoles, the MRL is stated to be the sum of all appropriate parents and metabolites 'expressed as sulphone'.

Where glucuronide or sulphate metabolites are known to be formed, often the initial stage of the analytical procedure involves aqueous homogenisation at a pH suitable for use of an appropriate enzyme preparation to liberate the parent drug (Harwood et al., 1980). This is almost universally done for the oestrogenic hormones such as the stilbenes, oestradiol, etc. and is suggested for the phenolic-type β-agonists of which salbutamol is the most frequently mentioned (Montrade et al., 1993). However, this procedure is seldom undertaken for chloramphenicol analysis, where it is less well recognised that the drug undergoes this metabolic change, particularly in kidney. Recent experiments in my own laboratory have shown a 14-fold increase in concentration of chloramphenicol in incurred kidney after treatment with a glucuronidase enzyme. No or little change was observed in incurred muscle or liver from the same animal. However, as kidney is the target organ and is most often analysed, it would appear sensible that the enzyme step should be introduced.

8.6 Methods of analysis

As has been seen, MRLs are generally in the μg/kg (ppb) range, with highest values in the organ tissues, medium in muscle and lowest in milk. Analysing to (and much better below) such concentrations in animal tissues and products poses very demanding problems for the analyst. However, with experience, new technology and new approaches, such problems are being overcome. The other major problem is the high number of samples and the fact that most are negative or drug free. There is therefore demand for high-throughput screening methods that are sufficiently sensitive to monitor the MRL and produce low incidence of false positive results and zero false negative results. Such methods are termed screening methods. Where positives are found, re-analysis is required using quantitative methods to determine which drug and how much of it is present and, finally, where MRLs are exceeded, confirmatory methods are required that unequivocally identify the analyte in question to allow legal proceedings to be taken. There are some true screening methods available, but mainly physicochemical methods are used for many drugs. A growing number of immunochemical methods are also becoming available.

8.6.1 Screening methods

Antibiotics are readily detectable using microbial inhibition tests; indeed Fleming's attention was drawn to penicillin by the basis of these tests, namely the failure of organisms to grow. Two basic approaches to these tests are available, one depending on colorimetric measurement, the other on measurement of zones of inhibition where organisms fail to grow. As microorganisms grow they consume nutrients from their surroundings and excrete substances that are generally more acidic or have undergone oxidative or reductive change. Such changes can be detected either by change of colour of a pH indicator such as bromocresol purple or by reduction of a redox indicator such as Brilliant Black. In both cases the initial purple colour changes to yellow either because of increased acidity or because of the appearance of reducing substances. A number of commercial kits based on this principle are marketed, some of which are listed in Table 8.6. Similar tests can readily be generated in the laboratory. The kits contain organism, nutrient media and colour reagent either as a desiccated pellet or in agar. The sample is added and the

Table 8.6 Selected list of microbiological inhibition tests using acid–base indicators

Test	Organism	Matrix
Delvotest P and Delvotest SP[a]	Bacillus stearothermophilus	Milk
Charm AIM test[b]	Bacillus stearothermophilus	Milk, tissues
Valio T101[c]	Streptococcus thermophilus	Milk and milk products
Charm Farm Test[b]	Bacillus stearothermophilus	Milk, meat, eggs, fish serum, urine
Accusphere[d]	Streptococcus thermophilus	Milk, dried milk

[a]Gist Brocade, The Netherlands.
[b]Charm Sciences, USA.
[c]Valio, Finland.
[d]International Dairy Federation, *Bulletin of the International Dairy Federation* no. 258, 2nd edition (Brussels, 1991), p 48.

system is incubated at 60–65°C for about 3 h as the organisms used are heat resistant (thermophiles). In the absence of antibiotic, the organism grows and a colour change occurs; in the presence of antibiotic the test colour remains constant. A useful review has been given by Boison and MacNeil (1995). Such tests are widely used in the milk industry where they allow high throughput with highly reliable results. For example, one highly automated laboratory operating 7 days per week, 24 hours per day can process on the order of 30 000 samples per week.

In a variation on this, the sample is placed in a well or the surface of an agar plate containing nutrient medium and seeded with an organism. The antibiotic is allowed time to diffuse out into the agar before incubation at about 30°C usually for 18–24 h. In the presence of antibiotic, a clear zone of inhibition surrounds the well or sample. These tests often use organisms such as *Bacillus subtilis* or *Micrococcus luteus*. Where wells are used, liquid samples such as milk, urine or blood are usually involved and, where the sample is placed on the surface of the agar, discs of tissue (Four Plate Test) or filter paper saturated with fluid (Dutch Kidney Test) or a cotton swab that has absorbed fluids (SWAB Test) are used. A number of such tests are listed in Table 8.7, some of which use multiple plates such as three, four or six. The different plates are at different pHs and contain different organisms or additives. Thus the Bogaerts and Wolf (1980) Four Plate Test, widely used in the EU, uses *B. subtilis* at pH 6.0, pH 8.0 and pH 7.2 in presence of 0.1% trimethoprim and *M. luteus* at pH 8.0. These plates are supposed to be particularly sensitive to penicillins, aminoglycosides, sulphonamides and macrolides, respectively, as well as other antimicrobials.

The Limit of Detection (LOD) of such tests is at or close to the MRL, as seen in Table 8.8. Positive results indicate the presence of inhibitory material, probably antibiotic, but not the nature or the quantity of it. Corry *et al.* (1983) summarised LODs for a number of these tests. The

Table 8.7 Selected list of microbiological inhibition tests using zones of inhibition

Test	Organism	Matrix
Four Plate Test (FPT)[a]	*B. subtilis, M. luteus*	Muscle, kidney, liver
Liver Animal Swab Test (LAST)[b]	*B. subtilis*	Urine
Calf Antibiotic and Sulpha Test (CAST)[c]	*B. megaterium*	Kidney, urine
Swab Test on Premises (STOP)[d]	*B. subtilis*	Kidney
Dutch kidney test[e]	*B. subtilis*	Kidney

[a]Bogaerts and Wolf (1980).
[b]Environmental Diagnostics Inc., *Performing The Live Animal Swab Test*, 1979.
[c]USDA, *Performing the Calf Antibiotics and Sulpha Test*, FSIS Meat and Poultry Program, 1984.
[d]Johnston *et al.* (1981).
[e]Nouws *et al.* (1988).

Table 8.8 Claimed limits of detection (µg/kg) for some microbiological inhibition tests

	Charm[a] Farm	Delvo P[a]	Delvo SP[a]	Disc[b]	FPT[c]
Penicillin G	3	2	2	5	20
Cloxacillin	40	20	20	50	–
Ampicillin	5	4	4	10	–
Tetracycline	50	200	200	980	100
Oxytetracycline	80	200	200	1000	–
Chlortetracycline	150	500	500	1000	–
Erythromycin	200	400	400	1000	100
Tylosin	50	100	100	1000	300
Gentamicin	–	250	250	460	–
Neomycin	–	1000	1000	800	200
Streptomycin	–	4000	4000	1000	500

[a]See Table 8.6.
[b]Ginn *et al.* (1982).
[c]Four Plate Test, see Table 8.7.

Four Plate Test (FPT) is prone to show many false positives from frozen pig and horse kidney samples. To overcome this, a permeable membrane is placed between the agar and sample to prevent the suspected high molecular mass inhibitory protein, suggested to be lysozyme, passing into the agar. A follow-up to positive FPT results is the use of high-voltage electrophoresis across an agar plate. Discs of meat are placed on the plate before performance of the electrophoresis. An overlayer of organism-seeded agar is applied and allowed to incubate. Clear zones of inhibition show the presence of antibiotic. The position in terms of distance and direction of movement of the antibiotic assists in determining its nature. Smithers and Vaughan (1978) made a considerable study of this using agar and agarose plates of pH6.0 and 8.0 with the organisms *B. subtilis* and *M. Luteus*.

8.6.2   Immunochemical assays

Immunoassays in various formats have been developed over the last 30 years. Initially they were used for macromolecules but it was found possible to use them for smaller molecules, particularly if they had a reasonably rigid structure. The process involves attaching the drug to a protein in a specific manner. This substance is termed an antigen. The product is injected into an animal and, because it is a foreign proteinaceous substance, the animal's immune system responds by producing an antibody to it. This antibody develops a conformation or shape that is like a lock to the antigen's key shape. With luck, part of the lock shape involves a space with the shape of the drug and in the presence of drug this shape can adsorb the drug. This coming together of drug and antibody can be used as an assay.

One format known as an ELISA (enzyme-linked immunosorbent assay) involves the antibody (Ab) in the presence of both the drug (D) and the drug linked to a protein that has an enzyme attached to it (DPE). The Ab is present in limited quantity and the D and DPE are allowed to compete for available antibody sites. They will attach in proportion to their concentration. Substrate is then supplied and the enzyme produces a colour that can be measured, e.g.

$$Ab + \frac{D}{DPE} \rightleftharpoons \frac{Ab.D}{Ab.DPE} + \frac{D}{DPE}$$

If no D is present, the maximum amount of DPE will be absorbed and a maximum colour will develop. In the presence of increasing D, the Ab.DPE decreases and therefore the colour developed under standard conditions decreases.

A number of such tests are commercially available, as listed in Table 8.9, and a greater number have been developed 'in house' but are not readily obtainable or may be of dubious worth. The main advantage is speed of action, with results obtained in minutes from the time of presentation of sample. For milk or urine, presentation can be direct but, for tissue, some form of aqueous extraction and possible clean-up is required. The other advantage is that the tests are generally very sensitive, often being capable of measuring to sub-ppb concentration and with the

Table 8.9 ELISA kits currently available from some European commercial sources

Randox (N. Ireland)	R Biopharm (Germany)	Tepnel (England)
β-Agonists	Clenbuterol	Sulphamethazine
Stilbenes	Diethylstilboestrol	Tetracyclines
Trenbolone	Zeranol	Clenbuterol
Nortestosterone	Trenbolone	Streptomycin
3×Sulphonamides	Nortestosterone	Chloramphenicol
Corticosteroids	Oestradiol	
Testosterone	Testosterone	
Chloramphenicol	Sulphamethazine	
	Tetracycline	
	Chloramphenicol	
	Streptomycin	
	Carazolol	
	Dexamethasone	
	Methyl testosterone	
	Gestagens	

capability of desensitisation by means of sample dilution. The main disadvantage is selectivity. At best, they will only determine the presence of one class of drug, e.g. β-lactams, and often they will be specific for a single drug. Nevertheless, such assays are being used for hormones (Harwood *et al.*, 1980; Jansen *et al.*, 1985) and several antibiotic classes at present.

A variation on this is the receptor assay. One particular form is available commercially in the Charm test. Here the antibody is replaced by inactivated microbial cells the surfaces of which have receptor sites capable of absorbing different classes of antibiotics: β-lactams, tetracyclines, aminoglycosides, sulphonamides, macrolides and chloramphenicol. Here the competition for the receptor sites is between the drug and a radiolabelled form of the drug. The cells are spun off and the radioactivity is counted in a scintillation counter. The radioactive counts are inversely proportional to the quantity of drug present. The benefits of the Charm test are that it identifies and quantifies the drug from among six of the more likely candidate classes. Its disadvantage is the high basic cost for set-up and the fact that six separate assays have to be performed. Nevertheless, this assay is widely used in the United States for milk testing and a reformatted form, Charm II, has been introduced for tissue assays. This involves aqueous extraction of the tissue followed by heat precipitation of protein before starting the assay. Some problems with kidney extracts were initially encountered but are believed to have been overcome. The Charm Corporation is actively trying to introduce the products in Europe at present.

Despite being quantitative in nature, both the Charm receptor assay and ELISAs must generally only be regarded as qualitative and therefore as screening tests.

8.6.3 Chemical screening tests

There is a wide spectrum of such tests used for veterinary drug residues varying in complexity and a detailed account cannot possibly be given here. However, a general description of the process with specific examples will be given. Any trace analysis usually involves the three stages of extraction, clean-up and determination. In the methods described above, some of these stages are missing. Thus, in the FPT the extraction step could be said to be dependent on the leachate from the tissue, followed by determination. In the ELISA and Charm there is much more clearly an extraction with aqueous buffer, a selection out or clean-up of the specific antibiotics, and a colorimetric or radiochemical determination. These stages tend to be combined and are achieved quickly. In more classical physicochemical approaches, the stages are more distinctly observable and separate, and the methodology is more time consuming.

8.6.3.1 Extractions

The extraction stage can be achieved by homogenisation of the tissue in aqueous or organic solvents, such as alcohol (Dimenna et al., 1986; Takasuki et al., 1986), ethyl acetate (Thomas et al., 1983; Porter et al., 1993), acetone (Goodspeed et al., 1978), acetonitrile (Arnold and Somogyi, 1985; Meetschen and Petz, 1991), or chlorinated solvents, such as dichloromethane, chloroform (Parks, 1982), or occasionally carbon tetrachloride, often in a tissue:solvent ratio of 1:5 or 1:10. Multiple extractions may be necessary. The use of chlorinated solvents is currently being reduced or replaced on environmental grounds. Such a procedure is crude as it extracts not only the desired compound but also many other co-extractives, most often in far greater abundance.

8.6.3.2 Clean-up

The clean-up stage is designed to remove selectively as many of the co-extractives as possible while retaining as much of the analyte as possible. Where organic solvents have been used, fat has been extracted from animal tissue and this is most readily removed by defatting using non-polar solvents such as petroleum or diethyl ether. This is often achieved by evaporation of the organic solvent, taking the residue up in aqueous phase, and adding an immiscible defatting solvent to the aqueous phase. This process is referred to as liquid–liquid extraction. The chlorinated solvents have the advantage that they are immiscible with aqueous phases and addition of dilute aqueous acid or alkali allows such liquid–liquid extraction directly. This approach is dependent on the analyte being a base or acid, respectively, so that the analyte favours partition into the aqueous phase. Fortunately, veterinary drugs often contain acidic or basic groups. Neutralisation of the aqueous phase allows a further liquid–liquid partition back into the organic phase. The use of repeated liquid–liquid extractions like this has become somewhat outdated as it is cumbersome and uses sizeable volumes of solvent that often are expensive and the disposal of which is regarded as environmentally undesirable. Nevertheless, sulphonamides used to be extracted in such a manner, as evidenced by the method of Thomas et al. (1983) in which the acidic and basic nature of these compounds was exploited by use of both basic and acidic phases to hold the sulphonamides while co-extractives were removed, before altering the pH to a more neutral pH 5.0 to allow the sulphonamides to return to an organic phase. Another and more recent use of liquid–liquid extraction as an approach to clean-up is for β-agonists (van Ginkel et al., 1992).

Where aqueous extraction has been used, the problem of fat extraction is less and the extract sometimes can be taken straight to the clean-up stage. Such an example is the method of Farrington *et al.* (1991) for tetracyclines. Yet another reason for having the residue in aqueous phase can be to allow enzymatic action to be used either to remove extracted protein or to allow production of parent drug from metabolite. In the first case, a proteinase such as subtilisin (Ossleton, 1977; Aureli *et al.*, 1995) may be used and, in the latter, a glucuronidase or sulphatase (Heitzman and Harwood, 1983; Montrade *et al.*, 1993).

More recent approaches use chromatographic procedures for clean-up. There are many of these and only some examples can be given. An early example is the clean-up used for the ionophore coccidiostats monensin, narasin and salinomycin. Here the drugs were extracted into carbon tetrachloride, which is very non-polar, and applied, after drying, to a silica chromatographic column (Dimenna *et al.*, 1986; Takasuki *et al.*, 1986). The slightly polar ionophores were retained in the column while the co-extracted fat passed through and, by modifying the eluent with a small percentage of methanol, the ionophores were partitioned into the mobile phase and selectively extracted from other co-extractives.

The use of reversed-phase, coated silica allows the opposite effect to be achieved. By applying an extract in reasonably polar solvent such as alcohol or acetonitrile, the less polar extractives remain on the column while more polar co-extractives pass through. Modification of the eluent by addition of less polar solvent selectively removes the analyte leaving behind less polar co-extractives. Such a procedure is used in the clean-up of extracts for analysis of tylosin (Chan *et al.*, 1994) and erythromycin (Stubbs *et al.*, 1986).

Cation exchange columns allow bases to be retained while anion columns allow acids to be selectively retained. Modification of the eluent to a more basic or more acid medium, respectively, then allows the retained substances to be released. An example of this is the method of Haagsma and van de Water (1985) for sulphonamides. The extract is applied to a cation exchange column, which retains the sulphonamides as bases, and these in turn are displaced by use of an ammoniated methanol.

A more specific chromatographic approach was used by Farrington *et al.* (1991) for separation of tetracyclines. Recognising the tetracyclines' affinity for complexing with divalent ions, they used Ca^{2+}-prepared Sepharose to remove tetracyclines selectively from the primary extract before eluting with the stronger metal chelater ethylenediaminetetraacetic acid (EDTA).

Clean-up by specific selective chromatography is further exemplified by the use of immunoaffinity columns (see Chapter 4). Here the antibodies, produced as for immunoassays and capable of selectively binding their analyte or analyte class, are attached to some inert column material, often a cellulose. Change from the initial aqueous eluting medium to an alcohol-containing eluent removes the selectively held drug or drug class. There are numerous examples of this technology (Hasmoot et al., 1990; Cooper and Shepherd, 1996).

Improvement in chromatographic materials through the 1980s has allowed such developments in clean-up methods to take place. This is because of the increased number of selective phases available attached to more uniformly sized particles. This has been further improved by the commercial introduction of solid phase extraction (SPE) cartridges of many of these chromatographic materials. These cartridges are small and uniformly packed, allowing excellent and consistent separations to be obtained using very small volumes of solvent.

8.6.3.3 Determination step

There are many determinative methods available. Most of these depend on analytical chromatographic separations with many forms of detection.

Some drugs can be detected on thin-layer chromatography (TLC) plates at extremely low concentrations and separations of drugs has been improved by use of high-performance TLC (HPTLC). Thus, sulphonamides can be observed as purplish spots through on-plate diazotisation using the Bratton–Marshall procedure (Parks, 1982) or through fluorescence by tagging with fluorescamine (Thomas et al., 1983). The ionophores monensin, narasin and salinomycin likewise can be observed by fluorescence developed by spraying with acidic vanillin (Takasuki et al., 1986).

Gas chromatography (GC) is also a useful determinative step for many drugs, particularly the less polar ones such as the hormones and β-agonists (Le Bizec et al., 1993; Montrade et al., 1993). To volatilise these substances they have to be derivatised after the extraction and clean-up stages, but thereafter they may be reproducibly separated and quantified using flame ionisation detection. More polar drugs often require increased or multiple derivatising steps, which renders them less useful for GC, but examples for sulphonamides and chloramphenicol are available using electron capture detection (Goodspeed et al., 1978; Arnold and Somogyi, 1985).

Liquid chromatography (LC or HPLC) is generally a more useful chromatographic determinative step, although the detectors here are less sensitive than for GC. Nevertheless, there are plentiful methods available for a wide range of drugs using ultraviolet, visible, fluorescence, diode-array or electrochemical detectors. The examples quoted involve penicillin (Boison and Keng, 1996), levamisole, benzimidazole (Marti et al., 1990) (UV), ivermectin (Tway et al., 1981; Li and Qian, 1996) (fluorescence), and promazine and β-blockers/tranquillisers (Rose and Shearer, 1992) (electrochemical).

8.6.4   Confirmatory methods

All of the above methods, although very specific because of their clean-up, chromatographic separation and detection procedures, are still considered no more than indicative that the compound is in fact that which is claimed. To confirm, two approaches are permitted. The first is to determine the analyte using a totally separate analytical method, ideally with different extraction, clean-up and determinative stages. If the same qualitative and quantitative results are obtained by both routes, it would appear conclusive that the compound identified is the one originally claimed. The second approach, and the one regarded as more desirable, is to use mass-spectrometry (MS). This technique could be used in isolation having recovered a drug residue from a determinative step that was non-destructive. However, linked GC-MS has been available for a good number of years now and this can be used with low-resolution benchtop equipment or high-resolution instruments (van Peteghen et al., 1987; Le Bizec et al., 1993; Batjoen et al., 1996).

LC-MS instrumentation is much newer and, practically, has only been routinely available for the past 5 years or so. For drug analysis this is probably the preferred method because LC is applicable for separation of a wider range of these more polar substances (van Vyncht et al., 1996; Bean and Henion, 1997). Mass spectrometry can be very selective, being capable of displaying the unique 'fingerprint' of a compound, including the molecular ion and unique patterns of breakdown ions. Because of this selectivity it is possible to use less demanding clean-up procedures, but a trade-off has to be made between this benefit and the likelihood that a dirty extract will more quickly foul the MS detector.

For complete identification, Commission Decisions 87/410, 89/610 and 93/256 suggest that for any compound up to four ions should be detected and their ratios should be within 10% for the EI (electron impact ionisation) mode. This is not always possible to achieve, but a molecular ion and one or two degradation ions would be regarded as reasonable identification criteria for any analyte. These Decisions also give criteria for other determinative methods such as LC, GC, TLC, immunoassay, etc. These Decisions also indicate the accuracy and precision that should be expected of a method of analysis. As might be anticipated, a multistep procedure, as most physicochemical methods are, involves losses, hopefully minor, along the way. Thus, less than total recovery of the analyte can be tolerated. Equally, even after thorough clean-up there may still be some co-eluting contamination that may cause apparent over-recovery of spiked analyte. The suggested tolerances permitted are as shown in Table 8.10, with less accuracy and precision expected the lower the concentration. Much of the information in Directives and the detailed methods of analysis being used in the EC are collated in a book edited by Heitzman (1994) on behalf of the EU.

8.7 Accreditation and quality assurance

It is becoming essential that laboratories undertaking analytical work should be accredited to standards such as BS 5750, ISO 9000 or EN 45001. This requires laboratories to submit to inspection by the national accreditation body (UKAS in the case of the United Kingdom), have the methodology and its performance agreed, and thereafter ensure that the standards laid down are maintained. This requires formal noting of all essential data associated with analytical work, keeping of training records of staff and maintenance and standardisation records for all equipment. All this information is subject to internal and external audit several times a year. In addition, it is desirable that laboratories take part in performance assessment schemes wherever possible to ensure that their analytical capability can be measured against other laboratories when analysing standard or reference materials. In the veterinary drug field, the Food Analysis Performance Assessment Scheme (FAPAS) provides such materials (Key et al., 1997).

8.8 Future developments

In terms of current technology, improved efficiency in veterinary drug analysis, and particularly surveillance schemes where large numbers of samples are involved, will occur. This can most readily arise from automation not simply of the end-determinative method but also of the clean-up stage. Some steps in this direction have already been taken but such developments are not widespread.

Table 8.10 Tolerances for methods of analysis at varying drug concentrations

		Precision	
True content	Recovery	Within lab	Between labs
≤1 µg/kg	50–120%	30%	45%
<1 µg/kg to 10 µg/kg	70–110%	20%	32%
<10 µg/kg	80–110%	15%	23%

The next most obvious development is that of multiresidue methods. Presently, either single drugs or drug classes (sulphonamides, tetracyclines, benzimidazoles, etc.) are analysed, but it should be possible to develop methodology that will cope with multiple classes of drugs. At least, common extraction and clean-up procedures should be the first goal. This would reduce a time-consuming aspect of the analytical methods even if the final extracts are dealt with separately. For example, many drugs exhibit basic properties and it should be possible to target this fact in extraction and clean-up.

LC-MS offers great potential to allow less extensive clean-up as well as a more universal end-determinative method. It certainly would allow screen positives to be taken on to a confirmatory method without passing through the physicochemical qualitative/quantitative stage.

Immunochemical advances both in the immunoaffinity clean-up stage and in more rapid screening phases offer considerable hope. After disappointment in the 1980s of the expectation that immunochemical screening methods would rapidly solve the drug analysts' problems, there are now several materials and kits available which offer practical advantage. One of the problems in this field is that the development costs for production of a robust product are high and often kits are rushed to the marketplace to recover investment before being fully developed to achieve reliable application. The kits are also expensive to purchase, particularly if sample numbers are insufficient to fully utilise the 96-well formats that most kits offer. Immunoaffinity chromatography clean-up procedures would appear to offer quicker returns.

Among new technologies that may bear fruit, critical fluid extraction and chromatography seem unlikely to be useful. The normal equipment available does not appear to cope satisfactorily with the polar drug molecules and, although at one time very much higher pressures than normal seemed to offer a means of extracting such compounds, this technology does not appear to have succeeded after a number of years. Capillary zone electrophoresis (CZE) in some form may offer hope. At present this technology lacks sensitivity, being capable at best of detecting at the ppm level, but its ability to cope with a wide range of polar compounds seems attractive. Whether the present format can be developed to detect concentrations down to the ppb level or whether some new detection mode is necessary is not clear. For example, it might be possible to link MS to CZE which may be sufficiently sensitive and selective to allow this approach to be utilised. If it were the case, this might reduce the need for clean-up stages with the electrophoresis supplying this stage. Time will tell.

References

Aerts, M.M.L., Beek, W.M.J., Keukens, H.J. and Brinkmann, U.A.T. (1988) Determination of residues of carbadox and some of its metabolites in swine tissue by HPLC. *Journal of Chromatography*, **456**, 105-119.

Arnold, D. and Somogyi, A. (1985) Trace analysis of chloramphenicol in milk, eggs and meat. Comparison of gas chromatographic method and RIA. *Journal of the Association of Official Analytical Chemists*, **68** 984-990.

Aureli, P., Ferrini, A.M. and Monnoni, V. (1995) Study of the effect of some proteolytic enzymes to improve the sensitivity of the microbial method for the detection of antibiotics and sulphonamide residues in milk. *Residues of Antimicrobial Drugs and Other Inhibition in Milk*. Brussels: International Dairy Federation, pp 201-202.

Batjoen, P., de Brabender, H.F. and de Wasch, K. (1996) Rapid and high performance analysis of thyrostatic drug residues in urine by GC-MS. *Journal of Chromatography A*, **750** 127-132.

Bean, K.A. and Henion, J.D. (1997) Direct determination of anabolic steroid conjugates in human urine by combined LC-tandem M.S. *Journal of Chromatography B*, **690** 65-75.

Blanchflower, W.J., Hewitt, S.A., Cannovan, A., Elliot, C.T. and Kennedy, D.G. (1993) Detection of clenbuterol residues in bovine liver, muscle, retina and urine using GC/MS. *Biological Mass Spectrometry*, **22** 326-330.

Bogaerts, R. and Wolf, F. (1980) A standardised method for the detection of residues of antibacterial substances in fresh meat. *Fleischwirtschaft*, **60** 672-673.

Boison, J.O. and Keng, L. (1996) Application of solid phase extraction technology for the multi residue analysis of penicillin residues in animal tissues by HPLC with UV detection. *Seminars in Food Analysis*, **1** 33-38.

Boison, J.O. and MacNeil, J.D. (1995) New test kit technology, in *Chemical Analysis for Antibiotics Used in Agriculture* (eds H. Oka, H. Nakazuwa, K.-I. Haroda and J.D. NacNeil), AOAC International, pp 77-119.

Chan, W., Gerhardt, G.C. and Salisbury, C.D.C. (1994) Determination by tylosin and tilomicosin residues in animal tissues by reversed phase liquid chromatography, *Journal of the Association of Official Analytical Chemists*, **77** 331-333.

Cooper, A.D. and Shepherd, M.J. (1996) Evaluation of a novel immunoaffinity phase for the purification of cattle liver extracts prior to HPLC determination of β-agonists, *Food and Agricultural Immunology*, **8** 205-213.

Corry, J.E.L., Sharma, M.R. and Bates, M.L. (1983) in *Antibiotics: Assessment of Antimicrobial Activity and Resistance* (eds A.D. Russell, L.B. Quesnel), Academic Press, London, pp 349-370.

Craigmill, A.L., Sundlof, S.F. and Riviere, J.E. (1994) *Comparative Pharmacokinetics and Residues of Veterinary Therapeutic Drugs*, CRC Press, Boca Raton, FL.

Dimenna, G.P., Creegan, J.A., Turnbull, L.B. and Wright, G.J. (1986) Determination of sodium salinomycin in chicken skins/fat by HPLC utilising column switching and UV detection. *Journal of Agricultural and Food Chemistry*, **34** 805-810.

Erasmuson, A.F., Scahill, B.G. and West, D.M. (1994) Natural zeranol (zearalanol) in the urine of pasture fed animals. *Journal of Agricultural and Food Chemistry*, **42** 2721-2725.

Farrington, W.H.H., Tarbin, J., Bygrave, J. and Shearer, G. (1991) Analysis of trace residues of tetracyclines in animal tissues and fluids using metal chelate affinity chromatography/ HPLC. *Food Additives and Contaminants*, **8** 55-64.

Fleming, A. (ed.) (1946) *Penicillin. Its Practical Application*, Butterworth, London.

Giera, D.D., Abdulla, R.F., Occolowitz, J.L., Dorman, D.E., Mertz, J.L. and Sieck, R.F. (1982) Isolation and identification of polar sulphamethazine metabolites from swine tissues. *Journal of Agricultural and Food Chemistry*, **30** 260-263.

Ginn, R.E., Case, R.A., Packard, V.S. and Tatini, S.R. (1982) Quantitative estimates of beta-lactam residues in raw milk around a reference standard, collaborative study, *Journal of the Association of Official Analytical Chemists*, **65** 1407-1412.

Goodspeed, D.P., Simpson, R.M., Ashworth, R.B., Shafer, J.W. and Cook, H.R. (1978) Sensitive and specific GLC-screening procedure for trace levels of 5 sulphonamides in liver, kidney and muscle tissue. *Journal of the Association of Official Analytical Chemists*, **61** 1050-1053.

Haagsma, N. and van de Water, C. (1985) Rapid determination of 5 sulphonamides in swine tissue by HPLC. *Journal of Chromatography*, **333** 256-266.

Harwood, D.J., Heitzman, R.J. and Jouquey, A. (1980) A radioimmunoassay method for the measurement of residue of the anabolic agent hexoestrol in tissues of cattle and sheep. *Journal of Veterinary Pharmacology and Therapeutics*, **3** 245-254.

Hasmoot, W., Ploun, M.E., Paulussen, R.J.A., Schult, R. and Huf, F.A. (1990) Rapid determination of clenbuterol residues in urine by HPLC with on-line automated sample processing using immunoaffinity chromatography. *Journal of Chromatography*, **519** 323-335.

Heitzman, R.J. (1994) in *Veterinary Drug Residues. Residues in Food Producing Animals and Their Products: Reference Material and Methods*, Commission of European Communities, Blackwell Scientific, Oxford.

Heitzman, R.J. and Harwood, D.J. (1983) Radioimmunoassay of hexoestrol residues in faeces, tissues and body fluids of bulls and steers, *Vet. Res.*, **112** 120-123.

Jansen, E.H.J.M., van Den Berg, R.W., van Blitterswijk, H., Both Miedema, R. and Stephany, R.W. (1985) A highly specific detection method for diethyl stilboestrol in bovine urine by radiommunoassay following HPLC. *Food Additives and Contaminants*, **2** 271-281.

Johnston, R.W., Reamer, R.H., Harris, E.W., Fayate, H.G. and Schwab, B. (1981) A new screening method for the detection of antibiotic residues in meat and poultry tissues. *Journal of Food Protection*, **44** 828-831.

Kennedy, D.G., Blanchflower, W.J., Hughes, P.J. and McCaughey, W.J. (1996) The incidence and cause of lasalocid residue in eggs in Northern Ireland. *Food Additives and Contaminants*, **13** 787-794.

Key, P.E., Patey, A.L., Rowling, S., Wilbourn, A. and Worner, F.N. (1997) International proficiency testing of analytical laboratories for foods and feeds from 1990-96: the experiences of the United Kingdom Food Analysis Performance Assessment Scheme. *Journal of the Association of Official Analytical Chemists*, **80** 895-899.

Korsrud, G.O., Boison, J.O., Papich, Yates, W.D.G., MacNeil, J.D., Janzen, E.D., McKinnon, J.J., Laundy, D.A., Lambert, G., Young, M.S. and Ritter, L. (1994) Depletion of penicillin G in tissues and injection sites of yearling beef steers dosed with benzathine penicillin G alone or in combination with procaine penicillin G. *Food Additives and Contaminants*, **11** 1-6.

Le Bizec, B., Montrade, M.P., Monteau, F. and Andre, F. (1993) Detection and identification of anabolic steroids in bovine urine by GC-MS. *Analytica Chimica Acta*, **275** 123-133.

Li, J. and Qian, C. (1996) Determination of ivermectin B_1 in biological samples by immunoaffinity column clean up and LC with UV detection. *Journal of the Association of Official Analytical Chemists International*, **79** 1062-1067.

McCalla, D.R. (1978) Nitrofurans, in *Mechanism of Actions of Antibacterial Agents* (ed. F.E. Halin), Springer-Verlag, Berlin, vol. 1, pp 176-213.

MacIntosh, A.I. and Neville, G.A. (1984) LC Determination of carbadox, desoxy carbadox and nitrofurazone in pork tissues. *Journal of the Association of Official Analytical Chemists*, **67** 958-962.

Marti, A.M., Mooser, A.E. and Koch, H. (1990) Determination of benzimidazole anthelmintics in meat samples. *Journal of Chromatography*, **498** 145-157.

MAVIS (1997) Medicines Act Veterinary Information Service. Veterinary Medicines Directorate, Woodham Lane, Addlestone, Surrey KT15 3NB, UK. Publishers, J.C. Alborough Ltd, Ipswich, UK.

Meetschen, U. and Petz, M. (1991) GC determination of penicillin in tissue and milk. *Zeitschrift für Lebensmittel-Untersuchung und -Forschung*, **193** 337.

Montrade, M.P., Le Bizec, B., Monteau, F., Siliart, B. and Andre, F. (1993) Multi-residue analysis for β-agonists drugs in urine of meat producing animals by GC-MS. *Analytica Chimica Acta*, **275** 253-268.

Nouws, J.F.M., Broex, N.J.G., den Hartog, J.M.P. and Driessies, F. (1988) The new Dutch Kidney test in practice. *Archiv für Lebensmittel Hygiene*, **39** 133-156.

Oka, H. (1995) Chemical analysis of tetracycline antibiotics, in *Chemical Analysis of Antibiotics Used in Agriculture* (eds W. Oka, H. Nakazuwa, K.-I. Harad and J.D. MacNeil), AOAC International, pp 333-406.

Ossleton, M.D. (1997) The release of basic drugs by the enzymic digestion of tissues in cases of poisoning. *Journal of the Forensic Science Society*, **17** 189-194.

Parks, O.W. (1982) Screening test for sulphamethazine and sulphathiazole in swine liver. *Journal of the Association of Official Analytical Chemists*, **65** 632-634.

Porter, S., Patel, R., Neate, S. and Osso, P. (1993) Determination of levamisole in liver by gas chromatography-mass spectrometry, in *Euroresidues II Proceedings. Residues of Veterinary Drugs in Food* (eds N. Haagsma, A. Ruiter and P.B. Czedick-Eysenberg), Veldhoven, pp 548-552.

Riviere, J.E., Craigmill, A.L. and Sundlof, S.F. (1991) *Pharmacokinetics and Residues of Veterinary Antimicrobials*, CRC Press, Boca Raton, FL.

Rose, M.D. and Shearer, G. (1992) Determination of tranquillisers and carazolol residues in animal tissues using HPLC with electrochemical detection, *Journal of Chromatography*, **624** 471-477.

Smithers, R. and Vaughan, D.R. (1978) An improved electrophoretic method for identifying antibiotics with special reference to animal tissues and animal feedingstuffs. *Journal of Applied Bacteriology*, **44** 421-329.

Stubbs, C., Haigh, J.M. and Kaufer, I. (1986) HPLC analysis of oleandomycin in serum and urine. *Journal of Chromatography*, **353** 33-38.

Takasuki, K., Suzuki, S. and Ushizawa, J. (1986) LC determination of monensin in chicken tissues with fluorometric detection and confirmation by GC-MS. *Journal of the Association of Official Analytical Chemists*, **69** 443-448.

Tarbin, J.A. and Shearer, G. (1992) HPLC determination of imidocarb in cattle kidney with cation exchange cleanup. *Journal of Chromatography*, **577** 376-381.

Thomas, M.H., Epstein, R.L., Ashworth, R.B. and Marks, H. (1983) Quantitative TLC chromatographic multi-sulphonamide screening procedure. Collaborative study. *Journal of the Association of Official Analytical Chemists*, **66** 884-892.

Tway, P.C., Wood, J.S. and Downing, G.V. (1981) Determination of ivermectin in cattle and sheep tissue using HPLC with fluorescence detection. *Journal of Agricultural and Food Chemistry*, **29** 1054-1063.

van Ginkel, L.A., Stephany, R.W. and Rossum, H.J. (1992) Development and validation of a multi-residue method for β-agonists in biological samples and animal feed. *Journal of the Association of Official Analytical Chemists*, **75** 554-560.

van Peteghen, C.H., Lefevre, M.F., van Haver, G.M. and Leenheer, A.P. (1987) Quantification of diethylstilboestrol residues in meat samples by GC-isotope dilution mass spectrometry. *Journal of Agricultural and Food Chemistry*, **35** 228-231.

van Vyncht, G., Preece, S., Gasper, P., Maghiun Rogister, G. and de Pauor, E. (1996) GC and LC coupled tandem MS for the multi residue analysis of β-agonists in biological matrices. *Journal of Chromatography A*, **750** 43-49.

Legislation cited

EC 90/2377. Regulation laying down a Community procedure for the establishment of maximum residue limits of veterinary medical products in foodstuffs of animal origin. Official Journal No. L224, 18.1.90, 1-8.

EC 86/469. Council Directive concerning the examination of animals and fresh meat for the presence of residues. Official Journal No. L275, 26.9.86, 36-46.

EC 81/602. Council Directive concerning the prohibition of certain substances having a hormonal action and of any substances having a thyrostatic action. Official Journal No. L222, 7.8.81, 32-33.

SI 82/626. The Medicines (Stilbenes and Thyrostats Substances) Regulation.

SI 85/649. Council Directive prohibiting the use of livestock farming of certain substances having a hormonal action. Official Journal No. L382, 31.12.85, 228-231.

SI 88/705. The Medicines (Hormone Growth Promoters) (Prohibition of Use) Regulation.

SI 88/849. The Animals and Fresh Meat (Hormonal Substances) Regulation.

EC/96/23. On measures to monitor certain substances and residues thereof in live animals and animal products. Official Journal No. L125, 23.5.96.

9 Polychlorinated biphenyls, dioxins and other polyhalogenated hydrocarbons as environmental contaminants in food

David E. Wells and Jacob de Boer

9.1 Historical perspectives

The first recording of the effects of polyhalogenated hydrocarbons (PHHs) was at the end of the nineteenth century during the production of caustic potash by the electrolysis of potassium chloride (Herxheimer, 1899). The workers who suffered from dermatitis related the cause to the nascent chlorine produced in the process. However, the most likely explanation was retrospectively identified as contamination from PHHs formed by the reaction of the chlorine and the pitch tar lining of the electrolytic cell. The first link was made during World War I when polychloronaphthalenes were used as additives in the rubber of the gas masks (Wauer, 1918). Since that time there have been numerous incidents of chloracne associated with occupational exposure to PHHs. Kimbrough (1980) has reviewed these incidents along with other chronic and acute cases relating to the PHH group of compounds (for relevant structures see Figure 9.1).

The widespread manufacture and use of PHHs in the 1960s and 1970s meant that, in addition to occupational exposure, there was the ready threat of severe environmental contamination close to the sites of manufacture and usage. As a consequence, there was concern both for the environment and for the food produced in those areas.

Some of the key incidents that triggered the intense interest in the PHHs as occupational, environmental and food contaminants, and, more importantly, human health issues are given in Table 9.1, together with a list of the compound groups discussed in this chapter.[1] Many of these incidents resulted from improper use or disposal with the result that animal feed, or human food, was contaminated. Alternatively, the material was released into the environment at relatively high concentrations.

---

[1] The prefix 'poly-' refers to the generic group of compounds with similar structure, but with a different degree or pattern of halogen substitution. Individual congeners are identified by their numbering system (Ballschmiter et al., 1985, 1992). While the numbering system for some groups of compounds has been relatively well established (e.g. CBs), others still need to be developed and used, e.g. CHBs (Oehme and Kallenborn, 1995). For abbreviations see Table 9.1. Relevant structures for the PHHs described in this chapter are given in Figure 9.1.

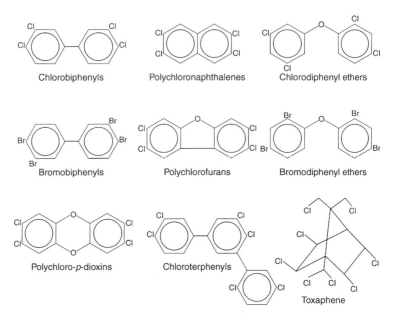

Figure 9.1 Examples of the chemical structures of PHHs covered in this chapter.

The early advantages of the PHHs in both agriculture and industry were quickly tempered by acute and chronic effects of these materials in the environment and in humans. During the late 1960s there was a growing awareness that the persistence, toxicity and rapid redistribution of these chlorinated compounds at sites remote from any production or use meant that low-level, long-term contamination of all compartments of the environment and the food chain was inevitable. The apolarity and lipophilicity of the PHHs also meant that all of these compounds were bioaccumulated in fatty tissue through the food web. Initially, manufacturers of these materials were reluctant to reduce production, treat effluent or admit to the causal links between the presence of these compounds through the food chain and the effects on the human body. As a result, it took almost a decade or more to reduce output and place restrictions on the use and disposal of these PHHs.

As environmental contaminants, PHHs and organochlorine pesticides were among the first group of organic compounds to gain public notoriety following the publication of Ratchel Carson's *Silent Spring* (Carson, 1962). The order of events preceding and subsequent to this publication have been mirrored on different occasions for other PHH contaminants. These events follow the sequential process of scientific investigation leading to public awareness, political response, environ-

Table 9.1 Main uses and sources of PHH, together with examples of incidents of acute and chronic exposure of PHHs to humans

Polyhalogenated hydrocarbons	Abbreviation	Main uses/sources	Incidents	References
Polychlorinated biphenyls	PCBs	Dielectric in capacitors and transformers Hydraulic fluids, heat transfer fluids, additives in paints, carbonless copy paper	Yusho Rice Oil contamination, Japan 1969 Yucheng, Taiwan 1979–1981 Hudson River—General Electric PCB contaminated effluent	Kuratsune (1980) (review) Kashimoto and Miyata (1982) (review) Horn et al. (1979)
Polychlorinated terphenyls	PCTs		Cow milk contamination from contaminated silage	Fries and Marrow (1973)
Polychlorinated dibenzo-p-dioxins	PCDDs	Impurities in PCP, PCBs Combustion product of PHHs	Seveso, Italy 1976, trichlorophenol plant out of control Vietnam 1962–1971 impurity in agent orange defoliant Missouri Soil contamination 1971	Reggiani (1980)
Polychlorinated dibenzofurans	PCDFs	Impurities in PCP, PCBs Combustion product of PHHs	Yusho Rice Oil contamination, Japan 1969 PCDFs as impurities	Kuratsune (1980) (review)
Polychlorinated naphthalenes	PCNs	Impurities in PCBs Additives to rubber gas masks, insulation for detonators Insulation in cables	Chloracne and acute yellow atrophy of the liver	Kimbrough (1980) (review)

Table 9.1 (Continued)

Polyhalogenated hydrocarbons	Abbreviation	Main uses/sources	Incidents	References
Polybrominated biphenyls	PBBs	Fire retardants	Michigan Cattle feed contamination with Firemaster BP6	Landrigan (1980)
Polybrominated diphenyl ethers	PBDEs	Fire retardants		WHO (1994) (review)
Polychlorinated diphenyl ethers	PCDEs	Impurities in chlorophenols		Becker et al. (1991)
Toxaphene (chlorobornanes)	CHBs	Insecticide on cotton	Inhalation of 'cotton dust' formulations	Deichmann (1973), Saleh (1991) (review)

PCP, pentachlorophenol. Other abbreviations as in the text.
Structures for the various PHHs are given in Figure 9.1.

mental monitoring and legislation and have been aptly described by Ashby (1978) as the 'chain reaction'. The initial investigation of an incident related to an environmental pollutant transfers from the scientific to the public domain once the initial findings are made available. Scientific facts, along with sound reasoning and appropriate caveats, are replaced for a while by public concern that is fuelled by pressure-group publicity. It is often these events that, in turn, create the political will to provide the resources necessary to support essential monitoring programmes, enact the required legislation and, where appropriate, provide the stimulus for further research (Figure 9.2). The effect of the 'chain reaction' has been manifest in a number of ways. First, governments throughout many developed countries have established systematic monitoring programmes to determine the extent of the contamination of these chemicals in foodstuffs and the dietary input of PHHs. Total diet studies or the Market Basket Programmes were started by the United States Food and Drug Administration (USFDA) in 1964 (Johnson et al., 1984) and total diet studies for organochlorine pesticides and PCBs were started in Japan in 1977 (Matsumoto et al., 1987). Similar programmes have ben mirrored in the United Kingdom by the Ministry of Agriculture, Fisheries and Food (MAFF), and by TNO Nutrition and Food Research Institute in The Netherlands. These food monitoring programmes developed as a direct result of scientific investigation, public pressure and political awareness and provide essential on-going

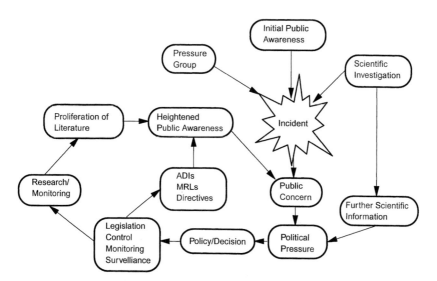

Figure 9.2 The 'critical chain' of events following an incident or publication of chemical contaminants in the environment. (After Ashby, 1978).

information on the concentrations of these contaminants in the key foodstuffs, giving an objective basis for public reassurance.

Initially, the studies included polychlorinated biphenyls (PCBs) and polybrominated biphenyls (PBBs) in the United States, and PCBs in Europe and Japan. By the mid 1980s there was a growing realisation, fuelled by incidents like Seveso in 1973, that there was a secondary source of contamination from PHHs via the incineration of materials containing polyhalogenated waste. The polychlorinated dibenzo-p-dioxins (PCDDs) and the polychlorinated dibenzofurans (PCDFs), and the 2,3,7,8-tetrachlorodibenzo-p-dioxins/furans (TCDD/Fs) in particular, were added to the list of deteminands for analysis in the market basket sample. Other PHHs such as chlorinated bornanes (CHBs; toxaphene) were included in the US food survey programme and in German monitoring programmes, but current information on the polyhalogenated diphenyl ethers (PCDEs and PBDEs), and polychlorinated naphthalenes (PCNs) in food have only been generated by monitoring or research studies of individual institutes.

The second major effect of the 'chain reaction' was an almost exponential growth of publications resulting from the numerous measurements of PHH contaminants in food and the environment. The shear volume of this activity stimulated the development of more reliable methods of analysis and the manufacture of stable and sensitive instrumentation capable of providing reliable measurements. The three key improvements in the field of instrumentation were the manufacture of the capillary gas chromatograph (GC) and, in particular, the stable GC oven capable of the necessary level of temperature control; the development of the fused-silica capillary columns that gave the much-needed level of resolution of the complex PHH mixtures; and computer software capable of sampling, integrating and quantifying the data.

Certified reference materials have been prepared and used to validate measurements, and external quality assurance/proficiency testing programmes were put in place to improve and assure the quality of the data provided for assessment. Proficiency testing schemes such as the Food Analysis Performance Assessment Scheme (FAPAS) operated by MAFF in the United Kingdom offer different types of foodstuffs for the determination of a broad range of additives and contaminants including PHHs. In addition to providing the regulatory authorities with data on these ubiquitous and persistent contaminants, monitoring programmes provide a level of public assurance on the safety of different foodstuffs.

One disadvantage to emerge from the extensive analytical work on the PHHs was the data-rich–information-poor syndrome. The measurement of individual compounds in each group of PHHs has the resultant effect of producing an extensive data set, but with limited interpretation of the likely toxic impact.

The original selection of chlorinated biphenyls (CBs), for example, was largely based on their persistence and relative abundance, and the ability to measure unequivocally the CBs in the specific matrices. The initial selection of the seven CBs made by the European Community Bureau of Reference in 1982 was based on these criteria and the analytical technology at that time. For most workers this selection has been a valuable starting position for general monitoring work, but is clearly insufficient when relating the measurement of CBs to toxicological effects. Much of the recent developments have centred around the determination on the non-*ortho*-chloro- and mono-*ortho*-chlorobiphenyls (McFarland and Clarke, 1989; Hess *et al.*, 1995), which are known to be toxic and to induce liver microsomal enzyme activity. These planar CBs, and the PCDDs and PCDFs, induce both arylhydrocarbon hydrolase (Ah) and ethoxyresorufin *O*-deethylase (E*R*OD), bind with high affinity to the cytosolic receptor protein and are iso-electronic with 2,3,7,8-tetrachloro-dibenzo-*p*-dioxin (TCDD). Both the toxic mechanism and the enzyme induction involve an initial binding of the CB, PCDD or PCDF to the same arylhydrocarbon (Ah) receptor.

An additivity model was developed to provide a link between possible cause and biological impact at low concentrations for a number of the toxic PHH compounds by relating their ability to bind to the Ah receptor with that of the 2,3,7,8-TCDD. The details of these developments have been reviewed elsewhere, along with the resultant Toxic Equivalency Factors (TEFs), for seven PCDDs, ten PCDFs (NATO/CCMS, 1988; Van Zorge *et al.*, 1989) and for three non-*ortho* CBs, eight mono-*ortho* CBs and two di-*ortho* CBs (Ahlbourg *et al.*, 1994). The summation of these compounds, in terms of their toxic equivalency (TEQ) to 2,3,7,8-TCDD, provides an indication of their contribution to the Ah binding and subsequent EROD induction.

The TEQs have also simplified the link in the 'chain reaction' between the analytical scientist, the regulatory authorities and the politician to provide a single value for the Tolerable Daily Intake (TDI) for a number of these related compounds.

9.2 Sources of contaminants

The production, properties and usage of each group of PHHs have been extensively reviewed and are well documented (Brinkman and de Kok, 1980; WHO, 1992, 1994a,b; Erickson, 1997). With the exception of the halogenated dioxins and furans, the bulk of all of these compounds have been manufactured as formulations for specific applications and in most cases are complex mixtures with varying degrees of halogenation, depending on the physicochemical properties required for the application. The primary sources of these compounds are in many cases interdependent, and examples are now discussed in turn.

Most manufactured PHHs are made from the chlorination (or bromination) of the parent hydrocarbon in the presence of an appropriate catalyst. Details have been reported elsewhere (Brinkman and de Kok, 1980; WHO, 1992, 1994a,b; Erickson, 1997). Also there are very few natural source of PHH and almost all of these materials are present in the environment from specific situations, e.g. toxaphene on cotton, from accidental release, or from permitted discharge.

9.2.1   Direct manufacture for use in open systems such as additives to plastics or in sealed systems such as transformers

Polychlorinated naphthalenes (PCNs) were manufactured in the United States (Halowaxes), in Germany (Bayer-Nibren waxes), in the United Kingdom (Seekay waxes) and in France (Clonacine waxes). Originally used for protective coatings on rubber products, PCNs were the earliest of the PHHs to be manufactured. As the awareness of their toxic properties became apparent, the uses focused mainly on the electrical industry, in cables and particularly in impregnated capacitors paper for automobiles.

PCBs and polychlorinated terphenyls (PCTs) are no longer used in open systems in most developed countries, e.g. as additives in paints. Manufacture and new applications have now been banned or severely curtailed. However, while the production of most chlorinated hydrocarbons has been terminated in most countries since the 1980s, PBBs and PBDEs are still manufactured and used as fire retardants for many synthetic, flammable materials. Small quantities of polybrominated terphenyls have been manufactured, mainly for trials to test their fire retardancy, but they have not been released commercially (Jamieson, 1977).

Apart from the Michigan disaster, the PBBs enter the environment at a relatively low concentration, primarily from leachates from landfill sites into the water course. PBBs are reported to be ~200 times more soluble in landfill leachate than in distilled water (WHO, 1994b). However, it is likely that the PBBs are associated with suspended particulates on which they readily adsorb. As with their chlorinated counterparts, their stability and persistence mean that they are distributed widely and are bio-accumulated through the food web to contaminate fish, avian and mammalian species. The congener patterns found, for example, in fish samples were different from commercial formulations, possibly owing to the photochemical debromination reactions following aerial transportation.

PBBs have been used as additives to plastics such as ABS (10% PBBs), polystyrene, polyesters, and polyamides, lacquers and polyurethane foam as fire retardants. In most applications these compounds were incorporated into thermoplastics and restricted to products that were not in contact with food. A full review of some 34 applications of PBBs has been made by Neufeld *et al.* (1977).

There are no reports that PBDEs occur in the natural environment, with the exception that some bromoaryl ethers do occur in marine organisms (Faulkner, 1990). PBDEs are manufactured by the bromination of diphenyl oxide with the highest percentage of congeners being the tetra (24–42%), penta (44–62%), hepta (43–44%), octa (31–35%), and deca (97%) BDE in any one formulation (WHO, 1994a). PBDEs have generally taken over for use as fire retardants, especially as additives in plastics in the electrical/computing industry and in furnishings. There are currently eight producers worldwide and the production in the European Union (EU) has been steadily growing from 8586 tonnes in 1975 to nearly 11 000 tonnes by 1986 (EBFRIP, 1990), with an annual consumption worldwide of over 40 000 tonnes (Arias, 1992). The main concern over the PBDEs as additives is that they are relatively easily leached from the plastics.

9.2.2 Impurities in manufacture and subsequent use of the primary product

PCDDs are present as impurities in pentachlorophenol (PCP) used as a wood preservative, and in herbicides such as agent orange which is used as a defoliant. PCNs and PBNs are present as impurities in PCB and PBB manufacture. In addition, PCDDs and PCDFs are present as impurities in the high-temperature chlorination manufacture of other PHHs that have widely differing applications. These include chlorophenols, chlorobenzenes and associated derivatives, hexachlorophene, 2,4-D and 2,4,5-T, PVC and epoxy resins (Jones and Alcock, 1996). Dioxins are also created by bleaching raw cellulose (paper and cardboard) with chlorine and during the Kraft paper process (Fiedler *et al.*, 1990). Unlike the PBDEs, the chlorinated homologues (PCDEs) are not manufactured directly for commercial use, but they have been found in the environment as a result of their occurrence as an impurity in technical chlorophenols produced primarily as wood preservatives (Paasivirta *et al.*, 1982).

9.2.3   Production through the disposal of other PHHs

PCDDs and PCDFs are formed in the incineration of material containing PCBs, PCNs and poly (vinyl chloride) (PVC) while the polybrominated dibenzodioxins (PBDDs) and polybrominated dibenzofurans (PBDFs) are produced from similar brominated homologues. Some compounds are formed in forest fires or in open fireplaces in houses and are inhaled by the occupants. Polybrominated flame retardants can be emitted from electronic circuit boards and cabinets of computers and TV sets and inhaled by the users of such equipment (WHO, 1994a; Sellström, 1996). Secondary sources of these PHHs come from leachates from landfill sites, incineration and combustion, waste water discharges, and sewage sludge disposal on land or at sea.

In most cases not only are the compounds themselves stable and persistent, but the sources provide a relatively continuous input into the wider environment. Cessation of manufacture or their use in closed systems may have reduced the volume of the source, but with the mass of these PHHs in landfill sites and in goods in homes and industry, there is a *virtual low level leak* into the wider environment. It is inevitable that most of the primary production of these PHHs will find its way onto the land or into the rivers and the sea. Recycling through vaporisation and aerial transport serves to dissipate these compounds to the more remote areas of the world.

PCDDs and PCDFs are not manufactured and have no commercial use. They are formed as by-products of incineration of waste products, combustion of sewage sludge and fires containing chlorinated materials such as PCBs, PVC, PCP, and PCDEs (Fiedler *et al.*, 1990). Most of these compounds and their brominated derivatives are formed in the incineration process when the temperature is not sufficient to completely destroy the parent compounds, e.g. PCBs. A similar process occurs with brominated flame retardants such as PBBs and PBDEs from which PBDDs and PBDFs are formed. Less research has been carried out on the latter category because, until recently, it has been assumed that the volume of PBDDs and PBDFs produced is much smaller than that of PCDDs and PCDFs. Dioxins and furans that are both chlorinated and brominated at the same time can occur as well.

PHHs can come from contaminated sediment in rivers following extensive flooding. Milk from an area in Rotherham, UK, was found to contain elevated levels of dioxins. The herd was grazing in a field close to the River Rother, which regularly floods in winter and deposits sediment from the adjacent River Doe Lea, the source of the contamination (MAFF, 1997b). This cyclic source of PHH contamination is less common since sediments are often regarded as a sink for these compounds.

Contamination can come from packaging prepared using chlorinated bleach material. The migration of PHHs through the wax coating is greater when the fatty foods, such as milk, are in direct contact with the container (Fürst *et al.*, 1992). Haywood *et al.* (1991) found that ~ 0.6 pg I-TEQ/kg/day[2] was added to the diet assuming ~ 500 ml of milk was consumed from particular brands of paper cartons.

9.2.4 Direct manufacture for use as pesticides

Although many PHHs have been banned or severely restricted in many developing countries, it has not stopped their manufacture and/or use in other areas. Significant quantities of toxaphene are still used in Mexico (de Boer, 1997), and the former USSR was manufacturing and using PCBs well into the 1990s (Ericksson, 1997).

The formulation of toxaphene is a very complex mixture resulting from the chlorination of α-pinene to form polychlorinated bornanes, bornenes and iso-camphenes (Muir and de Boer, 1995; Saleh, 1991). Although some 32 768 isomers are theoretically possible, the technical toxaphene manufactured, in practice, consists primarily of hepta- to nona-chloro substitution with some 6480 congeners possible. Initially, it was used as a direct replacement for DDT in some agricultural applications, but over 80% of the 410 000 tonnes produced was applied to the cotton fields (Saleh, 1991). These agricultural applications have been one of the main sources of these compounds into the wider environment. Unlike many of the other PHHs, toxaphene was used in the open environment where a significant proportion of the application was allowed to volatilise into the atmosphere and to be translocated worldwide and deposited especially in colder regions of the world (de Boer, 1997).

Production in the former Soviet Union continued through to the 1990s (Ivanov and Sandell, 1992). Since aerial translocation is one of the main mechanisms of global distribution, those countries still manufacturing PHHs remain a significant source of these contaminants for other nations that have placed strict controls over their use and emission.

The main source of these compounds, in terms of their entering the food chain, is via uptake on the land by animals eating the vegetation or animal feed contaminated at low levels or similarly through the aquatic food web to fish, shellfish or for some specific communities, to marine mammals.

---

[2] I-TEQ International Toxic Equivalence; see Ahlbourg *et al.* (1988, 1994).

9.3   Transfer through the food chain

In Western industrialised countries, 90% of the human intake of PHHs is via food consumption, and particularly through consumption of fish, meat and milk products (Huisman *et al.*, 1995). Fish consumption, although relatively small in volume compared to that of meat and milk products, can contribute up to 50% of the intake of PHHs. Some social groups may be exposed to higher PHH concentrations as a result of their particular diet. Inuit people from northern Canada are exposed to high PHH concentrations because a substantial part of their traditional diet consists of marine biota and bird eggs, both of which contain extremely high levels of PHHs. The same is true for people from the Får Öer Islands who consume whale meat. Total PCB concentrations in milk fat of Quebec Inuit women were similar to those in Beluga whales, while the profiles were similar to those found in polar bears (Dewailly *et al.*, 1993). Another group exposed to relatively high PHH concentrations are sucklings who are fed by breast milk. Owing to the biased diet and to the relatively high PHH concentrations in breast milk of women from Western industrialised countries, sucklings may quickly build up high PHH levels. The risks of these biased diets will be discussed later.

PHHs may enter the food chain in different ways. Toxaphene has been applied in cotton growing for many years in the southern states of the United States and in Central America. Although many countries have banned the use of toxaphene in recent years, it is still being used in several African and Asian countries (Voldner and Li, 1993, 1995). When toxaphene is sprayed on the cotton plantations, part of it immediately evaporates. These toxaphene residues can be transported in the atmosphere over very long distances and material that was used in the southern United States can finally precipitate in polar regions. This global fractionation and cold condensation may occur for other PHHs as well (Wania and Mackay, 1993). The effect of this global fractionation means that fish and marine mammals from the Arctic contain relatively high levels of PHHs, despite the fact that these compounds have never been used in these areas. A similar mechanism has been suggested to explain the toxaphene concentrations found in North Sea fish originating from the United States and central America (de Boer and Wester, 1993).

PCBs released from products containing these additives in waste incinerators will later be precipitated in other areas, often remote from the source. Cattle feed on the contaminated grass and finally the compounds reach man by ingestion of the meat or the milk. PCBs can also precipitate into water or reach the rivers from discharge through sewage or industrial waste water purification plants, which are normally unable to remove PHHs. Since the PCBs and related compounds are lipophilic, they migrate to more lipid-rich tissue in aquatic biota such as fish muscle or liver. Again, through consumption of the fish, the PHHs finally accumulate in man. Many of the so-called industrial contaminants such as PCTs, and polybrominated flame retardants show a very similar behaviour to PCBs.

Much of the dioxin contamination of food is related to waste incineration. The dioxins and furans formed in waste incinerators normally precipitate in an area near the installation, depending on the prevailing wind direction. PCDDs and PCDFs are taken up and enter the food chain through cattle grazing in that area and, through meat or milk consumption, are transferred to humans.

Information on the gastrointestinal absorption of PCDDs and other PHHs is not conclusive (Liem and Theelen, 1997), but it is estimated that the uptake of most PHHs is >90% (Boersma, 1996). Dulfer (1996) demonstrated that PCBs in fatty food are very well absorbed in an *in vitro* system that simulates human intestines. Data on gastrointestinal absorption with breast-fed babies suggests a nearly complete absorption (Jödicke *et al.*, 1992; Körner *et al.*, 1993; McLachlan, 1993; Abraham *et al.*, 1994; Dahl *et al.*, 1995; Liem and Theelen, 1997). PHHs are associated with fat in food, and thus dissolved in a fatty matrix. Absorption of lipids by breast-fed children is high owing to their ability to absorb intact lipids. In adults fat is not absorbed as whole lipids but as fatty acids and monoglycerides. This may explain a lower gastrointestinal absorption in adults. Nevertheless, there are indications of an absorption efficiency of >90%. For higher chlorinated compounds such as hepta- and octa-chlorinated dioxins and furans a lower absorption is expected, based on experiments with animals (van den Berg *et al.*, 1994). There is very little information on gastrointestinal absorption of PHHs other than PCBs, PCDDs and PCDFs.

9.4 Residence times/depuration rates

PHHs are known for their persistency. 2,3,7,8-TCDD, the most toxic dioxin, is only metabolised and excreted at 4.5% per year in humans (Boersma, 1996). Other PHHs may be even less susceptible to biotransformation. Most studies on elimination rates in humans have been carried out on PCDDs, PCDFs and PCBs, because of their well-known toxicity, but the half-lives of other PHHs may be expected in the same range. Table 9.2 shows an overview of half-lives of PCDDs, PCDFs and some PCBs in humans, determined by different authors. Clearly, there are still considerable differences between the results of the different

Table 9.2 Elimination half-lives of PCDDs, PCDFs and non-*ortho*-substituted PCBs in human adipose tissue, expressed in years

Compound[a]	Half-life (range) (years)
2,3,7,8-TCDD	5.8–9.6
1,2,3,7,8-PeCDD	8.6–15.7
1,2,3,4,7,8-HxCDD	8.4–19.0
1,2,3,6,7,8-HxCDD	3.5–>70
1,2,3,7,8,9-HxCDD	4.9–8.5
1,2,3,4,6,7,8-HpCDD	3.2–6.6
OCDD	5.6–6.7
2,3,7,8-TCDF	0.4
1,2,3,7,8-PeCDF	0.9
2,3,4,7,8-PeCDF	4.7–19.6
1,2,3,4,7,8-HxCDF	2.9–6.2
1,2,3,6,7,8-HxCDF	3.5–6.2
2,3,4,6,7,8-HxCDF	2.4–5.8
1,2,3,4,6,7,8-HpCDF	2.6–6.5
1,2,3,4,7,8,9-HpCDF	3.2
OCDF	<0.2
3,3′,4,4′-TeCB	0.1
3,3′,4,4′,5-PeCB	2.7
3,3′,4,4′,5,5′-HxCB	13.0

Data from Liem and Theelen (1997), Flesch-Janys *et al.* (1996), van den Berg *et al.* (1994), Michalek *et al.* (1992, 1996).
[a]T = tetra; Pe = penta; Hx = hexa; Hp = hepta; O = octa.

studies (Shiria and Kissel, 1996). Nevertheless it is clear that half-lives of most PCDDs, PCDFs and PCBs are of the order of several years.

There may be large differences in half-lives between congeners from one group of compounds. *In vitro* biotransformation experiments on toxaphene compounds with hepatic microsomes of marine mammals have shown that several toxaphene congeners are readily biotransformed, but others are very persistent (Boon *et al.*, 1998). In similar experiments almost no biotransformation was found for most PBB, PBDE and PCT congeners. Saleh *et al.* (1979) also reported a fast metabolism of several toxaphene congeners in different animals. Between 82% and 99% of a heptachlorobornane was biotransformed 72 h after an oral dose of 3 mg/kg in various test animals. Toxaphene is least extensively metabolised by mice and most extensively by monkeys. Humans seem to be able to metabolise many toxaphene congeners. The patterns found in human milk only show the presence of a selected number of congeners. Some of these congeners are, however, very persistent, and are found in many different environmental samples (Muir and de Boer, 1995). Fingerling *et al.* (1997) reported that particularly congeners with one chlorine atom at the carbon atoms in the ring in alternating *endo–exo* orientation were highly persistent, while congeners with geminal dichloro groups on the ring were rather labile, especially when the dichloro group was located at the C2.

9.5 Concentrations in food products

The reported concentration of PHHs in food products from various sources can require some care and normalisation if a meaningful comparison is to be made. Much of the difficulty stems from the development of the analytical techniques available for the measurement of these contaminants since the late 1960s. Most of the PHHs are complex mixtures that were only partially separated by packed-column gas chromatography (GC) with electron capture detection (ECD). This technique gave limited separation of the congeners and a *total* concentration for the sum of a group of PHHs, e.g. PCBs was made. As individual congeners became available and GC separation methods improved, the individual PHHs were reported, e.g. single PCDDs and PCDFs.

Some reports sum individual congeners in an attempt to make a comparison with the earlier data, while others report separate values. A further development focused on the measurement of the *toxic* compounds alone, i.e. those CBs that exhibit specific biological effects such as Ah and EROD induction. As a sequel, the large number of individual measurements are being normalised using the standardised TEF values agreed by the WHO (Ahlbourg et al., 1994) for CBs and the I-TEF (NATO/CCMS, 1988) for the PCDD/Fs to provide an additive TEQ for the normalised sum of all measured PHHs that have a known (and agreed) TEF value.

Currently, there are no agreed TEFs for most other PHHs, such as PBBs, PBDEs and CHBs and so most of the available data are either for individual congeners or for the sum of the specific PHH congeners. Safe (1992) has proposed TEFs for the mono- and non-*ortho*-PCDEs in a similar way to the PCBs. The toxic potencies as inducers of Ah and EROD are of similar magnitude to those of the CBs of similar structure (Howie et al., 1990).

Since all of the PHHs are primarily apolar in character and highly lipophilic the highest concentrations are found in fatty foods. In general, the relative concentrations are greater in fatty meats, fish and dairy produce than in vegetables, cereals and fruit.

The concentrations of PHHs are generally well correlated with the extractable lipid in the tissue and the data are often expressed as lipid weight in order to make a direct comparison between the contaminant levels in different biological compartments. The difficulty arises when the lipid values are not reported with the PHH data, making the comparison between studies, at best, difficult. Specific studies are usually internally consistent, but more widespread, global comparisons are more tenuous. The data available comes from numerous total diet studies (Tables 9.3 and 9.4), which have been confined almost exclusively to developed countries. Some of these total diet studies have focused on pesticide residues from direct application to the crops, but since the 1980s there has been an additional focus on PHH contamination. There are also many specific studies initiated to investigate a particular source of PHH contamination. In the United Kingdom, MAFF has monitored the level of PCBs and PCDD/Fs in cow milk taken form the area in the vicinity of a municipal incinerator (MAFF, 1997a). Similar extensive studies have also been conducted in The Netherlands, which are reviewed in comparison to studies in other countries (Liem and Theelen, 1997).

Data on PHH levels in food can also come from environmental monitoring programmes in which the information may be collated not only to estimate the intake levels but also to provide advice on a reduction of these levels. De Boer (1994) made a comparison of PHHs in the North Sea and the North Atlantic and found both temporal and spatial trends. CBs were highest in the southern North Sea and declined in concentration during the period of study, while the CHBs were consistently highest in the northern North Sea and the North Atlantic. The combination of environmental data and food surveillance data not only can provide supportive information on dietary inputs but can identify the reason for changes that occur in foodstuffs. Much of the data for the PBDEs and PCDEs are only available from research studies.

9.5.1   PCBs

Since the first reports on the measurement of PCBs by Jensen (1966) there has been a continuous expansion of the surveys in food and in total diet studies. A summary of these studies is given in Table 9.3 by country and categorised by concentrations of PHHs in meat, fish and shellfish, vegetables and fruit, cereals, dairy products and milk, Meat includes beef, veal, mutton/lamb, pork and poultry and the fish includes lean and lipid-rich species.

The main difficulty with a detailed comparison of such data is that most of the values are obtained by different methodologies based on either single congener analysis or different formulations. However, a broad interpretation can be made. McFarland and Clarke (1989) have identified that only 36 of the 209 congeners are relevant for any toxicological, environmental, food or risk assessment studies. In most

Table 9.3 Concentration of PCBs in foodstuffs from different countries. Values are μg/kg unless otherwise stated[a]

Country	Year of sampling	Fish	Shellfish	Meat	Dairy produce	Milk	Vegetables and fruit	Cereals	Fresh or lipid weight	Reference
Australia		55 0.22–720		11 <0.01–37	4.1 1.2–8.2	0.1–0.8	0.4 <0.01–0.85	0.62 0.22–1.5	Fresh	[1]
Canada	1985	3.2–21	2.3	0.1–0.6	0.1–3.4				Fresh	[2]
India	1989	<10–110		<10–40	<10–20	<10	<10	<10	Fresh	[3]
Indonesia	1991	2.0–3.8		2.9–4.5				0.34	Fresh	[1]
Papua New Guinea	1994	0.8–16		45–124	4.4				Fresh	[1]
Thailand	1989	2–19		2.3–12				0.22	Fresh	[4]
Slovakia	1993	2			2		1–18		Fresh	[5]
South Korea	1992	5.6–60		2.2–3.7					Fresh	[6]
Solomon Islands	1994	0.66–15		5.2–17					Fresh	[1]
United Kingdom	1988	3.9–33.6	0.9	0.27–0.96	0.44–1.3	0.27–0.31	0.06–0.48		Fresh	[7]

Table 9.3 (Continued)

Country	Year of sampling	Fish	Shellfish	Meat	Dairy produce	Milk	Vegetables and fruit	Cereals	Fresh or lipid weight	Reference
Vietnam		10 (3.1–24)	27 (6.6–59)	18 (7.3–29)	17 (11–22)			2.5 (0.95–3.4)	Fresh	[1]
Yugoslavia (polluted area)	1985	116000		60–160	700	1100				[8]
	1988	182000		40	40	60	2–220	50		
Yugoslavia (unpolluted area)	1989	50		50	50	11	<1			
Canada	as pg TEQ/g fat weight	2.9–13.6		0.94–1.3	0.88				Lipid	[9]
The Netherlands		12–158	40–79	1.6–25	1.8				Lipid	[10] [11]
		15–561								[11]

aValues are given as mean or range, except for Vietnam where both mean and range are available.

[1] Kannan (1994)
[2] Mes et al. (1989)
[3] Kannan et al. (1992)
[4] Tanabe et al. (1991)
[5] Veningerova et al. (1993)
[6] Kannan et al. (1997)
[7] Duarte-Davidson and Jones (1994)
[8] Jan and Admič (1991)
[9] Mes and Webster (1989)
[10] Leim and Theelen (1997)
[11] de Boer et al. (1993)

Table 9.4 Concentration of PCDDs and PCDFs expressed as I-TEQ in foodstuffs from different countries. Values are pg TEQ/g fat weight, except for vegetables (fresh weight) or as stated

Country	Location	Fish	Meat	Dairy produce	Milk	Vegetables and fruit	Cereals	References
Austria	Urban				20.1–69.5 1–2.1			Ris and Hagenmaier (1991), Ris (1993)
Belgium	Urban Rural				2.7–5.1 2.7–3.9			Van Cleuvenbergen et al. (1993)
Canada		8.5	0.19–0.56	0.84	0.07	0.05–0.07	0.05	Birmingham et al. (1989)
Denmark				0.5–2.2	2.6			Büchert (1988)
Germany	Rural	29–34	1.2–7.2	0.76–2.62	2.32			Fürst et al. (1990) Fürst et al. (1992) Frommberger (1991) Fürst and Wilmers (1995)
		8.1–39						
	Rural	31–43	0.9–2.6	0.61–1.75	0.8–3.2 1.9–6.5	0.015–0.23		Beck et al. (1989, 1990, 1994) Beck et al. (1990) Beck et al. (1990)
	Urban				5–26			
Japan		(0.87–0.33)						Takayama et al. (1991) (fresh weight)
The Netherlands	Rural	6.8–49	0.43–14	1.4–1.8	0.7–2.5	0.13		Liem et al. (1991), Liem and Theelen (1997) Traag et al. (1993) Liem et al. (1990, 1991)
	Urban				0.7–2 1.2–13.5			

Table 9.4 (Continued)

Country	Location	Fish	Meat	Dairy produce	Milk	Vegetables and fruit	Cereals	References
New Zealand					0.18–0.22			Buckland et al. (1990)
Norway		0.97–1.7	0.33	0.75	0.11			Biseth et al. (1990) SNT (1995) (fresh weight)
Russia		0.8–46	0.2–5	0.19–0.96	0.28–5			Schecter et al. (1990, 1992) (freshwater fish)
Sweden		2–10.4	0.76–0.86	0.44–1.4	0.44–1.4			Hallikainen and Vartiainen (1997)
Switzerland					0.7–3.28			Schmid and Schlatter (1992)
United Kingdom		0.48	0.21	0.21	0.08	0.04		MAFF (1992)
		0.21	0.08	0.16	0.06			MAFF (1995)
United States		2.5–12.5	0.2–0.89	0.8–1.5				Schecter et al. (1994a,b, 1995)
			0.53–1.10	0.68–0.85				Cooper (1995)

cases much of the earlier monitoring data was based on the seven congeners: CB 28, CB 52, CB 101, CB 118, CB 138, CB 153 and CB 180. More recent data on PCB residues in foods are being reported as WHO-TEQ values (Liem and Theelen, 1997). These values are not directly comparable with the earlier data in Table 9.3. The highly contaminated fish from a polluted site in the former Yugoslavia is a clear reminder of the need of surveillance on all food (Jan and Admič, 1991). Apart from these contaminated areas, the general level of PCBs in fish was highest in Australia and India and lowest in Indonesia.

The levels of PCBs in meat were relatively low in Canada and in the United Kingdom. Horse meat is not consumed in the United Kingdom and the levels of PCBs in these tissues are similar to those found in the livers of dairy animals. Dairy produce, including butter, cheese and eggs, constitute the next main source of PCBs in dietary uptake, with relatively high levels in these foods from India, Vietnam and former Yugoslavia. Most studies for PCBs in milk are now being undertaken by congener-specific analysis and are expressed as TEQ values along with the PCDDs and PCDFs in the total diet studies. (Liem and Theelen, 1997).

In some instances, restrictions can be placed on foodstuffs if they are regarded as unsafe to eat. In Germany, wild eels from the rivers and lakes became polluted with PCBs and were unsuitable for consumption (Karl and Lehmann, 1993). In such cases the market demand was met by importing the food from areas where the contamination is low enough to meet national regulations. This requires the market surveys to be extended to include the imported foods. The results of one such survey for eels in Germany are given in Table 9.5.

Table 9.5 Concentration of PCBs in eels imported into Germany[a]

Country	Mean concentration of PCB (μg/kg)
Australia	2
Canada	5
Denmark	75
German Baltic Sea	72
Ireland	15
Italy	47
New Zealand	25
Norway	39
Poland	76
Tunisia	7
Retail Store (unknown source)	56

[a] Karl and Lehmann (1993).

Bennet (1983) estimated that the mean daily intake of PCBs from food was between < 0.1 and 1.9 µg/person/day. Between 1982 and 1984 the FDA estimated the average human PCB intake was 0.042 µg/day. For a 70 kg person the average intake was 0.6 ng/kg/day and for infants and toddlers it was 1 µg/kg/day. Since the bans on manufacture and the restrictions on usage there has been a decrease in the daily intake so that by 1987–89 the daily intake was 50 ng/day or 0.7 ng/kg body weight/day. The estimated daily intake of PCBs and dioxins in different countries is given in Table 9.6.

9.5.2 PCDDs and PCDFs

PCDDS and PCDFs have been monitored intensely in foodstuffs in a number of different countries, particularly in Europe and the United States (Hallikain and Vartiainen, 1997; Liem and Theelen, 1997). Following the initial surveys, I-TEF values for the PCDD/Fs were established (NATO/CCMS, 1988; Van Zorge et al., 1989) and the data on these contaminants in food have been expressed as a summation of the total toxic equivalence of the PCDD/Fs. An overview of the mean values and/or the range of concentrations of PCDD/Fs in different foodstuffs is given in Table 9.4.

The intensity of activity to measure PCDD/Fs in foodstuffs was increased following the evidence, available in the 1980s, that elevated levels of these PHHs were emitted from waste incineration plants that subsequently contaminated the surrounding pastures used for grazing (MAFF, 1992, 1995; Liem and Theelen, 1997). The main route of dioxin exposure to food is via atmospheric deposition onto plants that are consumed by the animals or into the water and taken up by benthic organisms and fish (Farland et al., 1994).

Many of the data for PCDD/Fs in foodstuffs have been obtained from specific studies to investigate the transfer of these PHHs through particular routes of the food chain, such as the analysis of cow's milk (Liem and Theelen, 1997; MAFF, 1997a,b). These intense studies were undertaken because of the high bioconcentration of PCDD/Fs in the milk fat, the rapid transfer from pasture to milk and the wide cross-section of the population, especially babies and infants, who have a potentially high exposure to these compounds through the regular consumption of milk.

Cows' milk from the United Kingdom and Canada contains the lowest levels of PCDD/Fs, while levels in milk from urban areas of The Netherlands and Austria were very much higher than elsewhere in Europe. The spread of these data very much reflect the nature of the surveys and the contamination in local areas. The regulatory authorities do not use data from these specific studies to investigate particular

Table 9.6 Estimated total intake of PCBs and dioxins (assuming 60 kg person) in different countries

Country	Year of sampling	PCBs (µg/person/day)	Year of sampling	Dioxins TEQs (pg TEQ/person/day)	References
Australia	1990–92	6.9			Kannan (1994)
Canada	1985	0.09	1989	33.6–126	Davies (1990), Gillman and Newhook (1991),
				92–140	Birmingham et al. (1989)
Denmark			1988	290	Büchert (1988)
Finland	1983	14			WHO (1988)
Germany	1987	4.1	1989	93.5	Georgii et al. (1989)
					Fürst et al. (1990)
		93.5 [a]			Beck et al. (1989)
				23–96	Schery et al. (1995)
				44–67	Moser et al. (1996)
India	1989	0.86			Kannan et al. (1992)
Italy				260–480	Di Domenico (1990)
Japan	1977	3.3			Matsumoto et al. (1987)
				175	Miyata (1991)
				175	Theelen (1991)
The Netherlands				67	Liem and Theelen (1997)
Norway				90	Fæden (1991)
New Zealand	1982	54			Pickston et al. (1985)
Sweden	1975	7.5			Vaz (1995)
	1990	6.8		14–55	NEPB (1988)
Spain				81–142	Jiménez et al. (1996)
Thailand		1.5			Kannan (1994)
United Kingdom	1990	0.53		125	Duarte-Davidson and Jones (1994),
					MAFF (1992)
US	1980	0.45		116	Schaum et al. (1994)
	1987	1.6		18–192	Schecter et al. (1994a)
Vietnam		3.7			Kannan et al. (1992)

[a] Expressed as pg WHO-TEQ/day.

sources of contamination as the objective of the total diet studies is to determine the average uptake of PCDD/Fs.

Correlation of the concentrations in the local environment and levels in adipose tissue or human milk are poor because the source of the primary input is food that comes from various sources. In general, such correlations can occur when the main diet is of locally produced materials. Home-grown vegetables may be an important source for trace metals, but not for PHHs. Where there is a predominance of a particular lipid-rich food, e.g. fish or marine mammal tissue, then there is a greater risk of increased intake of PHHs related to the immediate location. Communities in rural areas are more likely to reflect the local environment, whereas the urban population will reflect the contents of the supermarket rather than the environment outside. Dietary habits of local or national communities greatly affect the balance of inputs. For example, horse meat and liver have relatively high levels of PCDD/Fs. Up to 14 pg I-TEQ/g has been reported in The Netherlands (Liem and Theelen, 1997), but is not directly relevant for countries where there is a low consumption of this type of meat. The TDI for dioxins in different countries is given in Table 9.6.

9.5.3 Toxaphene

Toxaphene usage in the United States was drastically reduced in the late 1970s before being banned in 1982 (Pollock and Kilgore, 1978). However, toxaphene residues were still present in food products and human adipose tissue 8 years after the ban (Saleh, 1991). However, interpretation of the early data is confounded by the difficulty in the analysis of these residues prior to the development of good methodology and calibration standards.

Toxaphene has not been used in western Europe (Voldner and Li, 1993, 1995), but it has been used extensively in the United States and in eastern Europe. Toxaphene has been found in North Sea fish (de Boer and Wester, 1993; Alder et al., 1997). The main route of toxaphene intake in western Europe is expected to be through fish consumption, although little is known on toxaphene levels in other food products. Relatively high toxaphene levels have been found in American food products, such as in parsley (2 mg/kg), jalapeño peppers (1–2 mg/kg), carrots, potatoes, cherry tomatoes and fresh bell peppers (0.5 mg/kg) (Luke et al., 1988). Gunderson (1988) reported that toxaphene was among the most frequently occurring residues found in total dietary foods during the period 1982–1984 in the United States and Muller et al. (1988) reported high concentrations of toxaphene in food from Nicaragua of between 1.1 and 50 µg/kg compared to < 10 µg/kg in European foods.

Most of the toxaphene enters the water column via aerial transport and deposition (de Boer, 1997) and since the octanol–water partition coefficient (log K_{ow}) for toxaphene has been estimated as 6.44 much of it is bioconcentrated in the biota. As a result, fish are one of the main dietary sources of toxaphene residues. Alder et al. (1997) determined three toxaphene indicator compounds in over 100 fish samples, which are summarised in Table 9.7 and the total toxaphene intake was estimated to be between ∼2.8 and 5.6 ng/kg body weight/day.

Extensive studies have also been undertake in the Great Lakes, with particular reference to the vulnerability of fishermen and their families who consume the catch from these waters (Table 9.8). From these data it is clear that there has been a significant decrease in the levels of the CHBs in the fish from the Great Lakes and therefore also a decline in intake of toxaphene by the fishermen. Most of the sources of contamination are indirect and so more clearly relate to the concentrations in the area where food is produced.

Table 9.7 Concentration of chlorobornanes (toxaphene) in fish as food[a]

Relative fat content	Marine species	Concentration (µg/g lipid weight)
High-fat	Herring	11.1–389
Medium-fat	Halibut	124–500
	Mackerel	19–82
	Redfish	130–786
	Sardine	8–50
Low-fat	Pollock, cod, hake, plaice, saithe	n.d.–367
	Freshwater species	
High-fat	Eel	8–30
Medium-fat	Trout	26–138
	Salmon	11–133

n.d. = not detected.
[a] From Alder et al. (1997).

Table 9.8 Concentration of toxaphene in trout and smelts from the Great Lakes[a]

Fish	Concentration (mg/kg lipid)	
	1982	1992/4
Trout	24–30	2.8–25
Smelts	10–13	1.1–3.1

[a]From Glassmeyer et al. (1997).

9.5.4 Polychlorinated naphthalenes

Recent data on the levels of PCNs in food is limited, although these materials have been in existence since the 1890s. The level of production was never as high as that of the PCBs, but their misuse and the poor disposal techniques in the first half of the twentieth century have meant that these compounds are widespread in the environment (Järnberg *et al.*, 1993). The concentration of PCNs in Baltic herring (Table 9.9) ranged from 0.024 µg/kg for tetra-CNs to 7.8 µg/kg for hexa-CNs (Falandysz *et al.*, 1997).

The studies on guillemot eggs by the same group showed a decline of PCBs from 270 µg/kg in 1972 to ~60 µg/kg in 1988 and a similar decline of the hexachloronaphthalene from 150 µg/kg in 1972 to 60 µg/kg in 1988. The PCNs were thought to be an impurity in the CBs. Many indicator studies such as these on the guillemot eggs often give a clearer indication of the temporal trends of PHHs than the data from the total diet studies.

9.5.5 Polychlorinated diphenyl ethers (PCDEs)

The PCDEs have been reviewed by Becker *et al.* (1991) and were first identified in Baltic salmon (Koistenen *et al.*, 1993) with levels of between 25 and 800 µg/kg fat. PCDEs were detected in human adipose tissue in Finland (Koistenen *et al.*, 1995b) as well as in fish from the region. PCDE contaminants in wood preservatives were considered to be the main source of these compounds. PCDEs have been detected in adipose tissue at concentrations of up to 2 µg/kg in a US census to establish the body burden of organic contaminants (Remmers *et al.*, 1990). Like other PHHs these compounds are relatively ubiquitous and will contribute to the overall toxic burden, particularly form leachates associated with chlorophenol-related waste sites (Becker *et al.*, 1991).

9.5.6 Polybrominated biphenyls

Almost all of the data available on PBBs in foodstuffs have been related to the cattle food contamination incident in Michigan in 1973 (WHO,

Table 9.9 Concentrations of PCNs in Baltic herring[a]

PCNs	PCN concentration (µg/kg)
Tetra-CNs	0.024–0.42
Penta-CNs	0.33–10
Hexa-CNs	0.19–7.8
Hepta-CNs	0.27–0.94

[a]From Falandysz *et al.* (1997).

1994a). Very high concentrations were found in milk and meat during that time and afterwards (Table 9.10). Since then there has been a limited amount of data available. Krüger (1988) analysed three congeners in cow's milk and found 0.025–0.053 µg/kg fat for BB 153, 0.001–0.007 µg/kg fat for BB 180 and 0.005–0.014 µg/kg fat for BB 187. Samples of fish were also analysed from Germany and from the Baltic (Table 9.10). A TDI for PBBs for the average 60 kg adult has been estimated at 0.002 µg/kg body weight from eating ∼100 g fish per day containing 20 µg/kg. A breast-fed infant would have an intake of 0.01 µg/kg body weight based on a mother's milk content of 2 µg/kg lipid.

9.5.7   Polybrominated diphenyl ethers (PBDEs)

The environmental impact of PBDEs, including in food, has been reviewed by the WHO (1994a). Currently, there is a limited amount of data available on PBDEs in food. Krüger (1988) detected PBDEs in cow milk (3 µg/kg) and in human milk (2.6 µg/kg). Stanley et al. (1991) identified PBDEs in 47 human samples ranging form birth to over 47 years. Most of the other information on PBDEs is from environmental studies. Tetrabromodiphenyl ethers (TBDEs) in cod from the southern North Sea showed a decrease from ∼700 µg/kg in 1980 to ∼200 µg/kg by 1993. Samples from the central and northern North Sea were constant at 200 µg/kg (de Boer, 1990, 1994).

Sellström et al. (1993) and Sellström (1996) found that PBDEs were widespread contaminants in both pristine and industrial areas of the Swedish environment and that the spatial trends were similar to those of PCBs in the same areas. The concentration of PBDEs in guillemot eggs, used for temporal trend studies, showed an increase from 0.2 µg/kg in 1970 to 450 µg/kg by 1989, which is a different trend shown from that elsewhere for these compounds (de Boer, 1990) and for PCBs. The concentrations of PBDEs in edible fish species such as herring ranged from 15 to 60 µg/kg (lipid weight) to 450 µg/kg lipid weight in the spring when lipid levels in the fish are low. The concentrations in trout were between 37 and 590 µg/kg lipid weight. This input into the food chain via the aquatic environment is most likely from sources remote from the Swedish rural environment.

9.6   Potential human hazards of consumption

9.6.1   PCDDs, PCDFs, PCBs

There is an extensive literature on the toxicology of PCDDs, PCDFs and PCBs (Anonymous, 1989; Hong et al., 1993; Kimbrough, 1995; Liem and

Table 9.10 Concentration of polybrominated biphenyls in food

Country	Year of sample	Fish (mg/kg)	Shellfish (mg/kg)	Meat (mg/kg)	Dairy (mg/kg)	Milk (mg/kg)	Reference
Michigan, USA	1974					44–900	Robertson and Chynoweth (1975)
	1974			up to 2700	1.4–15	2.8–270	Cordel et al. (1978)
	1974						Kay (1977)
	1975					1–13	Kay (1977)
	1975–6			n.d.–0.13			Cook et al. (1978)
		(µg/kg)				(µg/kg)	
Chippewa River, USA	1983	5300–15 000					Jaffe et al. (1985)
Germany	1988	6.3				0.05	Krüger (1988)
Baltic		30.5					Krüger (1988)

Theelen, 1997) and 2,3,7,8-TCDD, in particular, this being one of the most toxic compounds. A very low oral LD_{50} value of 0.6 μg/kg body weight was determined for the sensitive female Guinea pig (Liem and Theelen, 1997). Hong et al. (1989) reported that 500 ng/kg in the diet was lethal to Rhesus monkeys within 2 months. Large differences in sensitivity between test animals are reported (Pohjanvirta, 1991).

A wide variety of subchronic effects have been found in test animals. Decreased body weights, increased liver weights, altered haematology and pathological changes in liver, thymus and other lymphoid tissues, hair loss, and porphyria have been observed in different types of rats (Kociba et al., 1976; Liem and Theelen, 1997).

Holder and Menzel (1989) stated that 2,3,7,8-TCDD should be characterised as a multi-target carcinogen, which shows initiation by a hormonal mechanism and promotion of carcinogenic activities. Many studies on the carcinogenicity of PCDDs and PCDFs have been reported and tumour promotion activity has been frequently found (Flodström and Ahlbourg, 1989; Poland, 1991; Liem and Theelen, 1997). However, increased cancer mortality of humans was negligible in several of the well-documented cohorts made after the Seveso accident in which the local population were exposed to PCDDs and PCDFs following the accidental release at a chemical production site (Bertazzi, 1991; Manz et al., 1991; Fingerhut et al., 1991).

One of the biological responses most frequently studied following exposure to PHHs is the induction of the hepatic microsomal mixed-function oxidase system, cytochrome P450, which is essential for the oxidative detoxification of a broad variety of xenobiotic chemicals. Some of the enzymes of the P450 system can be induced by xenobiotics, which is often regarded as a strategy of the organism to cope with toxic chemicals. This type of enzyme induction is found in many organisms exposed to PCDDs, PCDFs and PCBs and also in humans who were exposed to PCDFs from PCB-contaminated rice oil (Lucier, 1991; Liem and Theelen, 1997).

Exposure of different test animals at varying doses to PCDDs, PCDFs and PCBs have induced neurobehavioural changes, reproductive and immunotoxicological abnormalities, reduced hepatic vitamin A storage, vitamin K deficiencies and dermal effects (Liem and Theelen, 1997). These biological effects have implicated PCDDs and related compounds in the deregulation of such processes when accumulated in the human body following consumption of contaminated food. However, clear effects such as pigmentation of skin and nails, follicular accentuation, hypersecretion of the Meibomian gland and a higher incidence of middle-ear disease in children were observed after two accidents in which people consumed PCB-contaminated rice oil (Chao et al., 1997). PCB levels in these children were as much as 14 µg/kg wet weight, and considerable amounts of dioxins were also found in the serum. Aylward *et al.* (1996) reported that humans were much less sensitive than rats to the carcinogenic effects of TCDD. Short-term effects following exposure to much lower doses from food may therefore be relatively small. Chronic, low level, effects from exposure to contaminants from food, may, however, not make themselves manifest until a number of years have elapsed. The effects of such a chronic exposure are not well known. The risks of dioxin exposure have led to relatively strict tolerance levels and values of PCDDs, PCDFs and PCBs in food (Tables 9.11 and 9.12). Different PCDD, PCDF and PCB congeners have a different potency to induce cytochrome P450. It is assumed that this potency is dependent on the structure and the planarity of the molecule. A system of toxic equivalency factors (TEFs) has been set up by the World Health Organisation (WHO) in which a potency of 1 is given to 2,3,7,8-TCDD, the strongest cytochrome P450 inducer. TEFs for other PCDD, PCDF and PCB congeners vary between 1 and < 0.000 005 (NATO/CCSM, 1988).

An extensive Dutch study on dioxins in duplicate diets over the period 1978–94 has shown decreases of dioxin levels in the foods. Dioxin levels in Dutch cow milk have also decreased over the last decade. The study on the dietary exposure of the general population in The Netherlands revealed a median daily intake of 2.3 pg dioxins, expressed as TCDD equivalents (TEQ) per kg body weight, in the early 1990s. Milk fat and related food items such as cheese, butter and beef contributed 50% to the total daily exposure of the general population. A major contributor to exposure from food items provided by the food industry (13–24% of the total exposure) is fish oil. The daily intake of 2.3 pg/kg body weight is clearly lower than the TDI of 10 pg/kg body weight per day advised by the WHO (1991a). Around 1.5% of the Dutch population is exposed to amounts of PHHs that exceed the TDI of 10 pg/kg body weight per day. A new exposure limit of 1 pg/kg body weight per day, proposed by the Dutch Health Council, is, however, exceeded by 93% of the population. Such a strict tolerance limit will also be exceeded in other countries, in particular where measures to reduce dioxin outputs from municipal waste incinerators have still to be taken. PCB levels in fish, which are decreasing only slowly in most places, are a significant contributor to the total human TEQ intake though food.

Some groups that have a special diet can be exposed to much higher TEQ levels. A human milk survey carried out in The Netherlands in 1993 (Liem and Theelen, 1997) showed that breast-fed children in The Netherlands were exposed to 200 pg/kg body weight per day TEQ. There were no observable negative effects on the development of young

Table 9.11 Maximum residue levels for PCBs in foodstuffs

Country	Effective date	All foodstuffs (μg/kg)	Fish (μg/kg)	Meat (μg/kg)	Dairy (μg/kg)	Milk (μg/kg)	Drinking water (μg/l)	Reference/comment
France	1983		2000					WHO (1992)
Germany	1988	8–600		1400				WHO (1992)
								Comprise CB28, 52,101 and 180 at 200 μg/kg and CB 138 and CB 153 at 300 μg/kg. Anonymous (1992) German Directive
Japan		200–3000						WHO (1992)
Netherlands		300–1000						WHO (1992)
Sweden	1983	50–2000						WHO (1992)
Switzerland		500–2000						WHO (1992)
United States	1981	200–2000	2000	30 000	300	1500		WHO (1992) FDA (1989)
Former Yugoslavia	1987		3000	2000	300	1000		Federal Ministry of Health (1987)
Drinking water European Union	1982						0.5	WHO (1992)

Table 9.12 Tolerable Daily Intakes (TDIs) as I-TEQs and Maximum Residue Levels (MRLs) in fish and milk and milk products for PCDDs and PCDFs in different countries

Country/agency	TDI/TEQ (pg I-TEQ/kg bw/day)[a]	MRL (pg I-TEQ/g) Fish	MRL (pg I-TEQ/g) Milk and milk products	References
WHO/EURO				
Canada	10	20		Ontario Ministry of Environment 1985 in Birmingham et al. (1989) Government of Canada (1990) Ryan et al. (1983)
Germany	1 (a) 1–10 (b) > 10 (c)		3 5	Schultz (1993) Bericht der Bund (1992) (a) Precaution (b) Range of Risk (c) Intervention
Japan	1			Masuda (1990)
The Netherlands	1		6	Health Council of the Netherlands (1996) Liem and Theelen (1997)
Nordic Council	5			NORD (1988)
Switzerland	10			Hallikainen and Vartiainen (1997)
Sweden	5			Ahlbourg et al. (1988)
United Kingdom	10		17.5	MAFF (1997a)
USFDA	0.0572	25		Vanden Heuvel and Lucier (1993)
USEPA	0.006			

[a] bw = body weight.

children, despite the high dioxin intake that takes place during a relatively short period. These children were monitored from birth onwards by a series of different development tests (Koopman-Esseboom, 1995). Consequently, Dutch authorities maintained the positive advice on breast milk feeding. Huisman *et al.* (1995) reported that the contribution of a special diet during pregnancy to the reduction of PCBs and dioxins in breast milk is relatively low. Decrease of exposure to PCBs and PCDDs of the fetus and the neonate probably requires long-term reduction of the intake of these contaminants. Substitution of normal cheese by low-fat cheese and the use of vegetable oils instead of fish oils in the preparation of food products by the food industry could contribute to a reduced intake of PCBs and PCDDs.

Anglers who eat fish taken from relatively polluted waters are another group of people at risk with regard to PCB and dioxin intake. Crane (1996) calculated an increase of an order of magnitude of carcinogenic risk due to recreational fishing in the Great Lakes, USA, and of an additional order of magnitude for people dependent on fish as their staple diet. She emphasised, however, that these estimates should be interpreted in the context of all the uncertainties associated with each step in the risk assessment. PCB concentrations in eel from the rivers Rhine and Meuse, determined in a Dutch eel monitoring programme for the benefit of anglers, indicate that the maximum eel consumption from those rivers should be less than 5 g per week to remain within the advised TDI of the Dutch Health Council of 1 pg TEQ per kg body weight per day (Anonymous, 1996a; de Boer *et al.*, 1997a). Most eels in The Netherlands originate, however, from cleaner waters, and most other fishes have lower fat contents and thus lower PCB and dioxin levels. Dar *et al.* (1992) studied the relation between fish consumption and reproductive effects in Green Bay, Wisconsin, USA. The only correlation between PCB concentrations and birth size of babies was positive.

9.6.2 Other PHHs

A number of other PHHs have toxicological effects very similar to PCBs and dioxins, although the degree of toxicity is normally lower than that of 2,3,7,8-TCDD. Compounds with a planar structure, such as PCNs and some PBBs, PBDEs and PCDEs, have a potency to induce cytochrome P450, and should therefore be included in the summation of the TEQ (Safe, 1984, 1992).

Most mechanisms of toxicity found for PBBs and PBDEs are similar to those of PCBs (Pijnenburg *et al.*, 1995). The concern over the toxic effects of these compounds on humans from the consumption of food containing these residues has not been so overt as for PCBs because, until recently, levels of PBBs and PBDEs have been lower than those of PCBs (Pijnenburg *et al.*, 1995). However, in contrast to PCBs, the PBBs and PBDEs are still produced in relatively high amounts. PBBs and PBDEs reach food products through waste incineration and water and air pollution and can accumulate in humans up to levels comparable to those of PCBs. However, little action has been taken by governments generally to place any restriction on these types of brominated flame retardants and to investigate the use of substitutes. The limited awareness of the levels and exposure of these brominated fire retardants has not created the same 'chain reaction' of events that has occurred with the chlorinated homologues. Yet these compounds are still manufactured and used extensively in electrical goods and synthetic textiles, giving a direct exposure in the home.

Toxaphenes (polychlorinated camphenes) are chlorinated cyclic aliphatic compounds. The toxicological effects are different from the aryl contaminants. Cytochrome P450 is not induced because the congeners cannot adopt a planar structure. The current discussion on the risks of toxaphene has focused on its carcinogenicity (Reuber, 1979). There is no doubting the mutagenicity of toxaphene (Hooper *et al.*, 1979) and the carcinogenicity of toxaphene is being reviewed within the framework of the European Union (de Boer, 1997). The congener concentration of a weathered toxaphene mixture found in fish has a considerably different pattern from that in the technical toxaphene standards, which could influence the level of toxicity. Reviews on the toxicology of toxaphene have recently been given (Saleh, 1991; Anonymous, 1996b, 1997). One group that is particularly at risk with regard to toxaphene intake through food are the Inuit people in northern Canada (Quebec). Fish and marine mammals are an important part of the traditional diet of the Inuit of northern Canada and Greenland. Breast milk of Inuit women contains 3-fold higher toxaphene levels (0.3 mg/kg on a lipid weight basis) than the milk of Dutch and Swedish women (Vaz and Blomkvist, 1985; Stern *et al.*, 1992; de Boer and Wester, 1993; Muir and de Boer, 1995).

9.6.3 Endocrine-disrupting chemicals

Many xenobiotic compounds are reported to disrupt endocrine development in wildlife and laboratory animals (Colborn *et al.*, 1993) and a number of these compounds are PHHs, including PCBs, PBBs, and 2,3,7,8-TCDD. Exposure to endocrine-disrupting chemicals is associated with abnormal thyroid function, decreased fertility and defeminisation and demasculinisation in fish and birds (Moccia *et al.*, 1981). Reproductive loss and early mortality have been observed in offspring of confined mink that were fed Great Lakes fish. PHHs have been linked to decreased male sperm counts and reproductive capacity. Furthermore, breast cancer in women may be related to human exposure to endocrine disruptors (Safe, 1997). Links have been made with PHHs and with diethylstiboestrol (DES), a synthetic oestrogenic compound used by physicians to prevent spontaneous abortions in women between 1948 and 1971. Daughters of these women have suffered from reproductive organ dysfunction, abnormal pregnancies, a reduced fertility, immune system disorders and periods of depression (Colborn et al., 1993). DES has been proposed as a model for other endocrine disruptors. Safe (1997) stated that humans are exposed to higher levels of food-derived hormone receptor antagonists in fruits and vegetables, which are usually associated with health benefits. Future studies are therefore required, focusing on identifying both xeno- and natural endocrine-active compounds and determining their hazards and risks.

9.7 Permitted concentrations

Guidelines to evaluate and establish the residue limits in food to asses the risk to human health have been developed (WHO, 1994c). Two terms have been used to describe permitted concentrations of chemicals in foodstuffs that may be potentially confusing. The first is Acceptable Daily Intake (ADI) which refers to permitted additives such as colourings or preservatives. These chemicals are usually knowingly added to the food as part of the processing and fulfil a specific function. Other chemicals such as the PHHs are present as a result of contamination of the foodstuff. They have no reason to be present and have no known benefit to the consumer. The term Tolerable Daily Intake (TDI) has been designated to describe the permitted concentrations of these compounds that would be consumed for a specific diet (WHO/IPCS, 1987).

Care should be taken when comparing literature data on the mean daily intake of PHHs since these values can be reported in different terms by different authors and countries. Some values are based on the weight of compound per kg body weight per day while others are based on weight of compound per person per day. The usual standard person is assumed to be 60 kg, although some reports also work on the basis of 70 kg. In addition, some regulatory levels are set as Maximum Residue Levels (MRLs) intrinsically permitted in the specified foodstuff (Table 9.11). These values for PCBs are only available for European countries, the United States and Japan and range from 8 to 3000 µg/kg. Some of these countries also have maximum residue levels for CBs in fish, meat, dairy produce and milk and the European Union has set a value of 0.5 µg/l for total PCBs in drinking water. Owing to the low water solubility of PCBs in purified water, this level is only likely to be exceeded if there are particulates or organic colloids present in the drinking water.

The estimated daily intake for PCBs expressed as μg per person per day, is given in Table 9.6. The highest estimate was that for New Zealand and then Finland. Both countries have a high intake of fish in the diet. Canada was among the lowest. In most dietary intake studies for PCBs, the standards used were industrial formulations (Sweden, Vaz (1995); Denmark, LST (1990); Finland, Kumpulainen (1988); Japan, Matsumoto *et al.* (1987); Canada, Davies (1988)).

No TDI has been proposed by the FAO/WHO (GEMS, 1988) for PCBs. Most comparison are currently made with the value of 1 μg/kg body weight per day proposed by the FDA (FDA, 1990a). A value of 0.6 μg/kg body weight per day has been proposed for PCBs by the Danish Institute of Toxicology under the National Food Agency (LST, 1990).

The tolerable daily intake for PCDD/Fs, expressed as pg I-TEQ per kg body weight per day set by different countries is given in Table 9.12. A number of countries have adopted the WHO/EURO intervention level of 10 pg I-TEQ per kg body weight per day. These are Canada, Switzerland, the United Kingdom and Germany. Germany has two other levels; 1–10 pg I-TEQ per kg body weight per day which is described as the range of risk and 1 pg I-TEQ per kg body weight per day which is the precautionary level. Some countries have also set specified levels for fish and for milk and milk products. Both Canada and the United States have set levels specifically for fish since there is a high consumption around the Great Lakes.

Following the exposure of PBBs in cattle feed, the USFDA set the MRL for PBBs in meat and dairy produces at 1 mg/kg and 0.1 mg/kg in eggs in May 1974. These values were revised down on November 1974 to 0.3 mg/kg in meat, milk and dairy produce, 0.5 mg/kg in eggs and 0.02 mg/kg in cows for slaughter (Kimbrough, 1980). No other data have been set for PBBs since that incident (WHO, 1994b).

Maximum residue limits for toxaphene have been set by the United States and Germany. In the United States the MRL was 5 mg/kg wet weight, but this value is no longer in force, while the German MRL is currently set at 0.1 mg/kg for fatty fish with > 10% extractable lipid and 0.1 mg/kg for lean fish with < 10% extractable lipid. It is unusual to have two MRLs set on such a basis and eventually a single value will be set on the basis of the sum of the three main CHBs (26, 50 and 62) in all species of fish. Canada has set a TDI of 0.2 mg per kg body weight per day.

The use of toxaphene has been banned in the United States, Canada, the United Kingdom, Sweden, Finland, Denmark, France, Germany, Switzerland, Hungary, Italy, Egypt, India, China and Algeria (Voldner and Li, 1993).

The US EPA is required, under the Toxic Substances Act (Sections 4 and 8) to identify all chemicals that may be contaminated with chlorinated or brominated dioxins or furans. The PBDEs used as fire retardants are known to contain traces of PBDD/Fs and these compounds are included under this legislation in the National Adipose Tissue Survey (Stanley *et al.*, 1991).

9.8 Specific methods of analysis and analytical problems

9.8.1 Analysis of PCBs

The PCBs were the first PHHs found in the environment and food. The analytical methods for most PHHs are based on the methods developed in the early 1970s for PCBs and organochlorine pesticides. These methods consist of the following steps: extraction, clean-up (fat separation), fractionation and GC determination (Wells, 1993; Hess *et al.*, 1995; Erikson, 1997).

9.8.2 Extraction

The most commonly used extraction techniques are suitable for a wide range of food matrices. The PHHs are lipophilic, so the extraction methods are based on the isolation of the lipid fraction from the sample matrix (Wells, 1988; Liem *et al.*, 1992). Soxhlet extraction has been used for the extraction of non-polar and semi-polar compounds from a wide variety of food matrices (Hess *et al.*, 1995). The size of the system can vary, but the more common configurations use ~100 ml of solvent and 5–50 g of sample. After maceration and homogenisation, the sample is ground with sodium sulphate to bind the water present in the sample. The dry powder is refluxed with the solvent for 4–18 h. It is essential to match the solvent polarity to the solute solubility. A combination of a non-polar solvent such as pentane or hexane with a medium-polarity solvent such as acetone or dichloromethane is recommended for an efficient extraction of the PCBs. Non-polar solvents alone do not extract all PCBs from low-fat, phospholipid, matrices (de Boer, 1988). Non-polar solvents by themselves may be sufficient for fatty matrices that contain only triglycerides. Saponification can be used prior to, or in conjunction with, solvent extraction. This is an exhaustive method, but, owing to the aggressive treatment, some more labile compounds, such as some higher-chlorinated PCBs, can lose chlorine atoms (de Boer, 1988; Wells and Echarri, 1992).

Other methods use column extraction (Huckins *et al.*, 1988; Schmidt and Hesselberg, 1992), blending with an Ultra Turrax and ultrasonic extraction (Jensen *et al.*, 1983; Jansson *et al.*, 1991) and supercritical fluid extraction (SFE) (Hawthorne, 1990). A supercritical fluid has a low viscosity, high diffusion coefficient, and, in the case of carbon dioxide, low toxicity and low flammability, which are all qualities superior to those of the organic solvents normally used. The advantage of carbon dioxide is that it has a low critical temperature (31.3°C) and pressure (72.2 atm). Other less commonly used fluids are nitrous oxide, ammonia, methane, pentane, ethanol, dichlorofluoromethane and sulphur hexa-fluoride (Hess *et al.*, 1995). A great advantage of SFE is that extracts are very clean and require only moderate additional clean-up. There are, however, limitations to this technique where lipids and PHHs are not separated (Bøwadt *et al.*, 1994). Initially samples with high fat contents were not extracted without the co-extraction of some lipids with the PHHs. The addition of different types of modifiers and the use of alumina, silica or silica with silver nitrate in the extractor have partially solved this problem for the most non-polar PHHs such as PCBs, PCDDs and PCDFs (van der Velde *et al.*, 1996; Yoo and Taylor, 1997). Another limitation with SFE is the small volume of the extractor, which can contain only a few grams of material. This is a considerable disadvantage when a higher sample mass is required for lean tissue samples. An evaluation of classic extraction techniques and SFE for the determination of PCBs has been made by Hess *et al.* (1995).

Microwave-assisted extraction (MAE) and accelerated solvent extrac-tion (ASE) have also been used. MAE uses electromagnetic wavelengths of 300–300 000 Hz. Most microwaves use a frequency of 2450 Hz. The electromagnetic field induces a rotation of molecules with dipole moments, but does not affect the molecular structure (Onuska and Terry, 1995). ASE is an extraction under elevated pressure and temperature. The solvent volume can be reduced because the solubility increases. Normally temperatures of 50–200°C are used with pressures of 110×10^5 to 140×10^5 Pa (Knowles *et al.*, 1995). ASE is relatively expensive, but shows more stable recoveries than MAE. No degradation of analytes was observed for either of the two techniques.

9.8.3 Clean-up

Destructive methods are mainly alkaline treatment (saponification) or oxidative dehydration (sulphuric acid treatment). Alkaline treatment is similar to the saponification used in conjunction with extraction, but is applied sequentially to the solvent extraction instead of applying it to the food matrix directly.

Concentrated sulphuric acid is mixed with the lipid extract (usually in pentane or hexane) and after mixing for some hours the organic layers are separated from the acid in a separation funnel (Harrad et al., 1992). Alternatively, sulphuric acid can be added to silica and the sample can be eluted from the silica chromatography column (Huckins et al., 1988; Jansson et al., 1993). Wells and Echarri (1992) found that a fine carbon layer formed during elution which tended to retain 2–3% of the PCBs present.

Non-destructive methods use solid-phase columns, gel permeation techniques and dialysis. Alumina columns are very effective and probably one of the most frequently used clean-up methods (de Boer, 1988). Columns of 10–20 g alumina have a fat capacity of $\sim$250 mg. This capacity may be too small for the removal of the larger quantities of lipids associated with the mass of sample used for the analysis of planar PCBs, PCDDs and PCDFs. Silica and Florisil columns are alternative adsorbents (Storr-Hansen and Cederberg, 1992; de Voogt et al., 1986; Wells et al., 1985). Gel permeation chromatography (GPC) or size exclusion chromatography (SEC) have also been used for lipid removal and the separation based on molecular size. SX-3 Bio Beads are used in most cases (Huckins et al., 1988; Beck and Mathar, 1985; Schwartz et al., 1993). The disadvantage is that it is very difficult to obtain a 100% separation of lipids from the PHHs. Often traces of the lipid that remain interfere with the measurement of the PHH, resulting in the need for further treatment or a second GPC elution. Dialysis techniques include the use of a polythene film of pore size $\sim$50 μm. Around $\sim$10 g of fat is dissolved in 15–20 ml cyclohexane and placed in the polyethylene tube. The PHHs migrate from the fat through the polythene tube to the cyclohexane surrounding the tube. Recoveries of more than 95% can be obtained (Huckins et al., 1990). A disadvantage may be the large volumes of solvent needed, because the solvent has to be renewed regularly.

9.8.4 Fractionation

Separation of PCBs and pesticides is normally required prior to the GC analysis, since the PCB chromatograms are already quite complex without additional pesticides. Silica gel columns of 1–2 g are frequently used to obtain such a pre-separation (de Boer, 1988). The PCBs and chlorobenzenes are eluted with a non-polar solvent (hexane or iso-octane) in the first fraction. Other absorbents, such as Florisil, may also be used (Harrad et al., 1992). The disadvantage of these gravity adsorption columns is that they are rather sensitive to moisture and the elution pattern has to be calibrated regularly. A pre-separation by HPLC on silica columns is more reproducible (Hess et al., 1995).

9.8.5 GC analysis

Gas chromatography (GC) is the most widely used method for the determination of PCBs because of their volatility. Electron capture detectors (ECDs) were used initially because of their high sensitivity for electronegative compounds, but mass spectrometric (MS) detection has become a routine method for measurement. Modern MS instruments are easy to operate, are more selective than the ECD, and have a better linear range. Packed columns that were used in the 1970s (Jensen, 1972) and early 1980s have been replaced by capillary columns, which offer a much better resolution (Jennings, 1978; Dandenau and Zerenner, 1989). Stationary phases of different polarities can be used for the determination of PCBs (Larsen et al., 1992; Larsen and Bøwadt, 1993; Ballschmiter et al., 1992). Non-polarity and medium-polarity phases generally offer a better resolution (Mullin et al., 1984; de Boer and Dao, 1989; de Boer et al., 1992).

There is, however, no single capillary column that is able to separate all 209 PCB congeners in one chromatogram. Therefore, more advanced techniques, such as the use of serial-coupled columns (Larsen and Bøwadt, 1993) and parallel-coupled columns (Storr-Hansen and Cederberg, 1992; Storr-Hansen, 1992), two-dimensional GC and, recently, comprehensive multidimensional GC (MDGC) have been developed. In two-dimensional GC the 'heart-cut' of co-eluting compounds can be made during elution on the first capillary column and transferred to a second capillary column of a different polarity (Schomburg et al., 1985; Duinker et al., 1988). This technique offers a complete separation of certain areas of the chromatogram from which the heart-cuts are made. The limitation of this technique is the number of heart-cuts which can be made during one run. As soon as more than three or four heart-cuts are being performed, peaks start to overlap again. Also, the analysis times of the heart-cut two-dimensional GC are relatively long. Nevertheless this technique has shown its merits in the determination of PCBs in food (de Boer and Dao, 1991). Comprehensive MDGC (or modulation MDGC; de Geus et al., 1997) is a new technique in which true three-dimensional chromatograms can be made (Phillips and Xu, 1995; de Geus et al., 1996). It is based on two capillary columns in series, connected by a retention gap (modulator), which can be heated very quickly in a reproducible way. Because peaks collected at the modulator are very highly focused before they enter the second column, an extremely high sensitivity can be obtained in combination with a very high selectivity.

9.8.6 Analysis of PCDDs, PCDFs and non-ortho-substituted PCBs

Most of the analytical methods for PCDD, PCDFs and non-ortho-CBs are comparable to those used for PCBs. However, concentrations of

PCDDs, PCDFs and planar PCBs are generally $\sim$1000-fold lower (ng/kg range) than those of PCBs (μg/kg range) (de Boer *et al.*, 1993) and it is difficult to measure these compounds simultaneously with the PCBs. An additional fractionation step is usually included in the clean-up procedure and MS is used for detection instead of ECD, which enables isotope-labelled standards to be used for the final detection to provide a better control on the recovery.

Three techniques are mainly used for the pre-fractionation of PCDDs, PCDFs and planar PCBs: use of gravity carbon columns (Huckins *et al.*, 1980; Tanabe *et al.*, 1987; Storr-Hansen and Cederberg, 1992); HPLC using graphitised carbon columns (Knox and Kaur, 1986; de Boer *et al.*, 1993); and HPLC using PYE (2-(1-pyrenyl)ethyldimethylsilylated) columns (Haglund *et al.*, 1990a).

There are several MS techniques that can be used for the determination of PCDDs, PCDFs and planar PCBs (de Jong *et al.*, 1989; de Jong and Liem, 1992). High-resolution instruments, using electron ionisation, are normally required for the determination of PCDDs and PCDFs. Low-resolution instruments, using negative chemical ionisation, may be used for the determination of planar PCBs.

9.8.7 *Analysis of other PHHs*

Most other PHHs can be determined using very similar methods to PCBs with some specific modifications of the method. Polychlorinated terphenyls have a very similar behaviour to PCBs. They are found in the same fraction after silica gel elution. Higher temperatures are required to elute them from GC columns because they have higher boiling points. PCT mixtures are very complex (over 6000 congeners). Congener-specific determination is still not possible. Total determinations, using the technical Aroclor mixtures as standards, have been used instead (Wester *et al.*, 1996). A disadvantage of such an approach is that neither information on toxic congeners nor on specific patterns can be obtained. Comprehensive MDGC may be a solution for this problem in the future.

PBBs and PBDEs are analytically analogous to the PCBs and PCTs. A specific problem is a higher instability of higher-brominated congeners during the analysis. This is caused by steric hindrance of the relatively large bromine atoms. PBBs and PBDEs can be determined by GC–ECD and GC–MS. MS is preferred because use can be made of the prominent bromine clusters in the MS spectra. PBBs normally elute together with PCTs, both on silica gel columns and, because of similar boiling points, in the GC analysis, so a selective detection, such as MS, is preferable. Most PBDEs are normally eluted in the silica gel fraction (Pijnenburg *et al.*, 1995).

Toxaphene is an even more complex mixture than PCTs. However, owing to weathering effects and biotransformation, fewer congeners are found in food products than for PCTs. Nevertheless the complexity is greater than that of PCBs. Congener-specific analyses have been performed, using both single-column GC and MDGC (de Boer et al., 1997b). Peak overlap is considered to be a serious risk in congener-specific analysis of toxaphene using single-column GC. Both ECD (Alder et al., 1995, 1997) and MS (de Boer et al., 1993) are used as detection techniques.

PCDEs are often found as an 'interference' during the PCDD and PCDF analysis. They can be separated from PCDDs, PCDFs and PCBs using different combinations of silica gel columns, alumina columns and HPLC techniques using PYE columns (Koistinen et al., 1995a,b; Kurz and Ballschmiter, 1995).

PCNs are compounds with a planar structure, comparable to PCDDs and PCDFs. They are often found in the same fraction as the PCDDs, PCDFs and planar PCBs after pre-fractionation over HPLC graphitised carbon or PYE columns (Jacobsson, 1994; Asplund, 1994; Haglund et al., 1990b). They can be determined in parallel with the PCDDs, PCDFs and planar PCBs, by scanning different masses using a selected-ion monitoring technique.

The occurrence of PHHs as complex mixtures has a considerable effect on the costs of a complete analysis of these contaminants in food. For most compounds highly sophisticated techniques are required and the clean-up methods are labour intensive. Multiresidue methods, in which several of these contaminants groups can be determined in parallel, are available, but it is often not possible without some compromise, resulting in lower recoveries of some compounds, and unresolved peaks in the chromatograms. Some examples of multiresidue methods can be found in Paasivirta et al. (1989) and Bergqvist et al. (1992).

9.8.8   Quality of analysis

During the last ten years specific attention has been given to the quality of the analysis. There have been several worldwide interlaboratory studies conducted for PCBs, PCDDs and PCDFs (de Boer et al., 1996). Between-laboratory standard deviations of laboratories with a better performance for these compounds are in the range of 10–25% (WHO, 1991b; de Jong et al., 1993; de Boer et al., 1996; de Boer and Wells, 1997). A few certified reference materials (CRMs) have become available during the last years, but many more are required (Law and de Boer, 1995).

For other PHHs the situation less clear. A worldwide interlaboratory study on toxaphene in fish oil showed between-laboratory standard deviations of ~50% for total toxaphene and of 50–>100% for toxaphene congeners (Andrews *et al.*, 1995). Interlaboratory studies for the other PHHs have not been reported. There is clearly a need to improve the quality of this type of analysis and the mutual comparability of laboratories working in this field.

9.9 Future perspectives

The 'chain reaction', described at the beginning of this chapter has resulted in the production of a vast amount of data on levels of PHHs in foodstuffs from a number of countries. Although there are inevitably differences due to sampling, analytical and transcription errors, it is clear that, away from local hot-spots, the concentrations of contaminants in food lie within a relatively small range. In most cases these concentrations are below the levels set by the regulatory agencies.

Long-term trends of PCB levels in human adipose tissue (Patterson *et al.*, 1994) and in the environment (Loganathan and Kannan, 1994), e.g. guillemot eggs (Bignert *et al.*, 1995), are clear examples of a decline in concentrations since the ban on use and manufacture, and control of disposal in the 1970s. The decline from these previously high levels has left the concentrations at relatively constant levels. In view of the diffuse, but continuous, sources of these PHHs it is likely that a further decline will be very slow, ~1–2% per annum. The effective control of emissions of PCDD/Fs from waste incineration plant also means that the future levels of these PHHs are likely to remain at low levels at sites remote from the immediate source. These environmental changes are reflected in the levels of these PHHs in foodstuffs. In addition to providing public reassurance, the analyses of these foodstuffs should continue to be used, in conjunction with toxicological studies, to provide further information on the long-term additive effects of these PHHs. Much of the effort in monitoring food for the presence of PHHs and the associated risk assessment to human health has been to provide a simplified guide by way of the TEQ values. The TEQ values have currently been used to provide the basis for setting legislative levels for the sum of these compounds in addition to relating them to toxicological effects.

In the future it may not be sufficient to include only the PCDD/Fs, PBDEs and the non-*ortho*- and mono-*ortho*-CBs in this additive model. Compounds such as PCNs, PBDD/F are present in food and the environment, probably in similar quantities to dioxins and non-*ortho*-CBs but are not measured regularly and do not have established TEF values. Clearly, there are distinct advantages in expressing the levels of these PHHs collectively in terms of their potential biological effect, particularly for the regulatory authorities. However, there are some disadvantages of such an approach.

Many of these PHHs exhibit mixed-function oxidase (MFO) action with both phenylbarbitol (PB) and methylacetylcholine (MC) type activity. Expressing the concentrations as TEQs alone overlooks the possibility of relating the levels of these and other congeners to alternative modes of biological activity. These PHHs have also been identified as endocrine disrupters (Colborn *et al.*, 1993) and have been shown to possess neurotoxic characteristics. Therefore they require a very different model of data reduction from that used to provide the TEQ information. Since the 1970s a considerable effort has been made to separate and measure individual PHHs and to provide unequivocal data on these compounds. It is, therefore, essential that all data are reported for individual congeners in addition to the reduced TEQ information, so that the values can be used for information other than for current regulatory purposes.

Most of the raw data produced by competent laboratories with a quality management system will provide information on the precision of these measurements. The calculation of the TEQ values involves the normalisation of each congener with its own TEF value and the final additive result is often not reported along with the estimated uncertainty of either the analytical variability or the uncertainty of the TEF values. In most cases the overall uncertainty of the TEQ is large (de Boer and Brinkman, 1994).

There are probably sufficient safety margins between the No Effect Level or Minimum Effect Level and the TDI, but it should be borne in mind that the calculated daily intake values are estimates and these estimates should be accompanied by a level of uncertainty. There is also an implicit assumption that the TEQ is a direct estimate of the potential health risk from intake of food. There is also no agreement that Ah-mediated effects are associated with any adverse health effects in humans (Stone, 1995). Many compounds other than PHHs induce CYPIA and these must be considered in the overall model. The overall additive values of TEQs from PHH measurements do not, on average, account for the overall equivalent biological toxicity as measured by EROD activity. Polycyclic aromatic hydrocarbons (PAHs) are present in foodstuffs and will bind to the Ah receptor and induce EROD activity. The Ah activities of PAHs are based on benzo[*a*]pyrene and are separate from the measurements for PHH. A comparison of the contribution of PCBs, PCDD/Fs and PAHs in sediment and biota to the EROD activity has shown that a correlation between the chemical and biological measurements is improved by the summation of all TEQs in *both* PHH and PAH to give a fuller account of the EROD activity (Hess, 1998; Hess *et al.*, 1998).

Silkworth and Brown (1996) have highlighted that there are a number of naturally occurring compounds such as plant flavanones and cooking spices that have a similar planar structure to PAHs and can bind to the Ah receptor. It is clear that the contribution to the TEQ from the PHHs currently measured is only part of the overall total input to the human exposure. In the future developments in the estimation of the risk assessment or impact of PHHs in humans there should be an evaluation of the additional sources of those compounds that bind to the Ah receptor whether they are from anthropogenic or natural sources. It is also essential to maintain the awareness of neurotoxicological effects of PHHs and the biological impact of other halogenated contaminants such as chlorinated paraffins that are used in large quantities, for which there is little information on residue levels in foodstuffs. The 'chain reaction' for many of these compounds has yet to become critical.

Acknowledgements

The authors acknowledge the assistance of Elizabeth Richardson and Philipp Hess in obtaining many of the sources used in this paper.

References

Abraham, K., Hille, A., Ende, M. and Helge, H. (1994) Intake and faecal excretion of PCDDs, PCDFs and PCBs in a breast fed and a formula fed infant. *Chemosphere*, **29** 2279-2286.

Ahlbourg, U.G., Håkansson, H., Wærn, F. and Hanberg, A. (1988) *Nordisk dioxinriskbedömning*, Nordic Council of Ministers, Copenhagen, Denmark, Miljørapport 1988:7.

Ahlbourg, U.G., Becking, G.C., Birnbaum, L.S., Brouwer, A., Derks, H.J.G.M., Freley, M., Golor, G., Hanberg, A., Larsen, J.C., Liem, A.K.D., Safe, S.H., Schlatter, C., Wærn, F., Younes, M. and Yrjänhekki, E. (1994) Toxic equivalency factors for dioxin-like PCBs. Report on a WHO-ECEH and IPCS consultation. *Chemosphere*, **28** 1049-1067.

Alder, L., Bache, K., Beck, H. and Parlar, H. (1995) Collaborative study on toxaphene indicator compounds (chlorobornanes) in fish oil. *Organohalogen Compounds*, **26** 369-374.

Alder, L., Beck, H., Khandker, S., Karl, H. and Lehman, I. (1997) Levels of toxaphene indicator compounds (chlorobornanes) in fish. *Chemosphere*, **34** 1389-1400.

Andrews, P., Headrick, K., Pilon, J.C., Lau, B. and Weber, D. (1995) An interlaboratory round robin study on the analysis of toxaphene in a cod liver oil standard reference material. *Chemosphere*, **31** 4393-4402.

Anonymous (1989) *Toxicological profile of 2,3,7,8-TCDD*. Agency Tox. Substances Disease Registry (88-23), US Public Health Service, US EPA, Washington, DC.

Anonymous (1992) German directive for allowable residues in food. Bundesgesetzbl I:1605.

Anonymous (1996a) *Dioxinen—polygechloreerde dibenzo*-p-dioxinen, dibenzofuranen en dioxine-achtige polychloorbifenylen, Dutch Health Council (Gezondheidsraad): Commissie Risico-evaluatie stoffen, Report 1996/10, Rijswijk, The Netherlands.

Anonymous (1996b) *Toxicological profile for Toxaphene (update)*, US Department of Health and Human Services, Public Health Service, Agency for Toxic Substances and Disease Registry, Atlanta, GA.

Anonymous (1997) *Nordic Risk assessment of toxaphene exposure*, Nordic Council of Ministers, TemaNord 1997:540, Copenhagen, Denmark.

Arias, A. (1992) *Brominated Diphenyl Oxides as Flame Retardants. Bromine Based Chemicals*, Consultant report to the OECD, Paris, France.

Ashby, E. (1978) *Reconciling Man with the Environment*, Oxford University Press, London.

Asplund, L. (1994) *Development and application of methods for determination of polychlorinated organic pollutants in biota*, PhD thesis, Stockholm University, Sweden.

Aylward, L.L., Hays, S.M., Karch, N.J. and Paustenbach, D.J. (1996) Relative susceptibility of animals and humans to the cancer hazard posed by 2,3,7,8-tetrachlorodibenzo-p-dioxin using internal measures of dose. *Environmental Science and Technology*, **30** 3534-3543.

Ballschmiter, K., Buchert, H., Class, Th., Krämer, W., Magg, H., Munder, A., Reuter, U., Schäfer, M., Swerev, R., Wittinger, R. and Zoler, W. (1985) Isomerenspezifische Bestimmung der Polychlordibenzodioxine (PCDD) und polychlordibenzofurane (PCFD) *Fresenius Journal of Analytical Chemistry*, **320** 711-717.

Ballschmiter, K., Bacher, R., Mennel, A., Fischer, R., Riehle, U. and Swerev, M. (1992) The determination of chlorinated biphenyls, chlorinated dibenzodioxins and chlorinated dibenzofurans. *Journal of High Resolution Chromatography*, **15** 260-270.

Beck, H. and Mathar, W. (1985) Bundesgesundheitsblatt, **28** 1-12.

Beck, H., Echart, K., Mathar, W. and Wittkowski, R. (1989) PCDD and PCDF body burden from food intake in the FGR. *Chemosphere*, **18** 417-424.

Beck, H., Echart, K., Mathar, W. and Wittkowski, R. (1990) Influences of different regional emissions and cardboard containers on the level of PCDD, PCDFs and related compounds in cow milk. *Chemosphere*, **21** 789-798.

Beck, H., Dross, A. and Mathar, W. (1994) PCDD and PCDF exposure and levels in humans in Germany. *Environmental Health Perspectives*, **102** 173-185.

Becker, M., Phillips, T. and Safe, S. (1991) Polychlorinated diphenyl ethers—a review. *Toxicology and Environmental Chemistry*, **33** 189-200.

Bennet, B.G. (1983) Exposure of man to environmental PCBs, and exposure commitment assessment. *Science of the Total Environment*, **29** 101.

Bericht der Bund (1992) *Länder Arbeitsgruppe Dioxine*, Federal Ministry for the Environment, Nature Conservation and Nuclear Safety, Bonn.

Bergqvist, P.-A., Bandh, C., Broman, D., Ishaq, R., Lundgren, R., Näf, C., Pettersen, H., Rappe, C., Rolff, C., Strandberg, B., Zebühr, Y. and Zook, D.R. (1992) Multi-residue analytical method including planar PCB, dioxins and other organic contaminants for marine samples. *Organohalogen Compounds*, **9** 17-20.

Bertazzi, P.A. (1991) Long term effects of chemical disasters. Lessons and results from Seves. *Science of the Total Environment*, **106** 5-20.

Bignert, A., Litzén, K., Odsjö, T., Olsson, M., Persson, M. and Ruetergårdh, L. (1995) Time - related factors influence the concentration of ΣDDT, PCBs and shell parameters in eggs of Baltic Guillimot (*Uria aalge*) 1961–1989. *Environmental Pollution*, **89** 27-36.

Birmingham, B., Gilman, A., Grant, D., Salminen, D., Boddington, M., Thorpe, B., Wile, I., Toft, P. and Armstrong, V. (1989) PCDD/PCDF multimedia exposure analysis for the Canadian population: Detailed exposure estimations. *Chemosphere*, **19** 637-643.

Biseth, A., Oehme, M. and Færden, K. (1990) Levels of PCDDs and PCDFs in selected Norwegian foods. *Organohalogen Compounds*, **1** 463-466.

Boersma, E.R. (1996) PCB's en dioxinen: effect op het jonge kind. *Tijdschrift Voeding*, **57** (6) 6-9.

Boon, J.P., Wester, P.G. and de Boer, J. (1998) The use of a microsomal *in vitro* assay to study phase I biotransformation of chlorobornanes (toxaphene®) in marine mammals and birds. Possible consequences of biotransformation for bioaccumulation and genotoxicity. *Comparative Biochemistry and Physiology*, in press.

Bøwadt, S., Johansson, B., Fruekilde, P., Hansen, M., Zilli, B., Larsen, B. and de Boer, J. (1994) Supercritical fluid extraction of polychlorinated biphenyls from lyophilized fish tissue. *Journal of Chromatography A*, **675** 189-204.

Brinkman, U.A.Th. and de Kok, A. (1980) Production, properties and usage, in *Halogenated Biphenyls, Terphenyls, Naphthalenes, Dibenzodioxins and Related Products* (ed. R.D. Kimbrough) Topics in Environmental Health **4**, Elsevier, Amsterdam, pp 1-40.

Büchert, A. (1988) *Dioxiner i danske levnedsmidler*, Publ. Levnedsmiddelstyrelsen No. 170, ISBN 87-503-7493-1.

Buckland, S.J., Hannah, D.J., Taucher, J.A. and Weston, R.J. (1990) The migration of PCDDs and PCDFs into milk and cream from bleached paperboard packaging. *Organochlorine Compounds*, **3** 223-226.

Carson, R. (1962) *Silent Spring*, Houghton Mifflin, Boston.

Chao, W.-Y., Hsu, C.-C. and Guo, Y.L. (1997) Middle ear disease in children pre-natally exposed to polychlorinated biphenyls and dibenzofurans. *Archives of Environmental Health*, **52** 257-262.

Colborn, T., vom Saal, F.S. and Soto, A.M. (1993) Developmental effects of endocrine-disrupting chemicals in wildlife and humans. *Environmental Health Perspectives*, **101** 378-384.

Cook, H., Chian, E.S., van der Weele, B.H. and de Jong, R.J. (1978) Histotoxic effects of PBBs in Michigan dairy cattle. *Environmental Research*, **15** 82-89.

Cooper, K.R., Fiedler, H., Bergek, S., Andersson, R. and Rappe, C. (1995) PCDDS and PCDFs in food samples collected in Southern Mississippi (USA). *Organohalogen Compounds*, **26** 51-57.

Cordel, F., Corneliussen, P., Jeinek, C., Hackley, B., Lehmann, R., McLaughlin, J., Rhoden, R. and Shapiro, R. (1978) Human exposure to PCBs and PBBs. *Environmental Health Perspectives*, **23** 275-281.

Crane, J.L. (1996) Carcinogenic risks associated with consuming contaminated fish from five Great Lakes areas of concern. *Journal of Great Lakes Research*, **22** 653-668.

Dahl, P., Lindström, G., Wiberg, K. and Rappe, C. (1995) Absorption of PCBs, PCDDs and PCDfs by breast fed children. *Chemosphere*, **30** 2297-2306.

Dandenau, R.D. and Zerenner, E.H. (1989) The invention of the fused silica capillary column: an industrial perspective. *Proceedings of the 10th International Symposium on Capillary Chromatography, Riva del Garda, Italy*, Huethig Verlag, Heidelberg, pp 5-11.

Dar, E., Kanarek, M.S., Anderson, H.A. and Sonzogni, W.C. (1992) Fish consumption and reproductive outcomes in Green Bay, Wisconsin. *Environmental Research*, **59** 189-201.

Davies, K. (1988) Concentrations and dietary intake of selected organochlorines, including PCBs, PCDDs and PCDFs in fresh food composites grown in Ontario, Cananda. *Chemosphere*, **17** 263-276.

Davies, K. (1990) Human exposure pathways to selected organochlorines and PCBs in Toronto and Southern Ontario, in *Food Contamination from Environmental Sources*. (ed. J.O. Nriago and M.S. Simmons), Wiley, New York, pp 525-540.

de Boer, J. (1988) Chlorobiphenyls in bound and non-bound lipids of fishes; comparison of different extraction methods. *Chemosphere*, **17** 1803-1810.

de Boer, J. (1990) PBDEs in Dutch freshwater and marine fish. *Proceedings of the 10th International Symposium Dioxin 90, Bayreuth, Germany*, **2** 315-318.

de Boer, J. (1994) Spatial differences and temporal trends of bioaccumulating halogenated hydrocarbons in the livers of cod (*Gadus morhua*) form the North Sea 1977–1992. Scientific Symposium on the 1993 North Sea Quality Status Report, 110–114, 18–21 April, Denmark, ISBN 87-7810-699-0.

de Boer, J. (1997) Toxaphene—recent developments in analysis and biomonitoring. *Organohalogen compounds*, **33** 7-11.

de Boer, J. and Brinkman, U.A.Th. (1994) TCDD equivalents of mono ortho substituted chlorobiphenyls. Influence of analytical error and uncertainty of toxic equivalent factors. *Analytica Chimica Acta*, **289** 261-262.

de Boer, J. and Dao, Q.T. (1989) The analysis of individual chlorobiphenyl congeners in fish extracts on 0.15 mm i.d. capillary columns. *Journal of High Resolution Chromatography*, **12** 755-759.

de Boer, J. and Dao, Q.T. (1991) Analysis of seven chlorobiphenyl congeners by multi-dimensional gas chromatography. *Journal of High Resolution Chromatography*, **14** 593-596.

de Boer, J. and Wester, P.G. (1993) Determination of toxaphene in human milk from Nicaragua and in fish and marine mammals from the North Eastern Atlantic and the North Sea. *Chemosphere*, **27** 1879-1890.

de Boer, J. and Wells, D.E. (1997) Chlorobiphenyls and organochlorine pesticides in fish and sediments—three years of QUASIMEME laboratory performance studies. *Marine Pollution Bulletin*, **35** 52-63.

de Boer, J., Dao, Q.T. and van Dortmond, R. (1992) Retention times of fifty one chlorobiphenyl congeners on seven narrow bore capillary columns coated with different stationary phases. *Journal of High Resolution Chromatography*, **15** 249-255.

de Boer, J., Stronck, C.J.N., Traag, W.A. and van der Meer, J. (1993) Non-ortho and mono-ortho substituted chlorobiphenyls and chlorinated dibenzo-*p*-dioxins and dibenzofurans in marine and freshwater fish and shellfish from The Netherlands. *Chemosphere*, **26** 1823-1842.

de Boer, J., van der Meer, J. and Brinkman, U.A.Th. (1996) Determination of chlorobiphenyls in seal blubber, marine sediment and fish: Interlaboratory study. *Journal of the Association of Official Analytical Chemists*, **79** 83-96.

de Boer, J., Pieters, H. and Dao, Q.T. (1997a) *Verontreinigingen in aal: monitorprogramma ten behoeve van de Nederlandse sportvissrij* 1996, RIVO-DLO report C048/97, IJmuiden, The Netherlands.

de Boer, J., de Geus, H.-J. and Brinkman, U.A.Th. (1997b) Multidimensional gas chromatographic analysis of toxaphene. *Environmental Science and Technology*, **31** 873-879.

de Geus, H.-J., de Boer, J. and Brinkman, U.A.Th. (1996) Multidimensionality in gas chromatography. *Trends in Analytical Chemistry*, **15** 398-408.

de Geus, H.-J., de Boer, J. and Brinkman, U.A.Th. (1997) Development of a thermal desorption modulator for gas chromatography. *Journal of Chromatography*, **767** 137-151.

de Jong, A.P.J.M., Liem, A.K.D., den Boer, A.C., van der Heeft, E., Marsman, J.A., van de Werken, G. and Wegman, R.C.C. (1989) Analysis of polychlorinated dibenzofurans and dibenzo-*p*-dioxins in human milk by tandem mass spectrometry. *Chemosphere*, **19** 59-66.

de Jong, A.P.J.M. and Liem, A.K.D. (1992) Gas chromatography-mass spectrometry in ultra trace analysis of polychlorinated dioxins and related compounds. *Trends in Analytical Chemistry*, **12** 115-124.

de Jong, A.P.J.M., Dross, A., Fürst, P., Lindström, G., Päpke, Ö. and Startin, J.R. (1993) Interlaboratory comparison study on dioxins in cow's milk. *Fresenius Journal of Analytical Chemistry*, **345** 72-77.

de Voogt, P., Klamer, J.C. and Govers, H. (1986) Simultaneous clean-up and fractionation of organochlorine compounds by adsorption chromatography. *Journal of Chromatography*, **363** 407-411.

Deichmann, W.B. (1973) The chronic toxicity of organochlorine pesticides in man. *Proceedings of the 8th Pesticide Symposium. Collective Papers of the Inter-American Conference on Toxicology and Occupational Medicine*, p 347.

Dewailly, E., Ayotte, P., Bruneau, S., Laliberté, C., Muir, D.C.G. and Norstrom, R. (1993) Inuit exposure to organochlorines through the aquatic food chain. *Environmental Health Perspectives*, **101** 618-620.

Di Domenico, A. (1990) Guidelines for the definition on Environmental action alert thresholds for PCDD/Fs. *Regulatory Toxicology and Pharmacology*, **11** 8-23.

Duarte-Davidson, R. and Jones, K.C. (1994) PCBs in the UK population: estimated intake, exposure and body burden. *Science of the Total Environment*, **151** 131-152.

Duinker, J.C., Schulz, D.E. and Petrick, G. (1988) Multidimensional gas chromatography with electron capture detection for the determination of toxic congeners in polychlorinated biphenyl mixtures. *Analytical Chemistry*, **60** 478-482.

Dulfer, W.J. (1996) *Partitioning and transport of PCBs over gastrointestinal epithelium*, PhD Thesis, University of Amsterdam, The Netherlands.

EBFRIP (1990) Information on PBDEs. Economic, technical and regulatory assessment of the European Brominated Flame Retardant Industry Panel. Response to questions submitted to the EU Brussels Rijswick (NL).

Erickson, M.D. (1997) *Analytical Chemistry of PCBs*, 2nd edn, CRC Press, New York.

Fæden, K. (1991) *Dioksiner i naeringsmidler*. SNT report 4 Norwegian Food Control Authority, Postboks 8187 Dep., 0034 Oslo, Norway.

Falandysz, J., Srrandberg, L., Bergqvist, P.A., Strandberg, B. and Rappe, C. (1997) Spatial distribution of PCNs in mussel and fish for the Gulf of Gdansk, Baltic Sea. *Science of the Total Environment*, **203** 93-104.

Farland, W.H., Schaum, J., Birnbaum, L., Winters, D. and Goldma, L. (1994) Status of dioxin related activity at the US EPA. *Organohalogen Compounds*, **20** 559-562.

Faulkner, D.J. (1990) Naturally occurring brominated compounds, in *Proceedings of the Workshop on Brominated Aromatic Fire Retardants, Skokloster, Sweden* (ed. L. Frij), Solna, National Chemicals Inspectorate, pp 121-144.

FDA (1989) Compliance Policy Guide 7141.01 Pesticide Residues in food or feed- Enforcement Criteria, Attachment B Action levels for unavoidable pesticide residues in food and feed commodities. FDA Washington, DC.

FDA (1990a) Action levels for residues of certain pesticides in foods and feed (4/17/90) Fed. Reg 55(74):14359-14263.

FDA (1990b) Food and Drug Administration Pesticide Programme Residues in Food 1990. *Journal of the Association of Official Analytical Chemists*, **74** 121A-141A.

FDA (1992) Action levels for poisonous and deleterious substances in human food and animal feed (8/92). Department of Health and Human Services, FDA, Washington, DC.

Federal Ministry of Health (1987) *Pravilnik, Official Journal of Yugoslavia* No. 79, 1842.

Fiedler, H., Hutzinger, O. and Timms, C. (1990) Dioxins: sources of environmental load and human exposure. *Toxicology and Environmental Chemistry*, **29** 157-243.

Fingerhut, M.A., Halperin, W.E., Marlow, D.A., Piacitelli, L.A., Honchar, P.A., Sweeney, M.H., Griefe, A.L., Dill, P.A., Steenland, K. and Surada, A.J. (1991) Cancer mortality in workers exposed to 2378-TCDD. *New England Journal of Medicine*, **324** 212-218.

Fingerling, G.M., Coelhan, M., Angerhöfer, D. and Parlar, H. (1997) Structure–stability relationship of chlorinated bornanes in the environment. *Organohalogen Compounds*, **33** 17-22.

Flodström, S. and Ahlbourg, U.G. (1989) Tumour promoting effects of 2,3,7,8-TCDD; effects of promoting exposure duration, administration schedule and type of diet. *Chemosphere*, **19** 779-783.

Flesch-Janys, D., Becher, H., Gurn, P., Jung, D., Konietzko, J., Manz, A. and Päpke, O. (1996) Elimination of PCDDs and PCDFs in occupationally exposed persons. *Journal of Toxicology and Environmental Health*, **47** 363-378.

Fries, G.F. and Marrow, G.S. (1973) Polychlorinated terphenyls as potential contaminants of animal products. *Journal of the Association of Official Analytical Chemists*, **56** 1002-1007.

Frommberger, R. (1991) PCDDs and PCDFs in fish from south-west Germany: River Rhine and Neckar. *Chemosphere*, **22** 29-38.

Fürst, P. and Wilmers, (1995) PCDD/F levels in dairy products 1994 versus 1990. *Organohalogen Compounds*, **26** 101-104.

Fürst, P., Fürst, C. and Groebel, W. (1990) Levels of PCDDs and PCDFs from the Federal Republic of Germany. *Chemosphere*, **20** 787-792.

Fürst, P., Beck, H. and Theelen, R.M.C. (1992) Assessment of human intake of PCDDs and PCDFs from different environmental sources. *Toxic Substances Journal*, **12** 1039-1048.

GEMS (1988) *Assessment of Chemical Contamination in Food*, Report on the results of the UNEP/FAO programme on health related environmental monitoring, Global Environmental Monitoring Systems, Monitoring and Assessment Research Centre, London.

Georgii, S., Brunn, H., Stojanowic, V. and Muskat, E. (1989) Fremdstoffe in Lebebsmitteln-Ermitteln einer täglichen Aufnahme mit der Nahrug II. Chlororganische pestizide in ausgewählten lebebmitteln, tagesrationenstationär verpflegter patienten sowie in säuglings-und kleinkinderrnhrung. *Deutsche Lebensmittel-Rundschau*, **85** 385-389.

Gillman, A. and Newhook, R. (1991) An update assessment of the exposure of Canadians to dioxins and furans. *Chemosphere*, **23** 1661-1667.

Glassmeyer, S.T., de Vault, D.S., Myers, T.R. and Hites, R.A. (1997) Toxaphene in Great Lakes Fish: a temporal, spatial and trophic study. *Environmental Science and Technology*, **31** 84-88.

Government of Canada (1990) Priority Substances + List Assessment Report No 1 PCDDs and PCDFs Ottawa, Ministry of Supply and Services Cat No EN 40-215/E.

Gunderson, E.L. (1988) FDA total diet study, April 1982–April 1984, dietary intakes of pesticides, selected elements, and other chemicals. *Journal of the Association of Official Analytical Chemists*, **71** 1200-1209.

Haglund, P., Asplund, L., Järnberg, U. and Jansson, B. (1990a) Isolation of toxic polychlorinated biphenyls by electron donor-acceptor high performance liquid chromatography using a 2-(1-pyrenyl) ethyldimethylsilylated silica column. *Journal of Chromatography*, **507** 389-398.

Haglund, P., Asplund, L., Järnberg, U. and Jansson, B. (1990b) Isolation of mono- and non-ortho polychlorinated biphenyls from biological samples by electron donor-acceptor high performance liquid chromatography using a 2-(1-pyrenyl)ethyldimethylsilylated silica column. *Chemosphere*, **20** 887-894.

Hallikainen, A. and Vartiainen, T. (1997) Food Control surveys of PCDDs and PCDFs and intake estimates. *Food Additives and Contaminants*, **14** 355-366.

Harrad, S.J., Sewart, A.S., Boumphrey, R., Duarte-Davidson, R. and Jones, K.C. (1992) A method for the determination of PCB congeners 77, 126 and 169 in biotic and abiotic matrices. *Chemosphere*, **24** 1147-1154.

Hawthorne, S.B. (1990) Analytical scale supercritical fluid extraction. *Analytical Chemistry*, **62** 633A-642A.

Haywood, D.G., Petreas, M.X., Goldman, L.R. and Stephens, R.D. (1991) Assessing the risks from 2,3,7,8,TCDDD/F in milk packaged in paper. *Chemosphere*, **23** 1551-1559.

Health Council of the Netherlands, (1996) Committee on Risk Evaluation of Substances/Dioxins Publication No 1996/10, The Hague.

Herxheimer, K. (1899) Uber Chloracne. *Münchener Medizinische Wochenschrift*, **46** 278.

Hess, P. (1998) *The determination and environmental significance of planar aromatic compounds in the marine environment*, PhD thesis, Robert Gordon University, Aberdeen, UK.

Hess, P., de Boer, J., Cofino, W.P., Leonards, P.E.G. and Wells, D.E. (1995) Critical review of the analysis of non- and mono-ortho chlorobiphenyls. *Journal of Chromatography, A,* **703** 417-465.

Hess, P., Robinson, C., Stagg, R. and Wells, D.E. (1998) *Planar Aromatic Compounds in the Marine Environment: Toxic Impact of PAHs PDDD/Fs and Non-ortho CBs in Marine Sediments from the Firth of Clyde, Scotland,* Fisheries Research Report.

Holder, J.W. and Menzel, H.M. (1989) Analysis of 2,3,7,8-TCDD tumour promotion activity and its relationship to cancer. *Chemosphere,* **19** 861-868.

Hong, R., Taylor, K. and Abonour, R. (1989) Immune abnormalities associated with chronic TCDD exposure to Rhesus monkeys. *Chemosphere,* **18** 313-320.

Hong, C.S., Bush, B., Xiao, J. and Qiao, H. (1993) Toxic potential of non-ortho and mono-ortho coplanar polychlorinated biphenyls in Aroclors, seals and humans. *Archives of Environmental Toxicology,* **25** 118-123.

Hooper, N.K., Ames, B.N., Saleh, M.A. and Casida, J.E. (1979) Toxaphene, a complex mixture of polychloroterpenes and a major insecticide, is mutagenic. *Science,* **205** 591-593.

Horn, E.G.L., Hetling, J.L. and Tofflemire, T.J. (1979) The problem of PCBs in the Hudson River System. *Annals of the New York Academy of Science,* **320** 591-609.

Howie, L., Dickerson, R., Davis, D. and Safe, S. (1990) Immunosuppressive and mono-oxygenase induction activities of PCDEs in C57BL/6N mice:quantitative structure activity relationship. *Toxicology and Applied Pharmacology,* **105** 254-263.

Huckins, J.N., Stalling, D.L. and Petty, J.D. (1980) Carbon-foam chromatographic separation of non-*o,o'*-chlorine substituted PCBs from Aroclor mixtures. *Journal of the Association of Official Analytical Chemists,* **63** 750-755.

Huckins, J.N., Schwartz, T.R., Petty, J.D. and Smith, L.M. (1988) Determination, fate and potential significance of PCBs in fish and sediment with the emphasis on selected Ah inducing congeners. *Chemosphere,* **17** 1995-2016.

Huckins, J.N., Tubergen, M.W., Lebo, J.A.M., Gale, R.W. and Schwartz, T.R. (1990) Determination, fate and potential significance of PCBs in fish and sediment with the emphasis on selected AHH inducing compounds. *Journal of the Association of Official Analytical Chemists,* **73** 290-293.

Huisman, M., Eerenstein, S.E.J., Koopman-Esseboom, C., Brouwer, M., Fidler, V., Muskiet, F.A.J., Sauer, P.J.J. and Boersma, E.R. (1995) Perinatal exposure to polychlorinated biphenyls and dioxins through dietary intake. *Chemosphere,* **31** 4273-4287.

Ivanov, V. and Sandell, E. (1992) Characterisation of PCB isomers in sovol and trichlorodiphenyl formulations by HRCG with ECD and HRGCMS techniques. *Environmental Science and Technology,* **26** (10) 2012-2017.

Jaffe, R., Stemmler, E.A., Eitzer, B.D. and Hites, R.A. (1985) Anthropogenic, polyhalogenated organic compounds in sedentary fish from Lake Huron and Lake Superior tributaries and embayments. *Journal of Great Lakes Research,* **11** 156-162.

Jakobsson, E. (1994) *Synthesis and analysis of chlorinated naphthalenes. Biological and Environmental implications,* PhD thesis, Stockholm University, Sweden.

Jamieson, J.W.S. (1977) *PCTs in the Environment EPS-3-EC-77-22,* Environmental Protection Service, Ottawa, Canada.

Jan, J. and Admič, A. (1991) PCB residues in food from a contaminated region of Yugoslavia. *Food Additives and Contaminants,* **8** 505-512.

Jansson, B., Andersson, R., Asplund, L., Bergman, Å., Litzén, K., Nylund, K., Reutergårdh, L., Sellström, U., Uvemo, U.-B., Casja, W. and Wideqvist, U. (1991) Multiresidue method for the gas chromatographic analysis of some polychlorinated and polybrominated pollutants in biological samples. *Fresenius Journal of Analytical Chemistry,* **340** 439-445.

Jansson, B., Andersson, R., Asplund, L., Litzén, K., Nylund, K., Sellström, U., Uvemo, U-B., Wahlberg, C., Wideqvist, U., Odsjö, T. and Olsson, M. (1993) Chlorinated and brominated persistent organic compounds in biological samples for the environment. *Environmental Toxicology and Chemistry*, **12** 1163-1174.

Järnberg, U., Lillemore, A., de Wit, C., Gradströrm, A., Haglund, P., Jansson, B., Lexén, K., Strandell, M., Olsson, M. and Jonsson, B. (1993) PCBs and PCNs in Swedish sediment and biota: Levels, Patterns, and Time Trends. *Environmental Science and Technology*, **27** 1364-1374.

Jennings, W. (1978) *Gas Chromatography with Glass Capillary Columns*, Academic Press, New York.

Jensen, S. (1966) Report of a new chemical hazard. *New Scientist*, **15** 612.

Jensen, S. (1972) The PCB story. *Ambio*, **1** 123-131.

Jensen, S., Reutergårdh, L. and Jansson, B. (1983) Analytical methods for measuring organochlorine and methyl mercury by gas chromatography. *FAO Technical Paper*, **212** 21.

Jiménez, B., Hernàndez, L.M. and Gonzàlez, M.J. (1996) Estimated intake of coplanar PCBs in individuals for Madrid (Spain) associated with an average diet. Studies on Spanish representative diets. *Organohalogen Compounds*, **30** 131-136.

Jödicke, B., Ende, M., Helge, H. and Neubert, D. (1992) Faecal excretion of PCDDs/PCDFs of a 3 month old breast fed infant. *Chemosphere*, **25** 1061-1065.

Johnson, R.D., Manske, D.D., New, D.H. and Podrebarac, D.S. (1984) Pesticides, metals and other chemical residues in adult total diet samples, 1976–77. *Journal of the Association of Official Analytical Chemists*, **48** 668-675.

Jones, K.C. and Alcock, R.E. (1996) *Dioxin inputs to the environment: a review of temporal trend data and proposals for the monitoring programme to detect past and future changes in the UK*, Report of a project commissioned by the Department of the Environment Ref EPG/1/5/53, Lancaster University, Lancaster, UK.

Kannan, K. (1994) *Food pollution by organochlorine and organotin compounds in tropical Asia and Oceania*, PhD thesis, Ehime University, Maysuyama, Japan.

Kannan, K., Tanabe, S., Ramesh, A., Subramanian, A. and Tatsukawa, R. (1992) Persistent organochlorine residues in foodstuffs from India and their implications on human dietary exposure. *Journal of Agriculture and Food Chemistry*, **40** 518-524.

Kannan, K., Tanabe, S., Giesy, J.P. and Tatsukawa, R. (1997) Organochlorine pesticides and PCBs in foodstuffs from Asian and Oceanic countries. *Reviews in Environmental Contamination and Toxicology*, **152** 1-55.

Karl, H. and Lehmann, I. (1993) Organochlorine residues in the edible part of eels from different origins. *Zeitschrift für Lebensmittel-Untersuchung und -Forschung*, **197** 385-388.

Kashimoto, T. and Miyata, H. (1986) Differences in Yusho and other kinds of poisoning involving only PCBs, in *PCBs and the Environment*, **3** (ed. J.S. Waid), CRC Press, Boca Raton, FL, pp 1-26.

Kay, K. (1977) PBBs Environmental contamination in Michigan 1973–1976. *Environment Research*, **13** 74-93.

Kimbrough, R.D. (1980) Occupational exposure, in *Halogenated Biphenyls, Terphenyls, Naphthalenes, Dibenzodioxins and Related Products* (ed. R.D. Kimbrough), Topics in Environmental Health **4**, Elsevier, Amsterdam, pp 374-397.

Kimbrough, R.D. (1995) PCBs and Human Health: An update. *Critical Reviews in Toxicology*, **25** 133-163.

Knowles, D.E., Richter, B.E., Ezzel, J., Höfler, F., Waddell, D.S., Gobran, T. and Khurana, V. (1995) Extraction of chlorinated compounds by accelerated solvent extraction (ASE). *Organohalogen Compounds*, **23** 13-18.

Knox, J.H. and Kaur , B.J. (1986) Structure and performance of porous graphitic carbon in liquid chromatography. *Journal of Chromatography*, **352** 3-25.

Kociba, R.J., Keeler, P.A., Park, C.N. and Gehring, P.J. (1976) 2,3,7,8-TCDD, results of a 13-week oral toxicity in rats. *Toxicology and Applied Pharmacology*, **35** 553-574.

Koistinen, J., Vuorinen, P.J. and Paasivirta, J. (1993) Content and origin of PCDEs in salmon from the Baltic sea, Lake Saimaa and the Tenojoki river in Finland. *Chemosphere*, **27** 149-156.

Koistinen, J., Koivusaari, J., Nuuja, I. and Paasivirta, J. (1995a) PCDEs, PCBs, PCDDs and PCDFs in black guillemots and white-tailed sea eagles from the Baltic sea. *Chemosphere*, **30** 1671-1684.

Koistinen, J., Mussalo-Rauhamaa, H. and Paasiverta, J. (1995b) PCDEs, PCDDs and PCDFs in Finnish human tissue compared with environment samples. *Chemosphere*, **31** 4259-4271.

Koopman-Esseboom, C. (1995) *Effects of perinatal exposure to PCBs and dioxins on early human development*, PhD thesis, Erasmus University, Rotterdam, The Netherlands.

Körner, W., Dawidowsky, N. and Hagenmaier, H. (1993) Faecal excretion rates of PCDDs and PCDFs in two breast-fed infants. *Chemosphere*, **27** 157-162.

Krüger, C. (1988) *Polybrominated biphenyls and polybrominated biphenyl ether- detection and quantification in selected foods*, Thesis, University of Münster, Germany.

Kumpulainen, J. (1988) Low levels of contaminants in Finnish foods and diet. *Annales Agriculture Fenniae*, **27** 219-229.

Kuratsune, M. (1980) Yusho, in *Halogenated Biphenyls, Terphenyls, Naphthalenes, Dibenzodioxins and Related Products* (ed. R.D. Kimbrough), Topics in Environmental Health **4**, Elsevier, Amsterdam, pp 374-397.

Kurz, J. and Ballschmiter, K. (1995) Isomer-specific determination of 79 polychlorinated diphenyl ethers (PCDE) in cod liver oils, chlorophenols and in a fly ash. *Fresenius Journal Analytical Chemistry*, **351** 98-109.

Larsen, B. and Bøwadt, S. (1993) HRGC separation of PCB congeners. The state-of-the-art, *Proceedings of the 15th International Symposium on Capillary Chromatography, Riva del Garda, Italy*, Huethig Verlag, Heidelberg, pp 503-510.

Larsen, B., Bøwadt, S., Tilio, R. and Facchetti, S. (1992) Congener specific analysis of 140 chlorobiphenyls in technical mixtures on five narrow-bore GC columns. *International Journal of Environmental Analytical Chemistry*, **47** 47-68.

Landrigan, P.J. (1980) General population exposure to environmental concentrations of halogenated biphenyls, in *Halogenated Biphenyls, Terphenyls, Naphthalenes, Dibenzodioxins and Related Products* (ed. R.D. Kimbrough), Topics in Environmental Health **4**, Elsevier, Amsterdam, pp 267-286.

Law, R.J. and de Boer, J. (1995) Quality assurance of analysis of organic compounds in marine matrices: application to analysis of chlorobiphenyls and polycyclic aromatic hydrocarbon, in *Quality Assurance for Environmental Monitoring* (ed. Ph. Quevauviller), VCH Verlag, Weinheim.

Liem, A.K.D. and Theelen, R.M.C. (1997) *Dioxins: chemical analysis, exposure and risk assessment*, PhD thesis, University of Utrecht, The Netherlands.

Liem, A.K.D., Hoogerbrugge, R., Koostra, P.R., de Jong, A.P.J.M., Marsman, J.A., Den Boer, A.C., Den Hartog, R.S., Groenemeijer, G.S. and Van 't Klooster, H.A. (1990) Level and patterns of dioxins in cow's milk in the vicinity of municipal waste incinerators and metal reclamation plants in the Netherlands. *Organohalogen Compounds*, **1** 567-570.

Liem, A.K.D., Hoogerbrugge, R., Koostra, P.R., Van der Velde, E.G. and de Jong, A.P.J.M. (1991) Occurrence of dioxins in cow's milk in the vicinity of municipal waste incinerators and metal reclamation plants in the Netherlands. *Chemosphere*, **23** 1675-1684.

Liem, A.K.D., Baumann, R.A., de Jong, A.P.J.M., van der Velde, E.G. and van Zoonen, P. (1992) Analysis of organic micro pollutants in the lipid fraction of foodstuffs. *Journal of Chromatography*, **624** 317-339.

Loganathan, B. and Kannan, K. (1994) Global organochlorine contamination trends: an overview. *Ambio*, **23** 187-191.

LST (1990) *Food Monitoring in Denmark. Nutrients and Contaminants 1983–1987*, Levedsmiddelstyrelsen, The National Food Agency of Denmark, Copenhagen, No. 195.

Lucier, G.W. (1991) Humans are a sensitive species to some of the biochemical effects of structural analogs of dioxin. *Environmental Toxicology and Chemistry*, **10** 727-735.

Luke, M.A., Masumoto, H.T., Cairns, T. and Hundley, H.K. (1988) Levels and incidences of pesticide residues in various foods and animal feeds analysed by Luke multi-residue methodology for fiscal years 1982–1986. *Journal of the Association of Official Analytical Chemists*, **71** 415-444.

MAFF (1992) *UK Dioxins Food*, Food Surveillance Paper No. 31, HMSO London.

MAFF (1995) *UK Dioxins in Food, UK Dietary Intakes*, Food Surveillance Information sheet No. 71, HMSO, London.

MAFF (1997a) *UK Dioxins and Polychlorinated Biphenyls in Food and Human Milk*, Food Surveillance Information sheet No. 105, HMSO, London.

MAFF (1997b) *Dioxins and PCBs in Cow's Milk from Farms Close to an Industrial Site in Rotherham*, Surveillance Information sheet No. 133, HMSO, London.

Manz, A., Berger, J., Dwyer, J.H., Flesch-Janys, D., Nagel, S. and Waltsgott, H. (1991) Cancer mortality among workers in chemical plant contaminated with dioxin. *Lancet*, **338** 95-96.

Masuda, Y. (1990) *Toxic Evaluation of PCBs and PCDFs by Yusho, Approach to Risk Assessment of PCDDs and PCDFs in Japan*, Discussion paper for WHO meeting, 29 November.

Matsumoto, H., Murakami, Y., Kuwabara, K., Obaba, H., Inada, C., Nishimune, T. and Tanaka, R. (1987) Average daily intake of pesticides and PCBs in total diet studies in Osaka, Japan. *Bulletin of Environmental Contamination and Toxicology*, **38** 954-958.

McFarland, V.A. and Clarke, J.U. (1989) Environmental occurrence, abundance and potential toxicity of PCB congeners. Considerations for congener specific analysis. *Environmental Health Perspectives*, **81** 225.

McLachlan, M. (1993) Digestive tract absorption of PCDDs, PCDFs and PCBs in a nursing infant. *Toxicology and Applied Pharmacology*, **123** 68-72.

Mes, J. and Webster, D. (1989) Non-orthochlorine substituted coplanar PCBs in Canadian adipose tissue, breast milk and fatty foods. *Chemosphere*, **19** 1357-1365.

Mes, J., Newsome, W.H. and Conacher, H.B.S. (1989) Determination of some specific isomers of polychlorinated biphenyl congeners in fatty foods of the Canadian diet. *Food Additives and Contaminants*, **6** 365-375.

Michalek, J.E., Tripathi, R.C., Caudill, S.P. and Pirkle, J.L. (1992) Investigation of TCDD half-life heterogeneity in veterans from operation Ranch Hand. *Journal of Toxicology and Environmental Health*, **35** 29-38.

Michalek, J.E., Pirkle, J.L., Caudill, S.P., Tripathi, R.C., Patterson, D.G. and Needham, L.L. (1996) Pharmacokinetics of TCDD in veterans from operation Ranch Hand, 10 year follow up. *Journal of Toxicology and Environmental Health*, **47** 209-220.

Miyata, H. (1991) Pollution with dioxins and related compounds in food and the human body. *Journal of Environmental Chemistry*, **1** 275-290.

Moccia, R.D., Fox, G. and Britton, A.J. (1981) Quantitative interlake comparison of thyroid pathology in Great Lakes coho (*Oncorrhynchus kisutsch*) and chinook (*Oncorrhynchus tschauritscha*) salmon. *Cancer Research*, **41** 2200-2210.

Moser, G.A., Schlummer, M. and McLachlan, M.S. (1996) Human absorption of PCDD, PCDFs and PCBs from food. *Organochlorine Compounds*, **29** 386-388.

Muir, D.C.G. and de Boer, J. (1995) Recent developments in the analysis and environmental chemistry of toxaphene with emphasis on the marine environment. *Trends in Analytical Chemistry*, **14** 56-66.

Muller, R., Lach, G. and Parlar, H. (1988) Comparative studies on the presence of toxaphene residues. *Chemosphere*, **17** 2289-2298.

Mullin, M.D., Pochini, C., McGrindle, S., Romkes, M., Safe, S. and Safe, L. (1984) High-resolution PCB analysis: synthesis and chromatographic properties of all 209 PCB congeners. *Environmental Science and Technology*, **18** 468-476.

NEPB (National Environmental Protection Board) (1988) The Swedish Dioxin Survey Status report, Special Analytical Laboratory, Solna, Sweden.

NATO/CCMS (1988) *International Toxicity Equivalency Factors (I-TEF) Method of Risk Assessment for Complex Mixtures of Dioxins and Related Compounds*, North Atlantic Treaty Organization, Brussels, Report No. 176.

Neufeld, M.L., Sittenfield, M. and Wolk, K.F. (1977) Market Input/Output studies, Task IV PCBs EPA 560/6-77-017 EPA Washington D.C.

NORD (1988) *Miljörapport 1988/7 Nordisk dioxinrisk bedömning*, Copenhagen, Nordic Council.

Oehme, M. and Kallenborn, R. (1995) A simple numerical code for polychlorinated compound classes allowing an unequivocal derivation of the steric structure I: PCBs and bornanes. *Chemosphere*, **30** 1739-1750.

Onuska, F.I. and Terry, K.A. (1995) Extraction of pesticides from sediment using a microwave technique. *Chromatographia*, **36** 191-194.

Paasivirta, J., Lahtiperä, M. and Leskijärvi, T. (1982) in *Chlorinated Dioxins and Related Compounds: Impact on the Environment* (eds O. Hutzinger, R.W. Frei, E. Merian and F. Pocchiari), Pergamon Press, New York, pp 191-200.

Paasivirta, J., Mäntykoski, K., Koistinen, J., Kuokkanen, T., Manilla, E. and Rissanen, R. (1989) Structure analysis of planar polychloroaromatic compounds in the Environment. *Chemosphere*, **19** 149-154.

Patterson, D.G., Todd, G.D., Turner, W.E., Maggio, V., Alexander, L.R. and Needham, L.L. (1994) Levels of non ortho, mono and di ortho substituted polychlorinated biphenyls, dibenzo-*p*-dioxins and dibenzofurans in human serum and adipose tissue. *Environmental Health Perspectives*, **102** (Suppl. 1) 195-204.

Phillips, J.B. and Xu, J. (1995) Comprehensive multidimensional GC. *Journal of Chromatography A*, **703** 327-334.

Pickston, N.L., Brewerton, H.V., Drysdale, J.M., Hughes, J.T., Smith, J.M., Love, J.L., Sutcliffe, E.R. and Davidson, F. (1985) The New Zealand diet: a survey of elements, pesticides, colours and preservatives. *New Zealand Journal of Technology*, **1** 81-89.

Pijnenburg, A.M.C.M., Everts, J.W., de Boer, J. and Boon, J.P. (1995) Polybrominated biphenyl and diphenyl ether flame retardants: analysis, toxicity, and Environmental occurrence. *Review of Environmental Contamination and Toxicology*, **141** 1-26.

Pohjanvirta, R. (1991) *Studies on the mechanism of acute toxicity of 2,3,7,8-TCDD in rats*, PhD thesis, National Public Health Institute, Department of Environmental Hygiene and Toxicology, Helsinki, Finland.

Poland, A. (1991) Receptor-mediated toxicity, reflections on a quantitative model for risk assessment, in *Banbury Report 35, Biological basis for risk assessment of dioxins and related compounds* (eds M. Gallo, R.J. Scheuplein and K.A. van der Heijden) Cold Spring Harbor Laboratory Press, pp 417-424.

Pollock, G.A. and Kilgore, W.K. (1978) Toxaphene. *Residue Reviews*, **69** 87-140.

Reggiani, G. (1980) Localised contamination with TCDD- Seveso, Missouri and other areas, in *Halogenated Biphenyls, Terphenyls, Naphthalenes, Dibenzodioxins and Related Products* (ed. R.D. Kimbrough), Topics in Environmental Health, Elsevier, Amsterdam, pp 303-371.

Remmers, J.C., Breen, J.J., Schwemberger, J., Stanley, J.S., Cramer, P.H. and Thornberg, K.R. (1990) Mass spectral confirmation of chlorinated and brominated diphenyl ethers in human adipose tissues, in *Dioxin; 90. Tenth International Symposium on Chlorinated Dioxins and Related Compounds, Bayreuth, Germany*, **2** (eds O. Hutzinger and H. Fredler), pp 347-350.

Reuber, M.D. (1979) Carcinogenicity of toxaphene: a review. *Journal of Toxicology and Environmental Health*, **5** 729-748.

Ris, A. (1993) Impact of PCDD/PCDF emissions of a copper reclamation plant: five years of experience with Environmental monitoring. *Organohalogen compounds*, **14** 23-26.

Ris, A. and Hagenmaier, H. (1991) Kuhmilch als Indikator für Dioxinimmissionen in der Nähe eines Emittenten. *VDI-Berichte*, **901** 863-872.

Robertson, L.W. and Chynoweth, D.P. (1975) Another halogenated hydrocarbon. *Environment*, **17** 25-27.

Ryan, J.J., Lau, P.Y., Pilon, J.C. and Lewis, D. (1983) 2,3,7,8 TCDD/F residues in Great Lakes commercial and sport fish, in *Chlorinated Dioxins and Dibenzofurans in the Total Environment II* (eds. G. Choudary, L. Keith and C. Rappe), Butterworth, Boston, pp 87-97.

Safe, S.D.P. (1984) Polychlorinated biphenyls (PCBs) and polybrominated biphenyls (PBBs): biochemistry, toxicology, and mechanism of action. *CRC Critical Reviews in Toxicology*, **13** 319-395.

Safe, S.D.P. (1992) Development, validation and limitations of toxic equivalency factors. *Chemosphere*, **25** 61-64.

Safe, S.D.P. (1997) *Dietary estrogen/antiestrogens dose, potency and health benefits*, Workshop on endocrine active compounds in food and environment—setting the research agenda, M and T Graduate School Environmental Chemistry and Toxicology, 11 April, Ede, The Netherlands.

Saleh, M.A. (1991) Toxaphene: chemistry, biochemistry, toxicity and environmental fate. *Review of Environmental Contamination and Toxicology*, **118** 1-85.

Saleh, M.A., Skinner, R.F. and Casida, J.E. (1979) Comparative metabolism of 2,2,5-*endo*,6-*exo*,8,9,10-hepatchlorobornane and toxaphene in six mammalian species and chickens. *Journal of Agriculture and Food Chemistry*, **27** 731-737.

Schaum, J., Cleverly, D., Lorber, M., Philips, L. and Schweer, G. (1994) Update analysis of U.S. sources of dioxin-like compounds and background exposure levels. *Organohalogen Compounds*, **20** 237-242.

Schecter, A., Fürst, P., Fürst, C., Groebel, W., Constable, J.D., Kolesnikov, S., Beim, A. Boldonov, A., Trubitsun, E., Vlasov, B., Hoang, Din Cau, Le Cau Dai and Hoang Tri Quynh. (1990) Levels of chlorinated dioxins, dibenzofurans and other chlorinated xenobiotics in food from the Soviet Union and the south of Vietnam. *Chemosphere*, **20** 799-806.

Schecter, A., Fürst, P., Fürst, C., Grachev, M., Beim, A. and Koptug, V. (1992) Levels of chlorinated dioxins, dibenzofurans and other chlorinated xenobiotics in food from Russia. *Chemosphere*, **24** 2009-2015.

Schecter, A., Startin, J., Wright, C., Kelly, M., Päpke, O., Ball, M. and Olson, J. (1994a) Dioxins in US food and estimated daily intake. *Chemosphere*, **29** 2261-2265.

Schecter, A., Startin, J., Wright, C., Kelly, M., Päpke, O., Lis, A., Ball, M. and Olson, J. (1994b) Congener specific levels of dioxins and dibenzofurans in US food and estimated daily dioxin toxic equivalent intake. *Environmental Health Perspectives*, **102** 962-966.

Schecter, A., Crammer, P., Bogges, K., Stanley, J. and Olson, J. (1995) Levels of dioxins, dibenzofurans, DDE and PCB congeners in pooled food samples collected at supermarkets across the United States. *Organohalogen Compounds*, **26** 125-128.

Schery, P., Mackrodt, P., Wittsiepe, J. and Selenka, F. (1995) Dietary intake of PCDD/F measured by the duplicate method. *Organohalogen Compounds*, **26** 147-150.

Schmidt, L.J. and Hesselberg, R.J. (1992) *Archives of Environmental Contamination and Toxicology*, **23** 37-44.

Schmid, P. and Schlatter, C. (1992) PCDDs and PCDFs in cow's milk from Switzerland. *Chemosphere*, **24** 1013-1030.

Schomburg, G., Husman, H. and Hübinger, E. (1985) Multidimensional separation of isomeric species of chlorinated hydroacarbons such as PCB, PCDD and PCDF. *Journal of High Resolution Chromatography*, **8** 395-400.

Schultz, D. (1993) PCDD/PCDF German policy and measures to protect man and the Environment. *Chemosphere*, **27** 501-507.

Schwartz, T.R., Tillitt, D.E., Feltz, K.P. and Peterman, P.H. (1993) Determination of mono and non-*o*'-chlorine substituted polychlorinated biphenyls in Aroclors and environmental samples. *Chemosphere*, **26** 1443-1460.

Sellström, U. (1996) *Polybrominated diphenyl ethers in the Swedish Environment*, ITM-report 1996 45 (Licentiate thesis), Stockholm University, Stockholm, Sweden.

Sellström, U., Jansson, B., Kierkegaard, A. and de Wit, C. (1993) PBDEs in biological samples from the Swedish Environment. *Chemosphere*, **26** 1703-1718.

Shiria, J.H. and Kissel, J.C. (1996) The Uncertainty in estimated half-lives of PCBs in humans: impact on exposure assessment. *Science of the Total Environment*, **187** 190-210.

Silkworth, J.B. and Brown, Jr., J.F. (1996) Evaluating the impact of exposure to Environmental contaminants on human health. *Clinical Chemistry*, **42** 8(B) 1345-1349.

SNT (1995) *Norway: Fremmedstoffer og smittestoffer-holder maten mål?*

Stanley, J.S., Crammer, P.H., Thornburg, K.R., Remmers, J.C., Breen, J.J. and Schwemberger, S. (1991) Mass spectral confirmation of chlorinated and brominated diphenyl ethers in human adipose tissue. *Chemosphere*, **23** 1185-1195.

Stern, G.A., Muir, D.C.G., Ford, C.A., Grift, N.P., Dewailly, E., Bidlemann, T.F. and Walla, M.D. (1992) Isolation and identification of two major recalcitrant toxaphene congeners in aquatic biota. *Environmental Science and Technology*, **26** 1838-1842.

Stone, R. (1995) Panel slams EPA's dioxin analysis. *Science*, **268** 1124.

Storr-Hansen, E. (1992) Comparative analysis of thirty polychlorinated biphenyl congeners on two capillary columns of different polarity with non-linear multi-level calibration. *Journal of Chromatography*, **558** 375-391.

Storr-Hansen, E. and Cederberg, T. (1992) Determination of coplanar polychlorinated-biphenyls (CB) congeners in seal tissues by chromatography on active carbon, dual column high resolution GC/ECD and high resolution GC/MS. *Chemosphere*, **24** 1181-1196.

Takaymara, K., Mityata, H., Mimura, M. and Kashimoto, T. (1991) PCDDs, PCDFs and coplanar PCBs in coastal and marketing fishes in Japan. *Japanese Journal of Toxicology. Environmental Health (Eisei Kagaku)*, **37** 125-131.

Tanabe, S., Kannan, N., Wakimoto, T. and Tatsukawa, R. (1987) Method for the determination of three toxic non-orthochlorinc substituted coplanar PCBs in Environmental samples at part-per-trillion levels. *International Journal of Environmental Analytical Chemistry*, **29** 199-213.

Tanabe, S., Kannan, K., Tabucanan, M.S., Siriwong, C., Ambe, Y., Tatsukawa, R. (1991) Organochlorine pesticide and PCB residues in foodstuffs from Bangkok, Thailand. *Environmental Pollution*, **72** 191-203.

Theelen, R.M.C. (1991) Modeling of human exposure to TCDD and I-EEQ in the Netherlands: background and occupational, in *Banbury Report 35, Biological basis for risk assessment of dioxins and related compounds* (eds M.A. Gallo, R.J. Scheuplein and K.A. van der Heijden), Cold Spring Harbour Laboratory Press, pp 277-290.

Traag, W.A., Van Rhijn, J.A., van der Spreng, P.F., Roos, A.H. and Tuinstra, L.G.M.Th. (1993) *Onderzoek naar polygechloreerde dibenzo-p-dioxinenb, dibenzofuranen en planaire gechloreende bifenylen in Nederlandse melk*, Report No 93.14, RIKILT-DLO.

van Cleuvenbergen, R., Schoeters, J., Bormans, R., Wevers, M., De Fré, R. and Ryman, T. (1993) Isomer-specific determination of PCDDs and PCDFs in Flemish cows milk. *Organohalogen Compounds*, **13** 27-30.

van den Berg, M., de Jongh, J., Poiger, H. and Olson, J.R. (1994) The toxicokinetics and metabolism of PCDDs and PCDFs and their relevance for toxicity. *Critical Reviews in Toxicology*, **24** 1-74.

Vanden Heuvel, J.P. and Lucier, G. (1993) Environmental Toxicology of PCDDs and PCDFs. *Environmental Health Perspectives*, **100** 189-200.

van der Velde, E.G., Hijman, W.C., Linders, S.H.M.A. and Liem, A.K.D. (1996) SEF as clean-up technique for ppt-levels of PCBs in fatty samples. *Organohalogen Compounds*, **27** 247-252.

Van Zorge, J.A., Van Wijnen, J.H., Theelen, R.M.C., Olie, K. and Van den Berg, M. (1989) Assessment of the toxicity of mixtures of halogenated dibenzo-*p*-dioxins and dibenzofurans by use of toxicity equivalence factors (TEF). *Chemosphere*, **19** 1881-1895.

Vaz, R. (1995) Average Swedish dietary intakes of organochlorine contaminants via foods of animal origin and their relation to levels in human milk. 1975–90. *Food Additives and Contaminants*, **12** 543-558.

Vaz, R. and Blomkvist, G. (1985) Traces of toxaphene components in Swedish breast milk analysed by capillary GC using ECD, electron impact and negative ion chemical ionization MS. *Chemosphere*, **14** 223-231.

Veningerova, M., Uhnák, M. and Prachar, V. (1993) Chlorinated pesticides and benzenes in food. *Fresenius Environmental Bulletin*, **2** 735-744.

Voldner, E.C. and Li, Y.F. (1993) Global usage of toxaphene. *Chemosphere*, **27** 2073-2078.

Voldner, E.C. and Li, Y.F. (1995) Global usage of selected persistent organochlorines. *Science of the Total Environment*, **160/161** 201-210.

Wania, F. and Mackay, D. (1993) Global fractionation and cold condensation of low volatility organochlorine compounds in polar regions. *Chemosphere*, **27** 2079-2094.

Wauer (1918) Gewerbliche Erkrankungen durch gechlorte Kohlenwasserstoffe. *Zbl. Gewerbehyg*, **6** 100.

Wells, D.E., Cowan, A.E. and Christie, A.E.G. (1985) Separation of organochlorine residues by adsorption chromatography prior to capillary gas chromatography. *Journal of Chromatography*, **328** 332-337.

Wells, D.E. (1988) Extraction, clean-up and group separation techniques in organochlorine trace analysis. *Pure and Applied Chemistry*, **60** 1437-1448.

Wells, D.E. and Echarri, I. (1992) Determination of individual chlorobiphenyls (CBs) including non-ortho, and mono-ortho chloro substituted CBs in marine mammals from Scottish waters. *International Journal of Environmental Analytical Chemistry*, **47** 75-97.

Wells, D.E. (1993) Current developments in the analysis of polychlorinated biphenyls (PCBs) including planar and other toxic metabolites in *Environmental Analysis: Techniques, Applications and Quality Assurance* (ed. D. Barcelo), Elsevier Science, Amsterdam, pp 113-148.

Wester, P.G., de Boer, J. and Brinkman, U.A.Th. (1996) Determination of polychlorinated terphenyls in aquatic biota and sediment with gas chromatography/mass spectrometry using negative chemical ionisation. *Environmental Science and Technology*, **30** 473-480.

WHO/IPCS (1987) *Principles for the Safety Assessment of Food Additives and Contaminants in Food*, Environment Health Criteria, **70**, WHO, Geneva.

WHO (1988) *European Regional Programme on Chemical Safety Results of Analytical Field Studies on Levels of PCBs, PCDDs and PCDFs in Human Milk*, Report on a WHO consultation, Copenhagen, 24–25 February.

WHO (1991a) *Consultation on Tolerable Daily Intake from Food PCDDs and PCDFs: Summary Report*, Bilthoven, The Netherlands, 4–7 December, EUR/ICP/PCS 030(S).

WHO (1991b) *Levels of PCBs, PCDDs and PCDFs in Human Milk and Blood: Second Round of Quality Control Studies*, Environment and Health in Europe, **37**, WHO, Copenhagen.

WHO (1992) *PCBs and PCTs Health and Safety Guide*, IPCS/UNEP/ILO Health and Safety Guide No. 68, WHO, Geneva.

WHO (1994a) *Brominated Diphenyl Ethers*, Environmental Health Criteria No. 162, WHO, Geneva.

WHO (1994b) *Polybrominated Biphenyls,* Environmental Health Criteria No. 152, WHO, Geneva.

WHO (1994c) *Assessing Human Health Risks of Chemicals: Derivation of Guidance Values for Health Based Exposure Limits*, Environmental Health Criteria, **170**, WHO, Geneva.

Yoo, W.J. and Taylor, L.T. (1997) Supercritical fluid extraction of polychlorinated biphenyls and organochlorine pesticides from freeze-dried tissue of marine mussel, *Mytilus edulis. Journal of the Association of Official Analytical Chemists*, **80** 1336-1345.

10 Polycyclic aromatic hydrocarbons, petroleum and other hydrocarbon contaminants

Colin F. Moffat and Kevin J. Whittle

10.1 Introduction

This chapter is concerned with the sources and occurrence of hydrocarbons of industrial and environmental origin in foods and beverages, and with the implications for human health. This very diverse group of compounds ranges from the simple gaseous normal (n)-alkane, methane, to the poly-condensed aromatic ring systems, the polycyclic aromatic hydrocarbons (PAHs), such as benzo[a]pyrene. The 'hydrocarbon family' is described in detail in section 10.2. Hydrocarbons are ubiquitous in nature and are found in air, water, soil and all living things and therefore in food. They exhibit a wide array of physical, chemical, biochemical and carcinogenic properties and include volatile compounds; compounds with significant aqueous solubility (although all are lipophilic); compounds that are virtually insoluble in aqueous media, are relatively persistent in the environment and can bioaccumulate in the food chain; compounds that may act as synergists or co-carcinogens or are well-established mutagens and carcinogens, and those that are not mutagenic; and compounds that are relatively reactive chemically and biochemically, such as the PAHs. The PAHs have been described as the largest class of known environmental carcinogens (Grimmer, 1983) and are probably the most widespread environmental contaminants.

Hydrocarbons occur naturally in air, water, soil and the biomass as a result of biosynthesis by plants, animals and microorganisms, and as a result of biodegradative processes. In the biomass, hydrocarbons are found at highest concentrations in lipid-rich tissues, and hence in waxes, oils and fats. In general, they tend to be fairly stable in the food chain, but this is conditioned by the type and structure of the hydrocarbon, the species, its metabolic activity, and the particular tissue. They also occur naturally as a result of combustion of vegetation, and as a result of diagenesis ultimately to produce carboniferous, bituminous, petroleum oil and gas deposits, from which hydrocarbons can seep back into the onshore and offshore environment through natural seeps and through exposure and erosion of deposit-bearing strata.

Human leisure, industrial and other commercial activities also introduce to the environment, via emissions to the atmosphere, discharges to rivers and water bodies, and disposal on land, a similar range of hydrocarbons from a diverse range of sources: combustion of vegetation and other organic waste matter, e.g. from municipal sources; combustion of petroleum and petroleum products and other fossil fuels used in domestic heating and industrial energy production as well as transport; wood burning; tobacco smoke; industrial chemical and polymer production; petroleum production, refining, bulk transport and spillage; and waste disposal. All of these may directly or indirectly contaminate at some stage of water supply or food production. Hydrocarbons are also added directly to food by smoke curing, broiling, roasting or grilling over open fires or charcoal, by the use of permitted food additives such as lubricants, solvents, propellants, glazing agents and protective coatings, and from packaging. In the strict sense, these additions are not environmental contaminants and, arguably, can be avoided, or are subject to legislative control. However, they need to be considered to place environmental and industrial contamination in perspective.

Thus, for hydrocarbons, there are multiple sources, both point sources and non-specific sources, and multiple exposure routes for man, via air, food and water, to a complex array of compounds with a wide range of biological properties and activity. Such sheer complexity of composition, coupled with multiple sources of input, not only to the food to be analysed but also to the analytical laboratory environment, provides difficult analytical problems, particularly at trace levels, for unequivocal identification, quantification and interpretation to reliably determine the contaminant origin of the analyte. Interestingly, hydrocarbons from different sources tend to show unique and robust fingerprints or patterns, so that, increasingly, it is possible to unravel the various source contributions to complex hydrocarbon analyses of aliphatic and aromatic fractions from food, water and other samples. The key question is, what are the implications for human health from exposure to these compounds in the diet?

The groups of most interest *vis-à-vis* food contamination are the volatile monocyclic aromatic hydrocarbons (MAHs): benzene (a well-known human carcinogen), toluene, ethylbenzene, and the xylenes; other petroleum-derived hydrocarbons and products that contaminate the food chain and are also a source of tainting, especially of aquatic food sources (Chapter 11), but particularly the PAHs, some of which are carcinogens, co-carcinogens, teratogens and mutagens. Table 10.1 lists the probable human carcinogens (IRIS, 1992) and their estimated relative potencies (EPA, 1993), where available. The 16 PAHs identified as the priority environmental pollutants in this group (EPA, 1995) are listed below with the abbreviations used elsewhere in the text. Studies of PAHs in food have used a variety of PAH groupings (section 10.3.3.2) although there

Table 10.1 Probable human carcinogens (IRIS, 1992) and estimated relative potencies (EPA, 1993)

Polycyclic aromatic hydrocarbon	Relative order-of-magnitude potency
Benzo[a]pyrene	1.0
Benz[a]anthracene	0.1
Benzo[b]fluoranthene	0.1
Benzo[j]fluoranthene	
Benzo[k]fluoranthene	0.01
Chrysene	0.001
Cyclopenta[cd]pyrene	
Dibenz[a,h]anthracene	1.0
Dibenzo[a,e]fluoranthene	
Dibenzo[a,e]pyrene	
Dibenzo[a,h]pyrene	
Dibenzo[a,i]pyrene	
Dibenzo[a,l]pyrene	
Indeno[1,2,3-cd]pyrene	0.1

has been a tendency to focus on those marked * and benzo[e]pyrene (Gilbert, 1994).

Naphthalene	NAP	Benz[a]anthracene *	BAA
Acenaphthene	ANP	Chrysene *	CRY
Acenaphthylene	ANY	Benzo[b]fluoranthene *	BBF
Fluorene	FLO	Benzo[k]fluoranthene *	BKF
Anthracene	ANT	Benzo[a]pyrene *	BAP
Phenanthrene	PHE	Dibenz[a,h]anthracene *	DBA
Fluoranthene *	FLU	Benzo[g,h,i]perylene *	BPE
Pyrene *	PYR	Indeno[1,2,3-cd]pyrene *	IDP

The reader is directed to a number of reviews of different aspects of hydrocarbons for further reading and source material (Whittle *et al.*, 1977; Lee *et al.*, 1981; Grimmer, 1983; Bjørseth and Ramdahl, 1985; Nevenzel, 1989; Seiber, 1990; Bartle, 1991; Gilbert, 1994; Tomaniová *et al.*, 1997)

10.2 The hydrocarbon family: chemistry and properties

Hydrocarbons are the least polar of natural compounds and are composed only of carbon (C) and hydrogen (H). Chemists have divided this diverse array of organic compounds into two broad classes: aliphatic hydrocarbons and aromatic hydrocarbons. The original meanings of the words 'aliphatic' (fatty) and 'aromatic' (fragrant) no longer have any significance. In this chapter, the term 'aliphatic' relates to open-chain saturated and unsaturated compounds and alicyclic hydrocarbons (aliphatic cyclic hydrocarbons) that contain at least one ring (Figure 10.1). Aromatic hydrocarbons are those that contain at least one benzene nucleus (Figure 10.2). Both saturated and unsaturated aliphatic side-chains can be bonded to a benzene nucleus and such hydrocarbons are collectively known as arenes.

As emphasised above, the hydrocarbon family is extremely diverse and complex, and associated with this is a nomenclature that has developed such that there are a number of synonyms and subgroups for each component group (Table 10.2). Several of the component groups

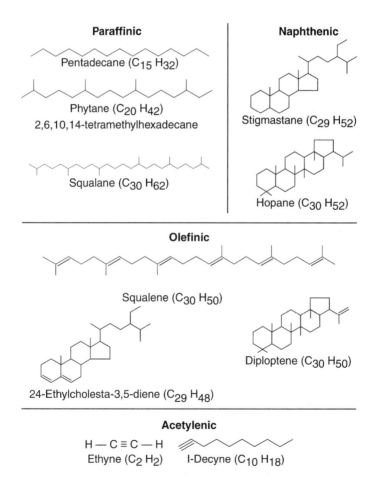

Figure 10.1 Structures for selected aliphatic hydrocarbons.

(A)   MONOCYCLIC AROMATIC HYDROCARBONS (MAHs)

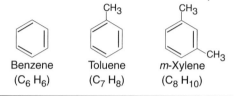

Benzene
($C_6 H_6$)

Toluene
($C_7 H_8$)

m-Xylene
($C_8 H_{10}$)

(B)   TWO –, THREE – AND FOUR – RING PAHs

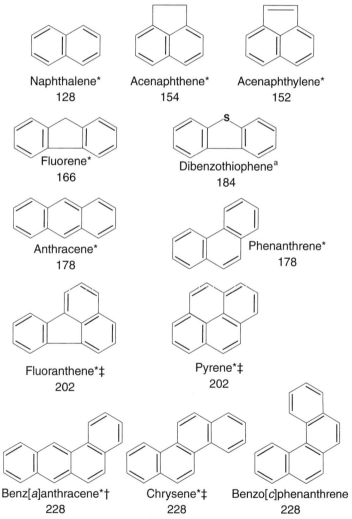

Naphthalene*
128

Acenaphthene*
154

Acenaphthylene*
152

Fluorene*
166

Dibenzothiophene[a]
184

Anthracene*
178

Phenanthrene*
178

Fluoranthene*‡
202

Pyrene*‡
202

Benz[a]anthracene*†
228

Chrysene*‡
228

Benzo[c]phenanthrene
228

(C)   **FIVE – AND SIX – RING PAHs**

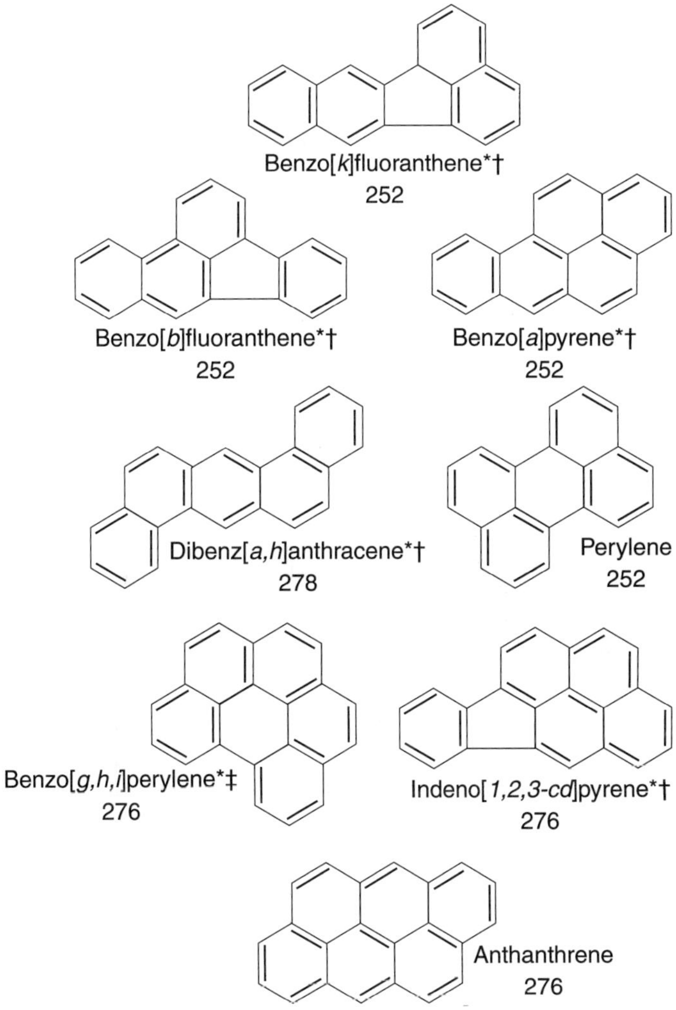

Benzo[*k*]fluoranthene*†
252

Benzo[*b*]fluoranthene*†
252

Benzo[*a*]pyrene*†
252

Dibenz[*a,h*]anthracene*†
278

Perylene
252

Benzo[*g,h,i*]perylene*‡
276

Indeno[*1,2,3-cd*]pyrene*†
276

Anthanthrene
276

Figure 10.2 Structures of (A) monocyclic aromatic hydrocarbons (MAHs), (B) two-, three- and four-ring polycyclic aromatic hydrocarbons (PAHs) and (C) five- and six-ring PAHs. The number below a PAH is the ion used for the determination of the PAHs by selected-ion monitoring mass spectroscopy. ªOften determined at the same time as PAHs.

*US EPA PAH; † classified as a possible human carcinogen by the International Agency for Research on Cancer; ‡ suspected as acting as a co-carcinogen. For additional structures , the ring numbering system and ring orientations see Lee *et al.* (1979).

Table 10.2 Summary of the hydrocarbon family, structures for which are presented in Figures 10.1 and 10.2

Hydrocarbon	Synonyms	General formula[a]/ abbreviation	Description	Examples
Aliphatic (open-chain compounds and those cyclic compounds that resemble the open-chain compounds)				
Paraffinic	Alkanes Paraffins	C_nH_{2n+2}	Saturated open-chain molecules that may or may not contain side-chains	Pentadecane ($C_{15}H_{32}$) Phytane ($C_{20}H_{42}$) Squalane ($C_{30}H_{62}$)
Naphthenic	Naphthenes Cycloalkanes Cycloparaffins		Saturated but contain at least one ring	Stigmastane ($C_{29}H_{52}$) Hopane ($C_{30}H_{52}$)
Olefinic	Olefins Alkenes	C_nH_{2n}[a]	Contain at least one non-aromatic double bond	Squalene ($C_{30}H_{50}$) 24-Ethylcholesta-3,5-diene ($C_{29}H_{48}$) Diploptene ($C_{30}H_{50}$)
Acetylenic	Acetylenes Alkynes	C_nH_{2n-2}[a]	Contain at least one triple bond	Ethyne (C_2H_2) 1-Decyne ($C_{10}H_{18}$)

Aromatic (contain at least one benzene nucleus)

Monocyclic	Alkylbenzenes Alkenylbenzenes Alkynylbenzenes Arenes[b]	MAH	Contain one benzene nucleus	Benzene (C_6H_6) Toluene (C_7H_8) Xylene (C_8H_{10})
Polycyclic	Polynuclear Polyaromatic	PAH	Contain a minimum of two benzene nuclei	Naphthalene ($C_{10}H_8$, two fused rings) Phenanthrene ($C_{14}H_{10}$, three fused rings) Pyrene ($C_{16}H_{10}$, four fused rings) Benz[a]anthracene ($C_{18}H_{12}$, four fused rings) Benzo[a]pyrene ($C_{20}H_{12}$, five fused rings) 3-Methylcholanthrene ($C_{21}H_{16}$)

[a] Where applicable: for the olefinic and acetylenic hydrocarbons, the general formula relates to an open-chain molecule containing only *one* double or triple bond respectively.

[b] The term arenes more generally relates to hydrocarbons that contain both aliphatic and aromatic units.

comprise a series of compounds for which each member differs from the next member by a constant amount. The family of alkanes forms such a homologous series, the constant difference being CH_2. Progression up the series of alkanes also results in an increasing number of isomers for each homologue; there are 3 isomeric pentanes, 75 decanes and 366 319 possible isomeric forms for the 20-carbon eicosane. Fresh crude oil contains, among other components, a vast array of hydrocarbons. The alkane homologous series in a fresh crude oil can extend to C_{40} and beyond, with little or no odd-chain preference. The n-alkane homologues are present at greater individual concentrations than the branched alkanes in a non-biodegraded fresh crude oil, although a decrease in the individual n-alkane concentration with increasing carbon number is generally observed. This results in the relatively typical pattern (Figure 10.3) that is obtained when a non-biodegraded fresh crude oil is analysed by gas chromatography with flame ionisation detection (GC-FID). Resolution of the individual branched isomers within a crude oil, by gas chromatography, is not practical. In addition, such oils contain a large number of naphthenic hydrocarbons. This combined complex array of hydrocarbons constitutes the unresolved complex mixture (UCM), which appears as a hump in the GC-FID chromatogram of either a biodegraded oil, a weathered oil (Figure 10.3) or so-called mineral hydrocarbons widely used in the production of food contact materials (Gough and Rowland, 1990; Castle et al., 1991; Grob et al., 1991a). The n-alkane patterns, and/or the presence of a UCM, can be indicative of petrogenic contamination of a food product. Thus, this complexity within one group of hydrocarbons can be a benefit to the analyst in identifying an environmental contaminant in food. The n-alkanes in edible oils show an odd-chain preference typical of terrestrial plants, with chain lengths of between C_{23} and C_{33} (McGill et al., 1993; Moffat et al., 1995; Cruickshank et al., 1997; Moffat and Cruickshank, 1997). Similarly, biogenic alkanes and alkenes of marine origin show a strong odd-chain preference, but the dominant chain lengths are C_{15}, C_{17}, C_{19} and C_{21} (Nevenzel, 1989). The alkanes of insects tend to show odd-carbon predominance and mono-, di- and trimethyl-branched components are common (Blomquist and Jackson, 1979). In contrast, petroleum hydrocarbons show little or no odd-chain preference and so the Carbon Preference Index (CPI) can be used to differentiate biogenic from petrogenic hydrocarbons:

$$CPI_{x \to y} = \left(\sum_{n=y}^{n=x} \%HC_{odd} \bigg/ \sum_{n=y}^{n=x} \%HC_{even} \cdot (D/N) \right) \qquad (10.1)$$

where $HC =$ hydrocarbon; $D =$ number of even-numbered homologues

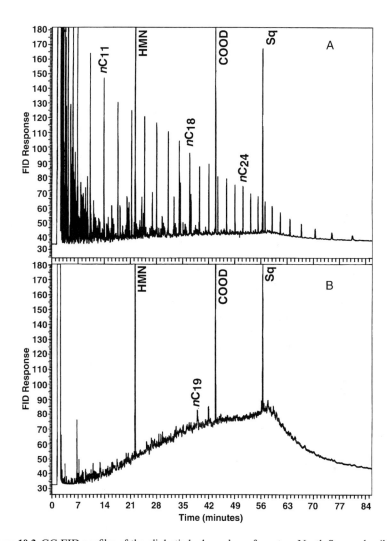

Figure 10.3 GC-FID profiles of the aliphatic hydrocarbons from two North Sea crude oils: (A) Forties Field crude oil and (B) Captain Field crude oil. The Forties oil profile contains a typical *n*-alkane pattern for a non-biodegraded oil. In contrast, there is no definitive *n*-alkane profile in the chromatogram of the aliphatic hydrocarbons from the Captain oil, a biodegraded oil. There is, however, an unresolved complex mixture (UCM) which shows a bimodal distribution. Heptamethylnonane (HMN), chlorooctadecane (COOD) and squalane (Sq) are added as internal standards. Specific *n*-alkanes are identified using abbreviations of the form nC$_{11}$ (undecane).

between chain length x and y inclusively; N = number of homologues in this range.

Of the polycyclic aliphatic hydrocarbons, the pentacyclic triterpanes and steranes have been shown to be particularly useful for the identification of crude oils, mainly in environmental samples (Brakstad and Grahl-Nielsen, 1988; Grahl-Nielsen and Lygre, 1990). This is related to the fact that cyclic hydrocarbons are more resistant to biodegradation than the open-chain compounds (Oudot *et al.*, 1998). Selective determination of the triterpanes and steranes, sometimes termed 'biomarkers', from the vast array of cyclic hydrocarbons is achieved using selective ion monitoring (SIM) gas chromatography–mass spectroscopy (GC-MS). Retene (1-methyl-7-isopropylphenanthrene) is a known marker for softwood combustion (Ramdahl, 1983; Halsall *et al.*, 1997) and, in general, the petrogenic PAHs tend to show higher proportions of alkyl-substituted to parent compounds while the pyrogenic PAHs tend to show higher proportions of the parent compounds.

Physical properties such as melting point, boiling point, vapour pressure and aqueous solubility depend on, among other things, molecular mass, structure and degree of unsaturation. Within a homologous series the molecular mass increases. This results in a progressive change in the physical state of the *n*-alkanes at standard temperature and pressure. The first four members of the *n*-alkane series (methane to butane) are gases. Pentane to heptadecane are liquids, and *n*-alkanes from octadecane onwards are solids (Walker *et al.*, 1996). The boiling points of individual *n*-alkanes present in a crude oil sample can range from 125.7°C for octane to 402°C for pentacosane and beyond for the larger molecules. This complicates the quantitative determination of these compounds from foods and necessitates the use of several internal standards. Great care also has to be exercised when concentrating extracts, so as to minimise losses. Similar complications are encountered with the PAHs; naphthalene is relatively volatile with a boiling point of 218°C, while chrysene has a boiling point of 448°C (Table 10.3). The analyst can, however, make use of such variability. Selective analysis of benzene, toluene and the xylenes and other substituted benzenes is possible using static or dynamic headspace sampling (Whittle *et al.*, 1997; Roose and Brinkman, 1998). Such techniques are rapid and avoid complicated extractions, but great care has to be taken to avoid adventitious contamination (Roose and Brinkman, 1998).

The environmental fate of organic pollutants, and whether or not they end up in the food chain, depends strongly on their distribution between different environmental compartments (Sabljiĉ, 1991). Thus the soil sorption coefficient (K_{om}), water (aqueous) solubility (WS) (May *et al.*, 1978; Sabljiĉ *et al.*, 1989), and the related octanol/water partition coefficient ($\log K_{ow}$) (Verschueren, 1983; Moore and Ramamoorthy, 1984; Güsten *et al.*, 1991), are fundamental parameters for assessing the extent and rate of dissolution of organic contaminants together with their fate, persistence and distribution. K_{ow} gives a measure of the

Table 10.3 Physical properties, including boiling point, aqueous solubility and the logarithm of the octanol/water partition coefficient ($\log K_{ow}$), for a selection of aliphatic and aromatic hydrocarbons

Compound	Molecular formula	Boiling point (°C)[c]	Aqueous solubility at 25°C (mg/l[b,c] or mg/kg[d])	$\log K_{ow}$[b,e,f]
Methane	CH_4	−164		
n-Octane	C_8H_{18}	125.7	0.66[b]	
Cyclooctane	C_8H_6	149		
1-Octene	C_8H_{16}	121.3	2.7[b]	
1-Octyne	C_8H_{14}	125.2		
n-Octadecane	$C_{18}H_{38}$	316.1	0.0075[c]	
Benzene	C_6H_6	80.1	1791[d]	2.13
Toluene	C_7H_8	110.6		2.69
m-Xylene	C_8H_{10}	139.1		3.20
Naphthalene	$C_{10}H_8$	218	31.69[d]	3.35
Anthracene	$C_{14}H_{10}$	340	0.0446[d]	4.50
Pyrene	$C_{16}H_{10}$	393	0.132[d]	5.00
Chrysene	$C_{18}H_{12}$	448	0.0018[d]	5.86
Benzo[a]pyrene	$C_{20}H_{12}$		0.003[b]	6.35

[a]CRC Handbook of Chemistry and Physics (1980); [b]Verschueren (1983); [c]Barrowman et al. (1989); [d]May et al. (1978); [e]Moore and Ramamoorthy (1984); [f]Güsten et al. (1991).

hydrophobicity of a chemical, i.e. its tendency to move from a polar liquid, in this case water, into a non-polar liquid that does not mix with water (Walker et al., 1996). K_{ow} for aromatic hydrocarbons varies by several orders of magnitude (Table 10.3), the MAHs having considerably lower K_{ow} values, and thus a greater hydrophilicity, than the four- and five-ring PAHs (Table 10.3). Naphthalene is slightly soluble in water (Table 10.3) and as such can be absorbed by finfish through their gills. The more hydrophobic PAHs tend to associate more with particulates. Filter-feeding molluscs are exposed to both water and suspended particulate material (SPM) and so are exposed to the full range of PAHs. As mentioned previously, the hydrophobic nature of hydrocarbons means that they tend to concentrate in lipid-rich tissues. Edible crab white meat (muscle) from reference sites around the Shetland Islands has been shown to contain between 1.8 and 13.0 ng/g wet weight PAHs (defined as two- to six-ring, parent and branched compounds) while the corresponding more lipid-rich brown meat contained between 6.6 and 48.8 ng/g PAHs (Topping et al., 1997). Larger differences were noted between scallop adductor muscle and the more lipid-rich gonad (Topping et al., 1997 and section 10.5.2).

The intestinal absorption of aromatic hydrocarbons is related to their solubility in water and, for the PAHs particularly, the 'hydrocarbon continuum' created by dietary lipids and their digestion products is very important (Laher and Barrowman, 1983; Barrowman *et al.*, 1989). The solubilities of benzene, naphthalene and benzo[*a*]pyrene in water are 1791 mg/kg, 31.69 mg/kg (May *et al.*, 1978) and 0.003 mg/l (Verschueren, 1983), respectively (Table 10.3). Thus, naphthalene is approximately 8000 times more soluble in water than benzo[*a*]pyrene. This contrasts with only 9 times greater solubility for the two compounds in trioleoylglycerol (Barrowman *et al.*, 1989).

Alkanes are less reactive than the PAHs. Aliphatic hydrocarbons undergo chiefly addition and free radical substitution; addition occurs at multiple bonds and free radical substitution at other points along the aliphatic chain. In contrast, aromatic hydrocarbons are characterized by a tendency to undergo ionic substitution. Furthermore, PAHs are more susceptible to chemical and biochemical transformation.

10.3   Determination of hydrocarbons in foods

10.3.1   General considerations

The concentration of hydrocarbons in foods can vary from trace amounts (e.g. < 0.1 ng/g individual PAH) to several percent. The hydrocarbons present in a food matrix will be a mixture of biogenic hydrocarbons, contaminant hydrocarbons and those resulting from processing of the food such as bleaching and refining of edible oils (Mennie *et al.*, 1994) or the use of γ-irradiation to preserve the food product (Kavalam and Nawar, 1969; Nawar, 1986). The aim, generally, of any analysis is to determine accurately the concentration of the relevant hydrocarbons in a representative sample of the food. In specific cases, such as the determination of steranes and triterpanes so as to identify a contaminant crude oil in food, the analysis is qualitative. All aspects of the analysis are important and include collection of the sample, suitable storage of the sample prior to the analysis, subsampling of the bulk sample, extraction, clean-up of the extract, chromatographic separation and detection, and data analysis. As the lower molecular mass aliphatic hydrocarbons, the MAHs and, to a lesser extent, the two- and three-ring PAHs are volatile, special care must be taken in handling the food product if these compounds are to be determined. Both PAHs and aliphatic hydrocarbons are ubiquitous in our environment and, if trace hydrocarbon analysis is being done, great care must be taken to avoid adventitious contamination. The extracting solvents can be hydrocarbons and must be routinely checked to ensure their purity as must the water used during extractions. Protocols must be in place to ensure rigorous cleaning of glassware.

Sodium sulphate used to dry extracts, column packings and Soxhlet thimbles must all be suitably treated to remove hydrocarbons. The minimum presence of any plastics, a strict regime for storage of samples, environmental control of the laboratory and assignment of all equipment to specific areas of analysis are further precautions taken to avoid contamination.

Very rarely in hydrocarbon analysis is there only one determinand, and thus the analyst is nearly always dealing with a group of compounds of varying volatility and solubility. A lengthy and involved extraction procedure coupled with a chromatographic separation by thin-layer chromatography (TLC), high-pressure liquid chromatography (HPLC), GC, or supercritical fluid chromatography (SFC) is a prerequisite of nearly all hydrocarbon analysis, making such analysis time-consuming and expensive. Screening of fish and shellfish samples by sensory assessment can assist in targeting of samples for chemical analysis, while ultraviolet fluorescence (UVF) spectroscopy can be used as an indicator of aromatic contamination. Immunochemical methods are being developed that should allow foods and environmental samples to be screened for PAHs (Mazet et al., 1997, and Chapter 4).

10.3.2   Volatile hydrocarbons

The volatile aromatic hydrocarbons (VAHs) include benzene, toluene, ethylbenzene and the xylenes (BTEX) together with propylbenzenes, butylbenzenes, trimethylbenzenes and naphthalene. An analysis may be restricted to the BTEX compounds, but these are often determined in conjunction with other volatile organic compounds (VOCs) including chlorinated aromatic hydrocarbons (e.g. dichlorobenzenes) and chlorinated aliphatic hydrocarbons (e.g. tetrachloromethane, trichloroethene). These compounds are particularly suited to headspace sampling, thus avoiding the lengthy extraction procedures often associated with hydrocarbon analysis. Jickells et al. (1990) simply homogenised foods with water and added a subsample of the homogenate, together with the relevant internal standard(s), to the headspace vial. Dynamic headspace or purge-and-trap techniques, which are less matrix dependent than static headspace techniques, can also be applied to VOCs in food. In this case, the VOCs are forced out of the tissue by heating in an oven under a flow of nitrogen, steam distillation or purging with an inert gas. The analytes are subsequently trapped, using either cryogenic or sorbent traps, and then desorbed into a GC (Roose and Brinkman, 1998). Solid-phase microextraction (SPME) has been used by Shirey (1995) and Nilsson et al. (1995) for the determination of VOCs, including the BTEX compounds, in drinking water. The SPME device was a fused-silica fibre coated with a layer of polydimethylsiloxane and housed in a stainless steel needle. The fibre was exposed to the water within a headspace vial and then transferred to the GC injector for thermal desorption. During studies on the tainting of Atlantic salmon by a water-soluble fraction of Flotta light crude oil, Heras *et al.* (1992) isolated the low-boiling aromatic hydrocarbons (BTEX compounds, naphthalene, methylnaphthalenes and dimethylnaphthalenes) from liver and muscle tissue using steam distillation.

Chromatographic separation of VAHs for quantitative analysis is generally done by capillary column GC (Jickells *et al.*, 1990; Roose and Brinkman, 1998; Schroers *et al.*, 1998; Yang and Lo, 1998). A variety of column lengths (10 to 75 m), film thicknesses (0.25 to 3.0 μm) and phases (non-polar, e.g. DB-1, HP-1; medium-polarity, e.g. DB-624; and polar, e.g. DB-Wax) have been used. A variety of manufacturers now produce columns designed specifically for the analysis of VOCs in accordance with USEPA methods 502.2, 524.2, 624, 8024 and 8260 and include HP-VOC, ZB-624, Rtx-502.2, DB-502.2, Rtx-Volatiles and VOCOL.

VAHs can be the only determinands or can be analysed with other VOCs; detection methods vary. Schroers *et al.* (1998) used photoionisation detector (PID), electron capture detector (ECD) and FID in series and these detectors have been used by other investigators (Shirey, 1995; Yang and Lo, 1998). Mass spectrometry (MS) does, however, have the advantage that deuterated internal standards can be used in conjunction with selective ion monitoring, making this a very powerful detection technique (Jickells *et al.*, 1990; Whittle *et al.*, 1997; Roose and Brinkman, 1998). Using a purge and trap technique coupled with GC-MS, Roose and Brinkman (1998) obtained a limit of detection for benzene, toluene and *m*- and *p*-xylene in marine biota of 80 pg/g. Dewulf *et al.* (1998) achieved a limit of detection of 4.8 ng/l for benzene and 2.7 ng/l for toluene in water, also using purge and trap with GC-MS.

10.3.3 *Polycyclic aromatic hydrocarbons (PAHs)*

Aliphatic hydrocarbons and PAHs are generally extracted at the same time and thus fractionation of the aliphatic and aromatic hydrocarbons is required and can be integrated into the clean-up stage of the analysis. TLC, HPLC, GC and SFC, with a variety of detection techniques, have all been used to determine PAHs in foods quantitatively and these are summarised, together with extraction and clean-up procedures, in Table 10.4.

Table 10.4 Summary of extraction, clean-up and analytical methods for the determination of PAHs in various food types

Extraction	Clean-up	Analysis	Food type	Reference(s)
General review		LC–fluorescence	Spinach, infant formula, corn oil, bran cereal	[1]
KOH saponification, sonication, Soxhlet, cyclohexane/caffeine–formic acid solution	Liquid–liquid chromatography, silica gel, HPLC	TLC HPLC–UV (254 nm) GC–FID, GC–MS	Meat, vegetable oils, smoked products, fish, plant tissue	[2]
Saponification, propylene carbonate extraction of a dried celite/sodium sulphate/food mixture, partitioning into DMSO following caffeine–formic acid complexation (fats and oils), Soxhlet	Liquid–liquid partition, liquid–solid chromatography, TLC, GPC	TLC GC–FID, GC–MS HPLC–UV HPLC–fluorescence	Oils and fats, smoked sausage, wheat grain, malt, sugar-beet fibre, lettuce, rye grain, cereal products, total diet samples	[3]
Ultrasonication with CHCl$_3$	GPC	HPLC–fluorescence	Smoked meat products	[4]
Saponification	SPE then silica	HPLC–UV	Beef	[5]
Saponification	Silica gel column	HPLC–fluorescence TLC–densitometry (336 nm)	Smoked meat, bacon, fish, eel, tea, flavours, herbs, spices, vegetable oils, roasted coffee, coffee extracts	[6]
Saponification	Liquid–liquid extraction,	HPLC–fluorescence	UK total diet sample	[7]
Saponification	Liquid–liquid extraction, HPLC	GC–MS	Fish and shellfish	[8] [9]

Table 10.4 (Continued)

Extraction	Clean-up	Analysis	Food type	Reference(s)
Saponification	KS and silica gel columns, GPC	GC-PID	Fish	[10]
Saponification	Liquid–liquid partition, column, chromatography on silica gel then Sephadex LH 20	GC-FID	Meat, poultry, smoked fish, yeast	[11]
Saponification	Liquid–liquid partition	HPLC–fluorescence (290 nm/430 nm)	Vegetables, fruit, pasta, rice, fish, milk, yogurt, cheese, meat, eggs, vegetable oils, butter, chocolate, wine, beer coffee, pizza	[12]
Soxhlet then saponification	Sep-Pak Florisil cartridge	HPLC-UV (254 nm) HPLC–fluorescence GC-MS	Smoked chicken	[13]
LC (silica gel) followed by in-line solvent evaporation	LC (aminosilane column)	GC-FID	Edible oils	[14]
Donor–acceptor complex chromatography (DACC)-HPLC	DACC-HPLC	HPLC–fluorescence	Edible oils	[15] [16]
LC (silica)	LC (silica)	GC-MS	Vegetable oils	[17]
Supercritical fluid extraction (CO₂– methanol) following freeze drying	SPE–alumina/silica gel 60 & C18	HPLC-UV (254 nm)	Smoked and broiled fish	[18]
Solvent (DCM) extraction	None	SFC-UV (254 nm) HPLC-UV (254 nm)	Water	[19]

Extraction	Cleanup	Analysis	Matrix	Reference
Liquid–solid extraction with Sep-pak tC18, SDB disks	None	HPLC–fluorescence	Water	[20] [21]
Solvent (cyclohexane) extraction	Liquid–liquid partition, column chromatography on silica gel then Sephadex LH 20	GC-FID GC-MS	Oils and fats	[11]
Solvent (cyclohexane) extraction	Liquid–liquid, silica column	GC-FID GC-MS	Edible oils	[22]
Solvent (cyclohexane) extraction	Liquid–liquid partition, GPC	GC-MS	Vegetable oils	[23]
Caffeine–formic acid complexation	Liquid–solid chromatography, TLC	GC-FID	Vegetable fats and oils	[24]
Homogenisation with trichlorotrifluorethane (TCFE)	Liquid–liquid partition	TLC–direct fluorimetry GC-FID	Ham	[25]

[1] Lawrence (1992) (review)
[2] Chen (1997) (review)
[3] Guillén (1994) (review)
[4] Cejpek et al. (1995) (review of saponification)
[5] Rivera (1996)
[6] Stijve and Hischenhuber (1987)
[7] Dennis et al. (1983)
[8] Webster et al. (1997)
[9] Topping et al. (1997)
[10] Lebo et al. (1991)
[11] Grimmer and Böhnke (1975)
[12] Lodovici et al. (1995)
[13] Chiu et al. (1997)
[14] Moret et al. (1996)
[15] Perrin et al. (1993)
[16] van Stijn et al. (1996)
[17] Vreuls et al. (1991)
[18] Järvenpää et al. (1996)
[19] Heaton et al. (1994)
[20] Fernandez et al. (1996)
[21] Harrak et al. (1996)
[22] Larsson et al. (1987)
[23] Speer and Montag (1988)
[24] Kolarovič and Traitler (1982)
[25] Morozzi et al. (1985)

10.3.3.1   Extraction and clean-up

The simplest medium from which to extract hydrocarbons is water. Mixing of water with a solvent such as cyclohexane (Staiken and Frank, 1979) or dichloromethane (Heaton *et al.*, 1994; Fernandez *et al.*, 1996; Law *et al.*, 1997) permits rapid extraction of the hydrocarbons. The solvent is dried over anhydrous Na_2SO_4 and the extract is concentrated by rotary evaporation or by using a Turbovap concentrator. Depending on the analysis, this extract can be analysed directly or the aliphatic hydrocarbons can be isolated from the aromatic hydrocarbons by solid-phase extraction (SPE) (Law *et al.*, 1997) or HPLC. SPE of PAHs from water has been investigated by Fernandez *et al.* (1996) and Zhou *et al.* (1996). Liquid–liquid extraction has been shown to give slightly better recoveries than SPE, but recoveries for SPE were generally in excess of 80% with detection limits of less than 0.1 µg/l for the three- to six-ring PAHs when analysed by HPLC-fluorescence (Fernandez *et al.*, 1996). Harrak *et al.* (1996) found that styrene-divinylbenzene copolymer SPE disks gave better recoveries than C18 extraction disks and achieved limits of detection of between 0.2 ng/l and 3.7 ng/l. SPME has been investigated for the isolation of PAHs from water (Eisert and Levsen, 1996) as has supercritical fluid extraction (SFE) (Janda *et al.*, 1996) but neither is currently being used extensively.

Concentration of the PAHs in high-fat tissues means that either the lipid must be destroyed or it must be extracted and the PAHs isolated from the resulting fat or oil. Extraction of PAHs from fats and oils is relatively straightforward if the materials are soluble in a solvent such as cyclohexane (Grimmer and Böhnke, 1975; Larsson *et al.*, 1987; Speer and Montag, 1988). Basically, the PAHs are extracted into DMF–water and then re-extracted with cyclohexane. Average recoveries for the four- to six-ring PAHs varied between 87% and 159% with lower recoveries for the three-ring PAHs (Larsson *et al.*, 1987). Alternatively, a caffeine–formic acid solution can be used to complex the PAHs. These are again re-extracted into cyclohexane with a resulting recovery of 68–95% (Kolàrovič and Traitler, 1982). Some form of clean-up is always required to remove traces of lipid and this can incorporate separation of the aliphatic and aromatic hydrocarbons or fractionation of the PAHs. Gravity column chromatography on silica gel or Florisil, followed by further column chromatography with Sephadex, gel permeation chromatography (GPC) on Biobeads S-X3 or TLC are commonly used. More recently, HPLC has been used directly to isolate PAHs from edible oils (Vreuls *et al.*, 1991; Moret *et al.*, 1996). On-line determination of PAHs in edible oils and fats by DACC-HPLC (Perrin *et al.*, 1993; van Stijn *et al.*, 1996) gave limits of quantification, using HPLC-fluorescence, of between 0.02 and 0.1 ng/g.

Protein-rich foods (fish, shellfish, meat, cheese, etc.) and insoluble fats require alkaline saponification to release the PAHs bound to food components. Saponification has also been used for tea, flavours, herbs, coffee, vegetables, cereals, fruit, pasta, rice, milk, yogurt, eggs, vegetable oils, fish oils and chocolate as summarised in Table 10.4. A comprehensive summary of the various saponification procedures and associated clean-up stages is presented in Cejpek et al. (1995). As with the determination of PAHs in water, fractionation of the aliphatic hydrocarbons from the PAHs can be incorporated into the clean-up (Hernández et al., 1995; Webster et al., 1997). Saponification is not used exclusively for the isolation of PAHs from the above foods. Morrozzi et al. (1985) homogenised ham with trichlorotrifluoromethane as the extracting solvent, boiled the homogenate for 30 minutes and then used liquid–liquid partition to clean-up the extract. Cejpek et al. (1995) applied ultrasonication in the presence of chloroform to extract the lipid and then used GPC to isolate the PAHs. Soxhlet extraction of lipid has also been applied to the determination of PAHs (see Table 10.4) and more recently SFE followed by SPE has been investigated for the extraction of PAHs from smoked and broiled fish (Järvenpää et al., 1996). Although microwave-assisted extraction of PAHs from soils has been reported (Barnabas et al., 1995), this would not appear to have been extended to food products.

10.3.3.2 Chromatographic analysis

The chromatographic method selected is, in part, related to the PAHs being determined. Capillary GC has the advantage of affording the greatest chromatographic resolution and, when coupled with mass spectroscopy, can be readily used for the complex determination of two- to six-ring parent and branched PAHs as required following a crude oil spill (Webster et al., 1997). GC-MS has the further advantage of enabling the use of deuterated internal standards for accurate quantification (Law and Klungsøyr, 1996), such standards being available to cover the relevant range of boiling points and solubilities. Any analysis, however, requires only to be 'fit for purpose' and if benzo[a]pyrene is the sole determinand then TLC with fluorimetry can be used (Morozzi et al., 1985). Many studies are restricted to the 16 PAHs that have been identified as 'priority pollutants' by the USEPA (Figure 10.2). Other groups use 12 PAHs, the USEPA PAHs excluding naphthalene, acenaphthylene, acenaphthene and fluorene, while Dennis et al. (1983) determined only the four- to six-ring PAHs from the EPA 16 and added benzo[e]pyrene. Lodovici et al. (1995) also determined only the four- to six-ring EPA PAHs but excluded indeno[1,2,3-cd]pyrene. As such their 'total PAH concentration' relates to nine PAHs. The current legislation[1] relating to drinking water in the United Kingdom requires only fluoranthene, benzo[b]fluoranthene, benzo[k]fluoranthene, benzo[a]pyrene, benzo[g,h,i]perylene and indeno[1,2,3-cd]pyrene to be determined (Statutory Instrument No. 1147, 1989). The term 'total PAH concentration' should, therefore, always be clearly defined and will vary markedly between studies.

HPLC analysis of PAHs can be selective and highly sensitive. Octadecyl (C18) columns are used almost exclusively with acetonitrile and water the most common solvents, sometimes in association with methanol. Many manufactures now produce specialised polymeric C18 packings for the analysis of PAHs. UV detection at 254 nm is commonly used but fluorescence detection allows a degree of selectivity through optimisation of the excitation and emission wavelengths. A variety of wavelength pairs have been quoted for the various PAHs. An example for each of the USEPA 16 PAHs, excluding acenaphthylene, is (excitation/emission) naphthalene, 221 nm/323 nm; acenaphthene 290 nm/323 nm; fluorene 265 nm/304 nm; anthracene, 251 nm/402 nm; phenanthrene 250 nm/348 nm; fluoranthene 286 nm/464 nm; pyrene 275 nm/374 nm; benz[a]anthracene 287 nm/388 nm; chrysene 267 nm/363 nm; benzo[b]-fluoranthene 254 nm/437 nm; benzo[k]fluoranthene 304 nm/411 nm; benzo[a]pyrene 295 nm/406 nm; dibenz[a,h]anthracene 297 nm/396 nm; benzo[g,h,i]perylene, 291 nm/409 nm; indeno[1,2,3-cd]pyrene 295 nm/496 nm (Beltrán et al., 1996). A typical single setting for PAHs is 290 nm/430 nm (Dennis et al., 1983; Lodovici et al., 1995). Minimum detectable amounts for individual PAHs determined by HPLC-fluorescence can be as low as 0.8 pg with limits of quantification in the range 0.02 ng/g to 3 ng/g.

Capillary gas chromatography is a reliable and robust analytical method. Non-polar phases and those of low to intermediate polarity are typically used for PAHs (Law and Klungsøyr, 1996) and again some manufacturers have produced columns to meet the requirements of the USEPA Method 8100 for PAHs. The FID offers excellent linearity, sensitivity and reliability but does not give the selectivity and increased sensitivity that is observed with the PID or ECD. GC-MS is a powerful combination for PAH analysis. Monitoring the total ion current yields a 'universal' detector, but selected ion monitoring (see Figure 10.2B and C for relevant ions) provides enhanced sensitivity and selectivity with detection limits in the range 0.1 ng/g to 1.0 ng/g (for relevant references see Table 10.4).

---

[1] Current EU legislation is likely to change before the end of 1998 and the United Kingdom will be required to adopt this revised legislation.

Thin-layer chromatography is a rapid and simple technique that has been applied, quantitatively, to the determination of selected PAHs (see Table 10.4). Oxidation of the PAHs during the analysis can be problematic but the technique has the advantage that separated components can be recovered for further analysis.

The separation, in 6 minutes, of the 16 USEPA PAHs has been achieved on a packed column using SFC with UV detection at 254 nm (Heaton et al., 1994). This was more rapid than for the equivalent HPLC analysis and has been applied to the determination of PAHs in water. SFC with UV, MS (chemical ionisation (CI) or charge exchange but not electron ionisation (EI)), PID, FID or Fourier transform infrared (FTIR) detection have been investigated for PAHs but primarily from environmental samples rather than from food (Mulcahey et al., 1994).

Capillary electrophoresis has been studied for the quantitative determination of PAHs. Efficient separation of anthracene, phenanthrene, chrysene, pyrene, benzo[a]pyrene and benzo[e]pyrene was achieved when using cyclodextrins as buffer modifiers with an analysis time of 18 minutes. The main advantage of this technique was the separation of isomeric forms of benzopyrene (Szolar et al., 1995).

10.3.4   Aliphatic hydrocarbons

Long-term chronic input of hydrocarbons to the marine environment from oil and gas exploration and production activities has resulted in the detection of petrogenic aliphatic hydrocarbons in dabs (*Limanda limanda*) caught in the vicinity of an oil platform (McGill et al., 1987) and in plaice (*Pleuronectes platessa*) caught in the vicinity of gas production platforms (Parker et al., 1990). Mineral oil[2] products used by the food industry can also contaminate foods and arise from the oils and greases used for lubricating machinery, releasing agents and blocking additives (Moret et al., 1996). Furthermore, mineral hydrocarbons used for the softening of jute and sisal fibres are present in jute bags used for the transportation of nuts, cocoa beans, coffee beans and rice and can migrate to the food and be carried through processing to the final product (Grob et al., 1991b; Nichol and Jickells, 1993). Concentrations of these hydrocarbons can exceed 100 mg/kg (Moret et al., 1996) and thus, although they are not regarded as being particularly toxic, quantitative

---

[2] The term 'mineral oil' means any hydrocarbon product, whether liquid, semi-solid or solid, derived from any substance of mineral origin and includes liquid paraffin, white oil, petroleum jelly, hard paraffin and microcrystalline wax (Statutory Instrument No. 1073, 1966). It can also be used as a collective term for liquid hydrocarbons that are obtained from crude oil or coal and are used as fuel, heating agents and lubricating oil (Levy, 1976).

determination of aliphatic hydrocarbons in food is important because of their reported ability to act as co-carcinogens (Bingham and Nord, 1977; Roy *et al.*, 1988).

10.3.4.1 Extraction and clean-up

Many of the comments relating to the extraction of PAHs can be reiterated for aliphatic hydrocarbons when determining those with molecular masses greater than that of undecane. Extraction of water with dichloromethane will isolate both aliphatic and aromatic hydrocarbons, but these can be readily separated as described in section 10.3.3.1. Extraction of hydrocarbons from edible oils can be achieved directly using silicic acid column chromatography but HPLC is required to remove traces of lipid (McGill *et al.*, 1993). For foods in general, the fat and associated lipophilic compounds can be extracted using methanol/dichloromethane (Mackie *et al.*, 1980; McGill *et al.*, 1987), ultrasonication in propan-2-ol/light petroleum (Rowland and Volkman, 1982), Soxhlet extraction (González-Barros *et al.*, 1997) or by simply mixing overnight in a suitable solvent (Grob *et al.*, 1991b; Nichol and Jickells, 1993). The aliphatic hydrocarbons can then be isolated from the lipid using SPE or HPLC, Grob *et al.* (1991b) using on-line coupled LC-GC. Alternatively, the extract can be saponified and the aliphatic hydrocarbons isolated by TLC (Rowland and Volkman, 1982) or SPE (Nichol and Jickells, 1993). Owing to the effects of oil spills on the marine environment, fish and shellfish are regularly studied. These products are often saponified directly, with column chromatography used to isolate the aliphatic from the aromatic hydrocarbons (Topping *et al.*, 1997), recoveries between 83% and 94% being obtained by Hernández *et al.* (1995) for hexacosane to triacontane. Newton *et al.* (1991) isolated diesel contaminants from fish by steam distillation. Saponification has also been used as the first step in the analysis of aliphatic hydrocarbons in edible oils (Bastić *et al.*, 1978) and foods generally (Di Muccio *et al.*, 1979), TLC being used to isolated the relevant hydrocarbons.

10.3.4.2 Chromatographic analysis

GC-FID is used almost exclusively for the determination of aliphatic hydrocarbons from foods. Indeed, the analysis of lubricating oils was one of the first applications of GC, but a katharometer was used for detection (Adlard and Whitham, 1958) Capillary columns (20–30 m length) coated with a 0.1–0.6 µm non-polar phase are generally used (more precise details can be obtained from the references in section 10.3.4.1), although Grob and Siegrist (1994) utilised a 3 m column when analysing mineral oil from jute bags. The GC-FID chromatograms of aliphatic hydrocarbons from petroleum or mineral oils are characterised by a UCM (Gough *et al.*,

1992) superimposed on which can be an *n*-alkane series. Quantification is complicated by the range of boiling points for the array of compounds present; several internal standards should be added and the most appropriate standard selected for the calculation. Approximately 5 pg of an individual *n*-alkane can be readily detected when using an FID, the detector response being linear up to 300 ng on-column.

Mennie *et al.* (1994) determined the concentration and composition of sterenes in edible oils, following bleaching and deodorisation, using GC-FID with a non-polar column. Compound identification required accurate mass measurement high-resolution mass spectroscopy. Hydrogenation greatly complicated the sterene profile and compound resolution was only achieved by using a phase of intermediate polarity (Moffat *et al.*, 1995).

Differentiation of biogenic from petrogenic hydrocarbons has been discussed in section 10.2. There is also a requirement to be able to determine the source of petrogenic contamination. Polycyclic aliphatic hydrocarbons, i.e. steranes and triterpanes, also known as biomarkers, are well suited to this purpose (Jones *et al.*, 1986; Grahl-Nielsen and Lygre, 1990; Glegg and Rowland, 1996). North Sea oils, for example, contain a marker compound not found in Middle Eastern oils (Figure 10.4). Furthermore, there are triterpanes that are unique to a specific North Sea oil (Moffat *et al.*, 1998). The steranes and triterpanes are extracted from fish or shellfish as for other hydrocarbons and are determined using SIM GC-MS.

10.3.5   Analysis of compounds used to assess exposure to PAHs

Assessment of the exposure of either humans or food-producing animals to PAHs is a challenging analytical problem. PAHs, along with other organic xenobiotics, undergo an oxidative transformation as the primary biochemical process in hydrocarbon detoxification. In addition, the carcinogenic potential of PAHs can only be realised following metabolic activation of the PAHs to electrophilic species that can then interact covalently with DNA (Dipple and Bigger, 1991; Cavalieri and Rogan, 1992; Carmichael *et al.*, 1996). Exposure to PAHs can thus be assessed through determination of the hydroxy metabolites in urine (Vyskocil *et al.*, 1997), bile and other tissues (Krahn *et al.*, 1993a,b), measurement of the induction of the relevant enzyme systems or determination of DNA adducts.

The determination 1-hydroxypyrene in urine requires enzymatic hydrolysis of the glucuronide and sulphate conjugates to produce the hydroxylated derivative. The 1-hydroxypyrene can then be isolated using reversed-phase solid-phase extraction (Buckley and Lioy, 1992) and the concentration of the metabolite determined by reversed-phase HPLC-

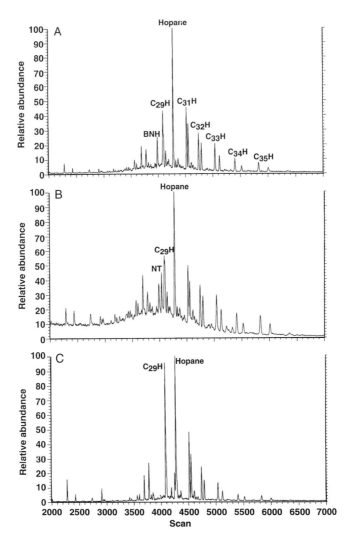

Figure 10.4 *m/z* 191 mass chromatograms of (A) Forties, (B) Captain and (C) Middle Eastern crude oil. The North-Sea marker bis-norhopane (BNH) can be seen in the chromatograms of the Forties and Captain oils both of which are North Sea oils. The homohopane doublets ($C_{31}H$ to $C_{35}H$) are present in all oils, while a novel triterpane (NT) is unique to the Captain oil.

fluorescence (excitation 254 nm/emission 388 nm). Recoveries in excess of 90% have been achieved using this methodology and a limit of detection of 0.023 ng/ml was established. The determination of 1- and 2-hydroxynaphthalene in urine, the concentration of which is regarded as being related to exposure by air, is achieved by GC-MS (Jansen *et al.*,

1996). The conjugated derivatives are first hydrolysed and the extract is purified using solid-phase extraction. The hydroxy compound is then derivatised with 3,5-bis(trifluoromethyl)benzoyl chloride prior to the GC-MS analysis using SIM (m/z 384).

Exposure of fish to PAHs can be assessed by determination of metabolites in bile. These metabolites are usually present in the form of phase-2 conjugates, covalently bound to glucuronic acid, glutathione or sulphate (Ariese et al., 1997). Either direct analysis of conjugated metabolites or analysis of the phase-1 (hydroxylated) metabolites after hydrolysis can be employed. Screening methods employing HPLC–fluorescence, and a GC-MS method for quantification of individual metabolites have been developed (Krahn et al., 1993a,b), the latter giving a limit of quantification in the range 7–110 ng/g wet weight of bile. Metabolites have also been determined in fish (Krone et al., 1992), rat (Chipman et al., 1981) and human (Selkirk et al., 1975) tissues, HPLC being the preferred analytical technique.

Many oil spills are marine based, affecting both farmed and wild fish populations. Although measurement of the levels of fluorescent aromatic residues in bile can be used as an indicator of exposure, the primary biological response of fish to PAH contamination is the induction of P4501A enzymes in the liver. Activation of this enzyme system can be assessed by the measurement of 7-ethoxyresorufin O-deethylase (EROD) activity. This technique has been used during investigations into the long-term impact of oil exploration and production activities in the North Sea on flat fish (Stagg et al., 1995) and to study the sub-lethal effects on farmed fish affected by the Braer oil spill in Shetland (Stagg et al., 1997). Induction of phase II enzymes (UDP-glucuronyltransferases and glutathione S-transferase) has been shown by Taysse et al. (1998) to occur in liver, spleen and head kidney of carp (Cyprinus carpio) following exposure to 3-methylcholanthrene. UDP-glucuronyltransferase activity is determined by incubation of the microsomal fraction with 4-nitrophenol and measurement of the absorbance at 400 nm, while glutathione S-transferase activity is measured by incubation of the cytosolic protein with 1-chloro-2,4-dinitrobenzene and measurement of the absorbance at 340 nm.

Following metabolic activation, PAHs can bind to DNA, forming the DNA adducts that are thought to have a crucial role in chemical carcinogenesis (Dale and Garner, 1996). Determination of DNA adducts in peripheral white blood cells by enzyme-linked immunosorbent assay (ELISA), with a limit of detection of 0.04 fmol adduct/µg DNA, showed a positive link between the consumption of char-broiled food and PAH–DNA adduct formation (Rothman et al., 1995). ^{32}P-Postlabelling analysis, with a limit of detection of 1 adduct/10^{10} nucleotides, showed there to be a dose-dependent response between the consumption of charcoal-broiled hamburgers and the levels of aromatic DNA adducts in mononuclear cells (van Maanen et al., 1994). ^{3}H-Label-HPLC, ^{14}C-label-HPLC, HPLC–fluorescence and GC-MS have also been used for the analysis of DNA adducts, but these methods are between one and three orders of magnitude less sensitive than the immunoassay and ^{32}P-postlabelling (Pfau, 1997).

10.3.6   Quality control

Good quality control is an essential aspect of ensuring that data released from a laboratory are fit for purpose. This can be achieved through applying internal quality control procedures (Thompson and Wood, 1995); through external accreditation by organisations such as the United Kingdom Accreditation Service (UKAS) or the British Standards Institution; by participation in interlaboratory calibration studies and the preparation of reference materials; and by participation in an international quality assurance scheme such as the Food Analysis Performance Assessment Scheme (FAPAS), Quality Assurance of Information for Marine Environmental Monitoring in Europe (QUASI-MEME) or AQUACHECK. QUASIMEME, for example, have conducted laboratory performance studies on PAHs in shellfish and have a development exercise for the determination of PAH metabolites in bile, while AQUACHECK provides proficiency testing materials for both the BTEX compounds and the PAHs appropriate to the UK Drinking Water Regulations. Two coconut oil reference materials have recently been certified for pyrene, chrysene, benzo[k]fluoranthene, benzo[a]pyrene, benzo[g,h,i]perylene and indeno[1,2,3-cd]pyrene (Luther et al., 1997). A full discussion of quality control procedures is outwith the scope of this chapter, but the nature of hydrocarbon analysis is such that regular blanks, assessment of recoveries, surrogate spikes, replicate analyses, determination of inter-batch variability, determination of sample variation and statistical control of instrumentation should all form an integral part of the analytical protocol.

10.4   The variety of sources and inputs of hydrocarbons

The major origins of hydrocarbons in the environment can be summarised as biogenic, petrogenic and pyrogenic, but petrogenic and pyrogenic sources can be both natural and anthropogenic. The sources contribute to both point and non-point (or diffuse) inputs.

- *Biogenic*: those produced naturally by the biomass of plants, animals and microorganisms on land and in the sea, and as a result of degradative processes on the biomass.
- *Petrogenic*: those naturally present in crude petroleum oils and gas, bituminous deposits and deposits of oil shales and sands, from which there are continuous localised natural oil seeps, exposure and erosion, both onshore and offshore, significant on a worldwide basis; those derived from petroleum and other fossil fuels as a result of man's activities giving rise to emissions, spillages or effluents during production and transportation of crude oil and gas both onshore and offshore, from the refining and petrochemical industries, from the use of petrochemical products for the manufacture of packaging materials, solvents, propellants and other permitted, mineral oil-based additives for industry, including the food industry, and in the dumping of wastes both onshore and offshore.
- *Pyrogenic*: those formed in natural combustion processes, mainly forest fires ignited by natural phenomena; those produced by man's activities such as combustion of oil and gas products and other fossil fuels, coal and peat, for industrial and domestic energy production, incineration of agricultural, industrial and municipal wastes, engine exhausts from the combustion of fuels for transport for commercial and leisure purposes; those produced by domestic wood burning, forest fires accidentally or maliciously initiated, or as a result of intentional forest clearances, such as the widespread fires burning out of control in the rain forests and underlying peat deposits of Borneo at the time of writing; those formed when oil wells burn out of control as happened in Kuwait on a massive scale during the Gulf War of 1991.

The relative importance of these inputs varies greatly from area to area, and from time to time, and can be influenced by specific major events such as spillages, droughts or volcanic eruptions. Thus, there is a very complex background of naturally derived and anthropogenic hydrocarbons ranging from methane to at least six-ring, condensed cyclic aromatic compounds in air, water and soil that provides a ubiquitous, low-level presence of a wide range of hydrocarbons in all foods.

10.4.1   Biogenic

Superimposed on the ever-present hydrocarbon background is a characteristic array of hydrocarbons actually produced by current biosynthesis in specific terrestrial or aquatic plants or insects, or acquired by animals from lower down the food chain, which of course applies also to those species used for food. It seems unlikely that biosynthesis is a significant source of PAHs. The cuticular waxes of plants have an aliphatic hydrocarbon fraction rich in the higher molecular mass n-alkanes, C_{25} to C_{33} (Dove and Mayes, 1991), with marked characteristic odd carbon number predominance. Such a distribution is regarded as typical of terrestrial inputs and is common to soil, sediments and terrigenous particulates that can be dispersed over thousands of kilometres by the wind. This type of distribution is also found in seed oils and the different patterns found in different seed oils are sufficiently robust to identify the species origin of the oils. Insect waxes also tend to show the characteristic odd-carbon predominance of the n-alkanes. Forests, in particular, contribute volatile hydrocarbons, rich in iso-prenoids, to the atmosphere, components of the so-called 'blue-haze' over densely forested areas. The biogenic hydrocarbons of the marine environment (Nevenzel, 1989) are particularly interesting, indicating the relatively stable nature of these hydrocarbons in their passage through the food chain. In the marine environment, aliphatic hydrocarbons, normal, iso-, anteiso-alkanes and isoprenoid alkanes such as pristane, and polyolefins such as n-$C_{19:5}$, n-$C_{21:5}$ and n-$C_{21:6}$, are produced by algae. These hydrocarbons are assimilated and conserved by the grazers at the next trophic level, e.g. the zooplankton, which also add to the complexity by dehydrating or decarboxylating phytol (a moiety of chlorophyll) producing a range of C_{19} and C_{20} branched olefins. In turn, these hydrocarbons are assimilated and deposited in the lipids of the edible flesh and other tissues of the pelagic, planktonivorous, oil-rich fish species at the next trophic level, such as herring (*Clupea harengus*), sardines and pilchards (*Sardinella*, *Sardinops*, and *Sardina* spp.), sprat (*Sprattus sprattus*) and anchoveta (*Engraulis ringens* and others) (Sargent and Whittle, 1981). These fish are not only available to species at the top of the marine food chain but are also used directly for human food and the production of edible marine oils and used for animal feeds in the form of fish-meal for silage.

There is ample evidence of close similarity between the tissue hydrocarbon array and dietary availability. The C_{19} olefinic hydro-carbons of herring livers were indicative of the principal zooplankton species in the same area (Blumer *et al.*, 1969). The hydrocarbons of herring muscle also reflected dietary composition (Mackie *et al.*, 1974), but the main difference between the hydrocarbons in muscle and liver was the odd-carbon, C_{25} to C_{33}, alkane predominance in the livers, which seems to be a normal characteristic of fish livers and the hepato-pancreas of shellfish. The alkane fraction of mackerel (*Scomber scombrus*) flesh was closely similar to the whole sprat on which it was feeding which, in turn, reflected that of its local zooplankton diet (Whittle *et al.*, 1974). The distinctive difference between the alkane composition of liver (hepato-pancreas) and muscle indicates the selective nature of the different tissues. The hydrocarbons of basking shark (a filter-feeder) liver were indicative of its recent food sources (Blumer, 1967), which probably accounts for the variation in hydrocarbon composition of oils and blubber oils from filter-feeding whales. Basking sharks and some other elasmobranchs also accumulate very large concentrations of squalene in the liver. The annual, total, global production of marine, biogenic hydrocarbons has been estimated to be of the same order as the estimated annual, global input of petroleum hydrocarbon contaminants to the marine environment from all sources, some 5 to 10 million tonnes.

10.4.2   *Petrogenic*

All crude oils (and other fossil fuels) contain similar molecular species amounting to tens of thousands of compounds ranging from gases to residues boiling above 350°C. They vary markedly in detailed composition from reservoir to reservoir and even during the lifetime of a single well on an oil-field. Nevertheless, most of the compounds (usually > 75%) are types of hydrocarbons in the classes *n*-alkanes, branched alkanes, cycloalkanes, triterpanes, aromatics, naphtheno-aromatics and PAHs with up to 10 condensed aromatic rings. The triterpanes (or hopanes) and steranes are useful diagnostic suites of compounds to help identify the source of a petroleum or petroleum product contaminant, but there are many diagnostic biomarkers available for this purpose that are also used to provide information on the natural origins of the fossil fuels (Philp and Xavier de las Heras, 1992), a concept first proposed in the 1930s. In addition, organosulphur compounds, acids, phenols, pyridine and pyrroles are present, and highly complex asphaltenes containing ten or more fused aromatic rings, some of which are heterocyclic, with aliphatic and naphthenic side-chains. Different crudes are enriched in different types of compounds, markedly affecting their properties. In the North Sea oil-fields, for example, Brent and Forties crudes are so-called light oils. Gullfaks crude, the cargo spilled from the oil tanker *Braer* in Shetland, is also a light crude but comes from a reservoir in which the crude was already highly biodegraded, and was low in *n*-alkanes but rich in aromatics, while Beatrice is a heavy, waxy crude. Distillates are also enriched in certain fractions and 'cracking' in the refining process introduces olefins, essentially absent in crudes. Fresh petroleum contamination is characterised particularly by alkyl substitution of the 2-, 3-, and 4-ring PAHs.

Table 10.5 World consumption of oil, gas and coal, 1997

	Oil	Natural gas	Coal
'Oil equivalents', million tonnes	3395	1977	2293
million tonnes	3395		3440
billion cubic metres		2196	
Percentage increase, 1972–1997	35	90	50

Source: *BP 1997 Statistical Review of World Energy* (BP, 1998).

10.4.3 Pyrogenic

Annual global consumption of oil, gas and coal for all purposes is some 9 gigatonnes and, overall, continues to increase (Table 10.5). The regional and national pattern of usage varies markedly (BP, 1998). Wood, peat and animal waste, which are important sources of fuel in some countries, are not reliably quantified. All combustion processes, whether natural, accidental, or as a result of man's industrial, municipal, commercial, domestic or pleasure activities, emit aliphatic and aromatic hydrocarbons most of which, apart from the volatiles, are associated with the micro particulate material in the emissions or subsequently become associated with particulates. These are transported, sometimes thousands of kilometres, with relatively little dilution, and are subsequently deposited by wet or dry precipitation. In boiler emissions, the PAH concentrations associated with fine particulates were 3 to 50 times higher than with coarser particles; the lower boiling PAHs to fluoranthene were mainly in the vapour phase, while the higher boiling PAHs were with the particulates (Bergström *et al.*, 1982). Greater formation of PAHs is favoured by incomplete combustion of organic matter in the range 400–900°C, particularly above 700°C (Bartle, 1991). In 1997 and 1998, massive forest fires, considered to be the worst on record, have occurred, for example in Brazil, Indonesia, Phillipines, Vietnam, Thailand, Australia and Mexico, resulting largely from the effects of logging and agricultural clearances combined with severe drought caused by the El Niño phenomenon in the Pacific. The contribution of these fires has not been quantified, however. PAHs are present in emissions from vehicle exhausts; industrial and domestic heating based on fossil fuels; wood-burning stoves; waste incineration; refining and cracking of petroleum; production of carbon blacks, creosote, charcoal, coal tar, carbon electrodes, and synthetic fuels; coal coking; use of carbon electrodes in aluminium and manganese alloy smelting; metal smelting; and production of electric power from fossil fuels. The typical pattern of PAHs from pyrolytic sources is a high proportion of parent compounds relative to their alkylated derivatives. Table 10.6 compares some of the large

Table 10.6 Differences in percentage composition of the aryl hydrocarbons of a crude oil and the combustion products of wood and coal (Massie *et al.*, 1985)

Hydrocarbons	Beryl crude oil	Combustion (coal + wood)
Naphthalenes	51.51	0.78
Biphenyls	3.70	0.08
Acenaphthalene/fluorenes	4.98	0.54
Dibenzofurans	0.91	0.10
Dibenzothiophenes	3.69	0.43
Phenanthrenes/anthracenes	32.99	42.93
Cyclopenta-3-ring PAHs	0.26	15.29
4-Ring PAHs	1.74	32.12
Cyclopenta-4-ring PAHs	0.03	3.50
Benzo[*e*]pyrene	0.043	0.81
Benzo[*a*]pyrene	0.006	0.91
Other 5-ring PAHs	0.12	1.86
Cyclopenta-5-ring PAHs	0.006	0.50
6-Ring PAHs	0.014	0.53
Benzo[*e*]pyrene/benzo[*a*]pyrene ratio	7.5	0.9

differences in aryl hydrocarbon composition between a crude oil and the combined combustion products of coal and wood and the difference between the respective benzo[*e*]pyrene/benzo[*a*]pyrene ratios. PAH inputs to water from all sources, natural and anthropogenic, were estimated to be greater than 80 000 tonnes/year (NRCC, 1983), so that the total global input is probably of the order of 100 000 tonnes/year, and annual global emissions of benzo[*a*]pyrene are probably in the range 5000 to 10 000 tonnes. In terms of human exposure, other significant pyrolytic sources include tobacco smoking, the smoking of foods, frying, and the broiling, grilling or barbecuing of foods over open fires, all of which can be considered to be within the realm of individual choice.

10.5   Transfer to and through the food chain

Various aspects of the principles of concentration in the food chain that ultimately affect human exposure to PAHs as environmental contaminants have been reviewed and investigated recently (Hattemer-Frey and Travis, 1991; Seigneur *et al.*, 1992; Butler *et al.*, 1993; Giordano *et al.*, 1994; EPA, 1995; Meador *et al.*, 1995).

10.5.1   Plants

Hydrocarbons in the air are either in the vapour phase or bound to the particles of a range of sizes present in the air. Benzo[*a*]pyrene concentrations range from less than $1\,ng/m^3$ in rural areas up to about $50\,ng/m^3$ in some urban areas. The main routes by which hydrocarbons from the air enter or deposit on plants (pasture, fruits, vegetables, cereals and other crops) are by direct deposition onto plant surfaces above ground and subsequent uptake by the foliage into or through the cuticular waxes, or by the foliage through the stomata from the air used for respiration. Volatiles such as benzene and toluene can be taken up by the foliage via the stomata in this way and can accumulate in the plant. If soil is contaminated with the more volatile compounds, volatilization from the soil can contribute to the foliar uptake. Direct deposition by wet or dry precipitation is an important pathway for the particle-bound PAHs, but some particles may be small enough to enter the stomata. Some deposited on the surface may transfer to the cuticular waxes and be transferred to the tissues, but some of the burden remains particle-bound on the surface and can be washed off or removed with the outer leaves or peel. It is well established that the PAH composition of the foliage derives from the surrounding air. Leafy vegetables such as lettuce and cabbage are the most exposed, while other crops, such as tomatoes, beans and peas, are less so. In root vegetables such as carrots there are similarities in composition between the foliage and the root core (Wild and Jones, 1992). PAH concentrations in leafy produce such as lettuce were found to be high near the source of PAHs, in this case 12 metres from a busy highway (Larsson and Sahlberg, 1982); the total PAHs were $102\,ng/g$ (benzo[a]pyrene $7.1\,ng/g$), declining to 15–27 and $0.2–1.6\,ng/g$ for the total PAHs and benzo[a]pyrene, respectively, some 50 metres from the highway. Washing removed on average 50% of the total PAHs but was more effective for phenanthrene (90%) than for benzo[a]pyrene (17%). Lettuce grown 0.5–1.5 km downwind from an aluminium smelter had concentrations of total PAHs and benzo[a]pyrene up to 922 and $18\,ng/g$, respectively. The dominant component from the smelter site was fluoranthene, whilst benzo[g,h,i]perylene was dominant in the samples from the highway site. In country areas, the median values for the total PAHs and benzo[a]pyrene were 12 and $0.1\,ng/g$, respectively (Larsson and Sahlberg, 1982). PAH levels are influenced substantially by the proximity of the source and the prevailing weather conditions. At harvesting, the amounts present in the foliage will be the balance reached between deposition, wash-off that occurs during precipitation, incorporation in the cuticular wax, cuticular penetration, uptake via the stomata and transfer to the plant tissues. Soil dust generated during the harvesting of crops and aeolar deposition during subsequent transportation also contribute to the burden.

Acute contamination can occur by aerosol as the result of a petroleum pollution incident. Following the grounding of the oil tanker *Braer* on the

Shetland coast in January, 1993, gale-force winds blowing onshore, driving rain and sea spray loaded with spilled oil combined to contaminate pasture and crops visibly with oil up to 17 km or more downwind from the wreck. In the worst-affected areas, concentrations of two- and three-ring PAHs (including methyl derivatives), which were the major constituents of the PAH in this crude oil, ranged up to about 270 ng/g wet wt vegetation 3 months after the grounding. The last areas were released for grazing 8 months later when levels there dropped to 40 ng/g or less (Milne et al., 1997).

High concentrations of hydrocarbons occur in soils, particularly those with a high organic matter content. PAHs in soil may be lost by abiotic degradation, volatilisation and biodegradation but these processes are structure specific. Potentially, plants can receive hydrocarbons directly from the soil by aqueous uptake via the root tissues, and by absorption from soil on the root surfaces. Those of most concern, the PAHs, are found in all soils from about 100 ng/g dry wt ranging over 3 orders of magnitude, with the higher concentrations in urban areas near point sources and in soils with high organic matter content (Wild and Jones, 1992). They have high octanol–water (K_{ow}) (Table 10.3) and organic carbon adsorption (K_{oc}) partition coefficients, are strongly associated with organic matter in the soil and are thought to be sequestered in nanopores in the soil matrix. In model systems, phenanthrene was sequestered and less than 10% was available for microbial mineralisation when it penetrated nanopores with hydrophobic surfaces (Nam and Alexander, 1998). PAHs can be expected to have low availability to plants via root uptake, especially in soils with high organic matter content owing to the greater adsorption potential. However, there is concern that some root crops such as carrots, beets, radishes, etc., particularly those with higher lipid contents, may have elevated PAH contents when grown in soils with high PAH concentrations, epecially where the concentrations of PAH have been elevated by the application of sewage sludge. This activity is increasing as dumping at sea is no longer permitted by a number of countries. PAHs enter waste waters and are subsequently removed during waste water treatment and concentrated in sewage sludges, which generally contain individual PAH compounds in the concentration range 1–10 mg/kg dry wt (Wild et al., 1990). The sludges are applied to soils as organic fertiliser and the higher molecular mass PAHs persist in the amended soils for many years. Carrots grown in sandy soils (organic carbon content about 1%) amended with sludge from an urban catchment (17 mg PAHs/kg dry wt) were analysed for 15 PAHs in foliage, root peels and root cores (Wild and Jones, 1992). Foliage was unaffected by the PAH loadings in the amended soils, since the PAH in the foliage came from the atmosphere. The PAH levels in the root peels increased with increasing PAH load in the soil, tending to level off at higher loadings, but were lower than the foliage. The cores were unaffected by the PAH concentrations in the soil, suggesting little or no transfer from the skin surface; 70% of the PAH burden was with the peel. The PAH compositions of the foliage and cores were similar, but different from that of the peel, which also had compositional differences with the soil. The core concentrations were less than 4.2 ng/g fresh wt (benzo[a]pyrene less than 0.14 ng/g), whereas concentrations of total PAHs were found in the foliage up to 55 ng/g.

Thus, PAHs in the above-ground foliage and probably in the core of the root crops are primarily derived from the atmosphere and largely independent of PAH concentrations in the soil, although there is some contribution from the soil to the surface tissues of the root which can be removed during peeling. Soil to plant bioconcentration factors were estimated to range from 0.0001 to 0.33 (Edwards, 1983).

10.5.2 Aquatic animals

Hydrocarbons enter the aquatic environment by direct discharge to rivers and water bodies from the petroleum industry and other industrial and municipal sources, by wet and dry deposition from the atmosphere, by land runoff, by accidental spillage and, in some countries, by dumping of waste and sewage sludge. Lower molecular mass members may be present in solution or highly dispersed, but the major proportion is associated with suspended organic matter and is particle-bound. The particles may be transported over considerable distances depending on size, nature and hydrographic conditions, ultimately sedimenting out and being incorporated in the sediments. In the aquatic environment, depending on behaviour and feeding patterns, aquatic animals (pelagic, demersal or benthic) are exposed not only to hydrocarbons in the dissolved or highly dispersed state but also in the organic-rich surface micro-layer. They are exposed to hydrocarbons incorporated in food organisms and on suspended or sedimenting particulates, from resuspended bottom sediments, from the interstitial waters of the sediments and from sediment ingested with food. In aquaculture, filter-feeding, bivalve shellfish mostly rely on natural sources of algae for food, but some fish species such as trout and salmon are fed on dry pelleted concentrated feeds, sometimes augmented, for example for salmon, with wet feed produced from comminuted fish offal derived from fish processing.

In the seas, the total PAH concentrations in unfiltered water in the open sea are of the order of 1 ng/litre or less; in enclosed seas such as the Mediterranean or the Baltic they may be 2 to 15 times higher. In coastal waters and estuaries the range may be 1 to 50 ng/l, but in industrial areas with particular sources of PAH concentrations may be of the order of 1 µg/l or even as high as 10 µg/l. The major portion (90%) of the higher molecular mass PAHs are associated with particulates and are ultimately deposited in the sediments. Law *et al.* (1997) found benzo[*a*]pyrene concentrations in unfiltered waters around the United Kingdom ranging from < 1 to over 900 ng/l. Compaan *et al.* (1992) estimated that benzo[*a*]pyrene concentrations in sea water were probably less than than 0.2 ng/l in the early nineteenth century and fluxes to sediments have increased 5–10-fold since the beginning of the twentieth century. Movement to cleaner fossil fuels and cleaner industrial processes may be contributing to a decrease from this source over the last decade or so. PAHs associated with particulates can be transported over tens to thousands of kilometres depending on particle size, currents, and other conditions affecting suspension and sedimentation, ultimately accumulating in the sediments. Benzo[*a*]pyrene in sediments in relatively clean areas is not usually more than 1 ng/g dry wt, but can commonly increase to 100 ng/g dry wt or more in industrial polluted areas. The Norwegian scale of quality for total PAHs in sediments indicates that > 20 µg/g dry wt is regarded as extremely polluted (Table 10.7), but concentrations can reach 500–600 µg/g dry wt close to industrial point sources (Naes *et al.*, 1995), and even an order of magnitude higher in the vicinity of large, heavily industrialised and municipal areas.

The hydrophobic character and low biodegradabilty of PAHs causes accumulation in organisms beyond their concentrations in the environment. Several processes are distinguished by which tissue concentrations of the hydrocarbon contaminants of concern can increase: bioconcentration, bioaccumulation and biomagnification, although these terms are not always used in strictly defined senses. Bioconcentration is the result of direct uptake by the tissues exposed to water: the gills, lamellae and skin. Bioaccumulation is the combined uptake from water and from ingested food, particulates or sediment through the gut lining to the tissues and, in the case of fish, from ingestion from drinking water. Biomagnification describes the situation in which species at higher trophic levels in the food chain or food web acquire higher tissue burdens of the contaminant hydrocarbons than their prey or food organisms at lower trophic levels. In inshore coastal areas, estuaries and rivers adjacent to industrial and

Table 10.7 Quality criteria for total PAH in polluted sediments (µg/g dry wt) (Naes *et al.*, 1995)

Degree of pollution				
Not or slightly	Moderately	Markedly	Severely	Extremely
< 0.3	0.3–2	2–6	6–20	> 20

municipal environments, aquatic animals used for food are exposed to elevated water, particulate and sediment hydrocarbon burdens and, depending on seasonal and weather conditions, significant fluctuations may occur in relative exposure at any locality. There are also considerable differences in the efficiency with which fish, bivalves and crustaceans are able to metabolise or detoxify and excrete the hydrocarbons of concern, the PAHs. Benzo[a]pyrene metabolism by aquatic organisms was ranked: fish > shrimp > amphipod crustaceans > clams (Varanasi et al., 1985). Consequently, PAHs have the potential to bioaccumulate in organisms such as crustaceans and bivalve molluscs that lack efficient mixed-function oxidase (MFO) mechanisms (the cytochrome P450 system, section 10.3.5), which in turn leads to the formation of the mutagenic and carcinogenic metabolites. Molluscs, crustaceans and fish are exposed to PAHs whether associated with particulates or not. Bioconcentration factors (BCF) with respect to water for the higher molecular mass PAHs in crustaceans and bivalves are generally in the range 3000–6000, and for the lower molecular mass PAHs are 1000 or less, but for fish that efficiently metabolise and excrete PAH metabolites BCF values are generally less than 500 (Eisler, 1987; Murray et al., 1991; Mackay et al., 1992; Hellou et al., 1994; Meador et al., 1995).

Although sediments are an important sink for PAH in aquatic systems, surface sediments can be mobilised and made available to aquatic organisms. Bioperturbation, resuspension and food-chain transfer are important processes mediated by benthic invertebrates, increasing the availability of sediment-associated PAHs that can be accumulated by bottom-dwelling invertebrates and fish (Varanasi et al., 1985; Eisler, 1987; Hellou et al., 1994). Approximate tissue-to-sediment ratios for benzo[a]pyrene were calculated to be 0.1 for clams and 0.05 for fish and shrimp (Varanasi et al., 1985).

Acute exposure to spilled crude oil, depending on the composition and characteristics of the oil, and the conditions of the spillage, in some circumstances can result in extremely elevated levels of PAHs in aquatic organisms used for human food, and loss or depuration of the contaminants towards background levels becomes an important issue. Contrary to some conclusions (Varanasi et al., 1989), PAH residue analysis in fish is an effective measure of petroleum contamination in some situations. Caged salmon in a salmon farm some 20 km away from the wreck of the oil tanker Braer, within 10 days of the incident, accumulated PAHs (mostly two- and three-ring parent and branched compounds) in the muscle tissue up to a maximum of 14 000 ng/g wet wt in the worst case, 450 times the average background value (30 ng/g) found in areas unexposed to the spilled oil. Concentrations returned to background levels in salmon flesh in all the exposed salmon farms after some 200 days (Whittle *et al.*, 1997). In line with established experimental results with other species, the order of rate of loss of aromatic components was: substituted benzenes > naphthalene > C_1-alkyl-substituted naphthalenes > C_2- to C_4-alkyl-substituted naphthalenes > phenanthrenes and anthracenes. The relative loss of naphthalene and substituted naphthalenes from salmon flesh has since been confirmed following exposure of salmon experimentally to Forties crude oil (Davis *et al.*, 1997). Initially, all commercial species of fish (demersal and benthic) and shellfish examined from within the established Exclusion Zone for fishing were found to have elevated levels of PAHs in the edible tissues, dominated by the two- and three-ring compounds (Topping *et al.*, 1997). Within 2 months, concentrations in all species of wild fish, of which the bottom-dwelling flatfish generally had the highest values, had decreased to within the range of reference samples (0.3–42.1 ng/g wet wt). Loss of accumulated PAHs from edible crabs was slower than from lobsters and was complete by 15 months. However, elevated PAHs have persisted in Norway lobster (*Nephrops norvegicus*), with animals obtained from the affected area during 1997 still containing higher PAH concentrations than those from reference sites, most likely as a result of continued exposure to oil sedimented out in the fine sediments of the habitat in which they burrow. As expected, the filter-feeding bivalve molluscs accumulated the highest concentrations of PAHs, up to 26 000 ng/g wet wt gonad tissue of queen scallops, some 130 times the mean background value. The entire suite of two- to six-ring compounds measured was found. In general, levels in lipid-rich gonad were some 5–10 times higher than muscle tissue. With the exception of mussels, levels in bivalve molluscs decreased to within the background range after 25 months. PAH concentrations in mussels (total two- to six-ring PAH, 13.7 ng/g wet wt soft tissue) transplanted into the Exclusion Zone consistently increased over a number of months up to 20 times the original level. This was more than could be accounted for in the apparent seasonal fluctuation at the reference site, which may be related to the normal seasonal changes in lipid content (Webster *et al.*, 1997).

10.5.3   Animals

Animals, for example beef and dairy cattle, are potentially exposed to hydrocarbons from forage (pasture, hay, silage, supplements and formulated feedstuffs), from drinking water, from the air by inhalation and from bedding materials. The target foods for human consumption are meat and meat products, various offal products and milk. None of 16 PAHs were detected in blood or livers of beef steers on sawdust or shredded newspaper beddings used to compare and assess the suitability of beddings from these waste materials (Comerford *et al.*, 1992). The total exposure from the forage, feeding stuffs and drinking water can be determined. Ingestion of soil along with the forage may be a substantial source since it is reported to range between 1% and 18% of the total feed intake on a dry weight basis depending on the make-up of the feed, which itself is of the order of 10 kg/day dry wt for beef cattle. Volatiles and hydrocarbons bound to particulates are present in the inhaled air, but it is not clear how comparatively efficient is absorption of the array of hydrocarbons from the inhaled air. However, MAHs are readily absorbed from the lungs and through the skin (Schroers *et al.*, 1998). Concentrations in the air are higher near point sources and animals closer to the urban environment are exposed to higher concentrations than in the rural environment. Only a proportion of the ingested hydrocarbons from all sources are absorbed via the gastrointestinal tract and the constituents of the array of hydrocarbons do not behave identically, but there is very little practical information on this aspect (Mayes *et al.*, 1986; Milne *et al.*, 1997).

On absorption, the lipophilic hydrocarbons will tend to acumulate to varying extents in the fatty tissues of the animal, being present in the meat, organs or offal products used for human food. The fat content of the meat depends *inter alia* on breed, diet and age at slaughter, being typically in the range 18–25%. Actual bioconcentration factors have to be determined experimentally for hydrocarbons of interest in different tissues, e.g. liver or muscle, but crude overestimates (because 100% absorption is assumed) of biotransfer from total exposure dose to beef can be made by correlation with octanol–water partition coefficients (K_{ow}) for individual compounds (Giordano *et al.*, 1994) based on animal feeding studies (Travis and Arms, 1988). Biotransfer factors obtained for benzene, toluene and benzo[*a*]pyrene were 3.3×10^{-6}, 1.3×10^{-5} and 2.9×10^{-2} (mg/kg fat)/(mg/day) respectively. Dairy cattle are larger animals than beef cattle and, depending on the husbandry practised, the feed combinations, feed consumption, soil ingestion and inhalation rates, exposure to hydrocarbons will be different. Hydrocarbons are transferred to milk, but the fat content of the milk, typically about 4%, is likely to be an important factor. Actual transfer from exposure has to be determined experimentally but, again, crude estimates can be made by correlation with the K_{ow} for a specific compound. The biotransfer factors calculated for dairy milk for benzene, toluene and benzo[*a*]pyrene were 1.0×10^{-6}, 4.3×10^{-6} and 9.1×10^{-3} respectively, roughly 3-fold less than the values for beef (Giordano *et al.*, 1994).

Ewes that were suckling single lambs (1 to 2 weeks old) and fed a pelleted, complete diet, were dosed orally with a mixture of two- to four-ring PAHs at a rate of 14.5 mg/day (Milne *et al.*, 1997), together with a faecal marker (hexatriacontane) for estimating faecal output (Mayes *et al.*, 1994). Faecal recoveries of the PAHs ranged widely from 2% to 84%. Naphthalenes appeared to be much more efficiently absorbed and/or metabolised in the gastrointestinal tract than three- and four-ring compounds, but the distribution of individual PAHs in the milk and tissues (liver, fat and muscle) did not reflect the differences found in the faecal recovery, suggesting greater metabolism of the naphthalenes. The total PAH concentrations in the muscle of the suckling lambs apparently increased about 10-fold towards 500 ng/g tissue over a period of 30 days from the time oral dosing of the ewes with PAH began (Milne *et al.*, 1997).

In recent experimental work in our laboratory, the tissue-specific uptake of petrogenic hydrocarbons was investigated in sheep. At the end of the 14-day exposure period, during which the animals consumed 1 kg feed/day that contained diesel oil at a concentration of 1.5%, all four tissues analysed—adipose tissue, liver, muscle and kidney—contained petrogenic hydrocarbons. The highest concentrations were measured in the adipose tissue (mean for $\sum n\text{-}C_{12}\text{-}n\text{-}C_{33}$ of 240.09 µg/g (SE 4.02, $n = 3$) relative to a mean control concentration of 18.87 µg/g (SE 1.86, $n = 12$)). There was an increasing preference for the shorter chain hydrocarbons in the order liver, adipose tissue, muscle, kidney. The half-life for depuration of the aliphatic hydrocarbons from the adipose tissue was estimated at 70 days. Lactating goats were fed for 21 days on a pelleted diet and on molassed-hay contaminated with the equivalent of 15 ml diesel oil/day/goat, which is rich in aliphatic hydrocarbons from C_{12} to C_{24} compared to the feeds, which were rich in C_{25} to C_{33}. In the milk, total alkanes (C_{12} to C_{33}, pristane and phytane) rose within 10 days from an average of 1300 ng/g to a range of 12 000 to 22 000 ng/g owing to the uptake of the aliphatics in the diesel oil, but rapidly declined to background concentrations a few days after oil-feeding stopped. Naphthenes were also found in the milk, evidenced by the presence of a UCM present in the chromatogram. The diesel oil was also rich in the two- and three-ring PAHs and the dibenzothiophenes. Naphthalenes rose on average from 1.6 ng/g in the milk to 5 ng/g after 7 days and had declined by day 25; three-ring compounds on average rose more slowly from 1.1 ng/g to 5 ng/g after 18 days and began to decline thereafter; relatively little change occurred in the dibenzothiophenes. There was some evidence that uptake of higher molecular mass PAHs into milk was enhanced in the later stages of oil-feeding.

Taking into account both inhaled air and drinking water, Giordano *et al.* (1994) and others concluded that in normal circumstances diet is the major source of exposure of hydrocarbons for animals.

10.6 Hydrocarbon concentrations in foods

10.6.1 Monocyclic aromatic hydrocarbons (MAHs)

Petroleum is the main source of these compounds and they are concentrated in the lighter distillate products of petroleum, such as aviation fuel, petrol (gasoline), particularly unleaded petrol, and diesel. They are widely used as solvents in paints, adhesives and degreasing agents, are present in polymers used for packaging, and are contaminants of styrene derived from petroleum. They are products of combustion processes, being present in vehicle emissions and tobacco smoke (Darrall *et al.*, 1998). The toluene/benzene ratio indicates the urban contribution to the non-methane volatile hydrocarbons in the atmosphere, since the major input is from vehicle exhausts (Yang and Lo, 1998). The higher the ratio, the greater the influence of the urban source: the ratio in Taiwan urban air was 3.75, compared with < 1.0 for rural air. Owing to their physicochemical properties they are widespread in the environment, in the outside air and in the workplace and home, and are present in marine and fresh waters. There is a paucity of information on levels in foods. The MAHs are readily absorbed via lungs and skin and, in some cases, exposure from oral ingestion may be important (Schroers *et al.*, 1998). Levels of MAHs in the air vary widely. For example, over the coast of Brittany concentrations ranged from 20 to 600 pg/ml (Boudries *et al.*, 1994), whereas over Southampton Water they ranged from 1 to 200 ng/ml (Bianchi and Varney, 1993). Not suprisingly, contamination of laboratory air causes problems for the analysis of these volatiles (Roose and Brinkman, 1998). Concentrations in coastal sea water were found to be of the order of 10 to 50 ng/l (Dewulf and Van Langenhove, 1995). Benzene is a genotoxic carcinogen (IARC, 1982), causally linked to leukaemias. Exposure of urban dwellers to benzene via the air, the main route of exposure, is of the order of 400 μg/day. The World Health Organisation set a limit of 10 μg/litre for benzene in drinking water. In 1998, some carbonated beverages were withdrawn from sale in the United Kingdom because they were found to contain benzene, 10–20 μg/litre, from contaminated carbon dioxide.

Concentrations in whiting and dab muscle, respectively, were 40 and 5 ng/g wet wt for total MAHs and 4 and 0.5 ng/g wet wt for benzene in fish caught off the Belgian coast (Roose and Brinkman, 1998). Fish can accumulate benzene in the tissues; one inshore species (not a food species) was reported to have concentrations up to 1000 ng/g (Ferrario *et al.*, 1985), but accumulation depends on species, fat content, type, extent and time scale of exposure and many other factors. Table 10.8 compares the MAHs measured by headspace analysis in the muscle of contaminated salmon

Table 10.8 Monocyclic aromatic hydrocarbons and other volatiles in farmed salmon muscle contaminated in the *Braer* oil spill (Whittle *et al.*, 1997)

	Benzene	Toluene	Other MAHs	Naphthalene	C_1–C_4 Naphthalenes[a]
				(ng/g muscle)	
Contaminated salmon	2.0	35.5	284.5	203.0	9425.0
Unexposed salmon	1.0	NA[b]	7.5	4.5	35.5
Salmon, retail packs	2.7	2.0	10.5	NA	NA

[a]The sum of the alkyl-branched (C_1 to C_4 chain length) naphthalenes.
[b]NA, not analysed.

(also strongly tainted) from a Shetland fish farm, 10 days after the wreck of the oil tanker *Braer*, with those in fish not exposed to the oil from the wreck, and with those in salmon in retail packs on sale in England at the time (Whittle *et al.*, 1997). Naphthalene and the C_1- to C_4-alkyl-substituted naphthalenes are included for comparison. Benzene concentrations were low even in the heavily contaminated fish (2 ng/g or less), but tended to be higher in the retail packs (2–5 ng/g). The degree of contamination was clearly shown by toluene, the substituted benzenes, naphthalene and the substituted naphthalenes. Tainted mackerel also had elevated concentrations of total MAHs, 160 ng/g muscle (benzene 4.4 ng/g), whereas total MAHs in lean species such as cod, haddock, whiting, saithe and ling from both exposed and unexposed areas were less than 10 ng/g muscle (benzene less than 2.5 ng/g); the total MAHs in the muscle of flat fish living on the bottom, plaice, lemon sole, dab and witch, ranged from 8 to 55 ng/g, with benzene less than 2 ng/g (authors' unpublished data).

Table 10.9 shows some data from the authors' laboratory giving the range of concentrations of MAHs (compared with naphthalene) found by headspace analysis in some vulnerable, packaged, fatty foods from a variety of retail outlets including petrol stations, road-side shops or stalls and supermarkets, products likely to absorb such compounds from the surrounding air (MAFF, 1996). The surfaces and core (or centre) of the products were analysed separately; concentrations in the latter were usually lower. The highest concentrations of MAHs were found in margarine. Low concentrations of benzene were detected in all samples up to 34 ng/g; 10% of products had benzene concentrations greater than 10 ng/g in the core. Toluene was detected in all samples up to 977 ng/g and was the most useful marker of contamination; 46% of products had toluene concentrations of 10 ng/g or less in the core; 21% were greater than 50 ng/g. Concentrations were consistently highest at the ends of the products (66% of all products), or on the surface where the wrapper

Table 10.9 Monocyclic aromatic hydrocarbons (ng/g) in retail samples of fatty foods[a]

Product (sample number)		Benzene	Toluene	Other MAHs	Naphthalene
Butter (32)	Surfaces	Tr–28	Tr–275	Tr–204	ND–140
	Core	Tr–15	Tr–189	ND–27	ND–53
Cheese (17)	Surfaces	Tr–18	Tr–109	Tr–158	Tr–82
	Core	Tr–13	Tr–55	ND–16	ND-13
Lard (18)	Surfaces	Tr–12	10–216	Tr–270	ND-56
	Core	Tr–14	Tr–118	Tr–18	ND-49
Margarine (14)	Surfaces	Tr–21	12–977	Tr–119	Tr–55
	Core	ND–16	Tr–274	ND–13	ND–43
Bacon (17)	Surfaces	ND–34	Tr–180	ND–44	ND–281
	Core	Tr–11	Tr–10	ND–Tr	ND–Tr
Sausage (15)	Surfaces	ND–34	Tr–122	Tr–19	ND–35
	Core	Tr	Tr–56	Tr–20	Tr–14
Sausage meat (1)	Surfaces	10	47	72	21
	Core	Tr	29	11	10

[a]Tr, less than 10 ng/g product; ND, not detected.

closed (20%) but, in some contaminated products, there was no significant difference between the surfaces and the core, suggesting relatively homogeneous contamination in these cases. Ethylbenzene or the xylenes were detected in all samples except bacon, up to 65 and 232 ng/g, respectively. Propyl- and butylbenzenes, also measured in these samples, were present at low concentrations in some products up to 42 and 24 ng/g, respectively. Naphthalene was found in 69% of samples, up to 281 ng/g. Overall, correlation with source was difficult, and location of the shops was not related to differences in concentration.

Packaging, cookware and cooking utensils in which polymeric materials are in contact with food can also contribute MAHs to the food by migration. Sometimes, contamination problems can occur (Haesook, 1990), but it is important to establish the background levels in the foods to assess migration. Background levels of benzene in foods used to assess migration of benzene from non-stick cookware and microwave susceptors during cooking were less than 5 ng/g product, the maximum found in fruit cake (Jickells *et al.*, 1993). Various MAHs are used as solvents in polymer and co-polymer production and as sealants in the production of food packaging materials. Thermoset polyester contains cross-linked styrene monomer and is used for a wide range of cooking utensils, cookware and packaging. Polystyrene and expanded polystyrene are widely used for food packaging and contain traces of residual styrene monomer; petroleum-derived styrene contains MAHs as impurities. Low concentrations of styrene can be found in packaged food products (Gilbert and Startin, 1983); commercial yogurts packed in polystyrene beakers had styrene concentrations in the range 2–11 ng/g (Linssen *et al.*, 1993), and styrene was present in cream packed in polystyrene tubs in the range 5–40 ng/g (Gilbert and Shepherd, 1981). The volatile contaminants can cause taints (Haesook, 1990; see also Chapter 11); in a tainted syrup, toluene was found ranging from 80 to 110 ng/g and styrene at 10 ng/g (Hollifield *et al.*, 1980).

Concentrations of benzene found by extraction or headspace analysis in various polymers used for packaging, non-stick coatings or susceptors are detailed in Table 10.10 (Jickells *et al.*, 1990; 1993). The high levels of benzene found in some of the thermoset polyester products originated by breakdown during polymerisation of the initiator *t*-butyl perbenzoate and could be avoided by the use of a non-aromatic initiator (Jickells *et al.*, 1990). In the non-stick coatings, benzene was attributed to the use of a phenylmethyl silicone component that had a benzene concentration of 360 mg/kg (Jickells *et al.*, 1993). Migration of benzene into cooked foods ranged from less than 10 to 90 ng/g from the use of thermoset polyester cookware (Jickells *et al.*, 1990), and was less than 2 ng/g from PTFE coatings on cookware or from microwave susceptors, (Jickells *et al.*, 1993).

10.6.2 *Polycyclic aromatic hydrocarbons in foods and beverages*

The presence of PAHs in foods is well documented, for example by Larsson (1982); McGill *et al.* (1982); Fazio and Howard (1983); Dennis *et al.* (1983, 1991); Vaessen *et al.* (1988); de Vos *et al.* (1990); Bartle (1991); Lijinsky (1991); Lodovici *et al.* (1995) and was most recently reviewed by Tomaniová *et al.* (1997). The data tend to focus on about 10 of the four- to five-ring compounds from fluoranthene to indeno[1,2,3-*cd*]pyrene (Figure 10.1, sections 10.1 and 10.3.3.2), including the potential carcinogens (Table 10.1, Figure 10.2), but highlighting benzo[*a*]pyrene. Hattemer-Frey and Travis (1991) summarised, for comparison with predicted values, the occurrence of benzo[*a*]pyrene in air, water, soil,

Table 10.10 Benzene concentrations in food contact materials (Jickells *et al.*, 1990, 1993)

Food contact material	Benzene concentration
Poly(vinyl chloride)	< 0.1 mg/kg
Expanded polystyrene	< 0.1 to 1.36 mg/kg
Polystyrene	< 0.1 to 1.7 mg/kg
Thermoset polyester retail cookware	0.3 to 84.7 mg/kg
Retail non-stick coatings based on PTFE[a]	< 1 to 50 μg/dm^2
Retail microwave susceptors	< 1 μg/dm^2

[a] Poly(tetrafluoroethylene).

sediments and a range of food types, but no data were included for vegetable oils. These published data, cited from 1986 and earlier, indicated that higher concentrations of benzo[a]pyrene were present in leafy vegetables, fish, beef and dairy products. The amount taken in via the diet as a whole is of course dependent on the detailed composition of the diet.

In total diet surveys, the diet is divided up into major food groups such as fruit, vegetables, cereals and cereal products (bread, biscuits, cake, etc.), meat and fish, sugar and sweets, oils and fats, and so on. Each group is analysed separately to assess the major contributions to dietary exposure. The UK diet surveys (Dennis *et al.*, 1983, 1991) indicated that the cereals group and the oils and fats group each contributed about one-third of the daily total PAH intake, which was estimated to be 3700 ng/person, and some 80% of the total daily dietary intake of benzo[a]pyrene (250 ng/person). Fruits, sugars and vegetables provided much of the balance. Margarine was the major source in the oils and fats group (contributing 70% of the benzo[a]pyrene, cf. MAHs in section 10.6.1), and edible oils contributed significantly to the cereals group. PAHs were low in retail fish and animal-derived oils and fats (e.g. average benzo[a]pyrene 0.06 ng/g butter, 0.36 ng/g lard), but higher and more variable in retail vegetable oils (average benzo[a]pyrene 1.29 ng/g), and margarine (average benzo[a]pyrene 1.68 ng/g). The average concentration of benzo[a]pyrene in chocolate was 0.21 ng/g. Higher concentrations of benzo[a]pyrene (maximum 2.2 ng/g) were found in cereal-derived products, puddings, biscuits and cakes, containing edible oils as an ingredient.

As expected, the average concentrations of total PAHs (parent and branched compounds) in smoked fish (245 ng/g), cheese (82 ng/g) and bacon meat (54 ng/g) were higher than in the corresponding unsmoked products (McGill *et al.*, 1982); concentrations in bacon rind were higher (average 168 ng/g) than in bacon meat. Concentrations higher than 100 ng/g were found in traditionally smoked products in which the product was exposed directly to the smouldering wood chips or sawdust. Concentrations of benzo[a]pyrene were up to 0.35 and 18 ng/g in unsmoked and smoked fish, respectively; up to 0.04 and 0.14 ng/g, respectively, in unsmoked and smoked bacon meat (without the rind); up to 0.03 and 1.62 ng/g in unsmoked and smoked cheese, respectively, and up to 0.26 ng/g for both smoked and unsmoked sausage, although concentrations are variable depending on the many factors influencing the conditions of smoking. A comparison of traditional and modern smoking processes for smoked fish products in Germany (Karl, 1997) found that on average the sum of seven carcinogenic PAHs (Table 10.1) in the edible portion of products smoked in traditional kilns was 9 ng/g product (benzo[a]pyrene, 1.2 ng/g), approximately ten times higher than products from modern kilns. In modern kilns, the mean concentration for the sum of 13 three- to five-ring PAHs was 54.2 ng/g; the seven carcinogenic PAHs, 1.5 ng/g; benzo[a]pyrene, 0.1 ng/g. The use of liquid smokes is effective in reducing the PAH concentrations of smoke-flavoured foods (Gomaa et al., 1993; Yabiku et al., 1993: Garcia Falcon et al., 1996). Cooking can also reduce the benzo[a]pyrene content of smoked sausages (Simko et al., 1993), the main factors being fat content, water-holding capacity and cooking time. Different methods of broiling or barbecuing over charcoal, preventing direct contact with the smoke, can also greatly reduce the levels of PAHs found in the food (Wu et al., 1997).

A comprehensive total diet study in the Netherlands, in which some 220 different food items were analysed (de Vos et al., 1990), estimated that the mean daily intake of total PAHs ranged from 5000 to 17 000 ng/ person, similar to previous results from the duplicate diet study of Vaessen et al. (1988). In the Dutch study, the carcinogenic PAHs contributed about half the intake, and the largest proportion came from sugar, sweets, cereals, oils, fats and nuts. Dennis et al. (1991) and Vaessen et al. (1988) noted that the concentrations of PAHs were similar in bread regardless of whether the ovens were gas-fired or electrically heated, indicating that combustion gases in gas-fired ovens did not contribute significantly to the PAH levels in bread, and no difference was found between crust, crumb, bread and dough. Higher levels of PAHs have been found in whole cereal or bran compared with milled cereal, but the higher fat content in the non-bread cereal-based products probably accounted for the higher PAH content (noted above) and for the higher PAH levels in infant formula powder (14.2 ng/g; benzo[a]pyrene 0.49 ng/g) compared with skimmed milk powder (3.3 ng/g; benzo[a]pyrene 0.11 ng/g) although, after reconstitution with water, the concentration in the infant feed would have been less than 0.1 ng/ml (Dennis et al., 1991).

In a food basket study of 33 items chosen from among the most common components of the Italian diet (Lodovici et al., 1995), the estimated daily total PAH intake was 3000 ng/person and the carcino-genic PAHs again contributed roughly half the daily intake, 1400 ng/ person. Table 10.11 summarises the individual concentrations of the nine PAHs measured in the 33 items (Lodovici et al., 1995), the total PAHs and the carcinogenic PAHs, and gives a reasonable impression of concentrations in a range of common foods. The meat, fish, eggs, cereals and some vegetables were cooked according to common Italian practice; the fish, meat and vegetables were cooked without added oil, salt and spices to avoid contamination, the beef, pork and rabbit items being fried. As expected, the highest total PAH concentrations were in barbecued beef and pork, and pizza baked in a wood-burning oven. The next highest were found among the leafy vegetable, chocolate, fruit, fried meat, cured meat and cereal product items, while the lowest were in potatoes, cooked fish, beverages and eggs. However, cereal and milk products, meat, vegetables and fruit contributed most to the dietary PAH intake as a consequence of the consumption pattern. Interestingly, Lodovici *et al.* (1995) also expressed the PAHs found in terms of benzo[*a*]pyrene equivalent potency for mutagenic and carcinogenic effects. For carcinogenic potency, the most significant factors are the combined concentrations of benzo[*a*]pyrene and dibenz[*a,h*]anthracene (Table 10.1) which, based on the data in Table 10.11, places barbecued beef, fried beef, apples, chocolate, peeled apples and squash in decreasing order of importance, and ranging from about 3 to 1 ng/g benzo[*a*]pyrene equivalents.

These surveys gave remarkably similar results in their overall estimates of exposure, ranging over a factor of about 5, and in the relative concentrations in the different types of foods, but differed according to the relative importance of the food items in the respective diets, e.g. consumption of oils and fats, coffee, cereals, sugar and sweets, and in the preparation and cooking of the products. For comparison, Table 10.12 summarises PAHs in uncooked products from locally reared animals in Kuwait (Husain *et al.*, 1997) and indicates generally elevated levels (taking into account the detection limit of 0.5–1.0 ng/g), of the order of those expected in smoked foods (McGill *et al.*, 1982) or in vegetables from contaminated areas (Larsson and Sahlberg, 1982).

Edible oils are clearly an important source of PAHs in the diet, either in food preparation or as food ingredients. Table 10.13 shows some unpublished data from the authors' laboratory for a range of two- to six-ring PAHs in retail samples of margarine and vegetable oil, fish oil supplements, and samples of ready-to-eat and dehydrated recipe baby foods. The vegetable oils included maize, rapeseed, sunflower, soyabean and an unspecified vegetable oil. Taking rehydration into account, the dehydrated baby food products had higher PAH concentrations, but the sum of the carcinogenic PAHs would be less than 0.4 ng/g in the rehydrated product. On average, the major contributors to the PAHs in each food group were the two- and three-ring naphthalenes and phenanthrenes, which accounted for 59%, 65%, 93% and more than 94% of the total PAHs in the margarines, vegetable oils, fish oils and baby foods, respectively. The range of values in each group show the variability from product to product, in which one or two products had concentrations some one or two orders of magnitude greater than the others analysed in the group. Overall, the average concentrations of the four- to six-ring PAHs in each group were similar to those reported by Dennis

Table 10.11 Concentrations (ng/g product) of polycyclic aromatic hydrocarbons (PAHs[a]) in foods (modified after Lodovici et al., 1995)

Food product	FLU	PYR	BAA*	CRY*	BBF*	BKF*	BAP*	DBA*	BPE	PAH[b]	CPAH[c]
Vegetable											
Cauliflower	2.45	ND	0.07	0.24	0.03	ND	0.01	ND	ND	2.79	0.34
Beet greens	5.41	0.69	0.25	3.66	0.92	0.14	0.10	ND	0.05	11.21	5.06
Squash	3.60	ND	2.39	1.65	0.42	0.10	0.45	0.29	0.01	8.91	5.30
Lettuce	0.10	0.43	0.96	0.84	0.12	0.08	0.01	0.06	0.02	2.62	2.06
Tomatoes	0.10	0.40	0.01	0.13	0.01	ND	ND	ND	ND	0.63	0.14
Potatoes	0.05	0.42	0.03	ND	0.02	0.01	ND	ND	ND	0.52	0.05
Fruit											
Apples	2.36	3.46	0.33	ND	0.26	0.08	0.53	1.23	0.02	8.27	2.43
Peeled apples	0.26	0.45	0.09	ND	0.30	0.20	0.06	0.99	ND	2.35	1.64
Citrus	0.05	ND	ND	1.48	0.08	0.03	0.03	ND	ND	1.67	1.62
Cereal											
Bread	0.85	ND	0.31	1.88	0.04	0.02	0.02	ND	0.01	3.12	2.26
Pasta	3.95	ND	0.03	1.88	0.04	0.02	0.02	ND	ND	5.93	1.98
Rice/corn	0.12	ND	0.06	0.52	0.03	0.03	0.02	0.07	ND	0.86	0.73
Cooked fish											
Trout	0.43	ND	0.14	0.91	0.13	0.01	0.03	0.11	ND	1.75	1.32
Cod	0.02	ND	0.12	ND	0.26	0.03	0.01	0.15	ND	0.58	0.57
Dried cod	0.38	ND	0.08	ND	0.03	0.02	0.03	ND	ND	0.53	0.15
Dairy											
Milk and yogurt	0.47	ND	0.68	ND	0.15	0.02	0.34	ND	ND	1.65	1.18
Cheese	0.46	ND	0.15	ND	0.09	0.02	0.01	0.26	ND	0.99	0.53
Meat											
Fried beef	1.00	0.01	2.22	ND	0.66	0.14	0.61	1.00	0.01	5.66	4.64
Fried pork	0.55	5.38	0.43	ND	0.65	0.04	0.04	0.20	ND	7.27	1.35
Fried rabbit	2.06	ND	0.13	1.18	0.03	0.02	0.02	ND	ND	3.43	1.37
Chicken	0.17	ND	0.18	ND	0.01	0.02	0.02	0.20	ND	0.60	0.43
Beef liver	0.29	ND	0.14	ND	0.24	0.05	0.03	0.16	ND	0.90	0.61
Cured meats	2.68	ND	0.62	ND	0.05	0.02	0.03	ND	ND	3.41	0.72
Eggs	0.10	ND	0.03	ND	0.44	0.01	0.02	ND	ND	0.59	0.49

Table 10.11 (Continued)

Food product	FLU	PYR	BAA*	CRY*	BBF*	BKF*	BAP*	DBA*	BPE	PAH[b]	CPAH[c]
Barbecued											
Beef	10.81	1.27	0.53	24.66	1.20	0.61	1.45	1.53	0.06	42.11	29.98
Pork	1.85	5.29	0.50	5.36	0.40	0.08	0.12	ND	0.01	13.60	6.46
Oils/fats											
Olive oil	0.17	ND	0.03	0.38	0.26	0.06	0.10	ND	ND	1.01	0.84
Butter	0.15	1.77	0.67	ND	0.02	0.03	0.02	0.02	ND	2.67	0.75
Chocolate	8.36	ND	0.51	ND	0.48	0.11	0.33	0.73	0.01	10.53	2.15
Beverages											
Wine	0.09	ND	ND	ND	0.04	0.01	0.01	0.06	ND	0.20	0.12
Beer	0.05	ND	0.11	ND	0.02	0.02	0.03	0.08	ND	0.31	0.26
Coffee	1.18	ND	0.10	ND	0.08	0.02	0.01	0.04	0.01	1.45	0.25
Pizza[d]	0.18	0.58	9.11	3.01	0.04	0.02	0.03	0.07	0.06	13.10	12.28

ND, not detected.
[a]Abbreviations used in this table are as shown in section 10.1.
[b]Sum of the nine PAHs.
[c]Sum of the six carcinogenic PAHs marked * above (see Table 10.1).
[d]Pizza cooked in a wood-burning oven.

Table 10.12 Average concentrations (ng/g product) of polycyclic aromatic hydrocarbons (PAHs[a]) in food products from animals reared in Kuwait (Husain et al., 1997)

Food product	PHE	ANT	FLU	PYR	BAA*	CRY*	BBF*	BKF*	BAP*	DBA*	IDP*	BPE	PAH[b]	CPAH[c]
Chicken														
Egg	18.53	29.73	1.19	5.55	4.46	5.57	3.49	4.53	7.49	4.75	8.73	1.25	95.27	39.02
Liver	47.43	100.00	10.50	14.05	3.63	6.36	6.00	2.00	3.25	ND	ND	ND	193.22	21.24
Sheep														
Liver	58.82	ND	15.91	15.91	3.00	9.73	7.25	3.00	1.78	ND	ND	ND	115.40	24.76
Kidney	40.45	ND	6.09	9.45	1.00	3.13	ND	1.9	1.00	ND	ND	3.00	66.02	7.03
Milk	10.03	ND	14.22	139.29	2.99	21.49	2.13	ND	1.60	ND	ND	ND	191.75	28.21
Goat														
Liver	32.14	ND	8.57	9.86	1.50	4.14	ND	2.14	1.33	2.00	2.00	ND	63.68	13.11
Kidney	38.71	ND	6.71	9.14	1.00	3.29	ND	1.86	ND	1.00	ND	ND	61.71	7.15
Cow														
Milk	3.05	0.50	3.39	35.52	2.42	8.64	3.12	ND	1.50	ND	ND	ND	58.14	15.68

ND, not detected.
[a] Abbreviations used in this table are shown as in section 10.1.
[b] Sum of the 12 PAHs measured.
[c] Sum of the seven carcinogenic PAHs marked * above (see Table 10.1).

Table 10.13 Concentrations (ng/g product) of polycyclic aromatic hydrocarbons (PAHs[a]) in baby foods, vegetable fats and oils, and fish oils

Product (samples)	NAPs	ANTs/PHEs	FLU	PYR	BAA*	CRY*	BBF/BKF*	BAP*	IDP*	BPE	PAH[b]	CPAH[c]
Baby foods (4)												
Ready-to-eat (2) mean	2.1	1.1	Tr	Tr	ND	Tr	Tr	ND	ND	ND	3.2	Tr
Dehydrated (2)[d] mean	18.3	31.0	0.6	0.5	0.2	0.5	0.7	0.2	0.2	0.2	52.4	1.8
Margarines, retail (6)												
Low	1.4	3.1	0.7	0.6	0.4	ND	0.2	ND	2.3	1.3		
High	20.0	23.6	2.5	1.9	2.0	3.6	5.4	1.6	5.3	3.5		
Mean	8.8	11.5	1.1	1.0	1.1	1.2	2.6	0.6	4.3	2.3	34.5	9.8
Vegetable oils, retail (12)												
Low	ND	2.3	ND	ND	0.2	0.5	ND	ND	3.1	0.9		
High	87.7	53.4	3.7	3.7	2.8	4.6	6.2	1.7	49.2	3.8		
Mean	34.7	16.8	2.2	1.4	1.2	2.2	2.5	0.5	16.0	2.3	79.8	22.4
Fish oils/supplements (7)												
Low	13.9	28.2	2.6	1.0	0.5	0.3	0.2	ND	ND	ND		
High	1204.0	362.4	44.9	21.5	6.4	20.5	20.0	2.8	8.5	4.2		
Mean	277.2	110.6	9.9	5.5	1.7	5.1	4.1	0.9	2.6	1.5	419.1	14.4

ND, not detected; Tr, less than 0.14 ng/g.

[a] Abbreviations used in this table are as shown in section 10.1.

[b] Sum of the naphthalenes, anthracenes, phenanthrenes (i.e. parent and branched compounds) and the nine other PAHs shown.

[c] Sum of the six carcinogenic PAHs marked * above (see Table 10.1).

[d] Note that the concentrations are given on a dry weight basis in this case.

et al. (1991) and confirm the importance of the edible oils as a source of PAHs in the diet and the importance of investigating the origin of the PAHs in the oils.

Analyses of 60 retail products of dietary supplements and health foods (unpublished data), for total PAHs from two- to six-rings, showed a very wide range of concentrations from 0.2 ng/g (garlic capsules) to nearly 8000 ng/g (mean 877 ng/g) in an evening primrose oil (EPO). Naphthalenes and phenanthrenes/anthracenes usually accounted for the major proportion of the PAHs, ranging from 28% to 99%; the proportion was 75% or more in 68% of samples. The sum of six carcinogenic PAHs ranged from undetected to 263 ng/g (the extremes of the range were found in different EPO products), but not all of these PAHs were detected in all products. Chrysene was found in most products; it was both the most abundant and found in the highest concentration (68 ng/g in an EPO). Benzo[a]pyrene was detected in only 23% of products; the maximum concentration found was 29.9 ng/g in an EPO. The wide variability in contamination found in these products is most likely to be due to the source material and the manufacturing process.

Oil seeds have low levels of PAHs at the point of harvest (Dennis *et al.*, 1991). Dust and soil raised during harvesting and transport to the processing unit probably contribute further contamination. Crude seed oils can occasionally show evidence of considerable contamination. For example, high levels of PAHs have been found in some crude rape seed oils (Sagredos *et al.*, 1988) and one was found to contain over 60 ng/g benzo[a]pyrene (Dennis *et al.*, 1991). Traditional smoke drying, as in the case of coconut oil, considerably enhances the PAH content (Sagredos *et al.*, 1988) of the crude oil and probably accounts for the high concentration of PAHs found in samples of dessicated coconut, 78.8 ng/g, to which fluoranthene and pyrene contributed about 70% (Dennis *et al.*, 1991).

The drying of rape seed showed no increase in any PAH concentrations when cold or electrically heated air was used (Dennis *et al.*, 1991). Combustion gas drying had no effect on the larger PAHs but resulted in average increases within the range 41–126% for fluoranthene, pyrene and chrysene that were not related to reductions in moisture content of the seeds. Control of combustion conditions during direct drying is clearly an important factor with respect to contamination with PAHs during drying. These PAHs and benz[a]anthracene were reduced by up to a factor of 5 when the crude oils were refined (Dennis *et al.*, 1991). Deodorisation during refining mainly removed these lighter PAHs, whereas active charcoal treatment was required to reduce the load of the five- and six-ring PAHs (Thomas, 1982; Larsson *et al.*, 1987; Sagredos *et al.*, 1988). Recently, Cejpek *et al.* (1998) investigated in detail the influence of stages of production on the PAH concentration of rapeseed oil. They confirmed that PAHs from soil and dust particulates contaminate rape seeds during harvesting, transport and storage and concluded that the PAHs were concentrated in the cuticular layer of the seeds, but prolonged silo-storage of 30 weeks or more before processing resulted in a reduction of about 50% in the total PAH concentration of the crude oil. Deodorisation followed by alkali-refining were the most effective steps in reducing PAH contamination of the refined oil; bleaching had only a small effect. As unrefined cold-pressed, native oils became more widely available, Speer et al. (1990) assessed the PAH content of some native oils, safflower, sunflower, maize germ, olive, sesame, linseed and wheat germ. Overall, the range was wide (0.7–124 ng/g), with wheat germ and olive oils contributing to the upper part of the range. Olive oils are not steam treated, a process that would decrease the light PAHs, and are subject only to cleaning by decantation, centrifugation and filtration. Pupin and Toledo (1996) studied a range of olive oils available in the Brazilian market and found benzo[a]pyrene in all oils up to 164 ng/g. An imported canned product, 'smoked oysters in oil', was reported by Speer et al. (1990) to have unacceptably high concentrations of benzo[a]pyrene in the oyster meat (10 to 12 ng/g) and in the oil (75.8 ng/g) compared with the unsmoked product (0.2 ng/g).

10.7 Human exposure to polycyclic aromatic hydrocarbons

Estimates have been made in a number of countries of human exposure to PAHs based on food consumption patterns and extrapolation from published analytical data on food commodities, or analyses of total diets, or analyses of food baskets, or analyses of the major constituents of the diet (Suess, 1976; McGill et al., 1982; Dennis et al., 1983; Vaessen et al., 1988; de Vos et al., 1990; Bartle et al., 1991; Hattemer-Frey and Travis, 1991; Butler et al., 1993; Lodovici et al., 1995). A number of the European studies were discussed in the previous section 10.6.2. In the United States, during a period of about 20 years, a number of estimates were made of daily dietary intake of benzo[a]pyrene. These ranged over three orders of magnitude: 2200 ng (Suess, 1976); 160–1600 ng (Santodonato et al., 1981); 50 ng (EPA, 1982); 30 ng from a total human environmental exposure study (Lioy et al., 1988); 2100 ng predicted by Hattemer-Frey and Travis (1991), and 140 ng estimated by Butler et al. (1993) from a human environmental exposure study, in which inhalation accounted for an additional 12 ng and drinking water 0.6 ng/day. These can be compared with 250 ng benzo[a]pyrene/person/day within a total dietary intake of PAHs (four and five rings) of 3700 ng/person/day in the

United Kingdom (Dennis *et al.*, 1983), and 1400 ng/person/day for 6 carcinogenic PAHs within a total dietary intake of PAHs (four and five rings) of 3000 ng/person/day in the Italian diet (Lodovici *et al.*, 1995). In the total diet study in the Netherlands, the intake of PAHs measured was 5000 to 17 000 ng/person/day, of which about 50% were carcinogenic PAHs.

Differences in dietary composition in different countries, differences in assessing the relative importance of dietary components, differences in the range of analytes measured, selection from published analytical data and differing methodology for estimation or prediction probably account for much of the apparent difference in the estimates. The studies show that dietary intake of the PAHs, and specific ones such as benzo[*a*]pyrene, is extremely variable depending on food habits, food source, preparation and cooking. The studies confirm that for the general population food is the major source of human exposure to PAHs, taking into account other sources such as drinking water, inhaled air (inside and outside the home) and cigarette smoking. Unusually high exposure in the workplace is most likely to be by inhalation or dermal contact and is not within the scope of this chapter. However, together with cigarette smoking, it is the source of most of the available direct information about carcinogenic risk to man from PAHs.

Among the general population, we can probably expect human dietary exposure to four- and five-ring PAHs to be within the range 1–25 µg/person/day, exposure to the 6 PAHs of concern to be within the range 0.1–12.5 µg/person/day, and exposure to benzo[*a*]pyrene to be within the range 0.01–2.5 µg/person/day. However, there is very little information about the extent to which benzo[*a*]pyrene in food is available and absorbed from the human gut, although this is thought to be strongly influenced by dietary components (Stavric and Klassen, 1994). Towards the upper end of the range of exposure, Hattemer-Frey and Travis (1991) estimated the health risk of dietary exposure to benzo[*a*]pyrene, assuming that 100% of the benzo[*a*]pyrene ingested is absorbed and that the cancer potency factor for oral exposure to benzo[*a*]pyrene is 11.5 $(mg/kg/day)^{-1}$ (EPA, 1986). They concluded, for the US population, that the excess lifetime cancer risk associated with a background exposure of 2.2 µg/person/day, 97% of which was estimated to come from the food chain, was a probability of 345 additional cancer cases per million persons, and that ingestion of this contaminant in food may pose a serious and unacceptable health risk. On the other hand, Butler *et al.* (1993) estimated that the total risk was of the order of 1×10^{-5} based on the average daily intake of 140 ng. They concluded that although this risk exceeded 1×10^{-6}, considered to be a *de minimis* level, it was ten times lower than the 'background' cancer risk calculated for other environmental contaminants such as benzene in indoor air, chloroform in tap water, and dioxins/furans in food (Travis and Hester, 1990).

As individuals we are all exposed to a vast range of organic xenobiotic chemicals through our occupation, environment and diet. The impact of PAHs relative to all the other chemicals is unclear. DNA adduct formation is a pertinent demonstration of exposure to PAHs from both diet and occupation (Rothman et al., 1993) but, although there is also a direct relationship between the extent of cigarette smoking and the number of DNA–benzo[a]pyrene adducts, the relationship between DNA–benzo[a]pyrene adducts and the occurrence of lung cancer is less well defined (Walker et al., 1996). The more potent carcinogenic PAHs have, however, been shown to react more extensively with adenine residues in DNA (Dipple and Bigger, 1991). Sjögren et al. (1996) concluded that bacterial mutagenicity best reflects the cancer initiation potency of PAHs, whereas the aromatic hydrocarbon receptor affinity reflects the promotive effect of some PAHs at the high doses applied in rodent carcinogenicity tests. They further concluded that initiation of carcinogenesis was provoked by reactive metabolites, while promotion was provoked by the parent PAHs.

PAHs, and many pesticides, have very short biological half-lives in most species but may, nevertheless, have long-term effects. It must, however, be concluded that reduction of human exposure to PAHs can only be beneficial and this can be achieved, in part, by reducing dietary exposure to PAHs through changes in cooking practices and reducing the overall concentration of these environmental contaminants in our food.

References

Adlard, E.R. and Whitham, B.T. (1958) Applications of high temperature gas-liquid chromatography in the petroleum industry, in *Gas Chromatography 1958* (ed. D.H. Desty), Butterworths Scientific, London, pp 351-368.

Ariese, F., Burgers, I., Oudhoff, K., Rutten, T., Stroomberg, G. and Vethaak, D. (1997) *Comparison of Analytical Approaches for PAH Metabolites in Fish Bile Samples for Marine and Estuarine Monitoring*, **R-97/9**, Institute for Environmental Studies, Vrije Universiteit, Amsterdam, The Netherlands.

Barnabas, I.J., Dean, J.R., Fowlis, I.A. and Owen, S.P. (1995) Extraction of polycyclic aromatic hydrocabons from highly contaminated soils using microwave energy. *Analyst*, **120** 1897-1904.

Barrowman, J.A., Rahman, A., Lindstrom, M.B. and Borgstrom. B. (1989) Intestinal absorption and metabolism of hydrocarbons. *Progress in Lipid Research*, **28** 189-203.

Bartle, K.D. (1991) Analysis and occurrence of polycyclic aromatic hydrocarbons in food, in *Food Contaminants—Sources and Surveillance* (eds C.S. Creaser and R. Purchase), Royal Society of Chemistry, Cambridge, pp 41-60.

Bastić, M., Bastić, Lj., Jovanović, J.A. and Spiteller, G. (1978) Hydrocarbons and other weakly polar unsaponifiables in some vegetable oils. *Journal of the American Oil Chemists' Society*, **55** 886-891.

Beltrán, J.L., Ferrer, R. and Guiteras, J. (1996) Determination of polycyclic aromatic hydrocarbons by HPLC with spectrofluorimetric detection and wavelength programming. *Journal of Liquid Chromatography and Related Technology*, **19** 477-488.

Bergström, J.G.T., Eklund, G. and Trzcinski, K. (1982) Characterization and comparison of organic emissions from coal, oil and wood fired boilers, in *Polynuclear Aromatic Hydrocarbons: Physical and Biological Chemistry, Sixth International Symposium* (eds M. Cooke, A.J. Dennis and G.L. Fisher), Battelle Press, Columbus, OH, 109-120.

Bianchi, A.P. and Varney, M.S. (1993) Sampling and analysis of volatile organic compounds in estuarine air by gas-chromatography and mass-spectrometry. *Journal of Chromatography*, **643** 11-23.

Bingham, E. and Nord, P.J. (1977) Cocarcinogenic effects of *n*-alkanes and ultraviolet light on mice. *Journal of the National Cancer Institute*, **58** 1099-1101.

Bjørseth, A. and Ramdahl, T. (eds) (1985) *Handbook of Polycyclic Hydrocarbons*, Marcel Dekker, New York.

Blomquist, G.J. and Jackson, L.L. (1979) Chemistry and biochemistry of insect waxes. *Progress in Lipid Research*, **17** 319-345.

Blumer, M. (1967) Hydrocarbons in the digestive tract and liver of a basking shark. *Science*, **156** 390-391.

Blumer, M., Robertson, J.C., Gordon, J.E. and Sass, J. (1969) Phytol-derived C19 di- and triolefinic hydrocarbons in marine zooplankton and fishes. *Biochemistry*, **8** 4067-4074.

Boudries, H., Toupance, G. and Dutot, D.L. (1994) Seasonal variations of atmospheric non-methane hydrocarbons on the western coast of Brittany, France. *Atmosphere and Environment*, **28** 1095-1112.

BP (1998) *BP 1997 Statistical Review of World Energy*, June 1998. The British Petroleum Company plc 1998, 40 pp.

Brakstad, F. and Grahl-Nielsen, O. (1988) Identification of weathered oils. *Marine Pollution Bulletin*, **19** (7) 319-324.

Buckley, T.J. and Lioy, P.J. (1992) An examination of the time course from human dietary exposure to polycyclic aromatic hydrocarbons to urinary elimination of 1-hydroxypyrene. *British Journal of Industrial Medicine*, **49** 113-124.

Butler, J.P., Post, G.B., Lioy, P.J., Walman, J.M. and Greenberg, A. (1993) Assessment of carcinogenic risk from personal exposure to benzo(a)pyrene in the total human environmental exposure study (THEES). *Journal of the Air and Waste Management Association*, **43** 970-977.

Carmichael, P.L., Stone, E.M., Grover, P.L., Gusterson, B.A. and Phillips, D.H. (1996) Metabolic activation and DNA binding of food mutagens and other environmental carcinogens in human mammary epithelial cell. *Carcinogenesis*, **17** 1769-1772.

Castle, L., Kelly, M. and Gilbert, J. (1991) Migration of mineral hydrocarbons into foods. 1. Polystyrene containers for hot and cold beverages. *Food Additives and Contaminants*, **8** (6) 693-700.

Cavalieri, E.L. and Rogan, E.G. (1992) The approach to understanding aromatic hydrocarbon carcinogenesis. The central role of radical cations in metabolic activation. *Pharmacology and Therapeutics*, **55** 183-199.

Cejpek, K., Hajšlová, J., Jehličková, Z. and Merhaut, J. (1995) Simplified extraction and cleanup procedure for the determination of PAHs in fatty and protein-rich matrices. *International Journal of Environmental and Analytical Chemistry*, **61** 65-80.

Cejpek, J., Hajšlová, J., Kocourek, V., Tomaniová, M. and Čmolík, J. (1998) Changes in PAH levels during production of rapeseed oil. *Food Additives and Contaminants*, **15** (5) 563-574.

Chen, B.H. (1997) Analysis, formation and inhibition of polycyclic aromatic hydrocarbons in foods: an overview. *Journal of Food and Drug Analysis*, **5** 25-42.

Chipman, J.K., Frost, G.S., Hirom, P.C. and Millburn, P. (1981) Biliary excretion, systemic availability and reactivity of metabolites following intraportal infusion of [^{3}H]benzo[*a*] pyrene in the rat. *Carcinogenesis*, **8** 741-745.

Chiu, C.P., Lin, Y.S. and Chen, B.H. (1997) Comparison of GC-MS and HPLC for overcoming matrix interferences in the analysis of PAHs in smoked food. *Chromatographia*, **44** 497-504.

Comerford, J.W. (1992) Heavy metal and hydrocarbon residues in tissue and blood of beef steers bedded on waste newspapers. *Bulletin of Environmental Contamination and Toxicology*, **49** 18-22.

Compaan, H., Smit, T., Law, R., Abarnou, A., Evers, E., Hünnerfuss, H., Hockstra, E., Klungsoyr, J., Holsbeek, L. and Joiris, C. (1992) Organic compounds, in *Background Concentrations of Natural Compounds in Rivers, Seawater, Atmosphere and Mussels* (ed. R.W.P.M. Laane), Report DGW-92.033, Tidal Waters Division, Rijkswaterstaat, Netherlands.

CRC Handbook of Chemistry and Physics (1980) (ed. R.C. Weast), CRC Press, Boca Raton, FL.

Cruickshank, P., Anderson, D.A. and Moffat, C.F. (1997) The concentration and composition of *n*-alkanes in crude edible oils, in *Pacific Oils 2000, The Proceedings of the International Conference on Plant Oils and Marine Lipids* (ed. C.J. O'Connor), Auckland, New Zealand.

Dale, C.M. and Garner, R.C. (1996) Measurement of DNA adducts in humans after complex mixture exposure. *Food and Chemical Toxicology*, **34** 905-919.

Darrall, K.G., Figgins, J.A., Brown, R.D. and Phillips, G.F. (1998) Determination of be-nzene and associated compounds in mainstream cigarette smoke. *Analyst*, **123** 1095-1101.

Davis, H.K., Moffat, C.F. and Shepherd, N.J. (1997) *A Study of Tainting of Fish by Chemically Dispersed Petroleum Products*, Report FD/62, A research contract commissioned and funded by the Marine Pollution Control Unit, Maritime and Coastguard Agency, Southampton, UK.

Dennis, M.J., Massey, R.C., McWeeny, D.J., Knowles, M.E. and Watson, D. (1983) Analysis of polycyclic aromatic hydrocarbons in UK total diets. *Food and Chemical Toxicology*, **21** 569-574.

Dennis, M.J., Massey, R.C., Cripps, G., Venn, I., Howarth, N. and Lee, G. (1991) Factors affecting the polycyclic aromatic hydrocarbon content of cereals, fats and other food products. *Food Additives and Contaminants*, **8** (4) 517-530.

de Vos, R.H., van Dokkum, W., Schouten, A. and de Jong-Berkhout, P. (1990) Polycyclic aromatic hydrocarbons in Dutch total diet samples (1984–1986). *Food and Chemical Toxicology*, **28** 263-268.

Dewulf, J. and Van Langenhove, H. (1995) Simultaneous determination of C_1 and C_2-halocarbons and monocyclic aromatic hydrocarbons in marine water samples at ng/L concentration levels. *International Journal of Environmental and Analytical Chemistry*, **61** 35-46.

Dewulf, J.P., Van Langenhove, H.R. and Van Der Auwera, L.F. (1998) Air/water exchange dynamics of 13 volatile chlorinated C1- and C2-hydrocarbons and monocyclic aromatic hydrocarbons in the southern North Sea and the Scheldt Estuary. *Environmental Science and Technology*, **32** 903-911.

Di Muccio, A., Boniforti, L., Palomba, A., Bernardini, M.P. and Delise, M. (1979) Idrocarburi saturi negli alimenti. Metodo d'analisi e valori riscontrati in alcuni alimenti per uso umano e in campioni da organismi unicellulari. *Annuario Istituto Superiore di Sanità*, **15** 525-540.

Dipple, A. and Bigger, C.A.H. (1991) Mechanism of action of food-associated polycyclic aromatic hydrocarbon carcinogenesis. *Mutation Research*, **259** 263-276.

Dove, H. and Mayes, R.W. (1991) The use of plant wax alkanes as marker substances in studies of the nutrition of herbivores: a review. *Australian Journal of Agricultural Research*, **42** 913-952.

Edwards, N.T. (1983) Polycyclic aromatic hydrocarbons (PAHs) in the terrestrial environment—a review. *Journal of Environmental Quality*, **12** 427-441.

Eisert, R. and Levsen, K. (1996) Solid-phase microextraction coupled to gas chromatography: a new method for the analysis of organics in water. *Journal of Chromatography A*, **733** 143-157.

Eisler, R. (1987) *Polycyclic Aromatic Hydrocarbon Hazards to Fish, Wildlife and Invertebrates: A Synoptic Review*, US Fish and Wildlife Services Biological Report 85(1.14), Patuxent Wildlife Research Center, Laurel, MD.

EPA (1982) *Exposure and Risk Assessment for Benzo(a)pyrene and Other Polycyclic Aromatic Hydrocarbons*, US Environmental Protection Agency, Volume 4, EPA/440/4-85/020, Office of Water Regulations and Standards, Washington, DC.

EPA (1986) *Superfund Public Health Evaluation Manual*, US Environmental Protection Agency, EPA/540/1-86/060, Office of Emergency and Remedial Response, Washington, DC.

EPA (1993) *Provisional Guidance for Quantitative Risk Assessment of Polycyclic Aromatic Hydrocarbons*, US Environmental Protection Agency, EPA/600/R-93/089, Office of Environmental Criteria and Assessment, Cincinnati, OH.

EPA (1995) *Guidance for Assessing Chemical Contaminant Data for Use in Fish Advisories Volume 1*, US Environmental Protection Agency, EPA/823/R-95/007, Office of Water, Washington, DC.

Fazio, T. and Howard, J.W. (1983) Polycyclic aromatic hydrocarbons in food, in *Handbook of Polycyclic Aromatic Hydrocarbons* (ed. A. Bjørseth), Marcel Dekker, New York, pp 461-505.

Fernandez, M.J., Garcia, C., Garcia-Villanova, R.J. and Gomez, J.A. (1996) Evaluation of liquid–liquid extraction and liquid–solid extraction with a new sorbent for the determination of polycyclic aromatic hydrocarbons in raw and finished drinking waters. *Journal of Agricultural and Food Chemistry*, **44** 1785-1789.

Ferrario, J.B., Lawler, G.C., DeLeon, I.R. and Laseter, J.L. (1985) Volatile organic pollutants in biota and sediments of Lake Pontchartrain. *Bulletin of Environmental Contamination and Toxicology*, **34** 246-255.

Garcia Falcon, M.S., Gonzalez Amigo, S., Lage Yusty, M.A., Lopez de Alda Villaizan, M.J. and Simal Lozano, J. (1996) Determination of benzo[a]pyrene in lipid-soluble liquid smoke (LSLS) by HPLC-FL. *Food Additives and Contaminants*, **13** (7) 863–870.

Gilbert, J. (1994) The fate of environmental contaminants in the food chain. *The Science of the Total Environment*, **143** 103-111.

Gilbert, J. and Shepherd, M.J. (1981) Headspace gas chromatography for the analysis of vinyl chloride and other monomers in plastic packaging and in foods. *Journal of the Association of Public Analysts*, **19** 39-49.

Gilbert, J. and Startin, J.R. (1983) A survey of styrene monomer levels in foods and plastic packaging by coupled mass spectrometry automatic headspace gas chromatogaphy. *Journal of the Science of Food and Agriculture*, **34** 647-652.

Giordano, M., Zale, R., Ruffle, B., Hawkins, E. and Anderson, P. (1994) Review of mathematical models for health risk assessment: V. Chemical concentrations in the food chain. *Environmental Software*, **9** 115-131.

Glegg, G.A. and Rowland, S.J. (1996) The *Braer* oil spill-Hydrocarbon concentrations in intertidal organisms. *Marine Pollution Bulletin*, **32** 486-492.

Gomaa, E.A., Gray, J.I., Rabie, S., Lopez-Bote, C. and Booren, A.M. (1993) Polycyclic aromatic hydrocarbons in smoked food products and commercial liquid smoke flavourings. *Food Additives and Contaminants*, **10** (5) 503-521.

González-Barros, S.T.C., Alvarez-Piñeiro, M.E., Smal-Lozano, J. and Lage-Yusty, M.A. (1997) Levels of aliphatic hydrocarbons in viscera of wolves (*Canis lupus* L) in Galicia (N.W. Spain). *Bulletin of Environmental Contamination and Toxicology*, **59** 534-547.

Gough, M.A. and Rowland, S.J. (1990) Characterization of unresolved complex mixtures of hydrocarbons in petroleum. *Nature*, **344** 648-650.

Gough, M.A., Rhead, M.M. and Rowland, S.J. (1992) Biodegradation studies of unresolved complex mixtures of hydrocarbons: model UCM hydrocarbons and the aliphatic UCM. *Organic Geochemistry*, **18** 17-22.

Grahl-Nielsen, O. and Lygre, T. (1990) Identification of samples of oil related to two spills. *Marine Pollution Bulletin*, **21** (4) 176-183.

Grimmer, G. (ed.) (1983) *Environmental Carcinogens: Polycyclic Aromatic Hydrocarbons*, CRC Press, Boca Raton, FL.

Grimmer, G. and Böhnke, H. (1975) Polycyclic aromatic hydrocabon profile analysis of high-protein foods, oils, and fats by gas chromatography. *Journal of the Association of Official Analytical Chemists*, **58** 725-733.

Grob, K. and Siegrist, C. (1994) Determination of mineral oil on jute bags by 20–50 µl splitless injection onto a 3 m capillary column. *Journal of High Resolution Chromatography*, **17** 674-675.

Grob, K., Lanfranchi, J.E. and Artho, A. (1991a) Determination of food contamination by mineral oil from jute sacks using coupled LC-GC. *Journal of the Association of Official Analytical Chemists*, **74** 506-512.

Grob, K., Artho, A., Biedermann, M. and Egli, J. (1991b) Food contamination by hydrocarbons from lubricating oils and release agents: determination by coupled LC-GC. *Food Additives and Contaminants*, **8** 437-446.

Guillén, M.D. (1994) Polycyclic aromatic compounds: extraction and determination in foods. *Food Additives and Contaminants*, **11** 669-684.

Güsten, H., Horvatiĉ, D. and Sabljiĉ, A. (1991) Modelling *n*-octanol/water partition coefficients by molecular topology: polycyclic aromatic hydrocarbons and their alkyl derivatives. *Chemosphere*, **23** 199-213.

Haesook, K.-K. (1990) Volatiles in packaging materials. *Critical Reviews in Food Science and Nutrition*, **29** 255-271.

Halsall, C.J., Barrie, L.A., Fellin, D.C., Muir, D.C.G., Billeck, B.N., Lockhart, L., Rovinsky, F.Y., Kononov, E. Y. and Patsukhov, B. (1997) Spatial and temporal variations of polycyclic hydrocarbons in the arctic atmosphere. *Environmental Science and Technology*, **31** 3593-3599.

Harrak, R. El., Calull, M., Marcé, R.M. and Borrull, F. (1996) Determination of polycyclic aromatic hydrocarbons in water by solid-phase extraction membranes. *International Journal of Environmental and Analytical Chemistry*, **64** 47-57.

Hattemer-Frey, H.H. and Travis, C.C. (1991) Benzo-a-pyrene: environmental partitioning and human exposure. *Toxicology and Industrial Health*, **7** (3) 141-157.

Heaton, D.M., Bartle, K.D., Clifford, A.A., Myers, P. and King, B.W. (1994) Rapid separation of polycyclic aromatic hydrocarbons by packed column supercritical fluid chromatography. *Chromatographia*, **39** 607-611.

Hellou, J., Payne, J.F., Upshall, C., Fancey, L.L. and Hamilton, C. (1994) Bioaccumulation of aromatic hydrocarbons from sediments: a dose–response study with flounder (*Pseudopleuronectes americanus*). *Archives of Environmental Contamination and Toxicology*, **27** 477-485.

Heras, H., Ackman, R.G. and MacPherson, E.J. (1992) Tainting of Atlantic salmon (*Salmo salar*) by petroleum hydrocarbons during a short-term exposure. *Marine Pollution Bulletin*, **24** 310-315.

Hernández, J.E., Machado, L.T, Corbella, R., Rodríguez, M.A. and Montelongo, F.G. (1995) *n*-Alkanes and polynuclear aromatic hydrocarbons in fresh-frozen and precooked-frozen mussels. *Bulletin of Environmental Contamination and Toxicology*, **55** 461-468.

Hollifield, H.C., Breder, C.V., Dennison, J.L., Roach, J.A.G. and Adams, W.S. (1980) Container derived contamination of maple syrup with methyl methacrylate, toluene, and styrene as determined by headspace gas–liquid chromatography. *Journal of the Association of Official Analytical Chemists*, **63** 173-177.

Husain, A., Naeemi, E., Dashti, B., Al-Omirah, H. and Al-Zenki, S. (1997) Polycyclic aromatic hydrocarbons in food products originating from locally reared animals in Kuwait. *Food Additives and Contaminants*, **14** (3) 295-299.

IARC (1982) *Monographs on the Evaluation of the Carcinogenic Risk of Chemicals to Humans*, Vol. 29, International Agency for Research on Cancer, Lyon, pp 93-148.

IRIS (1992) *Integrated Risk Information System*, US Environmental Protection Agency, Duluth, MN.

Janda, V., Fanta, J. and Vejrosta, J. (1996) Factors affecting SFE of PAHs from water samples. *Journal of High Resolution Chromatography*, **19** 588-590.

Jansen, E.H.J.M., Schenk, E., Engelsman, G.D. and van de Werken, G. (1996) Route-specific urinary biomarkers in the risk assessment of PAH exposure. *Polycyclic Aromatic Compounds*, **11** 185-192.

Järvenpää, E., Huopalahti, R. and Tapanainen, P. (1996) Use of supercritical fluid extraction-high performance liquid chromatography in the determination of polynuclear aromatic hydrocarbons from smoked and broiled fish. *Journal of Liquid Chromatography and Related Technology*, **19** 1473-1482.

Jickells, S.M., Crews, C., Castle, L. and Gilbert, J. (1990) Headspace analysis of benzene in food contact materials and its migration into foods from plastics cookware. *Food Additives and Contaminants*, **7** 197-205.

Jickells, S.M., Philo, M.R., Gilbert, J. and Castle, L. (1993). Gas chromatographic/mass spectrometric determination of benzene in nonstick cookware and microwave susceptors and its migration into foods on cooking. *Journal of the Association of Official Analytical Chemists International*, **76** 760-764.

Jones, D.M., Rowland, S.J. and Douglas, A.G. (1986) Steranes as indicators of petroleum-like hydrocarbons in marine surface sediments. *Marine Pollution Bulletin*, **17** 24-27.

Karl, H. (1997) Influence of the smoking technology on the quality of smoked fish regarding undesirable compounds, in *Developments in Food Science 38. Seafood from Producer to Consumer, Integrated Approach to Quality* (eds J.B. Luten, T. Börresen, J. Oehlenschläger), Elsevier Science, Amsterdam, pp 633-639.

Kavalam, J.P. and Nawar, W.W. (1969) Effects of ionizing radiation on some vegetable fats. *Journal of the American Oil Chemists' Society*, **46** 387-390.

Kolarovič, L. and Traitler, H. (1982) Determination of polycyclic aromatic hydrocarbons in vegetable oils by caffeine complexation and glass capillary gas chromatography. *Journal of Chromatography*, **237** 263-272.

Krahn, M.M., Ylitalo, G.M., Buzitis, J., Chan, S.-L. and Varanasi, U. (1993a) Rapid high-performance liquid chromatographic methods that screen for aromatic compounds in environmental samples. *Journal of Chromatography*, **642** 15-32.

Krahn, M.M., Ylitalo, G.M., Buzitis, J., Bolton, J.L., Wigren, C.A., Chan, S.-L. and Varanasi, U. (1993b) Analysis for petroleum-related contaminants in marine fish and sediments following the Gulf oil spill. *Marine Pollution Bulletin*, **27** 285-292.

Krone, C.A., Stein, J.E. and Varanasi, U. (1992) Estimation of levels of metabolites of aromatic hydrocarbons in fish tissues by HPLC/fluorescence analysis. *Chemosphere*, **24** 497-510.

Laher, J.M. and Barrowman, J.A. (1983) Polycyclic hydrocarbon and polychlorinated biphenyl solubilization in aqueous solutions of mixed micelles. *Lipids*, **18** 216-222.

Larsson, B. and Sahlberg, G. (1982) Polycyclic aromatic hydrocarbons in lettuce. Influence of a highway and an aluminium smelter, in *Polynuclear Aromatic Hydrocarbons: Physical and Biological Chemistry, Sixth International Symposium* (eds M. Cooke, A.J. Dennis and G.L. Fisher), Battelle Press, Columbus, OH, pp 417-426.

Larsson, B.K. (1982) Polycyclic aromatic hydrocarbons in smoked fish. *Zeitschrift für Lebensmitteluntersuchung und Forschung*, **174** 101-106.

Larsson, B.K., Eriksson, A.T. and Cervenka, M. (1987) Polycyclic aromatic hydrocarbons in crude and deodorised vegetable oils. *Journal of the American Oil Chemists' Society*, **64** 365-370.

Law, R.J. and Klungsøyr, J. (1996) The 1994 QUASIMEME Laboratory-performance studies: polycyclic aromatic hydrocarbons (PAH) in standard solutions. *Marine Pollution Bulletin*, **32** 667-673.

Law, R.J., Dawes, V.J., Woodhead, R.J. and Matthiessen, P. (1997) Polycylic aromatic hydrocarbons (PAH) in seawater around England and Wales. *Marine Pollution Bulletin*, **34** 306-322.

Lawrence, J.F. (1992) Determination of environmental contaminants in foods by liquid chromatography—an overview. *International Journal of Environmental and Analytical Chemistry*, **49** 15-29.

Lebo, J.A., Zajicek, J.L., Schwartz, T.R., Smith, L.M. and Beasley, M.P. (1991) Determination of monocyclic and polycyclic aromatic hydrocarbons in fish tissue. *Journal of the Association of Official Analytical Chemists*, **74** 538-544.

Lee, M.L., Vassilaros, D.L., White, C.M. and Novotny, M. (1979) Retention indices for programmed-temperature capillary-column gas chromatography of polycyclic aromatic hydrocarbons. *Analytical Chemistry*, **51** 768-773.

Lee, M.L., Novotny, M.V. and Bartle, K.D. (1981) in *Analytical Chemistry of Polycyclic Aromatic Compounds*, Academic Press, New York, pp 17-49.

Levy, R. (1976) Detection of mineral oil and paraffins in foods. *Die Nahrung*, **20** 773-775.

Lijinsky, W. (1991) The formation and occurrence of polynuclear aromatic hydrocarbons associated with food. *Mutation Research*, **259** 251-261.

Linssen, J.P.H., Janssens, A.L.G.M., Reitsma, H.C.E., Bredie, W.L.P. and Roozen, J.P. (1993) Taste recognition threshold concentrations of styrene in oil-in-water emulsions and yoghurts. *Journal of the Science of Food and Agriculture*, **61** 457-462.

Lioy, P.L.; Harkov, R., Waldman, J.M., Pietarinen, C. and Greenberg, A. (1988) The total human environmental exposure study (THEES) to benzo(a)pyrene: comparison of the inhalation and food pathways. *Archives of Environmental Health*, **43** 304-312.

Lodovici, M., Dolara, P., Casalini, C., Ciappellano, S. and Testolin, G. (1995) Polycyclic aromatic hydrocarbon contamination in the Italian diet. *Food Additives and Contaminants*, **12** (5) 703-713.

Luther, W., Win, T., Vaessen, H.A.M.G., van de Kamp, C.G., Jekel, A.A., Jacob, J. and Boenke, A. (1997) *The Certification of the Mass Fractions of Pyrene, Chrysene, Benzo[k]fluoranthene, Benzo[a]pyrene, Benzo[ghi]perylene and Indeno[1,2,3-cd]pyrene in Two Coconut Oil Reference Materials* (CRM 458 & CRM 459). European Commission, BCR Information, Reference Material, Report EUR 17545 EN.

Mackay, D., Shiu, W.Y. and Ma, K.C. (1992) *Illustrated Handbook of Physico-Chemical Properties and Environmental Fate for Organic Chemicals*, Vol. II, *Polynuclear Aromatic Hydrocarbons, Polychlorinated Dioxins and Dibenzofurans*. Lewis Publishers, Boca Raton, FL.

Mackie, P.R., Whittle, K.J. and Hardy, R. (1974) Hydrocarbons in the marine environment. 1. *n*-Alkanes in the Firth of Clyde. *Estuarine and Coastal Marine Science*, **2** 359-374.

Mackie, P.R., Hardy, R. Whittle, K.J., Bruce, C. and McGill, A.S. (1980) The tissue hydrocarbon burden of mussels from various sites around the Scottish coast, in *Polynuclear Aromatic Hydrocarbons: Chemistry and Biological Effects: 4th International Symposium* (eds A. Bjørseth and A.J. Dennis), Batelle Press, OH, pp 379-393.

MAFF (1996) Hydrocarbons in foods from shops in petrol stations and stalls or shops in busy roads. *MAFF Food Surveillance Information Sheet*, 98.

Massie, L.C., Ward, A.P., Davies, J.M. and Mackie, P.R. (1985) The effects of oil exploration and production in the northern North Sea: Part 1—The levels of hydrocarbons in water and sediments in selected areas, 1978–1981. *Marine Environmental Research*, **15** 165-213.

May, W.E., Wasik, S.P. and Freeman, D.H. (1978) Determination of the solubility behaviour of some polycyclic aromatic hydrocarbons in water. *Analytical Chemistry*, **50** 997-1000.

Mayes, R.W., Lamb, C.S. and Colgrove, P.M. (1986) The use of dosed and herbage n-alkanes as markers for the determination of herbage intake. *Journal of Agricultural Science*, **107** 161-170.

Mayes, R.W., Beresford, N.A., Lamb, C.S., Barnett, C.L., Howard, B.J., Jones, B.-E.V., Eriksson, O., Hove, K., Pederson, Ø. and Staines, B.W. (1994) Novel approaches to the estimation of intake and bioavailability of radiocaesium in ruminants grazing forested areas. *The Science of the Total Environment*, **157** 289-300.

Mazet, J.A.K., Gardner, I.A., Jessup, D.A. and Rittenburg, J.H. (1997) Field assay for the detection of petroleum products on wildlife. *Bulletin of Environmental Contamination and Toxicology*, **59** 513-519.

McGill, A.S., Mackie, P.R., Parsons, E., Bruce, C. and Hardy, R. (1982) The polynuclear aromatic hydrocarbon content of smoked foods in the United Kingdom, in *Polynuclear Aromatic Hydrocarbons: Physical and Biological Chemistry, Sixth International Symposium* (eds M. Cooke, A.J. Dennis and G.L. Fisher), Battelle Press, Columbus, OH, 491-499.

McGill, A.S., Mackie, P.R., Howgate, P. and McHenery, J.G. (1987) The flavour and chemical assessment of dabs (*Limanda limanda*) caught in the vicinity of the Beatrice oil platform. *Marine Pollution Bulletin*, **18** 186-189.

McGill, A.S., Moffat, C.F., Mackie, P.R. and Cruickshank, P. (1993) The composition and concentration of n-alkanes in retail samples of edible oils. *Journal of the Science of Food and Agriculture*, **61** 357-362.

Meador, J.P., Stein, J.E., Reichert, W.L. and Varanasi, U. (1995) Bioaccumulation of polycyclic aromatic hydrocarbons by marine organisms. *Review of Environmental Contamination and Toxicology*, **143** 79-165.

Mennie, D., Moffat, C.F. and McGill, A.S. (1994) Identification of sterene compounds produced during the processing of edible oils. *Journal of High Resolution Chromatography*, **17** 831-838.

Milne, J.A., Mayes, R.W., Smith, A. and Sinclair, A.H. (1997) The effects of the *Braer* oil on crops and sheep, in *The Impact of an Oil Spill in Turbulent Waters: The Braer* (eds J.M. Davies and G. Topping), The Stationery Office, Edinburgh, pp 63-72.

Moffat, C.F. and Cruickshank, P. (1997) The effects of processing on the concentration and composition of n-alkanes in edible oils, in *Pacific Oils 2000, The Proceedings of the International Conference on Plant Oils and Marine Lipids* (ed. C.J. O'Connor), Auckland, New Zealand.

Moffat, C.F., Cruickshank, P., Brown, N.A., Mennie, D., Anderson, D. and McGill, A.S. (1995) *The Concentration and Composition of n-Alkanes in Edible Oils*. A report to the UK Ministry of Agriculture Fisheries and Food, 17 Smith Square, London, UK.

Moffat, C.F., McIntosh, A.D., Webster, L., Shepherd, N.J., Dalgarno, E.J., Brown, N.A. and Moore, D.C. (1998) Determination and environmental assessment of hydrocarbons in fish, shellfish and sediments following an oil spill at the Captain Field. *Fisheries Research Services Report*, **9/98**.

Moore, J.W. and Ramamoorthy, S. (1984) *Organic Chemicals in Natural Waters*, Springer-Verlag, New York.

Moret, S., Grob, K. and Conte, L.S. (1996) On-line high-performance liquid chromatography–solvent evaporation–high-performance liquid chromatography–capillary gas chromatography–flame ionisation detection for the analysis of mineral oil polyaromatic hydrocarbons in fatty foods. *Journal of Chromatography A*, **750** 361-368.

Morozzi, G., Gambelunghe, C. and Manenti, R. (1985) Metodica analitica per la determinazione del benzo(a)pirene e di altri idrocarburi aromatici policiclici in alimenti affumicati o cotti. *La Rivista della Società Italiana di Scienza dell'Alimentazione*, **14** 351-356.

Mulcahey, L.J., Rankin, C.L. and McNally, M.E.P. (1994) Environmental applications of supercritical fluid chromatography. *Advances in Chromatography*, **34** 251-308.

Murray, A.P., Richardson, B.J. and Gibbs, C.F. (1991) Bioconcentration factors for petroleum hydrocarbons, PAHs, LABs and biogenic hydrocarbons in the blue mussel. *Marine Pollution Bulletin*, **22** 595-603.

Naes, K., Bakke, T. and Konieczny, R. (1995) Mobilization of PAH from polluted seabed and uptake in the blue mussel (*Mytilus edulis* L.). *Marine and Freshwater Research*, **46** 275-285.

Nam, K. and Alexander, M. (1998) Role of nanoporosity and hydrophobicity in sequestration and bioavailability: tests with model solids. *Environmental Science and Technology*, **32** 71-74.

Nawar, W.W. (1986) Volatiles from food irradiation. *Food Reviews International*, **2** 45-78.

Nevenzel, J.C. (1989) Biogenic hydrocarbons of marine organisms, in *Marine Biogenic Lipids, Fats and oils* (ed. R.G. Ackman), CRC Press, Boca Raton, FL, pp 3-71.

Newton, J.M., Rothman, B.S. and Walker, F.A. (1991) Separation and determination of diesel contaminants in various fish products by capillary gas chromatography. *Journal of the Association of Official Analytical Chemists*, **74** 986-990.

Nichol, J. and Jickells, S. (1993) Mineral hydrocarbons in chocolate and drinking chocolate. *Food Science Laboratory Report*, **93/26**.

Nilsson, T., Pelusio, F., Montanarella, L., Larsen, B., Facchetti, S., and Madsen, J.Ø. (1995) An evaluation of solid-phase microextraction for analysis of volatile organic compounds in drinking water. *Journal of High Resolution Chromatography*, **18** 617-624.

NRCC (1983) *Polycyclic Aromatic Hydrocarbons in the Aquatic Environment: Formation, Sources, Fate and Effects on Aquatic Biota*, NRC Associate Committee on Scientific Criteria for Environmental Quality, National Research Council of Canada, NRCC Publication No. 18981, Ottawa.

Oudot, J., Merlin, F.X. and Pinvidic, P. (1998) Weathering rates of oil components in a bioremediation experiment in estuarine sediments. *Marine Environmental Research*, **45** (2) 113-125.

Parker, J.G., Howgate, P., Mackie, P.R. and McGill, A.S. (1990) Flavour and hydrocarbon assessment of fish from gas fields in the southern North Sea. *Oil and Chemical Pollution*, **6** 263-277.

Perrin, J.L., Poirot, N., Liska, P., Thienpont, A. and Felix, G. (1993) Trace enrichment and HPLC analysis of PAHs in edible oils and fat products, using liquid chromatography on electron acceptor stationary phases in connection with reverse phase and fluorescence detection. *Fat Science and Technology*, **95** 46-51.

Pfau, W. (1997) DNA adducts in marine and freshwater fish as biomarkers of environmental contamination. *Biomarkers*, **2** 145-151.

Philp, R.P. and Xavier de las Heras, F. (1992) Fossil fuels. *Journal of Chromatography B*, **51** 447-473.

Pupin, A.M. and Toledo, M.C.F. (1996) Benzo[a]pyrene in olive oils on the Brazilian market. *Food Chemistry*, **55** 185-188.

Ramdahl, T. (1983) Retene—a molecular marker of wood combustion in ambient air. *Nature*, **306** 580-582.

Rivera, L., Curto, M.J.C., Pais, P., Galceran, M.T. and Puignou, L. (1996) Solid-phase extraction for the selective isolation of polycyclic aromatic hydrocarbons, azaarenes and heterocyclic aromatic amines in charcoal-grilled meat. *Journal of Chromatography A*, **731** 85-94.

Roose, P. and Brinkman, U.A.Th. (1998) Determination of volatile organic compounds in marine biota. *Journal of Chromatography A*, **799** 233-248.

Rothman, N., Poirier, M.C., Haas, R.A., Correa-Villasenor, A., Ford, P., Hansen, J.A., O'Toole, T. and Strickland, P.T. (1993) Association of PAH–DNA adducts in peripheral white blood cells with dietary exposure to polyaromatic hydrocarbons. *Environmental Health Perspectives*, **99** 265-267.

Rothman, N., Shields, P.G., Poirier, M.C., Harrington, A.M., Ford, D.P. and Strickland, P.T. (1995) The impact of glutathione *S*-transferase M1 and cytochrome P450 1A1 genotypes on white-blood-cell polycyclic aromatic hydrocarbon–DNA adduct levels in humans. *Molecular Carcinogenesis*, **14** 63-68.

Roy, T.A., Johnson, S.W., Blackburn, G.R. and Mackerer, C.R. (1988) Correlation of mutagenic and dermal carcinogenic activities of mineral oils with polycyclic aromatic compound content. *Fundamentals of Applied Toxicology*, **10** 466-576.

Rowland, S.J. and Volkman, J.K. (1982) Biogenic and pollutant aliphatic hydrocarbons in *Mytilus edulis* from the North Sea. *Marine Environmental Research*, **7** 117-130.

Sabljiĉ, A. (1991) On the prediction of soil sorption coefficients of organic pollutants from molecular structure: application of molecular topology model. *Environmental Science and Technology*, **21** 358-366.

Sabljiĉ, A., Lara, R. and Ernst, W. (1989) Modelling association of highly chlorinated biphenyls with marine humic substances. *Chemosphere*, **19** (10/11) 1665-1676.

Sagredos, A.N., Sinha-Roy, D. and Thomas, A. (1988) The determination, occurrence and composition of polycyclic aromatic hydrocarbons in oils and fats. *Fat Science and Technology*, **90** 70-81.

Santodonato, J., Howard, P. and Basu, D. (1981) Health and ecological assessment of polynuclear aromatic hydrocarbons. *Journal of Environmental Pathology and Toxicology*, **5** (1) 1-364.

Sargent, J.R. and Whittle, K.J. (1981) Lipids and hydrocarbons in the marine food web, in *Analysis of Marine Ecosystems* (ed. A.R. Longhurst) Academic Press, London, pp 491-533.

Schroers, H.-J., Jermann, E., Begerow, J., Hajimiragha, H., Chiarotti-Omar, A.-M. and Dunemann, L. (1998) Determination of physiological levels of volatile organic compounds in blood using static headspace capillary gas chromatography with serial triple detection. *Analyst*, **123** 715-720.

Seiber, J.N. (1990) Industrial and environmental chemicals in the human food chain, Part 2: Organic chemicals, in *Chemicals in the Human Food Chain* (eds C.K. Winter, J.N. Seiber and C.F. Nuckton), Van Nostrand Reinhold, New York, pp 183-219.

Seigneur, C., Venkatram, A., Galya, D., Anderson, P., Liu, D., Foliart, D., Von Burg, R., Cohen, Y., Permutt, T. and Levin, L. (1992) Review of mathematical models for health risk assessment—1. Overview. *Environmental Software*, **7** 3-7.

Selkirk, J.K., Croy, R.G., Whitlock, J.P. and Gelboin, H.V. (1975) *In vitro* metabolism of benzo [*a*] pyrene by human liver microsomes and lymphocytes. *Cancer Research*, **35** 3651-3655.

Shirey, R.E. (1995) Rapid analysis of environmental samples using solid-phase microextraction (SPME) and narrow bore capillary columns. *Journal of High Resolution Chromatography*, **18** 495-499.

Simko, P., Gergely, J., Karovicova, J., Drdak, M. and Knezo, J. (1993) Influence of cooking on benzo[*a*]pyrene content in smoked sausages. *Meat Science*, **34** 301-309.

Sjögren, M., Ehrenberg, L. and Rannug, U. (1996) Relevance of different biological assays in assessing initiating and promoting properties of polycyclic aromatic hydrocarbons with respect to carcinogenic potency. *Mutation Research*, **358** 97-112.

Stagg, R.M., McIntosh, A.D. and Mackie, P.R. (1995) The induction of hepatic monooxygenase activity in dab (*Limanda limanda* L.) in relation to environmental contamination with petroleum hydrocarbons in the northern North Sea. *Aquatic Toxicology*, **33** 245-264.

Stagg, R.M., McIntosh, A.D., Moffat, C.F., Robinson, C., Smith, S., Bruno, D.W. and Secombes, C.J. (1997) The sub-lethal effects of the *Braer* oil spill, Shetland Isles, Scotland, on farmed Atlantic salmon (Salmo salar) and the common dab (Limanda limanda), in *The Impact of an Oil Spill in Turbulent Waters: The Braer* (eds J.M. Davies and G. Topping), The Stationery Office, Edinburgh, pp 182-208.

Statutory Instrument No. 1073 (1966) *Food and Drugs Composition*, The Mineral Hydrocarbons in Food Regulations 1966.

Statutory Instrument No. 1147 (1989) *Water Supply*, Water Quality Regulations 1989.

Speer, K. and Montag, A. (1988) Polycyclische aromatische Kohlenwasserstoffe in nativen pflanzlichen Ölen. *Fat Science and Technology*, **90** 163-167.

Speer, K., Steeg, E., Horstmann, P., Kühn, Th. and Montag, A. (1990) Determination and distribution of polycyclic aromatic hydrocarbons in native vegetable oils, smoked fish products, mussels and oysters, and bream from the River Elbe. *Journal of High Resolution Chromatography*, **13** 104-111.

Stainken, D. and Frank, U. (1979) Analysis of Raritan Bay bottom waters for polynuclear aromatic hydrocarbons. *Bulletin of Environmental Contamination and Toxicology*, **22** 480-487.

Stavric, B. and Klassen, R. (1994) Dietary effects on the uptake of benzo[*a*]pyrene. *Food Chemistry and Toxicology*, **32** (8) 727-734.

Stijve, T. and Hischenhuber, C. (1987) Simplified determination of benzo(a)pyrene and other polycyclic aromatic hydrocarbons in various food materials by HPLC and TLC. *Deutsche Lebensmittel-Rundschau*, **83** 276-282.

Suess, M.J. (1976) The environmental load and cycle of polycyclic aromatic hydrocarbons. *The Science of the Total Environment*, **6** 239-250.

Szolar, O.H.J., Brown, R.S. and Luong, J.H.T. (1995) Separation of PAHs by capillary electrophoresis with laser-induced fluorescence detection using mixtures of neutral and anionic β-cylcodextrins. *Analytical Chemistry*, **67** 3004-3010.

Taysse, L., Chambras, C., Marionnet, D., Bosgiraud, C. and Deschaux, P. (1998) Basal level and induction of cytochrome P450, EROD, UDPGT, and GST activities in carp (*Cyprinus carpio*) immune organs (spleen and head kidney). *Bulletin of Environmental Contamination and Toxicology*, **60** 300-305.

Thomas, A. (1982) Über die Entfernung von Schadstoffen bei der Dämpfung von Speiseölen undfetten. *Fette Seifen Anstrichmittel*, **84** 133-136.

Thompson, M. and Wood, R. (1995) Harmonized guidelines for internal quality control in analytical chemistry laboratories. *Pure and Applied Chemistry*, **67** 649-666.

Tomaniová, M., Kocourek, V. and Hajšlová, J. (1997) Polycyklické aromatické uhlovodiky v potravinách. *Chemické Listy*, **91** 357-366.

Topping, G., Davies, J.M., Mackie, P.R. and Moffat, C.F. (1997) The impact of the *Braer* spill on commercial fish and shellfish, in *The Impact of an Oil Spill in Turbulent Waters: The Braer* (eds J.M. Davies and G. Topping), The Stationery Office, Edinburgh, pp 121-143.

Travis, C.C. and Arms, A.D. (1988) Bioconcentration of organics in beef, milk and vegetation. *Environmental Science and Technology*, **22** 271-276.

Travis, C.C. and Hester, S.T. (1990) Background exposure to chemicals: what is the risk? *Risk Analysis*, **10** (4) 463.

Vaessen, H.A.M.G., Jekel, A.A. and Wilbers, A.A.M.M. (1988) Dietary intake of polycyclic aromatic hydrocarbons. *Toxicological and Environmental Chemistry*, **16** 281-294.

van Maanen, J.M.S., Moonen, E.J.C., Maas, L.M., Kleinjans, J.C.S. and van Schooten, F.J. (1994) Formation of aromatic DNA adducts in white blood cells in relation to urinary excretion of 1-hydroxypyrene during consumption of grilled meat. *Carcinogenesis*, **15** 2263-2268.

van Stijn, F., Kerkhoff, M.A.T. and Vandeginste, B.G.M. (1996) Determination of polycyclic aromatic hydrocarbons in edible oils and fats by on-line donor–acceptor complex chromatography and high-performance liquid chromatography with fluorescence detection. *Journal of Chromatography A*, **750** 263-273.

Varanasi, U., Reichert, W.L., Stein, J.E., Brown, D.W. and Sanborn, H.R. (1985) Bioavailability and biotransformation of aromatic hydrocarbons in benthic organisms exposed to sediment from an urban estuary. *Environmental Science and Technology*, **19** 836-841.

Varanasi, U., Stein, J.E. and Nishimoto, M. (1989) Biotransformation and deposition of PAH in fish, in *Metabolism of Polycyclic Aromatic Hydrocarbons in the Aquatic Environment* (ed. U. Varanasi), CRC Uniscience Series, CRC Press, Boca Raton, FL, pp 93-150.

Verschueren, K. (1983) *Handbook of Environmental Data on Organic Chemicals*, 2nd edn, Van Nostrand Reinhold, New York.

Vreuls, J.J., de Jong, G.J. and Brinkman, U.A.Th. (1991) On-line coupling of liquid chromatography, capillary gas chromatography and mass spectrometry for the determination and identification of polycyclic aromatic hydrocarbons in vegetable oils. *Chromatographia*, **31** 113-118.

Vyskocil, A., Fiala, Z., Fialova, D., Krajak, V. and Viau, C. (1997) Environmental exposure to polycyclic aromatic hydrocarbons in Czech Republic. *Human and Environmental Toxicology*, **16** 589-595.

Walker, C.H., Hopkin, S.P., Sibly, R.M. and Peakall, D.B. (1996) *Principles of Ecotoxicology*, Taylor and Francis, London.

Webster, L., Angus, L., Topping, G., Dalgarno, E.J. and Moffat, C.F. (1997) long-term monitoring of polycyclic aromatic hydrocarbons in mussels (*Mytilus edulis*) following the *Braer* oil spill. *Analyst*, **122** 1491-1495.

Whittle. K.J., Mackie, P.R., Hardy, R. and McIntyre, A.D. (1974) *The Fate of n-Alkanes in Marine Organisms*. ICES CM 1974/E33, Fisheries Improvement Committee, Copenhagen.

Whittle, K.J., Hardy, R., Holden, A.V., Johnston, R. and Pentreath, R.J. (1977) Occurrence and fate of organic and inorganic contaminants in marine animals, in *Aquatic Pollutants and Biologic Effects with Emphasis on Neoplasia* (eds H.F. Kraybill, C.J. Dawe, J.C. Harshbarger, R.G. Tardiff), *Annals of the New York Academy of Sciences*, **298** 47-79.

Whittle, K.J., Anderson, D.A., Mackie, P.R., Moffat, C.F., Shepherd, N.J. and McVicar, A.H. (1997) The impact of the *Braer* oil on caged salmon, in *The Impact of an Oil Spill in Turbulent Waters:The Braer* (eds J.M. Davies and G. Topping), The Stationery Office, Edinburgh, pp 144-160.

Wild, S.R. and Jones, K.C. (1992) Polynuclear aromatic hydrocarbons uptake by carrots grown in sludge-amended soil. *Journal of Environmental Quality*, **21** 217-225.

Wild, S.R., McGrath, S.P. and Jones, K.C. (1990) The polynuclear aromatic hydrocarbon (PAH) content of archived sewage sludges. *Chemosphere*, **20** 703-716.

Wu, J., Wōng, M.K., Lee, H.K., Shi, C.Y. and Ong, C.N. (1997) Determination of polycyclic aromatic hydrocarbons in *Rougan*, a traditional Chinese barbecued food, by capillary gas chromatography. *Environmental Monitoring and Assessment*, **44** 577-585.

Yabiku, H.Y., Martins, M.S. and Takahashi, M.Y. (1993) Levels of benzo[*a*]pyrene and other polycyclic aromatic hydrocarbons in liquid smoke flavour and some smoked foods. *Food Additives and Contaminants*, **10** (4) 399-405.

Yang, K.L. and Lo, J.G. (1998) Volatile hydrocarbons (C6–C10) measurements at remote sites of Taiwan during the Pem-West A experiment (1991). *Chemosphere*, **36** (8) 1893-1902.

Zhou, J.L., Fileman, T.W., Evans, S., Donkin, P., Mantoura, R.F.C. and Rowland, S.J. (1996) Seasonal distribution of dissolved pesticides and polynuclear aromatic hydrocarbons in the Humber Estuary and the Humber Coastal Zone. *Marine Pollution Bulletin*, **32** 599-608.

11 Tainting of food by chemical contaminants
Peter Howgate

11.1 Introduction

11.1.1 Taint as a defect in food

The principal concerns about chemical contaminants in food[1] relate to toxic effects, i.e. the effects of contaminants on the health and well-being of consumers. However, contaminants that are not toxic, or not toxic at concentrations at which they might be present in food, can affect the nature of the food, particularly its flavour, and hence affect its acceptability. The contamination might then render the food unacceptable to the consumer, or might cause regulatory authorities to condemn the food as unfit for consumption, even though it might not be injurious to health. A wide variety of trace chemicals, of both natural and anthropogenic origin, can enter foods by a variety of routes. Inorganic chemicals are not known to have tainted foods. This chapter discusses tainting by organic chemical contaminants.

11.1.2 The nature of taints and tainting

The International Standards Organisation (ISO) glossary of terms used in the sensory evaluation of foods defines a 'taint' as 'an odour or flavour foreign to the product' (ISO, 1992a). The glossary distinguishes taints from off-flavours or off-odours, which are defined as atypical flavours or odours associated with deterioration or transformation of the product. The distinction is that taints are derived from materials present in the surroundings of the product; off-odours and off-flavours are produced by processes within the product. It is common usage in North America to apply the term 'off-flavour' both to taints and to off-flavours as defined above, and to assume that any taint adversely affects the flavour of the product (ASTM, 1989). However, it is useful, particularly in this context of food contamination, to maintain the concept of a taint as being derived from circumstances external to the food. The term is used in this chapter as defined in the ISO glossary. The ISO definition does not require that a taint necessarily be unpleasant, or that tainted food necessarily be unfit for consumption. While it is true that taints induced in foods by chemical

---

[1] Unless the context otherwise discriminates, 'food' here includes drinking water and beverages.

contaminants will usually be unpleasant, this need not be the case. Nevertheless, even if the contaminant improves the flavour, tainting should be considered an undesired effect of food contamination. Any procedure for assessment of taint should therefore seek to detect and measure a change in flavour rather than an impairment of flavour.

Taint is a sensory experience and can only be detected and measured by sensory procedures, by smelling or tasting the food presumed to be tainted. Chemical analytical procedures for trace chemicals are capable of identifying and quantifying very low concentrations of contaminants, and have made important contributions to identification of the origins of taints (Maarse, 1993). If the relationship between concentration of a chemical in the food and presence, or intensity, of taint is known, then analysis of that chemical should predict the presence and intensity of taint. However, such a process is only a prediction of tainting, not a direct measurement of it.

11.1.3   The sensory experience

The term 'sensory' is defined in an ISO standard (ISO, 1992a) as 'relating to the use of the senses'; sensory assessment is assessment of the properties of a substance using the senses. 'Sensory' in this context is preferred to 'organoleptic', which means relating to a property of a substance that is capable of being perceived by one or more of the sense organs, though many people use the latter as synonymous with the former. Humans, and other animals, have two mechanisms for detecting chemicals in food or in the environment: by taste in the mouth, and by odour in the nose.

Taste is perceived on the tongue, where taste buds respond to some types of chemicals. Four taste qualities—salt, sweet, acid, and bitter—corresponding to four different types of taste buds are commonly recognised, though there is some debate that this is a simplification, and there may be more. Some substances found in foods have an irritant effect in the mouth, which has a completely different physiological mechanism from that of taste.

Smell is detected by the olfactory sense organ situated in the nose that responds to chemicals in the vapour phase. The vapours can reach the olfactory organ through the nose when vapours from the substance are sniffed, or through the mouth and the passage connecting the mouth and the nose when the food is put into the mouth. The perception of an odour of a simple compound is the same in both cases, at least, qualitatively, but the overall perception of a food is not; perception of complex mixtures such as food assessed in the mouth is a combination of odour and taste. This combination is the flavour of the food.

Perception by taste is not as discriminating or as sensitive as perception by smelling, but taste can detect chemicals that do not have appreciable vapour pressures at the temperatures at which food would be evaluated and, consequently, would not be detected by odour. Chemicals that are completely ionised in foods would not be detected by odour, but might be detectable by taste. It is very rare for a taint to be detected by taste only; typically, taints are caused by chemicals detected by the olfactory system even if they are assessed as flavours in the mouth.

The olfactory organ is a very sensitive and discriminating detector of chemicals in the vapour phase. The great majority of actual or potential chemical pollutants in the environment that might contaminate foods are organic compounds and will have an odour. Experienced assessors[2] are capable of discriminating among thousands of different odours (though they cannot all be described and named). Some chemicals can be detected by smell at much lower concentrations than they can with sensitive analytical instruments such as gas chromatographs fitted with flame ionization detectors. The sense of smell is very important in evaluating taints in foods, and in assessing the odours of water or air that could be the vehicles for transferring the causative chemical to the food.

A few other terms relating to sensory analysis should be explained; formal definitions are given in the ISO glossary (ISO, 1992a). A *receptor* is the specific part of a sense organ—nose, tongue—that responds to a stimulus, and a *stimulus* is something that can excite a receptor. In the context of tainting by chemical contaminants, the chemical is the stimulus and the corresponding receptors are the taste buds on the tongue or the olfactory organ in the nose. If the stimulus is of sufficient intensity, the person subjected to a chemical stimulus perceives the stimulus and is aware of its effects as a taste or as an odour.

11.1.4   Relationship between perception and intensity of stimulus

It is the everyday experience of anyone who has experience of chemicals that they differ in the character and intensity of their odours and, for a chemical, the intensity of odour varies with concentration in the medium—air, water, food. Visualise an experiment in which an assessor is presented with a chemical over a range of concentrations and is required to state whether the stimulus can be perceived. (The experiment must be conducted so that the assessor gives an honest answer, but details of how this is done can be ignored for the moment.) At high

---

[2] An assessor is 'any person taking part in a sensory test' (ISO, 1992a), and this term is preferred over terms such as 'taster', 'judge' or 'panellist'.

concentrations the assessor will always detect the stimulus, and at low concentrations will never detect it. There is an intermediate range in which the assessor will sometimes detect the stimulus, and sometimes not. If the experiment is replicated often enough, the proportion of detections in this range will form an S-shaped curve when plotted against the intensity of the stimulus. When the intensity is plotted as the logarithm of the concentration, this ogive is approximated by the cumulative normal distribution (there are theoretical grounds for this approximation), or by a logistic of the form:

$$P = \frac{1}{e^{b(t-x)} + 1} \qquad (11.1)$$

where P is the proportion of detections, (or probability of detection), b and t are coefficients, and x is the logarithm of the concentration (Figure 11.1). At higher concentrations in this range, i.e. when the assessor can

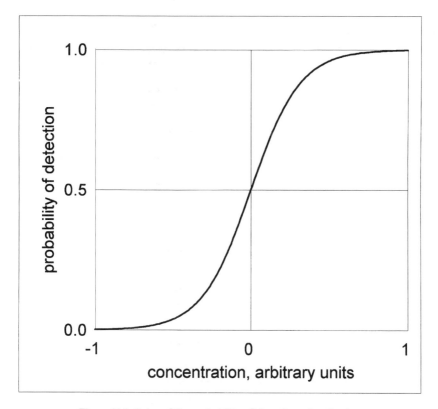

Figure 11.1 Ogive of the probability of detection of a stimulus.

almost always detect the stimulus, and above, an intensity can be assigned, e.g. weak, strong. This perceived intensity can be assigned a value on an intensity scale and, for an experienced assessor, the assigned score is linearly related to the logarithm of the concentration of the chemical. These are simplified accounts of the basic relationships between intensity of stimuli and perception. More details are given in textbooks of psychophysics, the study of the relationship between stimuli and sensory responses, such as those of McNicol (1972) and Torgerson (1958).

11.1.5 Detection thresholds and tainting thresholds

Reviews and other literature on tainting of foods often refer to the detection thresholds of tainting substances. Typically, in this literature, the threshold is equated with the minimum concentration of the chemical in the medium that can be detected by smell or taste. This is a simplistic view of the response to stimuli at low concentrations, as though there exists a sharp boundary between detection and non-detection; experiment shows a transition zone between these two states where there is partial detection. A threshold defined as the minimum concentration that can be detected would be towards the lower asymptote of the curve illustrated in Figure 11.1. The detectability of a chemical is better summarised by the concentration that produces a specified probability of detection away from the asymptotes. A common convention is to use the concentration that has a probability of detection of 0.5. This has several useful properties. It is the median value in a symmetrical frequency curve, and is the position of the maximum ordinate of a normal frequency distribution, of which the ogive of Figure 11.1 is the cumulative. Where the logistic of equation (11.1) is the mathematical model, the threshold is given by t. (When $x = t$, $e^{b(t-x)} = 1$, and $P = 0.5$.)

Chemicals differ in detectability, i.e. in the values of their thresholds. They also differ in the values of the slopes of the ogives, the change in detectability for a fixed increment in concentration. These two properties are measured by the parameters t and b, respectively, in the logistic curve of equation (11.1). They correspond to the mean and standard deviation of the distribution if the ogive is approximated by the cumulative normal distribution. The measured responses to a set of chemicals would then form a family of curves in Figure 11.1. A very wide range of detection thresholds has been reported, e.g. from 200 mg/l for cyclohexane in water to 3×10^{-11} mg/l for 4-butyl-5-propylthiazole in water. Typically, chemicals reported to cause tainting of food have detection thresholds in water below 0.1 mg/l. Thresholds are widely reported in the food, sensory and psychophysical literature, but it is often difficult to validate, or compare, results because of the variety of experimental procedures that have been used to determine them, and the variety of criteria that have been used to define the threshold. Only a minority of values have been obtained by procedures that allow the threshold to be estimated as a median response value. Where multiple values have been reported for a chemical, it is not uncommon to find that they differ by at least two orders of magnitude. There are simple compilations of threshold values (Fazzalari, 1978; Van Gemert and Nettenbreijer, 1977), but they do not list the methodologies or review the validity of values.

Threshold values for a chemical differ according to the medium in which the threshold is measured. Mostly, thresholds are measured in air or water. There are some reported values for thresholds in beverages and in milk (Fazzalari, 1978), but there are very few reported values of thresholds for chemicals in solid foods. For various reasons, it is to be expected that the detection threshold of a chemical in solid food would be greater than it is in water, that is, for detection a greater concentration is required in food than in water. A major factor is the lipid content of the food. Detection thresholds for chemicals are greater in lipids than they are in water, perhaps by up to three orders of magnitude, and the detection threshold of a chemical in a fatty food will be much greater than it is in water. Taints in foods arise predominantly from absorption of the contaminating chemical from the surrounding environment, air or water, (sections 11.3.2 and 11.3.4), and the tainting threshold is of more relevance for quantifying the potential of a chemical to taint. This is analogous to the detection threshold and is the concentration in the ambient air or water that will induce taint detectable with a probability of 0.5 in food that is exposed to the air or water. Measurement of tainting thresholds in the context of tainting of aquatic foods is discussed later in section 11.3.2.

Assessors also differ in their thresholds to a particular chemical, and in their change in response with change in concentration, and a family of curves can be constructed for a set of assessors tested with the same chemical. The difficulty in determining the distribution of sensitivities in a population is that a large amount of testing has to be carried out, in the order of 50 measurements over a range of concentrations, for a sufficiently accurate estimate of the threshold of an individual, and a large number of persons must be tested for a reliable estimate of the shape of the distribution. Consequently, only a few studies have been done on the distribution of detection thresholds of chemicals among populations and very few on chemicals that could cause taint from food contaminants. The few studies of distributions of sensory sensitivities to chemicals show that detection thresholds are approximately normally distributed in a population for most chemicals, but there are significant deviations from normality for some. The deviations range from skewing to polymodality (Reed *et al.*, 1995). This variation in sensitivity among individuals has important consequences for selection of assessors for analysis of taints, and for consideration of the impacts of taints in foods on consumers.

11.2 Measurement of taints

11.2.1 Appropriate procedures

It follows from the definition of taint that taints can only be evaluated directly by sensory methods, though chemical analytical methods are often used with sensory evaluations in investigations of taints and tainting. A variety of sensory methods are described in textbooks of sensory analysis of foods, and many of these are suitable for evaluation of taints. Kilcast (1993) recently reviewed sensory procedures used in evaluation of taints in foods, and Botta (1994) described sensory procedures for assessment of taint in fishery products.

Sensory evaluations of foods are made by individuals who differ in their sensitivities to organoleptic properties of substances and in their value judgements of the importance of sensory properties of foods to the acceptability of a food item. Because of these individual effects, sensory methods are often considered to be subjective, i.e. prone to biases because of personal whims. There is no doubt that individuals differ in their responses to chemical stimuli under given circumstances, but it is necessary to distinguish between several sources of variation. Some variation is due to intrinsic sensitivities to stimuli; this will show itself as variation in individual detection thresholds for substances, or in perceived intensities of fixed stimuli. These effects are intrinsic to the use of persons as measuring instruments, and need to be taken into account by the investigator when selecting assessors for assessment of taints. In a particular testing situation, individuals might differ because of the way they interpret and use the measuring procedure; the aim of selection and training of assessors is to minimise differences among assessors, that is to minimise personal biases.

Sensory analysts distinguish between 'objective' and 'subjective' sensory methods. In objective methods, the task required of the assessors focuses on properties of the product, and avoids any value judgements. Any terms used to characterise and quantify odours or flavours relate to properties—'musty', 'medicinal', 'petroleum', 'sulphide', 'weak', 'strong'—and not to personal responses to the odour or flavour, or to value judgements such as 'dislike', 'good' or 'bad'. Objective methods use panels of a few, typically 6–10, selected and trained assessors. Subjective methods measure affective responses to stimuli, that is they measure value judgements. The typical task for the assessors in subjective testing is to register like/dislike, or acceptability/unacceptability. People are expected to differ in value judgements, and subjective methods do not attempt to minimise, or to affect, these differences by experimental design or by training. Indeed, assessing the size and distribution of the differences among individuals, and among strata within populations, would likely be the purpose of the study. Typically, large panels of assessors are used in subjective testing to gain a sufficiently wide spread of opinions. Taint is defined in the ISO glossary in objective terms, a deviation from the normal odour or flavour of the product. Often, taints are subjectively unpleasant, but there is nothing in the definition of a taint that requires this to be the case. Taints should therefore be investigated by objective sensory methods. There can be a role for subjective methods, though they have very rarely been used in studies of tainting. The difference between objective and subjective methods can perhaps be summarised by posing two questions: is the product tainted and, if so, does it matter? To answer the first question, an objective method must be used; to answer the second question, a subjective method would be used. Several types of sensory methods are available for assessing taints in foods, but they can be grouped conveniently into two broad classes: difference tests and scalar tests. They are reviewed in sections 11.2.3. and 11.2.4, but selection and training of assessors is similar in both cases.

11.2.2 Selection and training of assessors

The general principles and practices for selection and training of assessors for sensory evaluations of foods are described in textbooks of sensory evaluation. Texts referred to earlier (Botta, 1994; Kilcast, 1993) provide descriptions in the context of taint testing, and further references. Selection, training and monitoring of assessors is summarised in an ISO standard (ISO, 1992b). It is most likely that investigations of incidents of, and studies of, tainting of foods will be conducted in factory quality control laboratories or in research or food testing laboratories that already have panels of experienced assessors. The important point for analysis of taints is that assessors should be familiar with the food under investigation and with the range of normal flavours, including off-flavours arising from inappropriate processing or storage.

Foods are complex materials with a wide range of sensory properties. Assessors for sensory testing of foods are generally selected for their abilities to evaluate the food as a whole, and the criteria for selection are based on their overall performance. Taints are usually caused by single chemicals, or perhaps a few sensorily potent chemicals if the contaminant is a mixture such as fuel oil. Individuals differ in their patterns of sensitivity to chemicals, but this variation is not too important when evaluating foods because, typically, several chemicals contribute to the flavour of a particular food. Initial selection of persons for inclusion in panels can include testing for sensitivity to chemicals found in foods but, in an individual, high sensitivity to one chemical does not guarantee high sensitivity to all chemicals. However, variations in sensitivity among individuals might be important in assessment of taints, and putatively tainted foods presented for evaluation should be assessed by a panel of several assessors. It will soon become apparent to the panel leader, as samples are evaluated, if there is a wide variation in sensitivities to the taint, and the leader might wish to use a subset of the panel comprising the more sensitive individuals. If the chemical nature of the taint is known, or suspected, the panel leader might wish to carry out some simple tests for sensitivity to the chemical. Measurement of individual thresholds is probably not justified, bearing in mind the amount of effort needed to measure these accurately. There are a few chemicals for which some persons are very insensitive, that is the distribution of sensitivities within a normal population is bimodally distributed, but there are no reports that the chemicals that have tainted foods are in this category. A panel of selected and experienced assessors, particularly if the assessors have been selected for high sensitivity, is not representative of the response of consumers to tainted foods; the panel is designed to detect the presence of taint, not its import. There is no doubt that a significant proportion of consumers will be as sensitive to the taint as members of the panel and, very likely, a small proportion may be even more sensitive but, unless extensive testing of a sample of consumers is carried out, it will not be possible to estimate what proportion will detect the taint or would find the taint unacceptable. In a commercial situation, the quality assurance manager would probably decide that even the suspicion of a taint would render the product defective in comparison with the specification for it and would withdraw the batch of product from use or distribution.

11.2.3   Assessment of taints by discriminative tests

In discriminative tests two or more samples are presented to the assessor, who is required to discriminate between them in some way. They are also referred to as difference tests—the term used in the ISO glossary—or comparison tests, but discrimination more properly describes the assessor's task. It is a well-established principle in psychophysics, and in sensory analysis (a practical application of psychophysics), that assessors can discriminate better between samples compared directly than between those assessed on separate occasions. Discrimination tests are more sensitive in detecting low levels of taint than are scalar methods (see section 11.2.4).

11.2.3.1 Pair comparison test

The simplest of the discrimination tests is the pair comparison test in which the assessor is presented with two samples and is required to discriminate between them. Typically, one sample is a reference and the other is a product in which a substance has been added or removed, or which has been treated or affected to increase or decrease the intensity of some property; the assessor has to select the sample of the pair that has more of the property. For contaminated material the question could be: which of the pair is more tainted or, which sample is least like the normal product? Reference and test samples should, as far as possible, differ only in the property of interest, though this requirement might be difficult to achieve in some circumstances. Raw materials, e.g. unprocessed crops or fish, can vary significantly in sensory properties among varieties and species and, within variety or species, with harvesting location or season. A particular processed food can vary in sensory properties arising from differences in raw material, variations in processing conditions, and effects of maturity or deterioration during storage. Experienced assessors will be familiar with the normal variation in sensory properties of a particular food product and should be able to discount these variations when evaluating the presence of possible taints. It is not always possible to obtain a suitable reference product, in which case the assessor is presented only with the test sample and asked to assess it for the presence of taint. This can be considered a pair comparison test but, in this case, the reference is a virtual sample based on the assessor's mental imagery of the sensory properties of the normal product.

The test and reference samples are presented to the assessors under code, and the order of presentation should be balanced to cancel out preferences for first or second sample assessed. The experimental design of pair comparison tests to avoid biases is described in texts of sensory methodologies. The panel leader decodes the results of the assessments and records for each assessor whether the test sample has been selected as being tainted, or having the stronger taint. The outcome from each assessment is a binary result: yes, a correct selection of the test sample, or no, an incorrect selection. The panel leader counts the number of correct selections and uses this number, perhaps expressed as a proportion of the total number of assessments, as the raw data for statistical evaluation. In a pair comparison test there is a probability of 0.5 that the test sample will be selected when there is no discernible difference between the two samples. The panel leader has to compare the observed proportion of successes against the probability of obtaining that proportion by chance in the event that there was no difference and the assessors were essentially guessing. (If the difference is negligible, some assessors might wish to record they could find no difference, but the test is best carried out by requiring the assessors to make a choice.) This probability is obtained from the binomial distribution, using the number of assessments and a probability of 0.5 from tables of the binomial expansion or, more conveniently, using the appropriate macro in a computer spreadsheet. The interested party can then decide whether this probability is low enough to warrant appropriate action. Texts on sensory analysis usually present tables of the minimum number of assessments that must be obtained from selected total numbers of assessments for statistical significance at given levels of confidence, usually $p < 0.05$.

The pair comparison method is widely used in sensory testing of foods, but there are few reports of its use in tainting studies, e.g. for aquatic food products (Jardine and Hrudey, 1988; Koning and Hrudey, 1992; Williams *et al.*, 1989). The procedure described above is an objective procedure, but it can be used as a subjective test. In this case, the assessors are asked which sample they prefer.

11.2.3.2   Triangle and three-alternative forced choice (3-AFC) tests
In these tests, the assessor is presented with three samples, two of which are the same. The two tests differ in the presentations of the two products—reference and test—in the set of three, and in the task set for the assessors. In the triangle test, the identical pair can be either reference or test; in the 3-AFC test, the pair is always the reference sample. In the triangle test, the assessors are asked to select the sample that is different from the other two; in the 3-AFC test, the assessor is asked to select the sample with the assigned property, knowing that only one has the property, the other two being references.

In the triangle test, the assessors are not instructed in, or trained to recognise, the nature of the difference in sensory properties between the test and reference samples; they may discriminate between the samples on the basis of any sensory property or combination of properties. This makes the test useful in cases of tainting when the nature of the taint is not known and it is not possible to train the assessors in detecting the causative chemical. The triangle test can be used with non-expert assessors because general training in sensory evaluation is not required, only adequate briefing in performing the test. However, it was found that training—repeated assessments of triangles comprising reference and test samples of the same nature—improves the ability of the assessors to detect small differences between the samples (Frijters *et al.*, 1982; McBride and Laing, 1979; O'Mahony *et al.*, 1988). As with the pair comparison test, a balanced design is used to avoid biases arising from the order of presentation of samples. In the triangle test, there are six combinations of triangles and orders of two products A and B: AAB, ABA, BAA, BBA, BAB and ABB. The triangle test is a popular test in sensory assessment of foods. Textbooks of sensory testing usually provide a full description of the test and of the precautions to be observed when conducting the test. There is an ISO standard (ISO, 1983) that has been used in several investigations of tainting of fish by pollutants (Carter and Ernst, 1989; Cook *et al.*, 1973; Davis *et al.*, 1992; Ernst *et al.*, 1989; Farmer *et al.*, 1973; Gordon *et al.*, 1980; Heras *et al.*, 1992, 1993; Howgate, 1987; Jardine, 1992; Mosse and Kowarsky, 1995; Redenbach, 1997; Williams *et al.*, 1989), and was incorporated in draft guidelines for determination of tainting thresholds of chemicals (ECETOC, 1987; Poels *et al.*, 1988).

Superficially, the triangle and 3-AFC tests are similar, but there are important differences between them, not least in the mental processes used by the assessor to select the odd sample. In the triangle test, the assessor does not know the nature of the difference or, at least, is not initially aware of it, and compares the sensory properties of the three samples in pairs. Referring to the samples as A_1, A_2 and B, with A_1 and A_2 the identical pair, the comparisons are A_1A_2, A_1B and A_2B. The assessor decides on the pair with the smallest difference and the third sample is the odd sample. The 3-AFC task is simpler. The assessor knows that the odd sample is the one with the target sensory property and scans the three samples for the one with the highest intensity of this property. (The 3-AFC test is so called because the assessor has three alternatives to choose from, and is required to make a choice.)

It has been observed that assessors are more frequently correct in selecting the odd sample in the triangle test when the odd sample is the test material—the 3-AFC design—than when the odd sample is the reference. Experimental findings, and fundamental considerations of the perception processes involved in the triangle and 3-AFC tests show that the 3-AFC test is superior to the triangular test in discriminating between samples (Ennis, 1990, 1993; Frijters, 1980, 1982, 1988). The pair comparison test described in the previous section also has a discriminatory power similar to that of the 3-AFC test and greater than the triangle test (Ennis, 1990, 1993; Geelhoed *et al.*, 1994; MacRae and Geelhoed, 1992). In practice, the samples in a triangle test are not evaluated exactly as described above. The test is presented as though the assessors have no previous knowledge of the sensory properties of the samples, but experienced assessors will almost certainly have knowledge of the reference material at least and, after a few trials in a tainting study, assessors will become aware of the different nature of the test material.

The strategy then adopted by the assessors converges towards that of the 3-AFC test, searching for the maximum intensity of the target sensory property. This drift from the triangle test strategy to that of the 3-AFC probably contributes to the training effect in the triangle test.

The 3-AFC test has been used only rarely in published investigations of taints in foods. Indeed, it is much less used in sensory evaluation of foods than is the triangle test despite its superiority in discrimination. Whitfield et al. (1997) used the 3-AFC test to investigate mustiness in food caused by the presence of 2,4,6-tribromoanisole, but they did not refer to their procedure by this name. Davis et al. (1992) used both the triangle and 3-AFC tests to investigate tainting of trout by diesel oil, but the data are insufficient to show that the latter test is superior. In later papers, Davis (1995) and Davis et al. (1995) used the 3-AFC test to study tainting of fish by petroleum.

It is important in these tests that the pair of references are identical. In discrete foods such as vegetables, fruit, fish or fillets of fish, it is not usually adequate simply to take portions of the unit. Triangle and 3-AFC tests are very sensitive, and in the author's experience can readily demonstrate sensory differences between portions of the same fillet of fish. Therefore, it is necessary to comminute, at least coarsely, both test and reference materials before sampling.

As in the pair comparison, the outcome of an evaluation is a binary result, a correct or incorrect selection of the odd sample. In the triangle and 3-AFC tests, there is a probability of one-third of correctly selecting the odd sample when there is no discernible difference between the test and reference, and the probability of obtaining the observed proportion of successes in the event that there is no difference should be calculated from the binomial distribution. Texts on the triangle test, including the ISO standard, have tables for determining the minimum number of correct selections that must be obtained for selected numbers of assessments for significant differences at, usually, probabilities of 0.05 and 0.01. However, it is probably more satisfactory in many situations to calculate the exact probability, and its confidence interval (Macrae, 1995), using functions available in computer spreadsheets.

11.2.3.3　Other discriminative tests

The literature on sensory testing of foods reports other discriminative tests, and there are variations on the basic pair comparison and triangle tests as just described. In the duo–trio test the assessor is presented first with a reference sample and then with a pair comprising test and reference samples. The assessor is required to select the test sample (ISO, 1991). Its discriminating power is no better than that of the pair-comparison test (Ennis, 1990, 1993) and has no advantages over it bearing in mind the extra complexity and cost of administering the test. It has not been reported for evaluation of taints. The 3-AFC test is just one of a family of m-AFC tests where m is 2 or more. (Effectively the pair comparison is a 2-AFC test.) Tests with m greater than 3 are very rarely used in sensory evaluation of foods, and have not been reported for use in evaluation of taints.

The pair comparison and triangle tests are sometimes extended to include supplementary questions on the score sheet. These may ask the assessor to indicate the nature of the difference observed, or their degree of liking for the selected sample. The nature of the difference is built into the pair-comparison test and it is not usual to extend it. Examples of the extended triangle test for taint evaluation are given by Farmer *et al.* (1973), Mosse and Kowarsky (1995) and Jardine (1992). The author does not favour the extended triangle test because it is unclear whether the results are valid. Often, only a proportion of assessors make correct selections and it is difficult to interpret the extended responses from those making incorrect assessments. Experimenters using extended triangle tests try to avoid this by using only results from assessors who correctly selected the odd sample. Even this approach is flawed because some of those correctly selecting the odd sample could have done so by a lucky guess. There are far better ways of obtaining the additional information using sensory tests designed for the purpose.

11.2.4 Scalar tests

11.2.4.1 Nature of scalar tests

The discriminative tests described above are best used to answer the question: is the product tainted? They are used when the level of taint is low, or only just discernible, and when only a proportion of the assessors are likely to detect it. When the product is definitely tainted, scalar methods can be used to investigate both the intensity and nature of the taint. Various types of scalar methods described in textbooks on sensory evaluation of food are summarised in ISO 6658–1985 (ISO, 1985a). Procedures used in investigations of tainting are those based on rating or on scoring. Both use ordered scales, i.e. scales running sequentially from a low to a high value, but differ in the nature of the data produced. Not all texts on sensory testing distinguish, or distinguish correctly, between rating and scoring, and to a large extent the difference is not of practical importance. ISO 6658 describes rating as a method of classification using categories on an ordered scale. Categories can be labelled by any system of labels, e.g. A, B, C, etc. but, when they are allocated numbers, the numbers do not have any mathematical significance; the concern is that the scale intervals may not be of equal size. Consequently, the data should not be subjected to arithmetic operations such as calculating means and variances; valid operations are restricted to those such as counting frequency of use of the numbered categories. Scoring is a form of rating using a numeric scale in which the intervals are of equal size; there is no restriction on the arithmetic operations that can be performed on the data. Discriminating between rating and scoring can be somewhat pedantic in practice. Assessors asked to allocate the position of a sample on a progressive scale appear to be able to subdivide the scale into approximately equal divisions. The literature on tainting of foods shows that investigators have treated any scales used as scoring scales, and there are very few examples of a scale being used as a rating scale.

11.2.4.2 Scoring
A variety of scales for specific purposes have been described in the literature. There are two main types (Figure 11.2). In numeric scales the intensity is scored by numbers on an ordered scale. Typically all points on the scale are described 'anchored' as in the example, but some systems

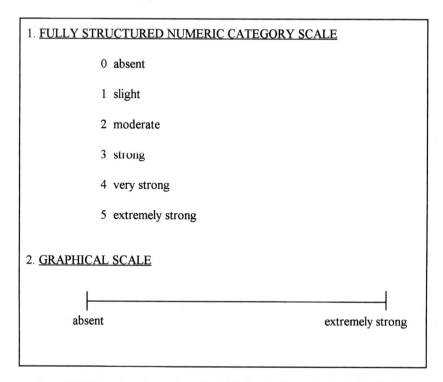

Figure 11.2 Examples of numeric and graphical scales for scoring intensity of taint.

have only some, usually alternate, scale points anchored (Martinsen *et al.*, 1992). The assessors assign numbers to a sample that best represents their perception of the taint intensity, and often assessors may use half-points. A graphical scale is typically 100 or 150 mm long and is anchored at the ends with the extremes of intensity. Sometimes a mid-point is marked and anchored. An assessor makes a mark on the line representing his or her perception of the taint intensity relative to the end positions, and the distance of this mark from the 'absent' end is used as the measure of intensity. Both types of scale have been used in taint assessment, though the numeric scale seems more popular (Branson *et al.*, 1979; Koning and Hrudey, 1992; Lindsay and Heil, 1992; Lockhart and Danell, 1992; Martin *et al.*, 1992; Martinsen *et al.*, 1992; McGill *et al.*, 1987; Münker, 1995; Parker *et al.*, 1990; Whittle *et al.*, 1995, 1997). Davis *et al.* (1992) used both scales for assessment of trout experimentally tainted by diesel oil and concluded that the two scales did not give systematically different results. Other published comparisons of numeric and graphical scales used for sensory assessment of foods do not show any marked advantage of one form over the other. Typically, the scores on either numeric or graphical scales are linearly related to the logarithm of the concentration in the test material of the substance giving rise to the flavour (Howgate *et al.*, 1977).

Data from scoring tests are processed by a wide variety of statistical methods. The simplest is to calculate the mean and variance of scores over the assessors; differences among samples are tested by analysis of variance (ANOVA) and multiple range tests. If several samples with a range of taint intensities are assessed by a single panel, the data can be subjected to two-way ANOVA with samples and assessors as effects. A significant F ratio of the assessor effect compared with the residual variance shows that one or more of the assessors deviates systematically from the rest of the panel. It must be accepted that assessors will differ from each other in scoring a given intensity of stimulus, and will demonstrate biases. There are several reasons why assessors can show consistent differences from each other but, for experienced assessors familiar with the scoring system, biases are intrinsic to the individual. Figure 11.3 shows the situation in which responses of four assessors are compared with the pooled response of all assessors in the panel over a range of concentrations. The responses, on average, show linear increases with logarithm of concentration, but the response lines of the assessors differ in location and slope. Assessor A, compared with the rest, is sensitive to low concentrations of the stimulus, i.e. has a low detection threshold, and his or her response changes rapidly with increase in concentration. This person is an effective assessor for evaluating this chemical in foods. Assessor B also has a low detection threshold, and is

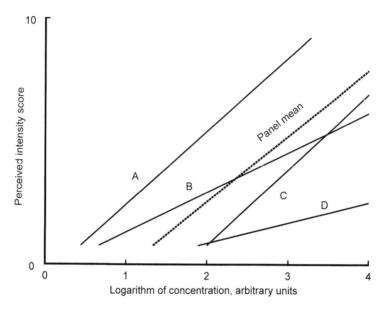

Figure 11.3 Illustration of possible relationship between perceived intensity of taint and concentration for members of a panel of assessors compared with the mean perceived intensity for the panel.

an effective assessor, especially if the samples being tested have a wide range of concentrations, including some in which the taint is extremely strong. Figure 11.3 shows linear responses but, in fact, they curve towards asymptotes at very low and very high concentrations, and assessor A's response is saturated at a lower concentration than B's. Assessors C and D have higher thresholds and again differ in slopes of responses. These assessors are not as effective as A and B in detecting and scoring low intensities of taints, though C is as sensitive and discriminating as B at higher concentrations. Differences in responses among assessors are inevitable in scoring systems. The differences are intrinsic to the individuals. Training, after initial familiarisation with the experimental procedure, will not bring about convergence. Biases contribute to the variance of the panel data for a sample. Total variance of scores for one sample from a panel of assessors has contributions from between-assessor variance (biases) and from within-assessor variance (random error). The size of the assessor deviations is obtained by comparing the assessors' means over the samples with the panel mean. It is quite common to find a significant assessor effect in the analysis of variance of data from a panel session, but only when deviations from panel means are persistent over panel sessions, and significant in the statistical sense, can they be considered as biases. The statistical treatment of data from scoring tests, and the estimation and examination of deviations of assessors, are given in the appendix of Shewan *et al.* (1953). In the example illustrated (Figure 11.3), a panel including assessors A, B, C and D, ANOVA of data over several samples will probably indicate significant biases for assessors A and D. Assessor D might show a significant bias if the samples tend to have low intensities of taint, but assessor B will not have a significant bias from the panel average.

Figure 11.3 also illustrates a difficulty with scoring for taints when the taint intensity is low. When the log concentration of taint substance is greater than 2 units, all of the assessors will almost certainly detect the taint with sufficient confidence to assign a score to its intensity. Below 2 units, assessors C and D will almost certainly not detect the taint, and below 0 units, none of the assessors should detect it. However, this does not mean that scores of zero will always be given to untainted samples. Even if the sample is not tainted, there could be doubt in an assessor's mind about the absence of taint. A taint is a flavour foreign to the product. In borderline cases an assessor has to decide whether the perceived flavour is within the natural range of flavours for that food product. Unprocessed food commodities and food products are not constant in sensory properties and there will be occasions when an assessor is unsure whether a particular flavour is within the normal range. The mental criterion of *difference from normal* established by an assessor in order to declare the presence of an effect differs among individuals, and within an individual from time to time. An assessor with a weak criterion for presence of taint will tend to decide that a taint is present in a sample that is not tainted, and allocate a score. In contrast, an assessor with a strong criterion, assessing the same sample, will tend to consider it not tainted and allocate a zero score. Because the scale is enclosed with a zero minimum, the latter type of assessor is not able to allocate corresponding negative scores to balance the positive scores of the former type. Weak criteria are encouraged by the testing situation. Assessors are aware, by briefing or by assumption, that they are evaluating possibly tainted samples, and this knowledge could influence the setting of the decision criterion in favour of declaring the presence of taint. This effect is sometimes referred to as the expectation bias.

In the typical procedure for scoring intensity of taint, these effects—a limit to the range of the scale, and expectation bias—produce a positive bias in the results, i.e. a taint score, even in the absence of a perceptible taint. The outcome across a panel of assessors is that low intensity scores are given to untainted samples. A typical example is given in Branson *et al.* (1979). Trout were exposed to water containing various levels of diphenyl oxide to determine the tainting threshold of that compound.

Two fish from each treatment, including a control, were assessed for taint on an intensity scale of 0 to 5 of unpleasantness of flavour by 12 assessors. The frequencies of scores of 0, 1, 2, and 3, for the 24 assessments of reference fish (same batch of fish but held in water to which no diphenyl oxide was added) were 15, 5, 2, and 2. Score 3 in the scale used is defined as 'definite off flavor, not objectionable'. Seven of the 12 assessors gave a non-zero score to one or both reference samples, and two assessors gave scores of 3. Other papers do not give sufficient data to allow examination of frequencies of scores, but the distribution of taint scores given to reference fish samples in Branson *et al.* (1979) is not untypical in the author's experience of managing panels assessing taints in fish.

There is an analogy in analytical chemistry. The panel of assessors can be considered as analytical instruments and the panel mean positive score given to reference samples as equivalent to the 'blank' in chemical analysis. Various experimental procedures have been used in the investigation of tainting to correct for the blank effect or to measure its value. In one experimental protocol one or more reference samples are identified to the members of the panel, and the test samples are compared with the reference. If a test sample is considered different from the reference, the assessor scores the intensity of taint. This strategy is not always successful because the reference(s) presented to the panel on that occasion might not be representative of the sensory properties of the type of product. It is sometimes difficult to obtain reference samples that are known not to be tainted. This can often be so if the reference does not come from the same batch of raw materials, or was not processed under the same conditions as the test. For example, McGill *et al.* (1987) and Parker *et al.* (1990), when investigating tainting of fish from around established gas platforms in the North Sea, presented reference samples gathered at locations well away from the platforms being surveyed. However, the sensory assessors considered that some of the samples from the reference sites were tainted. The panel in this case was very experienced in assessing fish, including the species comprising the test and references samples, and the reference samples were only a small subset of all the specimens of those species that the assessors had ever evaluated. They compared both reference and test samples against these virtual references and considered that some of the nominal reference samples were tainted.

Another approach is to include one or more reference samples coded and not identified to the assessors in the set of samples presented to the panel; the scores allocated to the hidden references are used to estimate the blank value. The experimenter determines whether the observed effect is significant by comparing the mean scores of test and reference samples.

Typical examples are ASTM (1989), Calbert *et al.* (1974), Parker *et al.* (1990) and Shumway and Palensky (1973). Significance of a difference is tested by statistical tests based on a normal distribution of scores, for example Student's *t*-test. The statistical justification for this is doubtful. The distribution of scores given to a sample, as the example quoted above shows, is not normally distributed, or even symmetrically distributed, at low values of mean intensity; there is a high proportion of zero scores. This comes from the effect referred to earlier that the scale is truncated at zero and there is no possibility of allocating negative scores to balance biased positive scores. Because of the problems of the validity of significance tests based on score values, one group of workers simply used the count of the number of assessors detecting taint at any intensity and compared effects as contingency tables using these frequencies (McGill *et al.*, 1987; Parker *et al.*, 1990). They also defined a criterion for declaring a sample tainted, i.e. when at least half of the panel members detected taint. This criterion was used in the monitoring of discharges of oily wastes and of salmon tainted by an oil spill (McGill *et al.*, 1987; Parker *et al.*, 1990; Whittle *et al.*, 1995, 1997).

A way of avoiding the problem of positive biases derived for the use of a simple intensity scale is to have a scale that is not limited by absence of taint. When assessors are required to score the intensity of taint, they are being asked to score their certainty that the sample is tainted, but they can only score in the one direction. Any slight certainty about the presence of taint can result in a score, but any slight certainty that the sample is not tainted cannot be allocated a score. This topic of certainty in judgements, either way, presence or absence of a property, is treated in psychophysics by the theory of signal detection (TSD) (Green and Swets, 1966; McNicol, 1972). The concepts could be applied to the detection of taints and, in the author's opinion, would be a fruitful area of investigation. A protocol in keeping with the ideas of TSD would be to ask assessors to rate their certainty of the presence or absence of taint on a bipolar scale running from 'certain the sample is not tainted' to 'certain the sample is tainted'. One end of the scale would be allocated a positive number and the other a negative number, with a zero value at the indifference point, 'not sure if the sample is tainted or not'. This would allow judgement criteria to be biased on either side of the indifference point, resulting in scores being at least symmetrically, and likely normally, distributed, with means for reference samples tending to zero with large numbers of observations. The only published example of the TSD approach being used in taint evaluation seems to be Linssen *et al.* (1991) in which the authors investigated tainting caused by styrene in packaging materials. They presented reference and test samples and asked assessors to rate certainty of the sample being either reference or test on a 4-point scale: standard, perhaps standard, perhaps not standard, and not standard. The data were processed by O'Mahony's short-cut signal detection procedure (O'Mahony, 1979, 1986, 1988; O'Mahony et al., 1979), which gives as a derived statistic an R-index. This ranges from 50% at a chance level, i.e. no difference, to 100% for certainty of difference.

11.2.4.3 Descriptive and profile methods

Descriptive methods, in which assessors describe the sensory attributes of the samples, provide only qualitative information and could be used in a preliminary examination of tainted foods. Profile and Quantitative Descriptive Analysis (QDA) methods are extensions of simple descriptive tests in which relevant sensory attributes of the material are listed and the intensities of the attributes are scored on an intensity scale (Einstein, 1991; ISO, 1985b). They are widely used in research into sensory properties of foods and in product development, and have been used to study off-flavours or taints in water (Krasner, 1988; Suffet and Mallevialle, 1993; Suffet et al., 1996) and fish (Johnsen et al., 1987). Profile methods provide a structured and formalised description of the flavour of food or water and have a role in monitoring products for changes in flavour, but there is little scope for their use in investigation of episodes of tainting.

11.3 Examples of tainting of foods

11.3.1 Potable water, beverages and drinks

Water is a good solvent for chemicals and the aqueous environment is the ultimate depository for a wide range of contaminants. Contaminants enter surface waters from runoff from the land, by deposition from the atmosphere, and from acute and chronic discharges of industrial chemicals. Contaminants can also enter underground aquifers by percolation through the overlying rocks to reappear in surface waters and boreholes. Contaminants can impart an odour to the water which might cause a nuisance, but their impact on the flavour of potable water (de Greef et al., 1983; Köster et al., 1981), beverages and drinks, is more relevant to the scope of this book. Persson (1983), Suffet and Mallevialle (1993), Suffet et al. (1996) and Wnorowski (1992) have reviewed odours in raw and potable waters. Most of the incidents are due to naturally occurring compounds such as the musty/earthy taints derived from the activities of microorganisms in soil or water (Maga, 1987). Chlorination of drinking water can result in naturally occurring chemicals being oxidised or chlorinated to derivatives with strong odours (Hrudey *et al.*, 1988).

Some taints in water are undoubtedly derived from industrial chemical contaminants. One identified industrial source is the paper pulp industry in Scandinavia and Canada (Cook *et al.*, 1973; Kenefick *et al.*, 1995a,b; Persson, 1983), and the tainting potential of the contaminants is considerably increased when chlorine is used in the process or for treatment of waste water (Paasivirta *et al.*, 1983). Discharges of contaminants from industrial sources can be odiferous of themselves, but can be made even more so by chlorination during treatment of potable water. The classic example of this is phenol, which has a detection threshold of the order of 10 mg/l; it is readily chlorinated in water and some of the chlorophenols have detection thresholds of the order of 0.01 mg/l (Suffet and Mallevialle, 1993). The chlorophenols, in turn can be methylated by bacterial action to chloroanisoles, which have even lower detection thresholds (Paasivirta *et al.*, 1983; Reineccius, 1991). Chlorophenols have medicinal, antiseptic-like odours, whereas chloroanisoles have musty odours. However, not all reports of medicine-like or disinfectant-like taints in chlorinated water arise from chemicals derived from industrial discharges; some breakdown products of organic matter naturally present in surface waters contain phenol groups that can be chlorinated to strongly odiferous substances. Pesticides were reported to taint water (Faust and Aly, 1964), but some of the examples in that paper could have been due to the vehicle in which the pesticide was suspended rather than the pesticide itself.

11.3.2 Aquatic food products

Contaminants in the aqueous environment can be taken up by animals inhabiting contaminated water. Other parts of this book describe the chemistry and pathways of these contaminants into aquatic biota used for food and their toxicity, but some contaminants can also impart taint to products. The previous section discussed taints in water and the general principle is that if the water body has an odour, then any fish[3] harvested from that water will also have an odour.

The flavour of fish varies for a variety of biological reasons, mostly connected with species, behaviour, season of capture, and fishing ground and, in the case of farmed fish, with cultural practices, even in the absence of any anthropogenic contaminants in the ambient water. A taint was

---

[3] 'Fish', here and in the subsequent text, unless specifically distinguished, includes crustacean and molluscan shellfish.

defined earlier as an atypical flavour, but there is some variation in the typical flavour of fish, both within and between species. At times it is not easy to distinguish between typical and atypical flavours in fish. However, persons who are experienced in tasting fish become accustomed to the normal range of flavours of a species and will readily be aware of flavours outside the typical range, which they will consider as taints. Though taints are usually considered as the result of contamination, fish can have unusual flavours due to naturally occurring processes. One particular natural taint that is sometimes erroneously attributed to anthropogenic contamination is a petroleum-like odour occasionally reported in several species of fish. Where it has been investigated (Ackman *et al.*, 1967, 1972) the odour has been attributed to dimethyl sulphide (DMS) which, in turn, is derived from a precursor, dimethyl-β-propiothetin, a component of marine and brackish water phytoplankton, and of organisms that graze the phytoplankton, (Andreae, 1980). Fish consume the phytoplankton or the grazing animals in their diet, and the dimethyl-β-propiothetin is converted in the fish into DMS. DMS is naturally present in sea water (Andreae and Raemdonck, 1983) and normally is present at very low concentrations in fish. It probably contributes to the typical flavour of species that graze phytoplankton, but the flavour becomes atypical when DMS is present in high concentrations, which can occur in some species at certain times of year, usually in spring when phytoplankton growth is high. Though the DMS taint is described at times as being like petroleum, experienced assessors can distinguish between the DMS flavour and true petroleum taint. Nevertheless, a naturally occurring DMS taint can cause problems for the regulatory authorities and quality controllers in the fish processing industry as it can give rise to complaints of petroleum contamination from consumers. Other sulphur compounds have been reported to give rise to taints in fish and shellfish (Whitfield, 1988). Again, the tainting compounds appear to have been formed in the fish from precursors present in the feed. These sulphur compounds have unpleasant flavours and are associated by the consumer with spoilage rather than with taint from chemical contamination.

A natural taint that occurs often in wild and in farmed fish from fresh or brackish waters, and already referred to earlier (section 11.3.1), is the musty, earthy flavour caused by compounds released by algae into the water. The causative chemicals have been shown to be geosmin and methylisoborneol (and possibly other related compounds) taken up by the fish, and there is now a large literature on the subject, and several reviews (e.g. Arganosa and Flick, 1992; Maga, 1987; Persson, 1983; Tucker and Martin, 1991). The causative compounds are not manufactured commercially in any appreciable quantities and the taint does not arise from an anthropogenic source, though eutrophication of waters by pollution can increase the likelihood and severity of the taint (Persson, 1982).

Fish and crustacean shellfish, and products, have been reported to have iodoform- or iodine-like taints. These flavours have been attributed to bromophenols (Anthoni et al., 1990; Bemelmans and den Braber, 1983; Boyle et al., 1992a; Münker, 1995; Whitfield et al., 1992) which are widely distributed in invertebrate species in the marine environment; it is assumed the animals pick up the flavour through their feed (Whitfield et al., 1992). In fact, Boyle et al. (1992a,b) have argued that low concentrations of bromophenols contribute to the typical flavour of marine, as distinct from freshwater, fish and shellfish. Though iodine- and iodoform-like flavours have an origin in natural processes in the marine environment, somewhat similar antiseptic-like taints have been attributed to the presence of chlorophenols. These do not seem to have a natural origin and are associated with anthropogenic pollution.

The most frequently reported incidents of tainting of fish and shellfish are petroleum taints arising from acute or chronic discharges of petroleum into the aquatic environment. Petroleum products are notorious for having strong odours and are clearly potential tainters if discharged into the aquatic environment. There is a considerable literature on the biological impacts of petroleum in the aquatic environment based on field studies and experimental investigations, but the literature on tainting by oil is rather sparse, probably less than 60 citations in all and some of these are multiple reports of the one incident. Tainting of aquatic food products by petroleum has been reviewed by, or included in reviews by, Connell and Miller (1981), Whittle (1978) and Ernst et al. (1987), and in reports issued by GESAMP (1977, 1993) on impacts of petroleum in the marine environment. It is rare that tainting of fish following spills of petroleum is systematically investigated and the data reported; in most cases, even in the event of massive spills (other than the Braer incident), descriptions of tainting are hardly more than anecdotal and there are rarely any formal laboratory evaluations of taint. To some extent, the scarcity of reports of tainting can be attributed to a low level of investigative effort but, even where the observers have specifically examined fish for tainting (e.g. Grainger et al., 1980; Mackie et al., 1978; Scarratt, 1980), any effect has been very small. Although large oil spills from shipping accidents are dramatic and often result in closure of fisheries by regulatory authorities, there is little evidence in the published literature, apart from the Braer accident, of large, or even significant, amounts of fish or shellfish being tainted. It is possible that wild populations of fish move away from incidents and do not become tainted. Stocks of sedentary species such as bivalve shellfish are at risk of being impacted by spills from groundings of tankers in coastal waters, and tainting of such shellfish has featured in reports of spills of petroleum or petroleum products (GESAMP, 1993). Aquaculture systems are also at risk; the fish cannot move away and high densities of fish are impacted. The *Braer* incident is a good example; salmon stocks in many salmon farms were tainted by the crude oil released in the accident (Davies and Topping, 1997). Whittle *et al.* (1997) provide an account of the extent of tainting among the farms, and of the uptake and loss of taint. Freshwater fish have been tainted by spills of petroleum. Trout in fish farms were tainted when vehicle fuel oil or heating oil was spilt into the ponds (personal communication); wild populations have also been tainted by fuel oil spills (Mackie *et al.*, 1972).

Fish have been tainted by chronic discharges of petroleum or petroleum-related chemicals into rivers, estuaries or inshore marine waters. An Australian group described tainting of mullet (*Mugil cephalus*) by a kerosene-like flavour and attributed its origin to petroleum wastes discharged in rivers (Connell, 1974). Krishnaswami and Kupchanko (1969) investigated tainting of rainbow trout in the Bow River, Alberta, Canada, and traced it to waste water discharges from a petroleum refinery. Jardine and Hrudey (1988) investigated the sensory properties of potential tainting compounds in waste waters from exploitation of oil sands. Nitta *et al.* (1965) studied experimental tainting of fish by waste water from three petroleum refineries and by muds collected near the waste outlet from one refinery. The effluent from one refinery produced strong tainting of fish after exposure of the fish for only 2½ hours. Fish in direct contact with the mud were tainted, but fish in water above the mud and not in contact with it were not tainted. Ogata and Miyake (1973) attempted to identify specific chemicals that could be tainting fish caught in the sea near petroleum industries. They reported that fish exposed to the effluent or caught in the vicinity of the refineries had an 'evil' or 'offensive' smell. Brunies (1971) reported that mussels (*Mytilus edulis*) harvested near the East Frisian Islands had a strong diesel oil flavour that rendered them unfit for consumption. It might be thought that exploration and production of oil and gas in the North Sea could potentially taint fish. However, surveys for tainted fish near platforms did not demonstrate significant tainting (McGill *et al.*, 1987; Parker *et al.*, 1990). The results are not entirely unequivocal because of the detection of taints in reference samples harvested from nominally pristine areas.

There are accounts of fish being tainted by discharges of industrial chemicals in freshwater systems. In continental Europe, discharges of phenolic and other compounds from the gas and coke industries have caused tainting of fish (Albersmeyer and von Erichsen, 1959; Bandt, 1955; Mann, 1951, 1969). The coal tar industry is much less important now than it was, but phenols are widely used as feedstock for many industrial chemicals and still pose a risk for tainting of fish in inland waters in the event of acute or chronic discharges (Funasaka *et al.*, 1975). As mentioned earlier, phenolic compounds are readily chlorinated in water treatment plants, and the odours of the chlorinated derivatives are far more intense that those of the parent phenols. Effluents from pulp and paper industries have long been known to taint fish (Calbert *et al.*, 1974; Cook *et al.*, 1973; Farmer *et al.*, 1973; Jardine, 1992; Whittle and Flood, 1977). Chemical analysis associated with sensory evaluation has identified a wide range of tainting chemicals (Berg, 1983; Heil *et al.*, 1989; Heil and Lindsay, 1990; Lindsay and Heil, 1992; Mikkelson *et al.*, 1996; Paasivirta *et al.*, 1983, 1987, 1992; Veijanen *et al.*, 1988). Some of the tainting compounds were chlorinated compounds formed in the chlorobleaching process. Paasivirta *et al.* (1987) showed that chlorinated compounds could be reduced by biological processes to the even more potent chloroanisoles and chloroveratroles. Fish have been tainted also by pesticides used in treatment of waterways (Folmar, 1980; Martin *et al.*, 1992).

It is possible that chronic tainting of fish, particularly shellfish, is more widespread than these reports would suggest. Various national and international agencies monitor the marine and freshwater environments for chemical and biological evidence of pollution, but there does not seem to be any programme of monitoring for possible impacts on the edibility of fishery products. The sparseness of the literature on tainting by chronic discharges might reflect more the level of resources devoted to its investigation rather than the rarity or importance of the effect. For example, Kuusi and Suihko (1983) summarised the results of tasting fish in Finland for 'off-flavours' during the period 1969 to 1981. A total of 1982 fish of 18 species were examined, coming predominantly from fresh or brackish water habitats. The sensory panel rated the flavour quality of the fish on a scale running from $10 =$ excellent to $0 =$ completely unfit for consumption. A rating of 5, described as 'noticeable off-taste', was taken as the level of acceptability. On this criterion, 35% of the fish were found unacceptable. Presumably, a higher proportion than this had some degree of tainting though not intense enough to render the sample unacceptable. Kraft pulp effluent was the most common source of taint, in 41% of the unacceptable samples.

In addition to observations of tainting of fish in the field, tainting has been studied under experimental conditions to gain some understanding of factors such as which chemicals are capable of inducing taint, the relationship between concentration in ambient water and intensity of taint, the components in complex mixtures such as petroleum that are responsible for the taint, and the kinetics of uptake and loss of taint. Petroleum and petroleum distillates are difficult materials to study because they are complex mixtures, and are variable in composition. Broadly, two procedures have been used to study tainting by petroleum: exposure in static tanks to which petroleum is added, and exposure of fish to petroleum, or to water-soluble fractions (WSF) of petroleum, in flow-through systems. The former is the simpler procedure and has been used in several studies, (Koning and Hrudey, 1992; Nitta et al., 1965; Ogata and Miyake, 1973, 1975; Ogata and Ogura, 1976; Ogata et al., 1986). These and other studies show that the intensity of tainting differs among crude oils and distillates, and that some can taint at concentrations of around 0.1 µl/l (ppm) of oil added to the tank. Water-soluble fractions of oils have been tested for tainting potential by the flow-through procedure (Carter and Ernst, 1989; Heras et al., 1992, 1993; Lockhart and Danell, 1992; Martinsen et al., 1992). Typically, the WSFs are prepared by stirring water, marine or fresh, with the oil or distillate at a ratio of 1 part of oil to a range of 100 to 300 parts of water, and recovering the water phase. The water phase is injected under controlled conditions into the water flowing through the exposure tanks. Again, these studies show that the different petroleum materials differ over a relatively narrow range of concentration in their capacities to taint, and can taint at concentrations down to around 0.5 mg/l of total hydrocarbons in the ambient water. The tainting threshold for these petroleum materials in a spill could well be lower than the values suggested by tests with WSFs because the WSF might not have fully extracted the tainting components. Davis (1995) and Davis et al. (1992) measured the tainting threshold of diesel oil by injecting the oil directly into the water supply flowing into the exposure tank and obtained tainting thresholds of 0.22 and 0.095 mg/l for two different oils. Fish can also become tainted by contact with oil-conta-minated sediments under experimental conditions (Howgate et al., 1977).

In some of these studies with petroleum, tainted fish were analysed for compounds potentially causing or contributing to the taint. Analyses showed the presence of a wide range of hydrocarbons in the flesh of tainted fish, similar to the pattern in the ambient water and in the original material. Only a small proportion of the compounds have been identified, usually the lower molecular mass alkanes, benzene and naphthalene derivatives, and lower molecular mass polycyclic aromatic hydrocarbons. All of the identified aromatic compounds individually have strong odours and have the potential to taint fish, but their collective presence in tainted fish does not indicate which compounds are the most important. Also, there could be other potent tainting compounds in the oil that were not identified. At present it is not known which chemicals in crude oils or derived fractions are critical for inducing taints in exposed fish.

Persson (1984) published a comprehensive review of the literature to that date on tainting of fish by chemicals under experimental conditions;

data on 45 chemicals are listed. Tainting thresholds ranged from 0.00002 mg/l (2-methylisoborneol) to 458 mg/l (aniline). In many of the cited publications, the conditions of exposure of the fish and of the sensory assessment of the exposed fish are poorly described, and it is often difficult to evaluate the validity of the data or to compare results. The largest set of comparable results was published by Shumway and Palensky (1973) for trout exposed in a flow-through system to 29 chemicals (mostly associated with pollution by the paper pulp industry) over a range of concentrations for an exposure period of 24 or 48 h. Tainting thresholds were estimated for 22 of the chemicals and found to range from 0.004 mg/l (2,4-dichlorophenol) to 95 mg/l (formaldehyde). The European Chemical Industries Environmental and Toxicology Centre sponsored a study of an experimental protocol for studying tainting of fish by exposure to chemicals in a flow-through system and the triangle test for evaluating the samples (ECETOC, 1987; Poels et al., 1988).

11.3.3   Agricultural products

Agricultural products can be contaminated by direct contact, e.g. sprays, or by absorption of chemicals from the air. However, the possibility of crops grown in the open being exposed to odoriferous air is small except in the case of oil-laden aerosols blown inland from stranded tankers and a few other situations, and published accounts of tainting of crops refer to taints arising from application of pesticides. Harries (1963) discussed the assessment of taints in fruit and vegetables and provided examples of tainting of crops by pesticides, particularly by hexachlorocyclohexane (HCH). HCH has several isomers and tainting is due to the β-HCH isomer. Tainting by HCH should be a thing of the past as it is much less used in agriculture than it was at the time of Harries' paper, and current preparations have only low levels of the β-isomer (Saxby, 1993a). HCH was the only pesticide for there was good evidence for causing taints in crops and recent reviews of tainting of foods do not refer to taints in crops, or refer to it only as a past problem (Baigrie, 1993; Land, 1989; Reineccius, 1991; Saxby, 1993a).

11.3.4   Processed foods

There is an extensive literature on off-flavours and taints in foods, applying mostly to processed food, discussed in detail by Saxby (1993b) and summarised in the review by Reineccius (1991). Much of the literature concerns off-flavours, that is atypical flavours generated within the food itself, rather than taints as atypical flavours arising by contamination of the food in some way. The major source of taints in processed foods is the ambient air in which the food is processed or stored. The same principle applies to tainting of foods as applies to aquatic foods—if the environment of the food has an odour, e.g. in a cold store, the food will pick up the odour. Reineccius (1991) reproduces a table of examples of airborne taints compiled from an earlier publication.

A common taint in foods is the medicinal, disinfectant odour of chlorinated phenols. This might come from the use of inappropriate disinfectants in the processing plant or store, but plant managers are well aware of taints from disinfectants and use odour-free materials. Tainting incidents typically originate from sources outside the premises. Another source of taints is fuel oils. Again, odours from spillages of fuel oil outside the premises can carry into food processing plants and stores and rapidly taint even wrapped products. Tainting has sometimes occurred from materials within factories such as paints and flooring compounds, and foods can become tainted from being in the proximity of other, strongly flavoured, foods. The water supply too can be a source of taints. Apart from the water having an odour that can be transferred to the food, traces of contaminants, which of themselves would not impart an odour at the concentrations present, can be chlorinated to more strongly odiferous compounds by the in-plant chlorination systems common in food processing plants. Wrapping and packaging materials can be sources of taints (Tice, 1993). Polystyrene is used both in containers and in wrapping film. Although polystyrene of itself does not have an odour, its monomer, styrene, does taint foods (Linssen *et al.*, 1991, 1993). The bulk polymeric materials used in wrapping films do not have odours but, apart from residual monomer and possible odoriferous impurities, the films are frequently prepared with other materials to improve their performance—plasticisers, stabilisers, adhesives in the case of laminates, for example—which might taint the enclosed food. The films might be overprinted for display purposes and the vehicles for the inks might have strong odours. Reineccius (1991) and Tice (1993) give examples of the potential or actuality of these additives in films for inducing taint in foods.

11.4 Implications for food control and food quality assurance

The fact that a food is tainted does not mean that it is unsafe to eat; the chemicals that have been identified as being responsible for taints in foods would not appear in the main to be toxic to humans at the levels at which they are apparent as odours or flavours. Where the possibility exists of a tainting chemical also being toxic, food legislation will control the maximum permitted level of the contaminant, though tainting might occur even below these levels.

Food legislation forbids the sale of food that is unfit for consumption, and food inspectors would consider the presence of a taint to cause a food to be unfit for consumption. Some regulations, e.g. those in the United States, specifically refer to tainting as a criterion for rendering a food unfit for human consumption. An inspection agency might also consider the presence of a taint as being indicative of possible contamination by toxic chemicals. For example, the presence of a petroleum taint might raise concerns about the concomitant presence of harmful chemicals such as polycyclic aromatic hydrocarbons. Condemning a food as unfit for consumption on the basis of a taint could cause problems for food inspectors. Taint has to be evaluated by sensory methods with a panel of experienced assessors, and testing laboratories in food control agencies might not be equipped for this purpose. Taint also has to be distinguished from off-flavours due to defects in processing or storage. Presence of off-flavours could also constitute a valid reason for declaring products as being unfit for consumption, but the control agency needs to be sure of the grounds for any condemnation. Where food is declared unfit because of contamination by chemicals, the agency must resort to established, if not validated, analytical procedures for confirming their presence, and regulations or guidelines for maximum permitted levels. This is not so in the case of taints. To the author's knowledge there are no formally established procedures for evaluation of taints in foods, and no validated (in the sense of collaboratively tested) procedures. Nor are there guidelines for the intensity of taint that might be permitted. Problems of detecting and quantifying low intensities of taint are discussed in section 11.2.4.2 above and the food inspector will have to make a judgement as to the impact a low level taint would have on typical consumers of the food. The inspector could take the view, quite correctly, that a sensory panel of experienced assessors in a testing laboratory situation will be more sensitive to taint than a group of consumers taken at random, and that a taint that is near the limit of detection for an analytical panel will not be detected by the large majority of consumers. However, it has to be accepted that there are bound to be some consumers in the population who will be more sensitive to the flavour of a particular chemical than the most sensitive assessor in the analytical panel.

There is a variety of national and international regulations and protocols covering the manufacture, use, storage, and transport of chemical substances designed to prevent discharge of dangerous chemicals into the environment, or at least to mitigate their impact. Chemical manufacturers, before a chemical is brought into use, or is stored or transported, have to lodge with competent authorities information relating to the environmental impact of the chemical, and this information forms the basis of any regulations or controls over its use, storage, or transport. Typically, the regulating authorities do not require the manufacturer to supply data on the sensory properties of the chemical, for example its detection threshold in air or water, that would give an indication of the chemical's potential to cause a nuisance or to taint food if discharged into air or water.

Some countries or groups of states have legislation protecting water or aquatic food products from tainting by contaminants. As examples, European Union (EU) Council Directive 80/68/EEC of 17 December 1979 on the quality of fresh waters (Commission of the European Communities, 1992) requires that fresh waters must not have phenolic compounds or petroleum hydrocarbons present in amounts that affect fish flavour; and Directive 79/923/EEC on the quality of shellfish water from which bivalve shellfish for human consumption are gathered requires that substances affecting the taste of shellfish be at lower concentrations than are likely to impair the taste of the shellfish (Commission of the European Communities, 1992). Typically, this legislation does not list specific compounds that have the ability to taint aquatic products. One international convention that includes sensory properties of chemicals within a system for categorising the impact of chemicals on the marine environment is the Convention for the Prevention of Pollution from Ships (MARPOL). Annexes II and III of this convention relate to prevention of pollution by chemicals carried in bulk or as packaged goods. Under the provisions of the convention, chemicals are categorised as to their impact on the marine environment, and the categories are used to regulate the requirements for carriage to prevent or minimise damage in the event of accident, and for possible discharge as tanker washings. Categorisation is based on a hazard profile for the chemical developed by a group of experts established for this purpose (GESAMP, 1989). The profile includes a criterion for a tainting threshold of 1 mg/l: that is, a chemical, or mixture, will be classed as a tainting chemical if it induces taint in fish when present in the ambient water at 1 mg/l or less. Unfortunately for the deliberations of this group, only a very small number of the 2000 or so substances on the GESAMP list have been tested for their ability to cause taint in fish and, as already stated, sensory properties and tainting potential are not properties that have to be measured by chemical manufacturers for their submissions to regulatory bodies relating to environmental impacts of chemicals. The Oslo and Paris Commissions and the signatories to the Convention for the Prevention of Marine Pollution and the Convention for the Protection of the Marine Environment of the North East Atlantic

(OSPAR) also take into account, *inter alia*, the potential of a chemical to cause taint, in their agreements on the use of chemical substances on, and discharge from, offshore installations. The Harmonised Manadatory Control System for the Use and Reduction of Discharge of Offshore Chemicals (PARCOM, 1996), in force for a trial period, includes a list of substances liable to cause taint. This list (Table 11.1) appears to consist of those chemicals classified as tainting substances by GESAMP referred to above. It is not a fully comprehensive list but is expected to be revised in 1999.

Table 11.1 Compounds liable to cause taint[a] (PARCOM, 1996)

Acrolein
Alkyl benzene
n-Amylbenzene
Amylmercaptan
Amyl thiol
Benzene
Butanethiol
Isobutylbenzene
n-Butylbenzene
sec-Butylbenzene
tert-Butylbenzene
Butylbenzenes (all isomers)
n-Butylbutyrate
Isobutyl isobutyrate
Butylphenols (liquid or solid)
p-tert-Butyltoluene
Calcium naphthenate
Carbolic oil
1-Chlorophenol
2-Chlorophenol
3-Chlorophenol
o-Chlorotoluene
Coal tar
Coal-tar naphtha
Cobalt naphthenate in solvent naphtha
Creosote (coal tar)
Creosote (wood tar)
Cresols (mixed isomers)
Cresylic acid, sodium salt solution
o-Dichlorobenzene
sym-Dichlorodiethyl ether
2,4-Dichlorophenol
Dichlorophenols (mixed)
2,4-Dichlorophenoxyacetic acid
2,6-Dichlorophenoxyacetic acid, diethanolamine salt
2,4-Dichlorophenoxyacetic acid, dimethylamine salt, 70%

Table 11.1 (Continued)

2,4-Dichlorophenoxyacetic acid, triisopropanolamine
1,1-Diethoxyethane
Diethylbenzene (mixed isomers)
Dimethyl sulphide
4,6-Dinitro-*o*-cresol
Diphenyl ether (diphenyl oxide)
Diphenyl oxide/biphenyl phenyl ether mixtures
Di-isopropyl benzenes
Ethane thiol
Ethyl acrylate
Ethyl chlorothioformate
o-Ethyl phenol
2-Ethyl-3-propylacrolein
Guaiacol
Heptylbenzene
n-Hexylbenzene
Laktane (light naphtha solvent, aromatics less than 25%)
Methanethiol
2-Methyl-4-chlorophenol
2-Methyl-6-chlorophenol
1-Methyl-2-ethylbenzene
1-Methyl-4-ethylbenzene
1-Methylnaphthalene
2-Methylnaphthalene
Methylnaphthalenes
Methylsalicylate
α-Methylstyrene
2-Methyl-5-ethylpyridine
1-Naphthol
2-Naphthol
Naphthalene
Naphthenic acids
Nitrocresols
Phenol (and compounds based on phenol)[a]
α-Pinene
Isopropylbenzene
Isopropyltoluene
Sulphide-containing compounds
Styrene (monomer)
Toluene
Tetramethylbenzenes (all isomers)
2,4,5-Trichlorophenol
2,4,6-Trichlorophenol
Triethylbenzene
1,2,3-Trimethylbenzene
1,2,4-Trimethylbenzene
1,3,5-Trimethylbenzene
Turpentine (wood)
Vinyltoluenes

Table 11.1 (Continued)

Xylene
2,3-Xylenol
2,4-Xylenol
2,5-Xylenol
2,6-Xylenol
3,4-Xylenol
3,5-Xylenol
Xylenols (mixtures)

[a]All the compounds listed above can cause taint at a concentration of 1 mg/l or less (GESAMP, 1989) except phenol. The list is not complete and is expected to be revised in 1999.

The detection threshold in water or in air is a good predictor of the potential for tainting of foodstuffs or foods exposed to contaminated water or air, but even these parameters have been measured for only a small number of industrial chemicals. Again, there are no generally accepted procedures for measuring detection thresholds and values of thresholds are dependent on the experimental procedures used to measure detection (see section 11.1.5).

Food manufacturers are well aware of the possibility of taints in foodstuffs, and of the consequences for consumer confidence in their products. Cummings *et al.* (1993) described one retailer's perspective on the subject. Quality assurance programmes require the sensory testing of raw materials and end-products for taints, and it is very uncommon for tainted processed food to enter the distribution chain. However, this does not necessarily mean the consumer will never experience tainted food, since tainting can occur during subsequent storage, distribution and retail sale.

References

Ackman, R.G., Hingley, H.J. and May, A.W. (1967) Dimethyl-β-propiothetin dimethyl sulfide in Labrador cod. *Journal of the Fisheries Research Board of Canada*, **24** 457-461.

Ackman, R.G., Hingley, H.J. and MacKay, K.T. (1972) Dimethyl sulfide as an odor component in Nova Scotia fall mackerel. *Journal of the Fisheries Research Board of Canada*, **29** 1085-1088.

Albersmeyer, W. and von Erichsen, L. (1959) Investigations into the effect of tar components in effluents. *Zeitschrift für Fischerei*, **8** 29-58.

Andreae, M.O. (1980) The production of methylated sulfur compounds by marine phytoplankton, in *Biogeochemistry of Ancient and Modern Environments* (eds P.A. Trudinger, M.R. Walter and B.J. Ralph), Springer-Verlag, Berlin, pp 253-259.

Andreae, M.O. and Raemdonck, H. (1983) Dimethyl sulfide in the surface ocean and the marine atmosphere. *Science*, **221** 744-747.

Anthoni, U., Larsen, C., Nielsen, P.H. and Christophersen, C. (1990) Off-flavor from commercial crustaceans from the North Atlantic zone. *Biochemical Systematics and Ecology*, **18** 377-379.

Arganosa, G.C. and Flick, G.J. (1992) Off flavours in fish and shellfish, in *Off-Flavours in Foods and Beverages* (ed. G. Charalambous), Elsevier Science, pp 103-126.

ASTM (1989) *Standard Practice for Evaluating an Effluent for Flavour Impairment to Fish Flesh*, ASTM designation: D3696-89. American Society for Testing and Materials, Philadelphia, PA.

Baigrie, B.D. (1993) Taints, in *Encyclopaedia of Food Science, Food Technology and Nutrition* (eds R. Macrae, R.K. Robinson and M.J.Sadler), vol. 7, Academic Press, London, pp 4504-4513.

Bandt, H-J. (1955) Damage to fisheries by phenolic effluents. *Wasserwirtschaft-Wassertechnik*, **5** 290-294.

Bemelmans, J.M.H. and den Braber, H.J.A. (1983) Investigation of an iodine-like taste in herring from the Baltic Sea. *Water Science and Technology*, **15** 105-113.

Berg, N. (1983) Chemical and sensory analysis of off-flavours in fish from polluted rivers in Norway. *Water Science and Technology*, **15** 59-65.

Botta, J.R. (1994) Sensory evaluation of tainted aquatic resources, in *Analysis of Contaminants in Edible Aquatic Resources: General Considerations, Metals, Organometallics, Tainting and Organics* (eds J.W. Kiceniuk and S. Ray), VCH, New York.

Boyle, J.L., Lindsay, R.C. and Stuiber, D.A. (1992a) Contributions of bromophenols to marine-associated flavors of fish and seafoods. *Journal of Aquatic Food Product Technology*, **1** 43-63.

Boyle, J.L., Lindsay, R.C. and Stuiber, D.A. (1992b) Distribution of bromophenols in salmon and selected seafoods of fresh and sea water origins. *Journal of Food Science*, **57** 918-921.

Branson, D.R., Blau, G.E., Alexander, H.C. and Peters, T.L. (1979) Taint threshold of diphenyl oxide in rainbow trout, in *Aquatic Toxicology, Proceedings of the Second Annual Symposium on Aquatic Toxicology* (eds L.L. Marking and R.A. Kimerle), ASTM Special Technical Publication 667, American Society for Testing and Materials, Philadelphia, PA, pp 107-121.

Brunies, A. (1971) Taste of mineral oil in mussels. *Archiv für Lebensmittelhygiene*, **22** 63-64.

Calbert, H.E., Dunnick, S.E. and Lindsay, R.C. (1974) Taste panel detection of pollution-related off-flavours in Flambeau River (Winsconsin) walleye pike. *Environmental Letters*, **7** 285-301.

Carter, J. and Ernst, R.J. (1989) Tainting in sea scallops (*Plactopecten magellanicus*) exposed to the water-soluble fraction of crude oil and natural gas condensate. *Proceedings of the Twelfth Arctic and Marine Oil Spill Program Technical Seminar, June 7-9, 1989, Calgary, Alberta*, Environment Canada, Ottawa, Ontario, Canada, pp 140-159.

Commission of the European Communities (1992) *European Community Environmental Legislation. Volume 7. Water*, Commission of the European Communities, Brussels.

Connell, D.W. (1974) A kerosene-like taint in the sea mullet *Mugil cephalus* (Linnaeus). I. Composition and environmental occurrence of the tainting substance. *Australian Journal of Marine and Freshwater Research*, **25** 7-24.

Connell, D.W. and Miller, G.J. (1981) Petroleum hydrocarbons in aquatic ecosystems — behaviour and effects of sublethal concentrations. Part 2. *CRC Critical Reviews in Environmental Control*, **11** 105-162.

Cook, W.H., Farmer, F.A., Kristiansen, O.E., Reid, K., Reid, J. and Rowbottom, R. (1973) The effect of pulp and paper mill effluents on the taste and odour of the receiving water and the fish therein. *Pulp and Paper Magazine of Canada*, **74C** 97-106.

Cummings, D.A., Swoffer, K. and Pascal, R.M. (1993) A retailer's perspective, in *Food Taints and Off-Flavours* (ed. M.J. Saxby), Blackie Academic and Professional, London, pp 236-243.

Davies, J.M. and Topping, G. (eds) (1997) *The Impact of An Oil Spill in Turbulent Waters: The Braer. Proceedings of a Symposium held at the Royal Society of Edinburgh 7–8 September 1995.* The Stationery Office, Edinburgh.

Davis, H.K. (1995) Depuration of oil taint and muscle pigment from fish. *Water Science and Technology,* **31** 23-28.

Davis, H.K., Geelhoed, E.N., MacRae, A.W. and Howgate, P. (1992) Sensory analysis of trout tainted by diesel fuel in ambient water. *Water Science and Technology,* **25** 11-18.

Davis, H.K., Shepherd, N. and Moffat, C. (1995) Uptake and depuration of oil taint from fish, in *Proceedings of the Second International Oil Spill Research and Development Forum, 23–26 May 1995,* International Maritime Organization, London, pp 353-361.

de Greef, E., Zoetman, B.C.J., van Oers, H.J., Köster, E.P. and Rook, J.J. (1983) Drinking water contamination and taste panel assessment by large consumer panels. *Water Science and Technology,* **15** 13-24.

ECETOC (1987) *Evaluation of Fish Tainting,* Technical Report No. 25, European Chemical Industries Environmental and Toxicology Centre, Brussels, Belgium.

Einstein, M.A. (1991) Descriptive techniques and their hybridisation, in *Sensory Science. Theory and Applications in Foods* (eds H.T. Lawless and B.P. Klein), Marcel Dekker, New York, pp 317-338.

Ennis, D.M. (1990) Relative power of difference testing methods in sensory evaluation. *Food Technology,* **44** (4) 114-117.

Ennis, D.M. (1993) The power of sensory discrimination methods. *Journal of Sensory Studies,* **8** 353-370.

Ernst, R.J., Ratnayake, W.M.N., Farquharson, T.E., Ackman, R.G. and Tidmarsh, W.G. (1987) *Tainting of Finfish by Petroleum Hydrocarbons,* Environmental Studies Revolving Funds Report No. 080, Ottawa, Canada.

Ernst, R.J., Ratnayake, W.M.N., Farquharson, T.E., Ackman, R.G., Tidmarsh, W.G. and Carter, J.A. (1989) Tainting of Atlantic cod (*Gadus morhua*) by petroleum hydrocarbons, in *Drilling Wastes* (eds F.R. Engelhardt, J.P. Ray and A.H. Gillam), Elsevier Applied Science, Barking, pp 827-839.

Farmer, F.A., Neilson, H.R. and Esar, D. (1973) Flavor evaluation by triangle and hedonic scale tests of fish exposed to pulp mill effluents. *Canadian Institute of Food Science and Technology Journal,* **6** 12-16.

Faust, S.D. and Aly, O.M. (1964) Water pollution by organic pesticides. *Journal of the American Water Works Association,* **56** 267-279.

Fazzalari, F.A. (1978) *Compilation of Odor and Taste Threshold Values Data,* ASTM Data Series DS 48A, American Society for Testing and Materials, Philadelphia, PA.

Folmar, L.C. (1980) Effects of short-term field applications of acrolein and 2,4-(DMA) on flavor of the flesh of rainbow trout. *Bulletin of Environmental Contamination and Toxicology,* **24** 217-224.

Frijters, J.E.R. (1980) Three-stimulus procedures in olfactory psychophysics: an experimental comparison of Thurstone-Ura and three-alternative forced-choice models of signal detection theory. *Perception and Psychophysics,* **28** 390-397.

Frijters, J.E.R. (1982) Expanded tables for conversion of a proportion of correct responses (Pc) to the measure of sensory difference (d′) for the triangular method and the 3-alternative forced choice procedures. *Journal of Food Science,* **47** 139-143.

Frijters, J.E.R. (1988) Sensory difference testing and the measurement of discriminability, in *Sensory Analysis of Foods* (ed. J.R. Piggott), Elsevier Applied Science, London, pp 131-154.

Frijters, J.E.R., Blauw, Y.H. and Vermaat, S.H. (1982) Incidental training in the triangular method. *Chemical Senses,* **7** 63-69.

Funasaka, R., Ose, Y. and Sato, T. (1975) Studies on the offensive odor of fish of the Nagara River. IV. Phenol as one of the offensive odor substances. *Journal of Hygienic Chemistry,* **21** 101-105.

Geelhoed, E.N., MacRae, A.W. and Ennis, D.M. (1994) Preference gives more consistent judgements than oddity only if the task can be modelled as forced choice. *Perception and Psychophysics*, **55** 473-477.

GESAMP (1977) IMO/FAO/UNESCO/WHO/IAEA/UN/UNEP Joint Group of Experts on the Scientific Aspects of Marine Pollution. *Impact of Oil on the Marine Environment*. GESAMP Reports and Studies No. 6, International Maritime Organization, London.

GESAMP (1989) IMO/FAO/UNESCO/WMO/IAEA/UN/UNEP Joint Group of Experts on the Scientific Aspects of Marine Pollution. *The Evaluation of Hazards of Harmful Substances Carried by Ships: Revision of GESAMP Reports and Studies No. 17*, Reports and Studies GESAMP 35, International Maritime Organization, London.

GESAMP (1993) IMO/FAO/UNESCO/WMO/IAEA/UN/UNEP Joint Group of Experts on the Scientific Aspects of Marine Pollution. *Impact of Oil and Related Chemicals and Wastes on the Marine Environment*, Reports and Studies GESAMP 50, International Maritime Organization, London.

Gordon, M.R., Mueller, J.C. and Walden, C.C. (1980) Effect of biotreatment on fish tainting propensity of bleached Kraft whole mill effluent. *Canadian Pulp and Paper Association, Transactions of the Technical Section*, **6** 1-7.

Grainger, R.J.R., Duggan, C., Minchin, D. and O'Sullivan, D. (1980) Fisheries-related investigations in Bantry Bay following the Betelguese disaster. *ICES C.M. 1980/E:54*, International Council for the Exploration of the Sea.

Green, D.M. and Swets, J.A. (1966) *Signal Detection Theory and Psychophysics*, Wiley, New York.

Harries, J.M. (1963) Some problems in the assessment of taints, in *Food Science and Technology. Proceedings of the First International Congress of Food Science and Technology, London, September 18–21, 1962, vol. III, Quality, Analysis and Composition of Foods* (ed. J.M. Leitch), Gordon and Breach, London, pp 325-332.

Heil, T.P. and Lindsay, R.C. (1990) Environmental and industrial factors relating to flavor tainting of fish in the upper Wisconsin river. *Journal of Environmental Science and Health*, **B25** 527-552.

Heil, T.P., Lane, N.A. and Lindsay, R.C. (1989) Sensory properties of thio- and alkyl-phenols causing flavour tainting in fish from the upper Wisconsin river. *Journal of Environmental Science and Health*, **B24** 361-388.

Heras, H., Ackman, R.G. and Macpherson, E.J. (1992) Tainting of Atlantic salmon (*Salmo salar*) by petroleum hydrocarbons during a short-term exposure. *Marine Pollution Bulletin*, **24** 321-315.

Heras, H., Zhou, S. and Ackman, R.G. (1993) Uptake and depuration of petroleum hydrocarbons by Atlantic salmon: effect of different lipid levels, in *Proceedings of the Sixteenth Arctic and Marine Oilspill Program Technical Seminar, Calgary, Alberta, 7–9 June, 1993*, Environment Canada, Ottawa, Ontario, Canada, pp 343-351.

Howgate, P. (1987) Measurement of tainting in seafoods, in *Seafood Quality Determination*. (eds D.E. Kramer and J. Liston), Elsevier Science, Amsterdam, pp 63-72.

Howgate, P., Mackie, P.R., Whittle, K.J., Farmer, J., McIntyre, A.D. and Eleftheriou, A. (1977) Petroleum tainting in fish. *Rapports et procès-verbaux des réunions. Conseil permanent international pour l'exploration de la Mer*, **171** 143-146.

Hrudey, S.E., Gac, A. and Daignault, S.A. (1988) Potent odour-causing chemicals arising from drinking water disinfection. *Water Science and Technology*, **20** 55-61.

ISO (1983) *ISO 4120-1983. Sensory Analysis—Methodology—Triangular Test*, International Standards Organization, Geneva.

ISO (1985a) *ISO 6658-1985. Sensory Analysis—Methodology—General Guidance*, International Standards Organization, Geneva.

ISO (1985b) *ISO 6564-1985. Sensory Analysis—Methodology—Flavour Profile Methods*, International Standards Organization, Geneva.

ISO (1991) *ISO 10399: 1991. Sensory Analysis—Methodology—Duo-Trio Test*, International Standards Organization, Geneva.

ISO (1992a) *ISO 5492-1992. Sensory Analysis—Methodology—Vocabulary*, International Standards Organization, Geneva.

ISO (1992b) *ISO 5496-1992. Sensory Analysis of Food. Part 9. Initiation and Training of Assessors in the Detection and Recognition of Odours*, International Standards Organization, Geneva.

Jardine, C.G. (1992) Public evaluation of fish tainting from pulp and paper mill discharges. *Water Science and Technology*, **25** 57-64.

Jardine, C.G. and Hrudey, S.E. (1988) Threshold detection values of potential fish tainting substances from oil sands wastewaters. *Water Science and Technology*, **20** 19-25.

Johnsen, P.B., Civille, G.V. and Vercellotti, J.R. (1987) A lexicon of pond-raised catfish flavour descriptors. *Journal of Sensory Studies*, **2** 85-91.

Kenefick, S.L., Brownleee, B.G., Perley, T.R. and Hrudey, S.E. (1995a) A chemical and sensory study of odour compounds in the Athabasca river, Alberta, Canada. *Water Science and Technology*, **31** 15-21.

Kenefick, S.L., Low, N.J., Hrudey, S.E. and Brownleee, B.G. (1995b) A review of off-flavour tainting of drinking water and fish by pulp mill effluents. *Water Science and Technology*, **31** 55-61.

Kilcast, D. (1993) Sensory evaluation of taints and off-flavours, in *Food Taints and Off-Flavours* (ed. M.J. Saxby), Blackie Academic and Professional, London, pp 1-34.

Koning, C.W. and Hrudey, S.E. (1992) Sensory and chemical characterization of fish tainted by exposure to oil sand wastewaters. *Water Science and Technology*, **25** 27-34.

Köster, E.P., Zoetman, B.C.J., Piet, G.J., de Greef, E., van Oers, H.J., van der Heijden, B.G. and van der Veer, A.J. (1981) Sensory evaluation of drinking water by consumer panels. *The Science of the Total Environment*, **18** 155-166.

Krasner, S.W. (1988) Flavor-profile analysis: an objective sensory technique for the identification and treatment of off-flavours in drinking water. *Water Science and Technology*, **20** 31-36.

Krishnaswami, S.K. and Kupchanko, E.E. (1969) Relationship between odor of petroleum refinery wastewater and occurrence of oily taste flavour in rainbow trout, *Salmo gairdneri Journal of the Water Pollution Control Federation*, **41** R189-R196.

Kuusi, T. and Suihko, M. (1983) Occurrence of various off-flavours in fish in Finland from 1969 to 1981. *Water Science and Technology*, **15** 47-58.

Land, D.G. (1989) Taints—causes and prevention, in *Distilled Beverage Flavour. Recent Developments* (eds J.R. Piggott and A. Paterson), Ellis Horwood, Chichester, pp 17-31.

Lindsay, R.C. and Heil, T.P. (1992) Flavor tainting of fish in the Upper Wisconsin river caused by alkyl- and thiophenols. *Water Science and Technology*, **25** 35-40.

Linssen, J.P.H., Janssens, A.L.G.M., Reitsma, H.C.E. and Roozen, J.P. (1991) Sensory analysis of polystyrene packaging material taint in cocoa powder for drinks and chocolate flakes. *Food Additives and Contaminants*, **8** 1-7.

Linssen, J.P.H., Janssens, A.L.G.M., Reitsma, H.C.E., Bredie, W.L.P. and Roozen, J.P. (1993) Taste recognition concentrations of styrene in oil-in-water emulsions and yoghurts. *Journal of the Science of Food and Agriculture*, **61** 457-462.

Lockhart, W.L. and Danell, R.W. (1992) Field and experimental tainting of arctic freshwater fish by crude and refined products, in *Proceedings of the 15th Arctic and Marine Oil Spill Programme Technical Seminar, Edmonton, Alberta, June 10–12, 1992*, Environment Canada, Ottawa, Ontario, Canada, pp 763-771.

Maarse, H. (1993) Analysis of taints and off-flavours, in *Food Taints and Off-Flavours* (ed. M.J. Saxby), Blackie Academic and Professional, London, pp 63-88.

Mackie, P.R., McGill, A.S. and Hardy, R. (1972) Diesel oil contamination of brown trout, *Salmo trutta. Environmental Pollution*, **3** 9-16.

Mackie, P.R., Hardy, R. and Whittle, K.J. (1978) Preliminary assessment of the presence of oil in the ecosystem at Ekofisk after the blowout, April 22–30, 1977. *Journal of the Fisheries Board of Canada*, **35** 544-551.

MacRae, A.W. (1995) Confidence intervals for the triangle test can give reassurance that the products are similar. *Food Quality and Preference*, **6** 61-67.

MacRae, A.W. and Geelhoed, E. (1992) Preference can be more powerful than detection of oddity as a test of discriminability. *Perception and Psychophysics*, **51** 179-181.

Maga, J.A. (1987) Musty/earthy aromas. *Food Reviews International*, **3** 269-284.

Mann, H. (1951) On the question of the influence of phenol on flavour. *Fischwirtschaft*, **1** 164-165.

Mann, H. (1969) Factors influencing the taste of fish. *Fette Seifen Anstrichmittel*, **71** 1021-1024.

Martin, J.F., Bennett, L.W. and Anderson, W. (1992) Off-flavor in commercial fish ponds resulting from molinate contamination. *The Science of the Total Environment*, **119** 281-287.

Martinsen, C., Lauby, B., Nevissi, A. and Brannon, E. (1992) The influence of crude oil and dispersant on the sensory properties of steelhead (*Oncorhynchus mykiss*) in marine waters. *Journal of Aquatic Food Product Technology*, **1** 37-51.

McBride, R.L. and Laing, D.G. (1979) Threshold determination by triangle testing: effects of judgemental procedure, positional bias and incidental training. *Chemical Senses and Flavour*, **4** 319-326.

McGill, A.S., Mackie, P.R., Howgate, P. and McHenery, J.G. (1987) The flavour and chemical assessment of dab (*Limanda limanda*) caught in the vicinity of the Beatrice oil platform. *Marine Pollution Bulletin*, **18** 186-189.

McNicol, D. (1972) *A Primer of Signal Detection Theory*, George Allen and Unwin, London.

Mikkelson, P., Paasivirta, J., Rogers, L.H. and Ikonomou, M. (1996) Studies on eulachon tainting problem: analyses of tainting and toxic aromatic pollutants, in *Environmental Fate and Effects of Pulp and Paper Mill Effluents* (eds M.R. Servos, K.R. Munkittrick, J.H. Carey and G.J. Van der Kraak), Sr. Lucie Press, Delray Beach, FL, pp 327-333.

Mosse, P.R.L. and Kowarsky, J. (1995) Testing an effluent for tainting of fish—the Latrobe Valley ocean outfall. *Water*, **22** 20-22.

Münker, W. (1995) Occurrence of off-flavour (tainting) in spring herring (*Clupea harengus*) from the Baltic Sea. *Information Fischwirtschaft*, **42** 202-209.

Nitta, T., Arakawa, K., Okubo, K., Okubo, T. and Tabata, K. (1965) Studies in the offensive odours in fish caused by wastes from petroleum industry. *Bulletin of the Tokai Regional Fisheries Research Laboratory*, **42** 12-26.

Ogata, M. and Miyake, Y. (1973) Identification of substances in petroleum causing objectionable odour in fish. *Water Research*, **7** 1493-1504.

Ogata, M. and Miyake, Y. (1975) Compound from floating petroleum accumulating in fish. *Water Research*, **9** 1075-1078.

Ogata, M. and Ogura, T. (1976) Petroleum components and objectionable odorous substances in fish flesh polluted by boiler fuel oil. *Water Research*, **10** 407-412.

Ogata, M., Miyake, Y., Fujisawa, K., Ogura, T. and Aramaki, M. (1986) Oily smell and oil components in fish flesh reared in seawater containing heavy oil. *Oil and Chemical Pollution*, **3** 329-341.

O'Mahony, M. (1979) Short-cut signal detection measures for sensory analysis. *Journal of Food Science*, **44** 302-303.

O'Mahony, M. (1986) *Sensory Evaluation of Foods: Statistical Methods and Procedures*, Marcel Dekker, New York.

O'Mahony, M. (1988) Sensory difference and preference testing: the use of signal detection measures, in *Applied Sensory Analysis of Foods* (ed. H.R. Moskowitz), CRC Press, Boca Raton, FL, pp 145-175.

O'Mahony, M., Kulp, J. and Wheeler, L. (1979) Sensory detection of off-flavours in milk incorporating short-cut signal detection measures. *Journal of Dairy Science*, **62** 1857-1864.

O'Mahony, M., Thieme, U. and Goldstein, L.R. (1988) The warm-up effect as a means of increasing the discriminability of sensory difference tests. *Journal of Food Science*, **53** 1848-1850.

Paasivirta, J., Knuutinen, J., Tarhanen, J., Kuokkanen, T., Surma-Aho, K., Paukku, R., Kääriäinen, H., Lahtiperä, M. and Veijanen, A. (1983) Potential off-flavour compounds from chloro-bleaching of pulp and chlorodisinfection of water. *Water Science and Technology*, **15** 97-104.

Paasivirta, J., Klein, P., Knuutila, M., Knuutinen, J., Lahtiperaa, M., Paukku, R., Veijanen, A., Welling, L., Vuorinenc, M. and Vuorinen, P.J. (1987) Chlorinated anisoles and veratroles in fish. Model compounds. Instrumental and sensory determinations. *Chemosphere*, **16** 1231-1241.

Paasivirta, J., Rantalainen, A-L., Welling, L., Herve, S. and Heinonen, P. (1992) Organo-chlorines as environmental tainting substances: taste panel study and chemical analyses of incubated mussels. *Water Science and Technology*, **25** 105-113.

PARCOM (1996) Oslo and Paris Commissions, PARCOM Decision 96/3 on a *Harmonized Mandatory Control System for the Use and Reduction of the Discharge of Offshore Chemicals*.

Parker, J.G., Howgate, P., Mackie, P.R. and McGill, A.S. (1990) Flavour and hydrocarbon assessment of fish from gas fields in the southern North Sea. *Oil and Chemical Pollution*, **6** 263-277.

Persson, P-E. (1982) Muddy odour: a problem associated with extreme eutrophication. *Hydrobiologica*, **86** 161-164.

Persson, P-E. (1983) Off-flavours in aquatic ecosystems—an introduction. *Water Science and Technology*, **15** 1-11.

Persson, P-E. (1984) Uptake and release of environmentally occurring odorous compounds by fish—a review. *Water Research*, **18** 1263-1271.

Poels, C.L.M., Fischer, R., Fukawa, K., Howgate, P., Maddock, B.G., Persoone, G., Stephenson, R.R. and Bontinck, W.J. (1988) Establishment of a test guideline for the evaluation of fish tainting. *Chemosphere*, **17** 751-765.

Redenbach, A.E. (1997) Sensory evaluation of fish exposed to pulp and paper mill effluent: a case study of methods used for environmental effects monitoring. *Water Science and Technology*, **35** 357-363.

Reed, D.R., Bartoshuk, L.M., Duffy, V., Marino, S. and Price, R.A. (1995) Propylthiouracil tasting: determination of underlying threshold distributions using maximum likelihood. *Chemical Senses*, **20** 529-533.

Reineccius, G. (1991) Off-flavors in foods. *CRC Critical Reviews in Food Science and Nutrition*, **29** 381-402.

Saxby, M.J. (1993a) A survey of chemicals causing taints and off-flavours in foods, in *Food Taints and Off-Flavours* (ed. M.J. Saxby), Blackie Academic and Professional, London, pp 35-62.

Saxby, M.J. (ed.) (1993b) *Food Taints and Off-Flavours*, Blackie Academic and Professional, London.

Scarratt, D.J. (1980) Taste panel assessments and hydrocarbon concentrations in lobsters, clams and mussels following the wreck of the Kurdistan, in *Scientific Studies During the Kurdistan Tanker Incident, Proceedings of a Workshop, June 1979, Bedford Institute of Oceanography* (ed. J.H. Vandermeulen), Report Series BI-R-80-3, Bedford Institute of Oceanography, Halifax, Canada, pp 212-227.

Shewan, J.M., MacIntosh, R.G., Tucker, C.G. and Ehrenberg, A.S.C. (1953) The development of a numerical scoring system for the sensory assessment of the spoilage of wet white fish stored in ice. *Journal of the Science of Food and Agriculture*, **4** 283-298.

Shumway, D.L. and Palensky, J.R. (1973) *Impairment of the Flavor of Fish by Water Pollutants*, EPA-R3-73-010, Office of Research and Monitoring, US Environmental Protection Agency, Washington, DC.

Suffet, I.H. and Mallevialle, J. (1993) Off-flavours in raw and potable water, in *Food Taints and Off-Flavours* (ed. M.J. Saxby), Blackie Academic and Professional, London, pp 89-121.

Suffet, I.H., Corado, A., Chou, D., McGuire, M.J. and Butterworth, S. (1996) AWWA taste and odour survey. *Journal of the American Water Works Association*, **88** 168-180.

Tice, P. (1993) Packaging material as a source of taints, in *Food Taints and Off-Flavours* (ed. M.J. Saxby), Blackie Academic and Professional, London, pp 202-235.

Torgerson, W.S. (1958) *Theory and Methods of Scaling*, Wiley, Chichester.

Tucker, C.S. and Martin, J.F. (1991) Environment-related off-flavours in fish, in *Water Quality in Aquaculture* (eds J.R. Tomasa and D. Brune), World Aquaculture Books, Baton Rouge, LA, pp 133-179.

Van Gemert, L.J. and Nettenbreijer, A.H. (1977) *Compilation of Odour Threshold Values in Air and Water*, National Institute for Water Supply, Voorburg, The Netherlands.

Veijanen, A., Paasivirta, J. and Lahtiperä, M. (1988) Structure and sensory analyses of tainting substances in Finnish freshwater environments. *Water Science and Technology*, **20** 43-48.

Whitfield, F.B. (1988) Chemistry of off-flavours in marine organisms. *Water Science and Technology*, **20** 63-74.

Whitfield, F.B., Shaw, K.J. and Walker, D.I. (1992) The source of 2,6-dibromophenol: cause of an iodoform taint in Australian prawns. *Water Science and Technology*, **25** 131-138.

Whitfield, F.B., Hill, J.L. and Shaw, K.J. (1997) 2,4,6-Tribromoanisole: a potential cause of mustiness in packaged food. *Journal of Agricultural and Food Chemistry*, **45** 889-893.

Whittle, D.M. and Flood, K.W. (1977) Assessment of the acute toxicity, growth impairment, and flesh tainting potential of a bleached kraft mill effluent on rainbow trout (*Salmo gairdneri*). *Journal of the Fisheries Research Board of Canada*, **34** 869-878.

Whittle, K.J. (1978) Tainting in marine fish and shellfish with reference to the Mediterranean Sea, *International Register of Potentially Toxic Chemicals. Data Profiles for Chemicals for the Evaluation of Hazards to the Environment of the Mediterranean Sea*, Vol II. IRPTC Data Profile Series No. 1, United Nations Environment Programme, Geneva, pp 89-108.

Whittle, K.J., Mackie, P.R. and Davis, H.K. (1995) Shellfish tainting—a means of monitoring petroleum-contaminated effluents. *Proceedings of the Royal Society of Edinburgh*, **103B** 127-135.

Whittle, K.J., Anderson, D.A., Mackie, P.R., Moffat, C.F., Shepherd, N.J. and McVicar, A.H. (1997) The impact of the *Braer* oil on caged salmon, in *The Impact of An Oil Spill in Turbulent Waters: The Braer. Proceedings of a Symposium held at the Royal Society of Edinburgh 7–8 September 1995* (eds J.M. Davies and G. Topping), The Stationery Office, Edinburgh, pp 144-160.

Williams, U.P., Kiceniuk, J.W., Fancey, L.L. and Botta, J.R. (1989) Tainting and depuration of taint by lobster (*Homarus americanus*) exposed to water contaminated with a No. 2 fuel oil: relationship with aromatic hydrocarbon content in tissue. *Journal of Food Science*, **54** 240-243, 257.

Wnorowski, A.U. (1992) Tastes and odours in the aquatic environment: a review. *Water SA*, **18** 203-214.

12 Risk assessment of environmental contaminants in food

Peter J. Abbott, Janis Baines, Simon Brooke-Taylor and Luba Tomaska

Abbreviations

ADI, acceptable daily intake; Codex, Codex Alimentarius Commission; FAO, Food and Agriculture Organisation; GAP, good agricultural practice; GEMS, Global Environmental Monitoring System; GLP, good laboratory practice; GMP, good manufacturing practice; IPCS, International Programme on Chemical Safety; JECFA, Joint FAO/WHO Expert Committee on Food Additives; MPC, maximum permissible concentration; ML, maximum level; MRL, maximum residue levels; NOEL, no-observed-effect level; OECD, Organisation for Economic Cooperation and Development; PTDI, Provisional tolerable daily intake; PTWI, Provisional tolerable weekly intake; UNEP, United Nations Environment Programme; WHO, World Health Organisation.

12.1 Introduction

The presence in food of environmental contaminants that may lead to adverse health effects is clearly undesirable and there is considerable effort on the part of governments and industry to minimise the level of contamination. Fundamental to the effective management of food contamination is an assessment of the potential health risks associated with specific contaminants. Assessment of risk is a complex matter, not only because of the poor understanding of the potential human health effects of some contaminants, but also because any contamination of the food supply is generally considered unacceptable by the broad community.

The challenge from a public health perspective is to ensure that the levels of unavoidable contaminants in food are well below the levels that may cause adverse health effects in any individuals and also to maintain contaminant levels as low as reasonably achievable without endangering the availability of an otherwise nutritious food source.

The range of potential environmental contaminants in food is very large and therefore some consideration needs to be given to focusing research, monitoring and regulatory activity on those contaminants for which effective public health outcomes can be achieved. In most cases, adverse effects as a result of acute (short-term) exposure to contaminants are already well managed, but for many contaminants there is low-level chronic (long-term) exposure of the human population. Assessing the risks associated with low-level chronic exposure is particularly difficult given the number of factors that may influence the outcome. In some cases, human exposure may occur via multiple routes and, for such contaminants, exposure via food needs to be considered in the context of the total level of exposure. For many contaminants, the paucity of the scientific information available can lead to an incorrect assessment of risk and subsequent public anxiety. Without further research, it is difficult to address many of these concerns.

This chapter discusses some of the issues that need to be considered when assessing the health risks associated with environmental contaminants in food, beginning first with the nature of risk in relation to food, then discussing each of the steps in risk assessment in relation to contaminants. Some options for risk management are also discussed to complete the chapter.

12.2   Concept of risk in relation to food

Risk can be considered as the probability of an undesirable outcome from a particular event. In relation to food, this is usually interpreted as the probability of an adverse health outcome that may occur either immediately or over the long term. Food, however, occupies an important place in our social structure, providing not only sustenance but also enjoyment, social intercourse, employment and trade. Thus, for many in the community, the perception of 'risk' in relation to food is relative and must be balanced against social, nutritional and other benefits associated with food. To the regulator, risk assessment involves the scientific assessment of health risks and, while it is performed independently of these other dimensions of risk, they may contribute to the subsequent risk management decisions. In communicating risk assessment decisions, the probability of adverse health effects needs to be considered in relation to the psychological, social, ethical and economic factors that can contribute to the perception of risk by an individual or by a community. These other dimensions of risk have been considered by Soby *et al.* (1994) and others (also see Chapter 13) but are considered only briefly in this chapter.

12.2.1   Public perception of risk

In general, the community has a poor understanding about health risks, and an unrealistic desire for a risk-free environment. For many individuals, the presence of low levels of environmental chemicals in food, some of which may carry an inherent, albeit low, level of risk, is unacceptable and, in this regard, the scientific view of health risks may be at odds with community expectations. Furthermore, in many cases, the low levels of risk associated with contaminants in food must be balanced against the health benefit of a nutritious and varied diet.

There are many factors that will influence whether a particular health risk is acceptable to an individual consumer. These factors have been discussed by a number of authors (BMA, 1987; Slovic, 1987; Sandman, 1989; Segal, 1990). Sandman identified a number of so called 'outrage factors' that influence an individual's perception of risk. Many of these factors are relevant to food contaminants and relate to such issues as whether the risk is voluntary (can it be avoided?), whether it applies to everyone (are the levels higher than elsewhere?), whether it is familiar (is it a new industrial contaminant?), whether the hazard is understood (is enough known?), and whether the real risk is understood (can the experts be trusted?). Plant-derived foods can contain relatively high levels of many unusual substances, some of which are potentially hazardous. The risks associated with these naturally occurring substances are often accepted by the majority of the population while minute levels of contaminants such as dioxins and polychlorinated biphenyls (PCBs) are deemed unacceptable. Although, on a purely comparative risk basis, this may seem illogical, it reflects the importance of social values and traditions that cannot be ignored when dealing with food safety matters. The scientist must work within this social framework, first to minimise health risks associated with contaminants, but also to achieve the levels of contamination that the community expects and considers acceptable.

12.2.2 Definition of terms

The terms used in risk analysis have been defined in different ways by different individuals and organisations. There appears, however, to be general agreement regarding the intent of these terms. The Codex Alimentarius Commission (Codex) is developing uniform definitions for these and other terms used by the Codex Committees. Some of the interim definitions, initially proposed by a Joint (FAO/WHO) Expert Consultation (FAO/WHO, 1995) and released following amendment by the Codex Executive Committee in June 1996 (Codex, 1996a), are shown below:

Hazard: A biological, chemical, or physical agent in, or condition of, food with the potential to cause an adverse health effect.

Risk: A function of the probability of an adverse health effect and the severity of that effect, consequential to a hazard(s) in food.

Risk analysis: A process consisting of three components: risk assessment, risk management and risk communication.

Risk assessment: The scientifically based process consisting of the following steps: (i) hazard identification, (ii) hazard characterisation, (iii) exposure assessment, and (iv) risk characterisation.

Risk management: The process of weighing policy alternatives in the light of the results of risk assessment and, if required, selecting and implementing appropriate control options, including regulatory measures.

Risk communication: The interactive exchange of information and opinions concerning risk among risk assessors, risk managers, consumers and other interested parties.

The interim Codex definitions for each of the steps in risk assessment are shown in that particular section in the text. A number of regulatory agencies consider risk communication as a part of the risk management strategy rather than as a separate component, since it is integral to the public acceptance of a particular risk management decision.

12.2.3 Risk analysis framework

Risk analysis can be divided into two distinct processes, namely, risk assessment and risk management (Figure 12.1). This chapter will focus on risk assessment, but some options for risk management in relation to environmental contaminants will also be considered later.

Risk assessment and risk management are regarded as two distinct activities. This distinction was noted by the US National Research Council (NRC, 1983). Risk assessment is essentially a scientific endeavour based on experimentation and observation, whereas risk management involves policy decisions based on a balance of scientific, social and economic considerations. By focusing on the scientific aspects of health

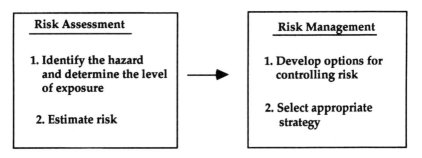

Figure 12.1 Framework for analysis of risk.

risks associated with environmental chemicals, it is possible to identify areas where further research is needed to provide a realistic assessment of risk, and also to give priority to those substances for which real reductions in food-related health risks can be achieved. At the same time, other legitimate concerns can be taken into consideration through an appropriate risk management and risk communication strategy that assesses factors such as the public's perception of risks and benefits. In some cases, where the risk is considered to be high, it may be possible to identify a measurable improvement in health outcomes resulting from particular risk management decisions.

12.3 Contaminants in food

12.3.1 Definition of a contaminant

Food contaminants are commonly considered to be those substances present in food that serve no technological function and whose presence may lead to adverse health effects. The Codex Alimentarius Commission defines a contaminant (Codex, 1995) as follows:

> Any substance not intentionally added to food, which is present in such food as a result of production (including operations carried out in crop husbandry, animal husbandry and veterinary medicine), manufacture, processing, preparation, treatment, packing, packaging, transport or holding of such food or as a result of environmental contamination. The term does not include insect fragments, rodent hairs and other extraneous matter.

Under this definition, the potential range of food contaminants is very large but has generally been restricted to synthetic chemicals, metals, mycotoxins and bacterial toxins. The synthetic chemicals include any industrial chemical or environmental contaminant. The definition stipulates "not intentionally added to food", and this excludes agricultural and veterinary chemicals, food additives and processing aids. A small number of substances classified as contaminants, such as copper, selenium and zinc, are also essential micronutrients although, for these substances, adverse health effects may be observed at high levels of dietary exposure. Certain pesticides, such as DDT, that are no longer intentionally added to crops but are still found in foods could also be regarded as contaminants.

The term 'environmental' in relation to contaminants generally refers to those substances that enter food as a result of environmental contamination of primary produce or that enter food during production, processing and storage. It generally excludes naturally occurring plant or marine toxins. It is debatable whether they are included under the Codex definition of a contaminant, although they will be encompassed within the scope of the Codex General Standard for Contaminants and Natural Toxins (Table 12.1) (Codex, 1995).

Table 12.1 Codex classification of food contaminants

- Metals, metalloids and their compounds
- Other elements and inorganic compounds
- Halogenated organic compounds
- Other organic compounds
- Mycotoxins
- Other microbial and food processing-related toxins
- Phytotoxins and other inherent naturally occurring toxins
- Radioactive isotopes

12.3.2   Priority chemicals for risk analysis

Given the large number of potential environmental contaminants in food, there is a need for national and international consensus with regard to which contaminants are a priority for a detailed assessment of potential risks. Some of the important factors to be considered when selecting priority foods and contaminants for risk assessment have been identified by the UN Food Contamination Monitoring and Assessment Program (GEMS, 1992) and are as follows:

- the potential for human risk posed by the contaminant, including the severity of possible adverse effects (e.g. carcinogenicity) together with information on human exposure;
- the frequency with which a food/contaminant combination is implicated;
- the importance of the food in the total diet (especially staple foods and human milk);
- the feasibility of measuring the level of the contaminant in a reliable manner in an adequate number of food samples;
- the persistence, ubiquity and abundance of the contaminant in the environment;
- the amount of contaminant being discharged into the environment by industry, agriculture and urban centres; and
- the economic importance of the food in international trade.

For any particular contaminant, information is generally available in some but not all of these areas. Ranking of contaminants for priority treatment is thus problematic and is generally biased by the availability of data. Often assessments are driven by necessity, either because of a particular contamination issue or because of the need to establish or review regulatory standards in relation to maximum levels in food.

Toxicity data can often be shared internationally because many environmental contaminants occur in all countries, but dietary exposure information generally needs to be considered on a regional basis and, in some cases, on a local basis, which can be both costly and time consuming. Individual countries may need to generate both toxicity data and exposure data for those contaminants that occur in particular geographical areas, such as some of the mycotoxins and bacterial toxins.

12.4 General aspects of risk assessment

12.4.1 Establishing a causal relationship

Assessing and managing the risk associated with contaminants depends, first, on an ability to predict the potential adverse health effects that may occur in humans as a result of exposure to contaminants and, second, on there being a direct relationship between the level of exposure and the effect(s) observed (i.e. a dose–response relationship). For most environmental contaminants, the low level of exposure means that potential adverse effects may be seen only after a long period of exposure. This can make it difficult to establish a causal relationship between the adverse effect and the contaminant and information on many different aspects related to the contaminant may be needed. These include its chemical and physical properties, the potential adverse effects, its rate of absorption, metabolism and excretion, the presence of a toxicity threshold, the level of exposure, the frequency of exposure and, perhaps most importantly, the mechanism of toxicity. Together, such information may be used to establish the potential risk to humans at different levels of exposure.

Establishing a safe level of exposure depends on determining that level of human exposure at which the observed effects are either known not to occur or can be predicted as unlikely to occur on the basis of current knowledge. Such a prediction often relies on the use of data extrapolation, i.e. making an assumption that cannot be directly inferred from the data available. Extrapolation is an essential component of risk assessment, and particularly so for environmental contaminants. There are basically three extrapolations that can be performed in risk assessment—high dose to low dose, short-term exposure to long-term exposure, and animal to human. High to low dose extrapolation may be used when the level of exposure is below the range of experimental doses; short-term to long-term exposure extrapolation may be used when only the results of short-term studies are available; and animal to human extrapolation will be used in the absence of direct studies on humans or of epidemiological data. The uncertainty that the use of these extrapolations brings to risk assessment is considered in later sections. Often, safety factors are used to make allowances for the uncertainties implicit in the assessment process.

12.4.2   Factors influencing the toxicity of contaminants

Many factors can influence the outcome of exposure to a contaminant, including the level of absorption and its disposition, metabolism and rate of excretion. These factors may in turn be influenced by the presence of other dietary components, the nutritional status of the recipient, as well as age and sex. While it may be possible to establish safe levels of exposure without a complete understanding of the toxicokinetics and metabolism of a contaminant, or indeed of the other factors that influence toxicity, information on these matters will enable a more accurate estimate to be made of the margin of safety between the level in the diet and the exposure level at which adverse effects are likely to occur. It will also enable the identification of particularly sensitive sub-populations. The group of contaminants most studied with regard to these modifying factors are the metals.

12.4.2.1   Nutritional status
The nutritional status of an individual can have a considerable influence on the toxicity of metals and other contaminants. Individuals with low iron status, for example, have increased susceptibility to zinc and copper. There is also evidence that cadmium toxicity is enhanced in communities where the diet is poor. Biological defence mechanisms are directly affected by nutritional factors such as energy, protein, lipids, vitamins and minerals (Parke, 1993). Thus, factors such as nutritional status need to be considered when establishing tolerable levels of exposure to metal contaminants for communities and for some individuals.

12.4.2.2   Speciation of metals
The speciation of metals provides several unique challenges for risk assessment. Speciation refers to the occurrence of metals in different chemical, physical or morphological forms. The total metal concentration in a food may not be useful in the determination of risk because different species of a metal can have different properties that influence solubility, bioavailability and, ultimately, toxicity. Therefore, ideally, each form of a metallic element should be considered individually. Inorganic arsenic, for

example, is rapidly absorbed and is toxic in mammals, whereas organic forms of arsenic are poorly absorbed and pose a low toxicity hazard. In contrast, inorganic mercury poses low toxicity to humans, whereas exposure to methylmercury can cause severe neurotoxic effects.

Some metals also interact with other components of the diet. Such interactions may change the chemical form of the metal; for example phytates may precipitate soluble zinc compounds and thereby prevent their uptake in the gut (Kargacin and Kostial, 1991). Conversely, dietary histidine may also change the chemical form of zinc in food, enhancing its absorption. Similarly, citrate is known to enhance aluminium absorption, and ascorbate has an impact on iron/copper absorption.

Metals in the diet may also interact with other metals, competing for uptake in the gut and causing deficiency. Exposure to high dietary levels of zinc over a period of time interferes with the uptake of copper by competing for transport molecules in the gut, and results in copper deficiency in the affected individuals.

The extent to which speciation of metals can be considered in human risk assessment is determined by the available data. Analytical methods to define different chemical species of metals in food are still being developed.

12.4.3   Use of thresholds in risk assessment

It is generally accepted that animals and humans can tolerate a certain level of exposure to chemicals; that is, there is a threshold of exposure below which adverse effects will not be observed. This view has determined the general regulatory approach to the assessment of risk from chemicals. For most chemicals, the identification of a no-observable-effect level (NOEL) coupled with an appropriate safety factor is used to determine an acceptable (tolerable or 'safe') level of exposure. Support for this approach is enhanced when the mechanism of toxicity of a particular contaminant is known. However, it is also recognised that where the mechanism of toxicity is unknown, it is difficult to demonstrate a clear threshold because of the statistical limitations of experimentation (Purchase and Auton, 1995). This is particularly apparent for carcinogenic chemicals that interact directly with DNA (so-called 'genotoxic carcinogens'). For these chemicals, the threshold of toxicity, if one exists, may be too low to be demonstrated experimentally. In the United States, mathematical modelling is used to calculate the exposure level likely to reduce the risk of cancer to an acceptable level ('virtually safe dose')—usually 1 case in 10^6 population (Lovell, 1988; Olin et al., 1995; Lovell and Thomas, 1996). For those carcinogenic chemicals that do not interact directly with DNA, a threshold of toxicity can often be identified based on the known mechanism of action, even if, as in some cases (e.g. dioxin), it is very low.

12.4.4   Steps in risk assessment

The general framework for analysis of risk described above has been used by a number of national governments and by international agencies (NRC, 1983; IPCS, 1987; ANZFA, 1996a) and by industry groups (ICME, 1996). The process of risk assessment in relation to chemicals can be divided into four distinct steps: hazard identification, hazard characterisation (or dose–response evaluation), exposure evaluation and risk characterisation (see Figure 12.2), each of which is considered below.

These steps have also been recognised by Codex as appropriate for consideration of the health risks associated with substances to be included in the proposed General Standard for Contaminants and Natural Toxins in Food (Codex, 1996b).

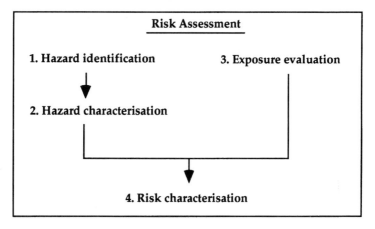

Figure 12.2 Steps in risk assessment.

12.5   Identifying hazards associated with contaminants

The Codex Alimentarius Commission (1996a) defines hazard identification as

> the identification of biological, chemical, and physical agents capable of causing health effects and which may be present in a particular food or group of foods.

12.5.1   Sources of data

The prediction of potential adverse effects caused by a chemical in humans is based on evaluation of data from a number of sources.

Toxicity data can be derived from *in vitro* studies, from the results of controlled toxicity tests in animals, from case studies of human or animal poisonings or, in more limited cases, from the results of epidemiological studies. Consideration of structure–activity relationships can also assist in identifying appropriate studies and possible mechanisms of toxicity. The outcome of the hazard identification process is to identify the major toxicological end-points and the dose levels at which toxic effects occur. These data in themselves are not sufficient for estimating risks, but can assist in identifying priority substances for further evaluation because of the serious nature of the toxicity end-point, e.g. cancer or birth defects.

Environmental chemicals present a particular concern with regard to hazard identification because, first, there is potentially a large number of such substances and, second, there is little incentive to conduct extensive toxicity testing when the level of human exposure is very low, as is the case for many such chemicals. Given this situation, it becomes important to identify which information is the most critical for hazard identification and to use the available information effectively. Increasingly, information on the structure of substances together with the results of short-term or *in vitro* studies is used to predict potential toxicity and focus testing on particular aspects that are likely to be critical in establishing a tolerable level of exposure for humans. While the most useful data are those that are most relevant to humans, data from all sources can contribute to an overall weight-of-evidence approach to hazard identification.

12.5.2   Animal toxicity data

Animal models used to predict the potential hazard of environmental chemicals are selected on the basis of their sensitivity for the particular biological parameter being measured. Thus, in assessing human safety, the conservative assumption is made that humans will be at least as sensitive as the most sensitive animal species. In some circumstances, the animal model may not be suitable or may be inappropriate for predicting effects in humans. Generally, some flexibility is required concerning the type of toxicity studies considered appropriate for assessing hazard and addressing the question of potential risk in particular circumstances.

The biological end-points for which toxicity information on contaminants in food is generally sought are outlined in Environmental Health Criteria No. 70 produced by the International Programme on Chemical Safety (IPCS, 1987). These normally include acute toxicity, reproductive toxicity, genotoxicity, embryotoxicity/teratogenicity, carcinogenicity, organ toxicity, and sometimes other specific end-points such as neurotoxicity, immunotoxicity, and where possible, behavioural and sensitisation studies. The most useful data are those obtained from toxicity tests performed according to internationally accepted protocols such as those published by the OECD (1993).

Another aspect of hazard identification is a consideration of the fate of the contaminant following ingestion. An understanding of its metabolism, distribution and excretion can assist in determining the mechanism of toxicity or, in some cases, the lack of toxicity. The nature of the food in which the contamination occurs can influence the bioavailability of the contaminant and thus the adverse health risk associated with consuming that particular food.

12.5.3   Human toxicity data

The use of data from epidemiological studies in risk assessment can, in some cases, overcome many of the problems inherent in the use of animal studies, such as species specificity, dose extrapolation and the influence of other dietary components. However, epidemiological data also have a number of shortcomings that limit their usefulness in assessing the risk associated with contaminants. These include the lack of a high-exposure group for most contaminants, the low sensitivity of contaminant–effect relationships, and the relatively long time interval between exposure and possible effect (de Hollander, 1996). There have, however, been a number of cases where a disease outbreak has been linked with a particular chemical exposure, e.g. methylmercury in fish (Tsubaki and Irukayama, 1977) and cadmium in rice (Hagino and Yoshioka, 1961).

The development of biomarkers for exposure as well as biomarkers that can be used as an early indicator of disease may improve the sensitivity of epidemiological data in the future (Crews and Hanley, 1995).

12.6   Evaluating dose–response relationships

The Codex Alimentarius Commission (1996a) defines hazard characterisation as

> the qualitative and/or quantitative evaluation of the nature of the adverse effects associated with biological, chemical, and physical agents which may be present in food. For chemical agents, a dose–response assessment should be performed. For biological or physical agents, a dose–response assessment should be performed if the data is obtainable.

Dose–response assessment is defined in the same document as

> the determination of the relationship between the magnitude of exposure (dose) to a chemical, biological or physical agent and the severity and/or frequency of associated adverse health effects (response).

12.6.1   Identifying critical effects

The relationship between the level of exposure and the effect(s) observed is fundamental to establishing the safety of chemicals in food. Hazard characterisation involves a consideration of the most significant toxicity end-points identified in either animal or human studies in relation to the exposure levels used.

The outcomes from a hazard characterisation may include:

- identification of the major toxicological end-points and the dose levels at which they occur;
- if there is a threshold, an estimate of the dose level below which the observed toxicity does not occur, i.e. the no-observed-effect level (NOEL);
- some understanding of the metabolism and kinetics of the substance in a mammalian system; and
- in some cases, information on the mechanism by which the chemical causes the observed toxicity.

The NOEL is generally the dose level just below the range of dose levels at which a dose–response relationship can be demonstrated for a particular adverse effect. The adverse effect should demonstrate a dose–response effect that is both statistically significant and also biologically plausible; that is, the contaminant could be capable of the effect. An understanding of the mechanism of toxicity can assist in this determination.

12.6.2   Establishing a tolerable level of intake

A tolerable level of intake for humans can usually be established on the basis of identifying a threshold for the adverse effects seen in animal or human studies. Generally, extrapolations are required to account for both the species used and the dosage at which the tests were conducted. Thus, the high doses used in animal experiments need to be interpreted in terms of the low doses to which humans are normally exposed. To compensate for uncertainty in the NOEL derived from animal experiments as a measure of safety for humans, the NOEL is generally adjusted by means of safety factors to arrive at a safe or tolerable level of intake for humans. The safety factor (or 'uncertainty factor') used may vary from 10 to 2000 depending on the confidence in the available data (Renwick, 1993) and, in some cases, the nature of the toxicity (Renwick, 1995). For chemicals intentionally added to food, such as food additives, a relatively conservative safety factor of 100 is generally used, which comprises a factor of 10 to account for potential inter-species differences between experimental animals and humans, and a factor of 10 to allow for a variation in response within the human population. If the NOEL is based on human data, a safety factor of 10 may be considered adequate. Environmental contaminants, however, cannot be regarded in the same way as food additives since they are not intentionally added to food and the levels necessarily cannot be adjusted to provide the same margin of safety that can be achieved with a food additive. Therefore, the issue of what is an appropriate safety factor needs to be balanced against available management options for the particular contaminant and the consequences of enforcing strict controls.

The reference value used to indicate the safe level of intake of a contaminant is the so-called 'tolerable intake', which can be calculated on a daily ('tolerable daily intake') or weekly ('tolerable weekly intake') basis. Reference values that define an acceptable level of exposure to a contaminant are established internationally by the Joint Expert (FAO/ WHO) Committee on Food Additives (JECFA). The tolerable intake is also generally considered to be 'provisional', since there is often a paucity of data on the consequences of human exposure at low levels, and new data may result in a change to the tolerable level. For contaminants that may accumulate in the body over time, the provisional tolerable weekly intake (PTWI) is used as a reference value in order to minimise the significance of daily variations in intake. This value generally applies to contaminants such as lead, cadmium and mercury. For contaminants that do not accumulate in the body, such as arsenic, the provisional tolerable daily intake (PTDI) can be used. The PTWI or PTDI are primary health standards that apply to total exposure, i.e. both food and non-food sources. The contribution from non-food exposure, therefore, has to be taken into account when comparing potential dietary exposure from food sources of a contaminant with the PTWI.

For those metals regarded as essential for human health, namely, selenium, zinc, iron, copper, chromium, manganese and molybdenum, it is clearly not appropriate to use a safety factor that will result in a PTWI that is below the level of intake considered essential or optimal for nutritional needs for maintenance of good health. The variety of forms of metals in foods and the variable toxicity of these different metal forms make establishing a single PTWI for each metal difficult and possibly not very useful in estimating risk.

12.7   Estimation of dietary exposure

The Codex Alimentarius Commission (1996a) defines exposure assessment as the qualitative and/or quantitative evaluation of the degree of the likely intake of biological, chemical and physical agents via food as well as exposures from other sources if relevant.

12.7.1 General issues

For contaminants, exposure evaluation involves estimating the level of human dietary exposure to a particular substance from individual foods, the whole diet and, where applicable, from other sources. Where dietary survey data are available, accurate dietary exposure evaluations for specified population groups can be made. In general, dietary exposure estimates for contaminants are based on an estimate of the level of the contaminant in particular commodities together with the known or anticipated consumption of foods containing the contaminant. In estimating risk, exposure estimates are then compared with the PTWI or PTDI for the contaminant.

The use of food consumption data appropriate for a particular situation is fundamental in dietary modelling. Rees and Tennant (1993, 1994) discuss various methods for estimating food chemical dietary exposure, emphasising that the type of food chemical and the purpose of the assessment will determine the most appropriate source of information on food consumption. Using a tiered approach to exposure assessment, initial dietary models will tend to be very conservative and are often used as part of a screening process. As the requirement for more accurate exposure assessment increases, so do the time and cost involved in collecting and analysing the data.

The methods for estimating dietary exposure to food chemicals in order of increasing cost are per capita consumption from food balance sheets, model diets, total diet surveys, individual food consumption surveys and duplicate diet surveys. The different approaches to estimating food chemical dietary exposures are shown in Figure 12.3 and are discussed in detail in later sections.

12.7.1.1 Use of data sets to estimate dietary exposure
One of the difficulties in estimating dietary exposure to contaminants is the need to combine data sets that are not immediately compatible. Data on contaminant levels are generally based on primary agricultural produce, while data on food consumption are generally based on processed or cooked foods. This difficulty can be overcome in two ways:

- by applying 'reduction or concentration factors' (as used for estimating pesticide residue dietary exposures) to apparent food consumption data (food balance sheet data) for commodities to

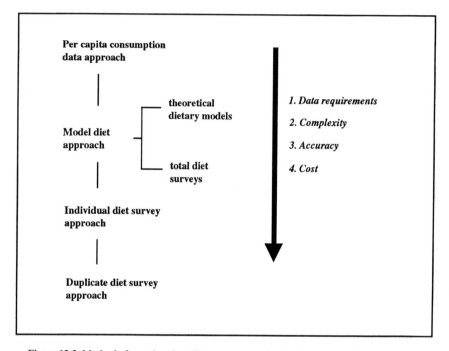

Figure 12.3 Methods for estimating dietary exposure in relation to complexity and cost.

account for potential losses or increases of contaminant concentrations through the food storage, processing and preparation stages;
- by using recipe information to disaggregate data on food consumption of processed foods derived from population dietary surveys to determine consumption of primary food commodities.

While both methods may be useful in some circumstances, the first method contains many uncertainties and the latter method is likely to be the one used for the majority of contaminants.

12.7.1.2 International guidelines

International guidelines for estimating dietary exposure of chemical contaminants have been prepared under the joint sponsorship of UNEP, FAO and WHO (GEMS, 1985, 1992, 1995). The guidelines give details of suitable sampling and analytical methods as well as examples of the use of food consumption data from different sources and suggest three basic approaches to determining dietary exposures:

- consideration of food consumption data together with data on contaminant levels in foods of major dietary importance generated from monitoring programmes;

- total diet studies (analysis of a 'standard' market basket), with foods usually prepared for customary consumption; and
- analysis of duplicate portions of food consumed by selected individuals.

The approaches considered below are, in general, consistent with these guidelines. Total diet studies can be considered as a special 'model diet' approach.

12.7.2   Specific approaches for estimating dietary exposure

12.7.2.1   Per capita consumption data approach
Using this approach, potential total dietary exposure to contaminants is estimated by combining per capita consumption data for each commodity with the level of contaminant in each commodity. Per capita consumption data are available for most countries and are derived from food balance sheets (FBS), in which food production (commercial and estimated home production), food imports, exports and estimated non-food use and waste are taken into account. The total amount of food (raw commodities) available for consumption is then divided by total population figure, estimating the dietary exposure to a contaminant as an average value per head of population, irrespective of age or gender. FBS data tend to overestimate actual consumption in industrialised countries because data represent food available for consumption only, with no account being taken of food wasted at a household level. In countries where subsistence farming provides an important food source, FBS data may underestimate food consumption. The data are also dependent on the availability of accurate data on population size, though this is less of a limitation than the accuracy of food supply data.

For some contaminants, determination of per capita consumption may not provide the detailed information required to provide a comprehensive assessment of risk because data cannot be provided for specific population subgroups. Furthermore, for many commodities there may be no available data on appropriate reduction or concentration factors with which to account for changes in the level of contamination when a raw commodity is processed. The per capita consumption method will have limited value when the contaminant occurs in a minor commodity because FBS data are normally for major food commodities only.

12.7.2.2   Model diet approach using population data
This method provides a more accurate estimate of contaminant exposure by constructing theoretical dietary models using food consumption data for populations from national dietary surveys or other sources combined with analytical data on contaminant levels in foods. The use of model

diets enables all sources of contaminants in the diet to be included in the analysis.

Using this method, the estimated total dietary exposure to a contaminant can be determined both for a person consuming at an 'average' level of consumption and at a high level of consumption. For the high-level consumer, it may also be important to distinguish, where possible, between a commonly consumed food (such as a staple food) and an occasionally consumed food, since the pattern of consumption for the high-level consumer is different for each.

For food consumption data derived from population surveys and model diets, the quality of such data or such a model is limited by the survey method and the representativeness of the sample. Data from national dietary surveys are usually based on records of daily diets (24-hour recall, one-day diary record, or weighed records). The general limitations of dietary methods used to collect dietary records for individuals have been extensively reviewed elsewhere (Block, 1989; Bingham, 1991; Sempos, 1992). Limitations depend on whether the dietary method was retrospective (recall, food frequency) or prospective (3–7-day diary records or weighed records). Both may include memory errors, reporting and coding errors and, for prospective methods, errors due to potential changes in dietary behaviour as a result of participation in the survey. Although food frequency surveys can provide useful information on patterns of food consumption over time, generally, the surveys do not provide adequate data on their own for dietary modelling purposes because there is insufficient detail on specific foods eaten and on the amount consumed.

There are also specific constraints on the use of food consumption data in dietary exposure assessments. Such constraints may be due to the methods applied to collect the data or to the ability to disaggregate the data to identify potential 'at risk' groups. In practice, cross-sectional dietary surveys are often the only food consumption data available on individuals. Short-term or cross-sectional dietary survey data are usually sufficient for use in acute exposure assessments. Food consumption data used in chronic dietary exposure assessments should represent the mean consumption of the whole population (consumers and non-consumers), to capture eating patterns over a lifetime of exposure (Petersen et al., 1994). Direct extrapolations from short-term food consumption data to lifetime food consumption are made by averaging daily food consumption over several age groups.

There are several limitations to using cross-sectional food consumption data to assess longitudinal food consumption in chronic exposure assessments. First, there is considerable variation in day-to-day consumption and, second, food consumption usually changes with age and over a given period of time. The extent of variation may change according to different population subgroups, in particular, for those groups where food choice is limited, for example, vegans. Third, it is difficult to account for the frequency of consumption of specific food items. Food frequency data may provide this information.

Average intake level consumer. The most appropriate measure of aver-age consumption for both a commonly consumed and an occasionally consumed food is the mean consumption for the whole population ('all respondents' in a national dietary survey) (Petersen *et al.*, 1994). The most common measure of the level of a contaminant in food is the median level in the surveyed food. The median is normally used in order to avoid the bias caused by a low number of samples and, for some contaminants, the skewed nature of the distribution curve.

The contribution of an individual commodity to the total average contaminant intake can only be considered in the context of an 'average' diet, i.e., average intake figures from all commodities containing the average (median) level of the contaminant. Thus:

Average dietary exposure to a = Mean consumption of the
contaminant from a specified   commodity × median
commodity   contaminant content for the
                                 commodity

and

Total average dietary exposure to a = $\sum$ Average dietary exposure to a
contaminant from all commodities   contaminant for each commodity

High-level consumer. The most appropriate measure of high consumption for a commonly consumed food is the 95th centile (by weight) food consumption for the 'consumers only' group (as identified in a national dietary survey), since an individual may consume such foods at a high level every day for a lifetime.

For occasionally consumed foods, however, the most appropriate measure of high consumption is considered to be the median intake level for 'consumers only', since consumption of such foods does not occur every day. However, care must be taken that the food is not consumed by a particular ethnic group or other population subgroups on a regular basis, in which case the intake level may be underestimated. The identification of an 'occasionally consumed food' should be based on food frequency data.

The contribution to the total contaminant dietary exposure from high consumption of a specific commodity can then be used in the dietary exposure estimate. Thus:

Contaminant intake from commodity A for the high consumer of this commodity = 95th centile consumption for the 'consumers only' group × median contaminant level for commodity A

For commodities that have been identified as occasionally consumed, it is considered more realistic to use the median consumption instead of the 95th centile level of consumption to represent high consumption over an extended period of time.

Contaminant dietary exposure from commodity B for the high consumer of this commodity = Median consumption for the 'consumers only' group × median contaminant level for commodity B

and therefore

Total contaminant dietary exposure for the high consumer of commodity B = (Total average dietary exposure of the contaminant from all foods – average contribution from commodity B) + high consumer dietary exposure for the contaminant from commodity B

The limitations of the model diet approach are as follows:

- The accuracy of the model is limited by the number of commodities for which data are available and the breadth and quality of the survey data.
- It is necessary to describe a diet in terms of the amounts of primary agricultural commodities using processing factors.
- Consideration needs to be taken of the difference in consumption pattern of a high consumer of a staple food compared to that of an occasionally consumed food.

12.7.2.3   Model diet approach using a total diet (market basket) survey
The use of total diet surveys to estimate dietary exposure to contaminants is a specialised model diet approach, where contaminant levels are measured in foods as consumed (prepared to a table-ready state) rather than in raw commodities. Total diet surveys can achieve the most accurate estimates of dietary exposure for contaminants found in the food supply and are commonly used to evaluate current risk management procedures. In both Australia (ANZFA, 1996b) and New Zealand (MoH, 1995), such surveys are conducted to estimate total dietary exposure to

pesticides, heavy metals and organic contaminants. These surveys are based on the approach recommended by GEMS (1992). Total dietary surveys have the following main limitations.

- The range of foods surveyed tends to be limited and single foods are used to represent the larger food group. Thus dietary exposure estimates may underestimate or overestimate potential exposures from that food group.
- Use of the mean of the contaminant levels found may overestimate the actual contaminant levels because of the skewed distribution often found in the samples.
- Assigning non-zero values when the contaminant is below the level of determination may also lead to an overestimate of dietary exposure.

Caution should be used when comparing data from different countries since different sampling procedures, sample sizes of foods analysed and sample preparation may make comparisons invalid.

12.7.2.4   Individual dietary record approach

A more accurate estimate of contaminant dietary exposure is made possible by using individual dietary records and body weight from national dietary surveys and combining this information with analytical data on contaminant levels. The limitations of national dietary survey data for chronic dietary exposure assessments have been mentioned in the previous section.

When individual dietary records are used, contaminant dietary exposures are first calculated for each individual, using that individual's body weight, before deriving population estimates of contaminant exposure, either for individual commodities or the total diet. These data can then be analysed by population subgroups, such as age and sex, and are particularly useful in the analysis of national dietary trends for various population subgroups and for studying long-term trends in the consumption of particular foods known to contain contaminants.

Thus, for a widely distributed contaminant, such as cadmium, which occurs in both staple and non-staple foods, major food sources containing the contaminant can be readily identified if appropriate survey data are available. Potential dietary exposures may be determined for the whole population or only for consumers of the foods of interest. Dietary exposure may also be determined for high consumers of individual foods that are either commonly or only occasionally consumed.

12.7.2.5   Duplicate diet survey approach

Using this approach, a duplicate diet of each individual in the survey is analysed for contaminants of interest. The duplicate diet method is considered the most accurate method available for determining the contaminant intake of individuals. It is not normally used, however, because of cost and time constraints. However, if surveillance indicates that particular subgroups of the population may be at risk, duplicate diet surveys may be needed for confirmation. Although high dietary exposures to potential contaminants for individuals can be estimated, the method cannot identify single food sources of contamination.

12.8   Characterising risk

The Codex Alimentarius Commission (1996a) defines risk characterisation as

> the qualitative and/or quantitative estimation, including attendant uncertainties, of the probability of occurrence and severity of known potential adverse health effects in a given population based on hazard identification, hazard characterisation and exposure.

12.8.1   Estimates of risk

Risk characterisation brings together information gained from previous steps and provides a practical estimate of the severity of the risk for a given population. Risk characterisation for a contaminant might be expressed as a margin of safety between the tolerable level of intake, based on the known hazard, and the known level of human exposure via the diet. The degree of confidence in the final estimation of risk depends on the level of uncertainty identified for each of the elements involved. Risk management strategies are then formulated on the basis of this determination.

12.8.2   Uncertainty in risk assessment

Acceptance of a degree of uncertainty is fundamental to an estimation of risk. The basis of this uncertainty is two-fold: first, there is uncertainty regarding the quantity and quality of the information upon which the risk estimate is made; second, there is uncertainty regarding the validity of the assumptions used in the estimation of risk, such as species and dose extrapolations. Together, these determine the degree of uncertainty that is associated with a particular risk estimation. Some of the factors that contribute to uncertainty in risk assessment are provided in Table 12.2, and are discussed in more detail other sections.

While there are inherent limitations in the risk assessment process, it provides a structured and logical approach to the determination of the risks associated with dietary exposure to environmental contaminants. The limitations can be minimised, and sometimes overcome, by careful consideration of the data available and by focused research that will provide data on specific matters where there is uncertainty.

Table 12.2 Factors that contribute to uncertainty in risk assessment

- Quality and quantity of the toxicity data
- Use of animal models
- Need for dose–response extrapolation
- Determination of a threshold dose level
- Lack of information on the target dose
- Basis for determination of safety factors
- Variability in the contaminant level in food
- Variability of individual diets over lifetime
- Availability of data on individual dietary exposure

12.8.3   Interpreting risk assessment data

While the methodology of risk assessment can in many cases provide an accurate estimate of risk, it can also be open to misinterpretation if the caveats that necessarily accompany the assessment are ignored. A fundamental issue for contaminants is the determination of the tolerable level of intake (either the PTWI or the PTDI). The nature of the toxicity end-point upon which the tolerable intake is established can influence consideration of the risk associated with exceeding this level. For a toxicological end-point that is associated with cumulative toxicity over a lifetime of exposure, e.g. in cadmium-induced renal toxicity, the risk of short-term exposure over the tolerable level of intake is minimal. On the other hand, for a toxicological end-point associated with more immediate hazard, e.g. enzyme inhibition induced by hydrocyanic acid or a teratogenic effect, exposure over the tolerable level of intake may be critical. The use of an additional safety factor to take account of the nature of the toxicity end-point has been suggested by Renwick (1995).

The risk associated with exceeding the PTWI is dependent on many factors, as discussed by Renwick and Walker (1993) and must be assessed on a compound-to-compound basis. Factors include the slope of the dose–response curve, the nature of the toxic end-point, the half-life of the contaminant following absorption, and the period of exposure. In some cases, the risk may also need to be considered against the availability of alternative food sources.

12.9   Options for risk management

12.9.1   General approaches

There are basically two approaches to managing the human health risks associated with the presence of contaminants in food:

- to control or reduce the levels of a contaminant in the food; or
- to control or reduce the level of consumption of contaminated food.

The first of these approaches is the most common and can be achieved by a variety of legislative methods. The second approach generally involves a public education programme and may be necessary when part of the food supply is found to be highly contaminated either as a result of undetected industrial activity or because of a naturally occurring high level of contamination.

12.9.2   Controlling contamination

Contamination of food with potentially hazardous substances can occur at all stages of food production. In many cases, the level of potential contamination with substances from environmental sources is low because of the existence of environmental and planning legislation that operates in conjunction with agricultural or industrial best practices. Control of the source of contamination, e.g. PCB concentrations in the environment, can often effectively control food contamination. Similarly, for mycotoxins, improvements in produce handling and storage may reduce levels in the final foods. Decontamination processes can also be effective for mycotoxins. It may not always be possible, however, to reduce either the range of foods in which a contaminant occurs or the contaminant concentration in a particular food, especially for contaminants that are ubiquitous to the environment, such as the metals cadmium and mercury. For both of these metals, there are naturally occurring levels in marine fish and other foods that are unrelated to industrial activities. Reduction of the level of the contaminant in food in these cases is not an option for risk management.

A generally recognised benchmark for control of contaminants is that the levels in food should be as low as reasonably achievable, i.e. the 'irreducible level'. The irreducible level is defined by JECFA as 'that concentration of a substance which cannot be eliminated from a food without involving the discarding of that food altogether, severely compromising the ultimate availability of major food supplies' (IPCS, 1987). Action to ensure that contamination of food is as low as reasonably achievable may be taken in a variety of ways, including the development of control measures at the source of contamination, the enforcement of both good agricultural practice (GAP) and good manufacturing practice (GMP) and, if necessary, the establishment of food standards that stipulate the maximum permitted concentrations of contaminants in particular commodities.

12.9.3   Maximum levels for contaminants

The maximum permitted concentration (MPC) or maximum level (ML) for a contaminant is the maximum concentration of a substance permitted in a food commodity. In assessing whether a maximum level is necessary for a particular contaminant, the effectiveness of existing controls for minimising contamination, and the potential for these controls to fail, need to be considered. Maximum levels for contaminants are normally set at the upper end of or above the range of contaminant levels normally found in the primary agricultural commodities. Maximum levels are established on the basis that sound production methods and natural resource management practices have been used throughout the food production system.

12.9.4   Carcinogenic contaminants

The issue of assessing the risk associated with chemicals in food that are capable of inducing cancer is particularly complex for a number of reasons. First, there is a widespread belief that there is no safe level of exposure for environmental carcinogens that operate through a genotoxic mechanism. This has led in some cases to extraordinary measures being taken to minimise exposure to some environmental contaminants. This approach to risk management is in sharp contrast to that taken with many other chemicals, which either occur naturally at relatively high levels in food or are present as a result of microbial contamination, and which are also capable of inducing cancer under certain circumstances (Scheuplein, 1992). The use of complex mathematical models to establish a 'safe' level of exposure to some environmental chemicals has possibly distorted their relative significance in contributing to the incidence of human cancers. Second, the factors influencing diet-related cancer are complex and cannot be related in a simplistic manner to the level of dietary exposure to chemicals that are capable of inducing cancer. The influence of other dietary factors, particularly macromolecules, on human cancer needs to be considered (Rowland, 1991; NRC, 1996). While it is appropriate to minimise exposure to carcinogenic chemicals, the extent to which this is necessary for a particular contaminant must be considered in the context of the potential benefits of the dietary source of the contaminant, as well as in relation to the anticipated decrease in incidence of human cancers.

The issues surrounding risk assessment and subsequent risk management of environmental carcinogens are the subject of wide and extensive debate largely focusing on the validity of the conservative assumptions and mathematical models used for low-dose extrapolation of risk (Clayson and Iverson, 1996). The outcome of this debate may be a more balanced perspective on carcinogenic risk (Moolenaar, 1996).

12.9.5   Controlling dietary exposure

In situations where a part of the food supply has become contaminated, the only option for risk management may be to encourage consumers to reduce the level of exposure to that particular food. This situation may also arise where particular subpopulations are found to be high consumers of a food that contains a naturally high level of a particular contaminant. This approach to risk reduction needs careful management, taking into account cultural and social norms.

12.9.6   Surveillance

Surveillance of contaminants in food provides the mechanism by which risk management decisions can be evaluated and modified if necessary and, as such, are an important part of the overall risk management process. Surveillance provides information on compliance with current regulations as well as surveying long-term trends in the level of contaminants in food.

12.10   Future directions

There is at present considerable interest in refining the methodology used in risk assessment in order, first, to focus resources on areas where positive public health outcomes can be achieved and, second, to define more clearly the margin of safety in particular situations of human exposure to contaminants by reducing the level of uncertainty in the data.

There are a number of areas where this might be achieved. In hazard identification and characterisation, for many chemicals, there is a poor understanding of the mechanisms of toxicity and, therefore, of the basis on which a threshold of toxicity might be identified. For contaminants, the usefulness of broadly focused toxicity studies is now limited and is being replaced by mechanistic studies on specific toxicological end-points, and, if possible, human data.

The recognition of the need for accurate dietary exposure information has probably been the most significant change in risk assessment in recent years. Refinement of the methodology used to estimate dietary exposure to contaminants and other chemicals in foods is likely to continue and to become more widely accepted and used in future.

The credibility and acceptability of risk assessment as a process for determining the health risks associated with contaminants in food will depend on its ability to provide positive health outcomes for the community. Confidence in the ability to achieve these health outcomes needs to be tempered by a realistic appraisal of the uncertainty in the risk assessment process, as well as the real costs of achieving positive health outcomes. This will, in the long term, lead to a better understanding by the community of the complexity of the relationship between food consumption and public health, and build public confidence in the food regulatory system.

References

ANZFA (1996a) *Framework for the Assessment and Management of Food-related Health Risks*, Australia New Zealand Food Authority, Canberra.

ANZFA (1996b) *The 1994 Australian Market Basket Survey*. Australia New Zealand Food Authority, Canberra.

Bingham, S.A. (1991) Limitations of the various methods for collecting dietary intake data. *Annals of Nutrition Metabolism*, **35** 117-127.

Block, G. (1989) Human dietary assessment: methods and issues. *Preventive Medicine*, **18** 653-660.

BMA (1987) Perception and acceptability of risk, in *Living with Risk*, Chapter 14. British Medical Association/Wiley, Chichester.

Clayson, D.B. and Iverson, F. (1996) Cancer risk assessment at the crossroads: The need to turn to a biological approach. *Regulatory Toxicology and Pharmacology*, **24** 45-59.

Codex (1995) *Draft Preamble to the Codex General Standard for Contaminants and Toxins*, Codex Alimentarius Commission, Alinorm 95/12A, Appendix VI.

Codex (1996a) *Terms and Definitions Used in Risk Assessment*, Codex Alimentarius Commission. Doc. CX/EXEC. 96/43/6, Annex 1, June 1996.

Codex (1996b) *Codex Risk Assessment and Management Procedures: Methods to Ensure Public Safety while Developing the Codex General Standard for Contaminants and Toxins in Food*, Codex Alimentarius Commission CX/FAC 96/15.

Crews, H.M. and Hanley, A.B. (eds) (1995) *Biomarkers in Food Chemical Risk Assessment*, Royal Society of Chemistry, London.

de Hollander, A.E.M. (1996) Risk assessment, risk evaluation and risk management, in *Food Safety and Toxicity* (ed. J. de Vries), CRC Press, Boca Raton, FL, pp 267-282.

FAO/WHO (1995) *Application of Risk Analysis to Food Standards Issues*, Report of the Joint FAO/WHO Expert Consultation, Geneva, Switzerland, 13–17 March 1995.

GEMS (1985) *Guidelines for the Study of Dietary Intakes of Chemical Contaminants*, Global Environmental Monitoring System, Joint UNEP/FAO/WHO Sponsorship, WHO, Geneva.

GEMS (1992) *Assessment of Dietary Intake of Chemical Contaminants*, Global Environmental Monitoring System (GEMS), Joint UNEP/FAO/WHO Sponsorship, Food Contamination Monitoring and Assessment Progam, UNEP, Nairobi.

GEMS (1995) *Guidelines for the Study of Dietary Intakes of Chemical Contaminants*, Global Environment Monitoring System (GEMS), Joint UNEP/FAO/WHO Sponsorship, WHO Offset publication No. 87, WHO, Geneva.

Hagino, N. and Yoshioka, K. (1961) A study of the etiology of itai-itai disease. *Journal of the Japanese Orthopaedic Association*, **35** 812-815.

ICME (1996) *Risk Assessment and Risk Management of Non-ferrous Metals* (draft) International Council on Metals and the Environment, Ottawa, Canada.

IPCS (1987) *Principles for the Safety Assessment of Food Additives and Contaminants in Food*, Environmental Health Criteria 70, International Programme on Chemical Safety, UNEP/FAO/WHO joint sponsorship, WHO, Geneva.

Kargacin, B. and Kostial, K. (1991) Toxic metals: influence of macromolecular dietary components on metabolism and toxicity, in *Nutrition, Toxicity and Cancer* (ed. I. Rowland) CRC Press, Boca Raton, FL, pp 197-222.

Lovell, D.P. (1988) Risk assessment of chemicals, in *Experimental Toxicology: The Basic Issues* (eds D. Anderson and D.M. Conning), Royal Society of Chemistry, London, pp 414-435.

Lovell, D.P. and Thomas, G. (1996) Quantitative risk assessment and the limitations of the linearised multistage model. *Human and Experimental Toxicology*, **15** 87-104.

MoH (1995) *The 1990/1991 New Zealand Total Diet Survey*, parts 1–4, ESR: Ministry of Health, Wellington Science Centre, Wellington.

Moolenaar, R.J. (1996) A perspective on carcinogen risk assessment. *Regulatory Toxicology and Pharmacology*, **23** 241-243.

NRC (1983) *Risk Assessment in the Federal Government*, A report of the Committee of the Institutional Means for Assessing Risks to Public Health. National Research Council, National Academy Press, Washington, DC.

NRC (1996) *Carcinogens and Anticarcinogens in the Human Diet. A Comparison of Naturally Occuring and Synthetic Substances*, National Academy Press, Washington, DC.

OECD (1993) *Guidelines for Testing of Chemicals*, 2nd edn, OECD Publications, Paris.

Olin, S., Farland, W., Park, C., Rhomberg, L., Scheuplein, R., Starr, T. and Wilson, J. (eds) (1995) *Low-dose Extrapolation of Cancer Risks—Issues and Perspectives*, International Life Science Institute, ILSI Press, Washington, DC.

Parke, D. (ed.) (1993) The importance of diet and nutrition in the detoxification of chemicals, in *Food, Nutrition and Chemical Toxicity* (eds D. Parke, C. Ioannides and R. Walker), Smith-Gordon, London, pp 1-16.

Petersen, B.J., Chaisson, C.F. and Douglass, J.S. (1994) Use of food intake surveys to estimate exposure to nonnutrients. *American Journal of Clinical Nutrition*, **59** (suppl) 240S-244S.

Purchase, I.F.H. and Auton, T.R. (1995) Thresholds in chemical carcinogenesis. *Regulatory Toxicololgy and Pharmacology*, **22** 199-205.

Rees, N.M.A. and Tennant, D.R. (1993) Estimating consumer intakes of food chemical contaminants, in *Safety of Chemicals in Food Chemical Contaminants* (ed. D.H. Watson), Ellis Horwood, Chichester.

Rees, N.M.A. and Tennant, D.R. (1994) Estimations of food chemical intake, in *Nutritional Toxicology* (eds F. N. Kotsonis, M. Mackey and J. Hjelle), Raven Press, New York.

Renwick, A. (1993) Data-derived safety factors for the evaluation of food additives and environmental contaminants. *Food Additives and Contaminants*, **10** 275-305.

Renwick, A. (1995) The use of an additional safety or uncertainty factor for nature of toxicity in the estimation of acceptable daily intake and tolerable daily intake values. *Regulatory Toxicology and Pharmacology*, **22** 350-261.

Renwick, A. and Walker, R. (1993) An analysis of the risk of exceeding the acceptable or tolerable daily intake. *Regulatory Toxicological and Pharmacology*, **18** 463-480.

Rowland, I. (ed.) (1991) *Nutrition, Toxicity and Cancer*, CRC Press, Boca Raton, FL.

Sandman, P.M. (1989) Hazard versus outrage in the public perception of risk, in *Effective Risk Communication: The Role and Responsibility of Government and Non-government Organisations*. (eds V.T. Corvello, D.B. McCallum and M.T. Pavlova), Plenum Press, New York, pp 45-49.

Scheuplein, R.J. (1992) Perspectives on toxicological risk—an example: foodborne carcinogenic risk. *Critical Reviews in Food Science Nutrition*, **32** 105-121.

Segal, M. (1990) Determining risk. *FDA Consumer*, June, 7-11.

Sempos, C.T., Briefel, R.B., Flegal, K.M., Johnson, C., Murphy, R.S. and Wotechi, C.E. (1992) Factors involved in selecting a dietary survey methodology for national nutrition surveys. *Australian Journal of Nutrition and Dietitics*, **49** (3) 96-101.

Slovic, P. (1987) Perception of risk. *Science*, **236** 280-285.

Soby, B.A., Simpson, A.C.D. and Ives, D.P. (1994) Managing food-related risks: integrating public and scientific judgements. *Food Control*, **5** 9-19.

Tsubaki, T. and Irukayama, K. (eds) (1977) *Minamata Disease*, Elsevier Science, Amsterdam.

13 Safety standards for food contaminants

Farid E. Ahmed

13.1 Introduction

Food is a complex mixture of nutrient and anutrient substances. Among the nutrient substances are macronutrients (e.g., carbohydrates, fats and proteins) and micronutrients (e.g., vitamins and minerals). Anutrient substances include naturally occurring substances (e.g., hormones and natural pesticides), and other substances added during food processing, or as a result of contamination of the food supply (Kotsonis *et al.*, 1995).

In order to determine whether consumption of any food may pose a risk due to the presence of deleterious environmental contaminants, it is essential to determine dietary intake of food with a reasonable degree of certainty, followed by estimating the actual intake of a particular contaminant in that food (as a measure of exposure), and then deciding whether a contaminant may pose a toxicological risk at that particular level (risk characterization) (Ahmed, 1991). Thus, dietary intake studies of food are an essential first step in determining food safety.

Chemical contaminants in food may arise from industrial pollution, from agricultural practices, or from food preparation and storage processes. Moreover, some contaminants may originate from natural geological formations, or from fungal contamination. These contaminants are likely to enter the human food supply and present a potential health hazard, as well as impede trade in affected foods (UNEP/FAO/WHO, 1988).

This chapter describes the surveys performed in the United States to estimate dietary intakes of various foods and elucidate their limitations, and critically reviews data on the distribution of various contaminants in the food supply. Because consumption of fishery products is increasing worldwide, and because they tend to accumulate certain chemicals in their aquatic environment and concentrate them through bioaccumulation (Ahmed, 1991), public health hazards due to organic contaminants of seafoods are examined.

The problems of food contamination are global. American studies are compared with studies performed in other countries. The importance of harmonizing standards globally to achieve the safety of the food supply, and the economic benefits to exporting countries that follow safety practices enabling them to participate in international food trade are discussed. Health and economic impacts of consuming contaminated food as well as comparison with other societal risks are reviewed, and the role of various government regulations, industry and consumer advocate activities in reducing contamination of the food supply and protecting public health considered.

In addition to chemical contamination of food, environmental intoxication can also result from plants, animals, and/or microorganisms. Although naturally occurring microbial food contaminants are considered a bigger threat to the world's general health than synthetic environmental contaminants, and most food-related illnesses are due to microbial contamination (Ahmed, 1992a,b; Kotsonis *et al.*, 1995; Miller, 1997; WHO/FAO, 1995), aetiological agents often associated with food-borne intoxications are outside the scope of this chapter.

13.2   US dietary data and their limitations

13.2.1   Group and individual dietary intake data

In most countries, average dietary intake is estimated from national food supply data or from national surveys of food intake by households or individuals. Food supply is estimated by adding the quantity of food imported to the quantity produced within a country and then subtracting the sum of food exported, destroyed by pests during storage, or put to non-food use such as production of industrial alcohol. The final figure is divided by the total population to obtain the average per capita food availability. The results are estimates of food that disappear into wholesale and retail markets; they fail to account for food wasted before consumption, food fed to pets, and home-grown foods if they are not included in the production data. In household food inventories, food consumed is estimated by recording the difference between inventories of foods available at the beginning and end of the study period (usually one week) and accounting for food purchased or brought into the house. Average per capita intake is estimated by dividing total household food intake by the number of people in that home (NRC, 1989).

Per capita intakes by different age–sex groups in the United States are provided through national surveys conducted by various federal government agencies using methods to assess individual intakes of food and dietary recalls, both of which include diet histories and food frequency questionnaires. These surveys are discussed below.

13.2.2   US national surveys

13.2.2.1   Historical perspective on food
The diet of Paleolithic hunter-gatherers is believed to have contained no dairy products or cereal grains, and to have consisted of approximately

35% meat and 65% plant food, mainly vegetables and fruits (Eaton and Konner, 1985). Two important major changes in the food supply occurred. The first major change happened around 10 000 BC, when people changed their lives from nomadic to agrarian; they began consuming dairy products and cereal grains. The second important change occurred with the onslaught of the Industrial Revolution of the 1800s. Industrialization enhanced production of variety in foods to meet the needs of people (Tannahill, 1973). A marked change in the quantity and quality of food occurred when large numbers of people left farming; in 1800, approximately 95% of the US population consumed minimally processed foods grown and produced on their own farms, but, by 1900, only 60% of people remained as farmers. Today, nearly the entire US population is dependent on others to produce and distribute food. The construction of railroads across the United States in the mid-1880s changed the character of the food supply, as this mode of transportation facilitated shipping of food and enhanced the advent of refrigerated railcars and trucks. In addition, innovations in food processing resulted in availability of canned, frozen, fermented, dehydrated and synthetic foods (Hampe and Wittenberg, 1964).

13.2.2.2 Surveys conducted by the US Department of Agriculture (USDA)

Since 1935, the USDA's Human Nutrition Information Service (HNIS) has conducted the Nationwide Food Consumption Surveys (NFCS) approximately every 7 to 10 years. NFCS focuses on food use of households and the dietary intakes and patterns of individuals. Data from 1935 to 1955 did not take into account food waste, or how food is distributed among household members. Since 1965, data on intake by individuals have been collected (Pao et al., 1982). Continuing Surveys of Food Intakes of Individuals (CSFII) have been conducted by the USDA since 1985, providing information on men and women 19 to 50 years of age, having either low or regular income, and on their children 1 to 15 years of age (USDA, 1985; 1986a,b; 1987a,b,c; 1988). CSFII have been studied by the USDA in 1985, 1986, 1989, 1990 and 1991 (Kotsonis et al., 1995).Today, the USDA estimates per capita use of foods or food groups by dividing total available food by the population of the entire country, as is done in many other countries (Pao et al., 1982; UNEP, 1992).

13.2.2.3 Surveys conducted by the US Department of Health and Human Services (DHHS)

The DHHS studied low-income people in 10 states during 1968 to 1970, the so-called 'Ten-State Nutrition Survey' (DHEW, 1972). In the biennial 'Food Label and Package Survey', a statistically representative sample of packaged food products was studied for information on ingredients and the extent of nutrient labelling (Woteki, 1986). The 'Total Diet Study' conducted by the DHHS's US Food and Drug Administration (FDA) is the only national study conducted in the United States for estimating average intakes of pesticides, toxic substances, radionuclides and industrial chemicals; it also provides estimates of dietary intakes of certain essential elements such as iodine, iron, sodium, potassium, copper, magnesium and zinc. The objective of the study was to determine safe intake levels of industrial chemicals and pesticides consumed by humans over a lifetime, without risk of developing cancer. The extent to which selected age–sex groups (males and females aged 6 to 11 months, 2 years, 14 to 16 years, 25 to 30 years, and 60 to 65 years) are exposed to harmful agents and to essential minerals in the diet can be estimated from this survey (Gunderson, 1988). Four times a year, 234 different food items representing US diets were purchased in grocery stores across the country and analysed in the FDA laboratories for contaminants. Although, in the diets looked at in this study, no pesticides have been found above tolerance levels as far back as 1961, the FDA's laboratory detection methods were not sensitive, and the number and kind of samples taken were not comprehensive enough either to determine their presence or to assess their deleterious effect on human health (Ahmed, 1991; NRC, 1989).

The National Health and Nutrition Examination Survey (NHANES) was conducted by the National Center for Health Statistics (NCHS) to monitor the overall nutritional status of the US population through health and medical histories, dietary interviews, physical examinations and laboratory studies. NHANES I was conducted between 1971 and 1974; NHANES II between 19776 and 1980, and the Hispanic HANES (HHANES) between 1982 and 1984 (Madans et al., 1986). NHANES III, which began in 1988, follows up people through their lives, surveying them at regular intervals (NRC, 1989).

13.2.3 Coordination of nutrition monitoring in the United States

13.2.3.1 National Nutrition Monitoring System (NNMS)
As a result of a 1979 Congressional directive, the USDA and DHHS developed in 1981 a Joint Implementation Plan for a National Nutrition Monitoring System (NNMS). In 1983, a Joint Nutrition Monitoring Evaluation Committee (JNMEC) was established to coordinate the survey methods used by the two departments to report their findings. The first report was issued in 1986. It provided food intake data derived from the 1977–78 NFCS, and information on nutritional status based on biochemical analyses from NHANES II (1976–80). USDA and DHHS

coordinated their survey methods, to publish results in a timely fashion, to conduct NFCS more frequently and to add longitudinal epidemiological aspects to data collected in NHANES (NRC, 1989).

13.2.3.2 Coordinated State Surveillance System (CSSS)

The nutritional status of high-risk children and pregnant women is monitored by the Centers for Disease Control (CDC) in Atlanta, Georgia, for information obtained from service delivery programmes operated by selected state and metropolitan health jurisdictions; data such as the prevalence of overweight, underweight, retarded growth and anaemia among high-risk children, and data on anaemia, abnormal weight changes, fetal survival, birth weights and infant feeding practices for pregnant women are included (Ahmed, 1991; NRC, 1989). A summary of food constituents reported by US national surveys is presented in Table 13.1.

13.2.4 Limitations and comparability of US data

13.2.4.1 Limitations of food intake assessment methods

Assessing dietary intake of humans accurately is a very difficult task. Each of the assessment methods has its weakness. The choice of method depends on whether assessment pertains to average intake of a group or to habitual intakes of individuals within a group, the level of detail desired (e.g., food groups, foods or nutrients) and the accuracy desired in determining food consumption. In addition, costs, burden on respondents, availability of resources and trained personnel, and availability of assessment methods are other compounding factors.

The most accurate data methods of assessment are largely free of random and systemic errors such as: sampling errors, non-response bias, reporting errors relative to day-to-day dietary variability, interviewer bias, and errors due to use of food consumption tables (NRC, 1989). Each method (e.g. food records, 24-hour recall, diet histories and food composition tables) has its strengths and weaknesses, and none is suitable for every purpose. Consequently, biological markers are becoming useful tools to validate assessment of food intakes. Examples are the 24-hour urinary nitrogen excretion as an estimate of protein intake, toenail levels of selenium to assess selenium intake, and adipose tissue concentrations of fatty acids to assess types of fatty acids consumed (Crews and Hanley, 1995; NRC, 1989).

13.2.4.2 Limitations of food composition data

The major repository of nutrition composition data for individual foods in the US is USDA's Nutrient Data Bank (NDB). Nutrient information

Table 13.1 Food constituents reported by US national studies[a]

Nutrient or food constitutent	Historical food supply[b]	House-hold	Indivi-dual	1985 CSFII[d]	I	II	Total diet study[f]	Food compo-sition[g]
	Data source							
	1977–78 NFCS[c]				NHANES[e]		Total diet study[f]	Food compo-sition[g]
Water	−	−	−	−	−	−	−	−
Energy (kcal)	+	+	+	+	+	+	−	+
Protein, total	+	+	+	+	+	+	−	+
Amino acids	−	−	−	−	−	−	−	+
Carbohydrates, Total	+	+	+	+	−	+	−	+
Sugars	+	−	−	−	−	−	−	−
Lipids								
Total fat	+	+	+	+	−	+	−	+
Saturated fat	+	−	−	+	−	+	−	+
Oleic acid	+	−	−	−	−	+	−	+
Total monoun-saturated fat	−	−	−	+	−	−	−	+
Linoleic acid	+	−	−	−	−	+	−	+
Total; polyun-saturated fat	−	−	−	+	−	−	−	+
Cholesterol	+	−	−	+	−	+	−	+
Minerals								
Calcium	+	+	+	+	+	+	+	+
Phosphorus	+	+	+	+	−	+	+	+
Magnesium	+	+	+	+	−	−	+	+
Iron	+	+	+	+	+	+	+	+
Iodine	−	−	−	−	−	−	+	−
Sodium	+	−	−	+	−	+	+	+
Potassium	+	−	−	+	−	+	+	+
Copper	−	−	−	+	−	−	+	+
Zinc	+	−	−	+	−	−	+	+
Manganese	−	−	−	−	−	−	+	−
Selenium	−	−	−	−	−	−	+	−
Chromium	−	−	−	−	−	−	+	−
Fibre, crude	+	−	−	−	−	−	+	+
Dietary	−	−	−	+	−	−	−	+
Alcoholic beverages	−	+	+	+	−	+	+	+

Modified from NRC (1989), with permission.

+, Data reported; −, data not reported.

[a]Table based on information from USDA (1987a) and Woteki (1986).
[b]USDA's food supply data indicating disappearance of food into consumer channels.
[c]Nationwide Food Consumption Survey (USDA, 1984).
[d]Continuing Survey of Food Intakes of Individuals (USDA, 1985, 1986b).
[e]National Health and Nutrition Examination Survey I (1971–74) and II (1976–80).
[f]Total Diet Study of Food and Drug Administration.
[g]Woteki (1986).

in this bank are derived from studies in the scientific literature, unpublished reports from federal government and university laboratories, studies contracted by USDA and data from industry for foods bearing nutrition labels. Often, NDB data pertain to foods as purchased, not as consumed, and do not account for losses during preparation and processing. Moreover, little is known about the quality of some nutrients and non-nutrients in foods (e.g. fibres) because of inadequate analytical methods. Thus, there are large gaps in food composition data. Although data on nutrients in commodities are available, there is little information on highly processed or manufactured foods such as snack foods, baked products, convenience foods, restaurant meals, fast foods and frozen dinners. In addition, there is limited knowledge regarding the amount of some vitamins (e.g. B_6, E, pantothenic acid) and minerals (e.g. zinc, copper, magnesium, manganese, chromium and selenium) in foods (Beecher and Vanderslice, 1984). Biases may result when inappropriate analytical methods are used, or when a food item is incorrectly identified owing to a coding error assigned to it when surveys are gathered (NRC, 1989).

13.2.5 Trends in per capita intake and consumption of foods in the United States

Dietary patterns in the United States have changed considerably since the turn of this century. Extensive changes in sources of calories, the composition of foods, consumption of specific food groups and eating patterns (e.g. snacking, eating away from home and diets different from those consumed by the average American). Trends in per capita food consumption are presented in Table 13.2. There is a tendency for increased consumption of meat, poultry, seafood and low-fat milk, and decreased consumption of whole milk and eggs. Intake of total fats and saturated fats has generally been increasing. Consumption of fresh fruits and vegetables and potatoes has decreased, while there has been an increase in consumption of the processed varieties. Grain products registered a decrease in their consumption, whereas white bread is, by far, the favoured bread, and sweet baked products (e.g. cookies, cakes) are also popular. There is also an increase in consumption of sweeteners and carbonated soft drinks (NRC, 1989).

As for seafood, consumption of finfish and shellfish in the United States in 1989 was estimated from landing information contained in National Marine Fisheries Services (NMFS) publication *Fisheries of the United States, 1989* (NMFS, 1990). Data show that consumption of all seafoods in the United States totaled 15.9 lb of edible meat per person (or about 20 g/day). A Committee of the US National Academy of Sciences (NAS) estimated per capita consumption for US and imported products of about 18 lb (Ahmed, 1991). In contrast, the US Environmental Protection Agency (EPA) has used a per capita consumption rate of only 6.5 g/day to represent the rate of consumption of commercially and recreationally caught finfish and shellfish from estuaries and freshwater, for the purpose of estimating risk from contaminants (EPA, 1989). Another source of value of the percentage of fish consumed is the USDA's NFCS. The average 50th and 95th centile consumptions rate for finfish and shellfish were 37 g/day and 128 g/day, respectively, and for finfish and other—canned, dried and raw—38 and 132 g/day (Pao *et al.*, 1982). A significant share of the US catch is taken recreationally. Consumption data by USDA and NMFS suggest that 20% of all fresh and frozen seafood consumed in the United States may now be attributed to non-commercial landings and distribution (Ahmed, 1991). A number of these surveys showing variations in total consumption among various government agencies in the United States and in methods of expression of seafood consumption is shown in Table 13.3. Overall, US consumers ate almost 23% more seafood in 1989 than they did 10 years earlier because they consider them to be an excellent source of high-quality protein, to have low content of saturated fat, and to be a good source for many important nutrients (Ahmed and Anderson, 1994).

The US population is ageing; going into the next century, the fastest growing groups will be those aged 45–54, and those over age 85. By the year 2000, the proportion over age 65 will be the same throughout the country as today's proportion in Florida. An ageing population means decreased discretionary spending and more demands for healthful and nutritious foods. Minorities are also growing in the United States; by the turn of the century, one-quarter of the population will be considered as such. As the number of working women and single dwellers increases, the consumer base continues to change, and the time spent in the kitchen decreases. The consumer's demand for convenience, gourmet foods, ethnic items, better nutrition, and other services is increasingly evident in the food service and retail food industries (Ahmed, 1991). Convenience stores, fast-food restaurants, specialty food service outlets, and prepared items in the supermarket (e.g. value-added products, ready-to-eat items, delicatessen and microwavable entrees) were the food industry's response. All these changes (demographic, ethnic, quality of life and new kinds of food consumed) have put pressure on government agencies to protect consumers by increased monitoring of the food service industry, and by more vigilant inspection of foods imported into the United States (Ahmed and Anderson, 1994; NRC, 1985).

Table 13.2 Trends in per capita food consumption in the United States[a]

Foods	Pounds (lb)/year			Foods	Pounds (lb)/year		
	1909–13	1967–69	1985		1909–13	1967–69	1985
Meat, poultry, and fish				**Vegetables**			
Beef	54	31	79	Tomatoes	46	36	38
Pork	62	51	62	Dark green and yellow[g]	34	25	31
Poultry	18	15	19	Other			
Total[b]	171	221	224	Fresh	136	87	96
				Processed	11	35	29
Eggs	37	40	32	Total	227	183	194
Dairy products				**Potatoes, white**			
Whole milk	223	232	122	Fresh	182	67	55
Low-fat milk	64	44	112	Processed	0	15	28
Cheese[c]	5	15	26	Total[h]	182	82	83
Other[d]	28	100	86				
Total[e]	339	440	450	Dry beans, peas, nuts, and soybeans	16	16	18
Fats and oils				**Grain products**			
Butter	18	6	5	Wheat products	216	116	122
Margarine	1	10	11	Corn products	56	15	7
Shortening	8	16	23	Other grains	19	13	26
Lard and beef tallow	12	5	4	Total	291	144	155
Salad and cooking oil	2	16	25				
Total[f]	41	54	67				

Fruits			
Citrus	17	60	72
Noncitrus			
Fresh	154	73	87
Processed	8	35	34
Total	179	168	193
Sugar and sweeteners			
Refined sugar	77	100	63
Syrups and other sweeteners	14	22	90
Total	91	122	153
Miscellaneous[i]	10	17	14

From NRC (1989). Used with permission.

[a]Based on unpublished data from R. Marston, USDA, 1986.

[b]Also includes veal, lamb, mutton, and offal.

[c]Product weight, not calcium equivalents.

[d]Includes cream; canned, evaporated, and dry milk; whey; yogurt; ice cream and other frozen desserts.

[e]Total given in calcium equivalents, in which cheese and other dairy products are expressed in terms of fluid whole cow's milk having the same quantity of calcium as the milk product in question.

[f]Totals may not appear accurate owing to rounding.

[g]Includes sweet potatoes.

[h]Data before 1960 not comparable with data after 1960.

[i]Includes coffee, tea, cocoa and spices.

Table 13.3 Consumption rate for commercial and recreational fisheries in the United States

Source	Consumption rate (g/day)
Commercial[a]	
EPA	6.5
NMFS[b]	20
USDA	
50th centile	37
95th centile	128
Recreational	
NMFS	3–5
Washington State Survey[c]	
50th centile	23
90th centile	54
Puget Sound Survey, 1986–87[d]	
Average	12.3
Highest consumer	95.1
Los Angeles Harbor Sports Fishing Survey, 1978[e]	
Median	37
90th centile	225
EPA recommended consumption rate[e]	
50th centile	30
90th centile	140
EPA Supplemental guidance for Superfund[f]	
Recreational fishing	54
Subsistence fishing	132

From Ahmed and Anderson (1994). Used with permission.
[a]Per capita consumption rate.
[b]Commercially caught marine and freshwater finfish and shellfish.
[c]Commencement Bay Survey (EPA, 1990).
[d]Source: Ahmed (1991).
[e]Source: EPA (1990).
[f]Source: EPA (1991a).

13.3 Environmental chemical contaminants of food

Over ten million chemical entities are known and around several hundred thousand are now in common use worldwide. Only a few of these chemicals have been fully characterized in terms of their potential toxicity, especially their long-term effects on humans. In addition, many naturally occurring chemicals known to be toxic to animals have not been fully assessed (Ahmed, 1991, 1992c; Jelinek, 1992).

A major cause of food contamination is the pollution of air, water and soil. Emissions from industry and vehicle exhausts are a common cause of air pollution, and dangerous air-borne elements such as lead can be deposited onto, and absorbed into, fruits, vegetables and cereal crops

(Costa and Amdur, 1995; Menzer and Nelson, 1980). Food may also be contaminated during processing and storage (e.g., aflatoxins and polychlorinated biphenyls (PCBs)) (Ahmed, 1992a,b,c).

Industrial and domestic waste is often discharged into water in which some harmful chemicals and microorganisms are broken down by biological and/or chemical processes. However, a large amount of untreated waste and certain chemicals cannot be detoxified, and contaminants therefore remain in the water, from which they are absorbed into the ecosystem and enter the food supply. Water can also become contaminated as rain water passes through contaminated soil and drains into rivers and lakes. Soil and plant contamination often occurs as a result of industrial or mining activities which produce poisonous wastes, particularly if the waste is not stored carefully, or is disposed of near agricultural land (Menzer and Nelson, 1980; UNEP, 1992).

Other common sources of agricultural contamination are fertilizers and pesticides that are deposited on crops and may build up in the soil during extended use. As air, water, soil, plants and animals are linked by a complex web of natural processes, contamination of any one element will likely affect the others (Menzer and Nelson, 1980). Drugs used in animal husbandry can also contaminate food (Ahmed, 1992b). A summary of contaminants affecting major foods is presented in Table 13.4.

13.3.1 Historical background and controversies in US regulations

The idea that ingestion of food is associated with certain risks has been around for centuries. The Greeks and Romans noted certain practices aimed at adulteration of both foods and beverages, and both civilizations had inspectors to prevent use of illicit additives in wine (Hutt and Hutt, 1984). Both the Bible and the Koran contain references to potential risks associated with certain foods, and instructions on how to prevent these hazards. Many of the current food laws can be traced to early British statutes such as the thirteenth-century 'Pillory and Tumbrel' statute. It remained in effect in England for about 600 years and specified penalties and punishments for food adulterers. However, penalties for adulteration were not common owing to lack of analytical methods to detect these practices. In eighteenth-century England and America, food adulteration was a regular occurrence for the purpose of achieving some economic advantage; occasionally, the adulterant would be highly toxic (Hutt and Hutt, 1984).

The first US food adulteration regulation appeared in 1641 in the Massachusetts Colony to ensure the wholesomeness of beef and pork. However, there was almost no comprehensive federal legislation to regulate either additives or contaminants in food until the twentieth

Table 13.4 Classes of chemical contaminants affecting major foods

Production-related	Processing-related	Aquaculture-related[a]
Agricultural chemicals Diethylstilboestrol Pesticides Fertilizers	*Unavoidable contaminants* Chlorinated hydrocarbons, nitrates, nitrites, *N*-nitroso compounds	*Chemotherapeutic drugs* Sulphonamides, antibiotics, nitrofurans
Environmental chemicals Inorganic trace metals: lead, mercury, cadmium, arsenic, selenium, copper, etc.	*Storage* Dehydration, polymerization, free-radical formation	*Chemicals for treating* *parasites* Formaldehyde
Organic contaminants: PCBs, PBBs, dioxins, furans	*Additives* Colours, flavours, anti- oxidants, sweeteners	*Chemicals introduced* *through construction* *materials*
Food-borne mycotoxins Aflatoxins, trichothecenes ochrotoxin, zeralenone, penicillinic acid, patulin, ergot alkaloids	*Irradiation by-products* Free radicals Radiological products	*Water treatment* Copper compounds
	Metal fragments	*Hormones*
Pharmaceuticals Antibiotics Hormones Drugs	*Packaging* PVC Acrylonitrile	*Products of chlorination,* *bromination and* *iodination*
		Residues of ozonation
	Heating/cooking/smoking Heterocyclic amines PAHs Chlorine	*Melanosis prevention* Sulfites

Derived from Ahmed (1991) and NRC (1985).
[a]Pertains to fishery products.

century. When federal legislation was finally enacted, control was through interstate commerce, rather than through regulation of production by control measures adopted by the federal government. Individual states took the initiative and enacted their own laws, and by 1900 most states had regulations designed to protect their inhabitants from adulterated foods (Luckey, 1968).

The Food and Drug Act was enacted in 1906 after disclosures of insanitary conditions in meat packing plants and findings of poisonous preservatives and dyes in food. The act prohibited interstate commerce of misbranded and adulterated foods, drinks and drugs. In addition, the Meat Inspection Act (Public Law 59-242), passed at the same time, strengthened requirements for sanitary conditions in packing houses and required inspection of meat for interstate commerce. In 1938, the Federal Food, Drug and Cosmetic Act (FFDCA: 21 U.S.C. 301 *et seq.*), which contained new provisions relative to food was passed by Congress. This Act provided for tolerances for unavoidable poisonous substances;

authorized standards of identity, quality, and fill of containers for foods; authorized factory inspection; and added court injunction to previous sanctions of seizure and prosecution. The Act has been administered by the FDA, which was organized in 1938 under the USDA; in 1940 it was transferred to the Federal Security Agency, and this agency became the Department of Health, Education and Welfare in 1953 (DHEW), which later changed its name to DHHS in the 1980s. The Act's regulations affect about 60% of the food produced in the United States; the remaining 40% is under state regulations, mostly tailored after federal legislation. Congress passed the Miller Pesticide Amendment in 1956 to streamline procedures for setting safety limits for pesticide residues on raw agricultural commodities. This was the first congressional action requiring pre-market evaluations of safety of chemicals added to food (Section 408).

In 1958, the Food Additive Amendment, known as the 'Delaney Amendment', became law (Public Law 85-929) (Section 409). This prohibits the use of new food additives until the sponsor establishes safety, and the FDA issues regulations specifying conditions of use. If the FDA finds the additive to be safe, a regulation (called an 'order') is issued permitting its use. This regulation may place a limit, or 'tolerance', on the amount that may be used, and specifies any other conditions necessary to protect public health. The 1958 Amendment explicitly states that no additive may be permitted in any amount if tests show that it produces cancer when ingested by man or animal, or by other appropriate test; it also specifies that only the smallest amount necessary to produce the intended effect may be permitted. If the additive cannot satisfactorily be shown to be safe, its use is prohibited. FFDCA applies equally to substances added directly to foods, and to substances likely to contaminate food accidentally, as for example during processing and packaging. It also applies to methods of irradiating foods for preservation, and covers any residues that may carry over into meat, milk or eggs as a result of use in animal feed. The Color Additive Amendment was enacted in 1960 to allow the FDA to establish regulations pertaining to safe use of colour additives in foods, drugs and cosmetics, and to require manufacturers to perform the necessary tests to establish safety (Kilgore and Li, 1980).

The FFDCA bears the presumption of safety of natural foods based on the history of their use, and the understanding that the consumption of food is essential and that eating of particular foods is deeply rooted in tradition (Section 402(a)(1)). On the other hand, safety standards for contaminants or additives that may render foods harmful are more stringent and require higher standards of safety. Because neither the law, nor the FDA or USDA regulations explicitly define what is a safe additive, an operational definition for safety of food ingredients, agreed upon by all concerned scientists and regulators, centres on both the nature of the substance and its intended conditions for use. This is because no substance is inherently safe, only the level at which it is present in food makes it either safe or not; the level is, of course, determined by the intended conditions of use and limitations of use of such substance. When the uncertainty about the risk of the additive is small compared with the uncertainties attending the food itself, the standard of reasonable certainty of no harm to the individual is presumed satisfied (Kotsonis et al., 1995).

Additives have been divided generally into two major categories on the basis of the degree of consumption and similarity to food. (1) Additives that are virtually indistinguishable from food (e.g. fats, oils, starches, sugars, proteins, amino acids, alcohols, etc.): these substances are considered to be direct additives according to the Code of Federal Regulations (CFR) (21 CFR 172), or as generally recognized as safe (GRAS) substances (21 CFR 182 and 184), and may be added in substantial amounts to foods, even to levels of more than 1% of dietary intake. (2) Indirect additives, which are substances added at much smaller amounts, usually below 1% and in many cases below 1 mg/kg (ppm), such as flavour enhancers and processing aids, and may also migrate into food from packaging. Foods containing unavoidable contaminants are not automatically banned because such foods are subjects to section 406 of FFDCA which allows FDA to set limits on the quantity of these contaminants by regulations (tolerances), or by informal action levels that do not have the force of the law but have the advantage of offering greater flexibility than that provided by tolerances established by regulators (Kotsonis et al., 1995).

It is important to point out that the ban on use of saccharin in the 1970s according to FFDCA Delaney clause was stayed by Congress (NRC, 1978). Moreover, the FDA was legally challenged by industry when it banned the use of the sweetener cyclamate in the late 1970s according to the same clause of that law (Ahmed and Thomas, 1992).

The Federal Insecticide, Fungicide and Rodenticide Act (FIFRA) was originally passed in Congress in 1947 as a labelling statute that put all pest control products, initially insecticides, fungicides, rodenticides and herbicides, under the jurisdiction of USDA. Conditions were set for the production, storage or transportation of agricultural commodities in raw or natural states (Section 201 (q)), as well as for registration of products with the USDA if its use was intended for interstate commerce or for export (FIFRA [7 USC 136a et seq.]). Amendments in 1959 and 1961 added nematicides, plant growth regulators, defoliants and desiccants to

FIFRA jurisdiction, plus the authority to deny, suspend or cancel registration of products, although it gave registrants the right to appeal. Neither the 1947 statute nor subsequent amendments required consideration of effects on either human health or the environment. In 1972, FIFRA was reorganized and the administrative authority was turned over to the newly formed EPA. The Act, and subsequent amendments, 1975, 1978, 1980 and 1984, defines registration requirements, impact studies, label specifications, use restriction, the establishment of tolerances for pesticide residues on raw agricultural products, and the responsibility to monitor pesticide residues in food to ensure that the benefits of using a pesticide were weighed against the risks to humans and/or the environment.

To cancel registration of a pesticide, the EPA must consider the impact of its action on the production and prices of agricultural commodities, retail food prices and the economy. In the mid 1970s, after pressure from activists, the EPA cancelled the registration of several pesticides owing to their deleterious effect on the environment and the availability of alternative control methods (Worobec and Ordway, 1989). Nevertheless, to suspend or cancel a pesticide, the EPA must employ a tortuous mechanism fraught with procedural obstacles that may account for the decline in enforcement actions (McCabe, 1989).

Pesticides intended for use on food crops may either be granted tolerance or exempted from tolerance under FFDCA. There are two types of tolerance: one permits the presence of pesticide residues in or on raw agricultural commodities, and the other permits such residues in processed food. A pesticide cannot be registered for use in the United States unless tolerances have first been granted to cover residues that are expected to remain in or on food. Tolerances for raw food are granted by the EPA under 40 CFR 180 on a commodity-by-commodity basis. Under FFDCA, two sections apply to the setting of tolerances: section 408 governs tolerances for pesticide residues in or on raw agricultural commodities; section 409 governs tolerances for residues that are concentrated in processed foods. Under section 408, tolerances are to be set at levels deemed necessary to protect public health, and because there is no Delaney anti-cancer provision in this section, tolerances for pesticides that are carcinogenic can be established. However, because pesticide residues in processed foods are considered to be food additives, they are subject to the requirements of section 409, which includes the Delaney clause. Thus, although the FDA retains the basic responsibility for monitoring pesticide residue levels and for seizure of foods not in compliance with the regulations under section 409, the EPA under the same section is responsible for evaluating the safety of pesticide residues that are not food additives, and is also responsible for setting tolerances listed under 40 CFR 185. As with section 408-based tolerances, food additive pesticide tolerances are granted on a food-by-food basis. In addition, the USDA continues to be the agency responsible for monitoring pesticides (and other chemicals) in meat and poultry products (Ahmed, 1992b; NRC, 1985).

The overlap of regulatory authority by separate federal government agencies in the United States has often led to ambiguities, different rules and different standards for tolerance setting for environmental contaminants, and has impacted on the effectiveness of food safety regulations in the United States (Ahmed, 1991; Ecobichon, 1995; Rodricks and Rieth, 1991).

Before 1992, the EPA granted food additive tolerance for carcinogenic pesticides if the risk was negligible, i.e. a lifetime chance of less than one cancer in a million individuals. However, a court ruling in 1992 (*Les vs Reily*) struck down the *de minimis* risk standard and forced the agency to the strict language of the Delaney clause (Kotsonis *et al.*, 1995).

During the 1960s, as hundreds of new chemicals were being introduced into the United States with potential to reach the environment through use or disposal, it was felt that legislation should control their disposal into the environment. Thus, the Toxic Substance Control Act (TSCA) legislation was enacted in 1976. Under TSCA (15 USC 2, 401 *et seq.*), the EPA Administrator was given a broad array of tools and options to regulate chemical substances and mixtures that may present an unreasonable risk of injury to human health or to the environment, and to take action against them under Section 5(e) if they cause imminent hazards. TSCA also addresses existing chemicals under Section 6, which gives the Administrator broad authority to prohibit or limit production, and to impose labelling or other requirements for existing chemicals if the manufacture, processing, distribution, use or disposal of that chemical generates an unreasonable risk; the determination of which requires consideration of both risks and benefits. Prohibitions, bans and other restrictions may take place in several ways, including restriction on the manufacture, processing, distribution, use and disposal of hazardous compounds. Finally, TSCA Section 6(e) prohibits the manufacture, processing, distribution in commerce or use of PCBs, except in a totally enclosed manner, unless a specific activity will not present an unreasonable risk of injury to health or to the environment. However, any person may petition for an exemption from the ban, which may be granted upon a finding that no unreasonable risk of injury to health or the environment would result, and that good-faith efforts had been made to develop a chemical substitute for PCBs.

Thus, while TSCA provides ample authority to deal with toxic substances under Sections 5 and 6, and to regulate new and existing chemicals, problems such as data deficiencies, resource deficiencies, lack of expert staff, and the Act's cumbersome procedures hamstrung most efforts to effectively exercise the Act's authority (Gaynor, 1977). In fact, only a small fraction of the chemicals that fall under the regulatory framework of TSCA have been adequately evaluated for hazard to the environment and human health. Since its inception in 1979, not one industrial chemical has been banned from use by TSCA. The most ambitious effort towards phasing-out an environmental pollutant under TSCA was the regulation of PCBs. However, after almost 15 years of regulatory activity, use of this ubiquitous environmental pollutant has not been completely banned (Foran, 1991).

The Clean Water Act (CWA) (33 USC 1, 251 *et seq.*) was originally written and then amended by Congress in 1987 to control the discharge of toxicants from point (e.g. industrial) and non-point (e.g. agricultural) sources into the water. This Act has failed to meet its mandate because: (1) it relied on dilution, accommodation for tolerability and detectability of toxic pollutants at the point of discharge, all factors that lead to biomagnification (i.e. uptake through the food chain) of toxicants; and (2) it has almost no regulation of non-point sources, which result in discharge of pollutants to surface waters. The only mechanism that will ensure that persistent bioaccumulative toxic pollutants are not discharged from point and non-point sources is either to eliminate the use of these substances entirely or to maintain their use within a completely closed system. The CWA has no authority to ban a pollutant, and generally lacks the authority, as currently implemented, to ban discharges of toxic pollutants (Foran, 1991).

13.3.2 Regulations and databases in other countries

Most nations now have laws regulating some aspects of food additives, sanitation, and protection of the environment. Early laws were often modelled after the British Statute of 1875. More recently, some emerging nations have followed the regulatory framework of developed countries (Ecobichon, 1995; Kilgore and Li, 1980). Nevertheless, an additive whose use is legal in one country may be prohibited in another (Ahmed and Thomas, 1992). In a few developing countries, there are still no legislations to curb adverse effects due to consumption, use and disposal of chemical contaminants (Forget *et al.*, 1993).

In the United Kingdom, the Food Safety Act of 1990 and its subsequent amendments/modifications updated and brought together all food legislation into a single comprehensive act, and incorporated some EU legislation requirements (MAFF, 1997). At the time of writing, the Ministry of Agriculture, Fisheries and Food (MAFF) is the lead department on food standards, chemical safety, food labelling and food technology; the department also plays an important role in promoting the economic interests of the constituent industries. On the other hand, the Department of Health (DOH) leads on issues of food hygiene, microbial surveillance and nutrition through regional public health laboratories and local environmental health authorities. Both MAFF and DOH are advised by a series of expert committees on food safety and, on enforcement activities, by the Local Authority Coordinating Body on Food and Trading Standards (LACOTS), which acts as the United Kingdom's single liaison body on food problems in the EU. On the other hand, responsibility for water safety lies mainly with a separate department, the Department of the Environment (James, 1997).

The major weaknesses in the present system include: (1) perceived conflict between food safety issues and economic interests of regulated industries; (2) fragmentation and lack of coordination between the different bodies involved in setting food policy and in monitoring food safety; and (3) uneven enforcement standards from one authority to another (James, 1997). In January 1998, the UK Government published proposals for a new Food Standards Agency.

Various countries differ in how they address the issues of food and safety standards, with respect to basic remit and the division of responsibility between different components, and many of these countries are in the process of restructuring these activities in response to consumer worries and pressures. For example, the Danish National Food Agency, which has recently moved from the Ministry of Health to the Ministry of Food, Agriculture and Fisheries, and the Swedish National Food Administration, which is part of the Ministry of Agriculture, have broad remits to legislate, and they strive to separate regulatory issues from economic interests in both Nordic countries. The Australian and New Zealand Food Authority has a much narrower remit, and is presently concerned with development of food standards and codes of practice to protect public health and promote trade, but not with issues related to food protection. However, it has industry representation. In Canada, the present organization is concerned with inspection and enforcement; it is hoped that all food inspection and quarantine services will be consolidated into one agency—the Canadian Food Inspection Agency—in the near future. The underlying culture and the political process, and the pressure exerted by consumers in the various countries, tend to influence the scope and role of the respective regulatory agencies (James, 1997).

The European Commission Scientific Committee for Food (SCF) is responsible for making inventory of food consumption data and for writing directives for EU member countries. Although there are many databases in the 15 member states describing food consumption habits, there is little harmonization among the data and, like the American example, they also have their potential limitations (Stephen Pugh, Ministry of Agriculture, Fisheries, and Food Consumers and Nutrition Policy Division, London; and Peter Wagstaff, SCF, Brussels personal communications, 1997). The Nordic example illustrates European limitations.

In Nordic countries, a Nordic working group has been formed to plan cooperation among Nordic countries to produce and harmonize databases on contaminants. The Nordic Council of Ministers provided in 1995 financial means for the project 'Intake of Nutrients and Non Nutrients' to make a survey of existing systems and methods for monitoring the intake of nutrient additives, contaminants and natural toxins. Although Denmark, and to a certain degree, Sweden, have databases on non-nutrients, Iceland and Norway do not yet have such data. All Nordic countries have carried out extensive dietary studies; they have, however, chosen different approaches for the conduct of these studies, making it difficult to compare studies. There is also a need to harmonize methods used to estimate intake of non-nutrients, as Nordic countries differ widely in the principles and methods for making such estimates (EC, 1996a,b).

13.3.3 International efforts and their limitations

Several international agencies are attempting to solve the problems of international trade and set up criteria to evaluate the safety of food. The *Codex Alimentarius* (Latin phrase meaning food law or code) has been developed by an international commission established in 1962 when the United Nations (UN) Food and Agriculture Organization (FAO) and World Health Organization (WHO) recognized the need for international standards to guide the world's growing food demands, to protect the health of consumers, and to facilitate international trade. Since its establishment, the Codex Alimentarius Commission (CAC) has produced sets of standards, principles and guidelines (Joint Standards Programmes) that include 237 food commodity standards and 41 hygienic and technological practice codes for all the principal foods, whether processed, semiprocessed or raw. It has evaluated the safety of over 7000 food additives and contaminants, and set more than 3200 maximum residue limits for pesticides. As of 1994, Commission membership included 146 countries. The standards contain requirements for foods, aimed at ensuring for the consumer a sound wholesome food product that is free from adulteration, correctly labelled and preserved. Upon promulgation of standards, each member state provides a measure to (1) accept them fully, (2) accept them with specified deviations, or (3) target an acceptance to specific foods or food groups (Miller, 1997). The first publication of the Codex was in 1981, and the second in 1995 (FAO, 1995). Revisions to the second publication include updates to take into account decisions made by the 21st through 29th sessions of the CAC (FAO, 1997a,b).

A Joint 'Food Contamination Monitoring and Assessment Programme' under the UN Environmental Programme (UNEP) in conjunction with FAO and WHO was established in 1976 with 13 countries participating. The number of countries had risen to 40 in 1992. A 'Food Programme' is part of the UN systemwide Global Environmental Monitoring System (GEMS), which also includes components for air and water contamination monitoring, as well as total human exposure assessment. The main objective of the Food Programme (GEMS/Food) is to inform governments, the Codex Alimentarius Commission, other relevant institutions and the public about levels and trends of contaminants in food, their contribution to total human exposure and significance with regard to public health and trade. A major activity of the Programme is to compare food contamination monitoring data from different countries for worldwide presentation, synthesis and evaluation (Jelinek, 1992). Participating countries have submitted data to GEMS on the concentrations of 19 specified contaminants in more than 1200 individual foods and in the typical diets of adults and children (UNEP, 1992). Most data submitted by the participating institutions to the Programme have been concerned with levels of priority chemicals in foods of major dietary importance, and a significant portion of the data concerns the estimated total dietary intake of those contaminants selected for further comparison with toxicologically available standards, or tolerance intake levels (UNEP/FAO/WHO, 1988).

13.3.4   Dietary intake of contaminants

13.3.4.1   Definitions and standards
Estimates of dietary intake of chemical contaminants are carried out periodically in various countries to estimate the dose (intake) of the contaminant. The intake data are combined with hazard and dose–response information to derive a characterization of the risk for that contaminant. In the United States, the original approach to determining the acceptable dose of exposure to food contaminants, namely the safety factor approach, was introduced in the mid-1950s in response to legislative guidelines to the FDA to interpret animal toxicity data. The term acceptable daily intake (ADI), represents a level of daily intake which, if not exceeded, or if exceeded only rarely in the life of an

individual, should permit an insignificant risk of carcinogenesis. An ADI is determined by dividing the no-observable-effect level (NOEL) established in chronic feeding studies in animals by some safety factor (usually from 10 to 1000) that presumably reflects uncertainties inherent in the extrapolation process (NRC, 1980). On the other hand, the term reference dose (RfD) is reserved for those toxicity values derived for evaluating non-carcinogenic health effects. Both ADI and RfD are established similarly, although the RfD is derived using a more strictly defined EPA method (EPA, 1991a), and both are expressed in milligrams (mg) of contaminant per kilogram (kg) of body weight (bw). Because certain environmental contaminants have a cumulative toxic effect, the tolerable intake of such substances is expressed on a weekly basis to allow for daily variations in intake levels; hence the term Provisional Tolerable Weekly Intake (PTWI) was developed by the Joint FAO/WHO Expert Committee on Food Additives (JECFA). The term 'Provisional' expresses the tentative nature of JECFA evaluation, and 'Tolerable' signifies permissibility rather than acceptability for the intakes of trace contaminants that have no necessary function in food, in contrast to those of permitted pesticides or food additives (Jelinek, 1992).

There are no international health-based residue limits for pesticides in food; instead CAC has established Maximum Residue Limits (MRL) or Extraneous Residue Limits (ERL) for pesticide residues in a variety of foods. MRL is defined as the maximum concentration of a pesticide residue resulting from its use according to Good Agricultural Practices (GAP) on a food, agricultural commodity, or animal feed. A Codex ERL refers to a pesticide residue arising from an environmental source (including former agricultural uses) other than its use directly or indirectly on a commodity. Both MRL and ERL are expressed in mg of pesticide residue per kg of commodity (UNEP/FAO/WHO, 1988).

13.3.4.2 Estimates of adult intake

Three basic approaches were used by the GEMS/Food Programme to determine intake of contaminants in food since 1980: (1) analysis of total dietary studies (e.g. market basket surveys), (2) analysis of food consumption data, together with data on contaminant levels in foods of major dietary importance generated from monitoring programmes, and (3) analysis of duplicate portions of food actually consumed by selected individuals (Jelenik, 1992). Table 13.5 summarizes the main characteristics of dietary intake studies performed in different countries. Countries participating in GEMS/Food submitted data on the median, mean and 90th centile concentrations of designated contaminants in individual foods and in total diets. These data often permitted a reasonable evaluation of median/mean levels, but seldom the 90th centile

Table 13.5 Characteristics of various national studies on intake of contaminants

	Australia	Austria	Canada	Fipland	Guatemala	Hungary	Ireland	Japan	Netherlands	New Zealand	Sweden	Switzerland	UK	USA
Description of consumer	Males, 20–34 years	Adult males and females	Average person	Male	Adult	Institutional, public catering	General adult population	Average person	Adult	Active young men	Males 23–50 years	Adult	Average person	Males, 16–19 years
Body weight (kg)	70	70[a]	70[a]	70	55	57	70[a]	~50	60	70	70	60	~60	69.1
Amount of food consumed per day (g)	2274	1645	1034	840	2400	1900	2495	2000	1900	3306	1733	~1900	1460	2913
Type of diet	Market basket[b]	Duplicate portion	Market basket (12)[c]	Individual food	Duplicate portion	Duplicate portion	Major foods	Market basket (14)[c]	Market basket (9)[c]	Market basket (8)[c]	Duplicate portion	Duplicate portion	Market basket (15)[c]	Market basket (12)[c]
No. of food items included	46	NA[d]	120	13	42	10	16	90	NA	NA[d]	~60	NA	115	123
Alcoholic beverages	Yes	Yes	NA[d]	No	No	No	No	Yes	No	Yes	Yes	No	No	No
Drinking water	No	Yes	No	No	Yes	No	Yes (1 litre)	Yes (600 ml)	Yes	No	No	No	No	Yes
Preparation of diet	Cooked and uncooked	Cooked and uncooked	Cooked and uncooked	Uncooked	Cooked and uncooked	Cooked	Cooked and uncooked	Cooked	Cooked and uncooked	Prepared for consumption, no cooking	Cooked	Cooked	Cooked	Cooked and uncooked

From UNEP/FAO/WHO (1988). Used with permission.

[a]Estimated weight.

[b]Individual food items, rather than composites, were analysed.

[c]Figures in parentheses indicate the number of composites per basket.

[d]Information not submitted by the Collaborating Centre.

values. Most data submitted to GEMS/Food were representative of the whole country. However, in a few countries, only selected sites were monitored. Nineteen contaminants were covered by the GEMS/Food project including selected pesticides, industrial chemicals and naturally occurring toxins. Pesticide residues in food were of priority concern, and the available data cover eight of the organochlorine pesticides most prevalent in the environment (e.g. DDT, aldrin, dieldrin) and five organophosphorus pesticides (e.g. malathion, parathion, diazinon, fenitrothion and methyl parathion). Industrial chemicals studied include PCBs, lead, cadmium and mercury. Among naturally occurring substances were aflatoxins on crops that have undergone stress conditions or have been stored improperly. Drinking water was included in some national monitoring programmes, as high levels of contaminants in water can affect dietary intake levels substantially. Some foods were monitored both before and after cooking, since contaminant levels can be altered by cooking and certain new chemicals may be produced (Sugimura et al., 1986). Similarly, fresh and canned foods were compared for contaminants, as processing is another potential source of contamination.

It is evident that the composition of the diet, the consumer of the diet, preparation for analysis, total weight of the diet and the study approach vary widely from country to country and from study to study. Review of dietary intake of food chemical contaminants on a global basis is hampered by factors such as (1) variation of study approach among countries (e.g. different diet compositions, different consumers, different methods for preparation and analysis of diet), (2) inadequacy and often lack of data from some developing countries, the disparity in nature of dietary surveys, and the number of years necessary to carry them out, and (3) inadequate assurance of the quality of some data. It is impossible to draw comparisons between countries, and it may only be possible to establish indications of the nature, levels and dietary intake trends within a country (Jelinek, 1992; UNEP/FAO/WHO, 1988).

13.3.4.3 Estimates of infant intake

Estimates of the dietary intake of contaminants in breast-fed infants were obtained from human milk data. During the first 3 months of life, an infant consumes daily about 120 g of human milk/kg bw. The volume consumed per unit weight decreases with increasing age. Multiplying the concentration (µg/kg or µg/l) of the contaminant by 0.12 gives the approximate intake of the contaminant in µg/kg bw/day. Alternatively, an acceptable level of contaminant in human milk may be estimated by dividing the ADI by 0.12 (WHO, 1985). The reported levels of certain contaminants in human milk result in estimates of intake by breast-fed infants that exceed toxocologically acceptable intake levels. Acceptable intakes are, however, usually derived on the basis of lifetime exposures, whereas intakes of contaminants from human milk by infants is limited to a few months in a lifetime. It is not known whether such intakes are detrimental to the child's physical or mental development. The problem, however, is not about breast feeding because a child taken off breast feeding will continue to acquire the pesticide through diet, just as his mother acquires and passes it to him through milk. To combat this problem, either societies must develop compounds that are easily execreted or they must control the use of the most harmful chemicals (WHO, 1985).

For assessment purposes, data were compared with standards established by CAC. Levels of pesticide residues in food were compared with MRL. Some countries have set their own standards for acceptable levels of contaminants in a number of foods. The standard criterion for pesticide consumption is the ADI. Whenever possible, monitoring data for dietary intake studies were compared with acceptable or tolerable levels recommended by JECFA (Jelinek, 1992).

Some institutions that submitted data to GMS/Food carried out studies to evaluate the performance of analytical methods employed, or used standard check samples, such as the US National Bureau of Standards reference material, to validate their results. In many cases, the quality of the analytical results was checked by participation in international collaborative studies. Some analytical quality assurance tests have indicated that results from a number of laboratories in some countries were unsatisfactory (FAO/WHO, 1988). These inadequacies, however, did not invalidate the conclusions on trends. In addition to UN studies, other studies appear in the open literature on surveys of contaminants in the food supply. Important environmental organic contaminants, as opposed to natural toxicants, affecting the food supply are discussed briefly below.

13.3.4.4 Polychlorinated biphenyls (PCB)

PCBs are industrial fluids with good thermal and chemical stability widely introduced into world markets in the 1930s. They found applications as dielectric fluids in transformers and capacitors, as heat transfer and hydraulic fluids, as plasticizers in paints, copying papers, adhesives, sealants and plastics, and were employed in the formulation of lubricating and cutting oils (Van der Kolk, 1984). Drastic restrictions on their production and many uses occurred since the early 1970, and their use was prohibited in the United States in 1977 (Rodricks and Rieth, 1991).

Contamination of rice oil with PCBs caused large-scale intoxications in Japan in 1968 and in Taiwan in 1979 (WHO, 1987a). Limited evidence of carcinogenicity was obtained from the incident in Japan (Kuratsune,

1976), as well as from occupational studies of exposed populations (Brown and Jones, 1981).The toxicological literature on the subject is vast, although adverse effects observed in humans are limited to certain dermal lesions and a relatively mild and reversible damage to the nervous system and liver. In addition, several lines of epidemiological evidence suggest adverse effects on reproduction, and on the offspring of exposed mothers, and excess cancers of the liver. The International Agency for Research on Cancer concluded that there is limited evidence for the carcinogenicity of PCBs in humans (IARC, 1987). But, there is considerable evidence of PCB carcinogenicity in experimental animals and short-term tests, including development of tumours and production of chromosomal aberrations (NRC, 1989), in addition to other toxic effects such as developmental delays and effects on cognitive functions (Jacobson et al., 1990; Yu et al., 1991). Manifestations of toxicity and their dose–response characteristics depend upon animal species, sex, and the particular set of congeners to which the animals were exposed (Rodricks and Rieth, 1991).

Occupational studies performed on a group of German workers exposed to tetrachlorodibenzodioxin (TCDD) and all PCBs/polychlorinated dibenzofurans at a pesticide-producing plant in Hamburg reported an increased rate of death from ischaemic heart disease (BIBRA, 1996).

No tolerable intake of PCBs to man has been established by the Codex (Jelinek, 1992). The FDA has suggested a consumption maximum of 1 µg/kg bw/day for adults (Swain, 1988). The EPA has estimated the upper limit on cancer risk per unit dose (potency) to be 7.7 per mg/kg bw/ day. This figure was derived by applying a simple linear dose–response model to the tumour incidence data from a study in rats using a particular congener (Aroclor 1260) (EPA, 1991a). The 'potency' represented the *extrapolated* slope of the linear dose–response relationship hypothesized to hold in the low dose region. At low doses, the excess lifetime cancer risk rises by a factor of 7.7 for each increase in lifetime average daily dose of 1 mg/kg bw/day. This potency factor, however, does not apply at high doses (those that yield lifetime risks above about 1/100) that correspond to a lifetime daily dose much less than 1 mg/kg bw/day, in the neighborhood of 0.0014 mg/kg bw/day (risk cannot exceed 1).

Risk assessors of PCBs were faced with a troublesome problem because commercially produced products are mixtures of dozens of chlorinated biphenyls, individual compounds being distinguished by the number and position of chlorine atoms on the biphenyl ring. In theory, 209 different isomers and congeners are possible; commercial products typically contain 70–100 of these compounds (Van den Huvel and Lucier, 1993).

PCBs were manufactured in the United States under the commercial name Aroclor; the major commercial products were called Aroclor 1242,

Aroclor 1248, Aroclor 1254, Aroclor 1260 and Aroclor 1016 (a purified form of 1242). The last two digits of the number represent the approximate weight per cent chlorine in the PCB mixture; thus 1242 has a relatively low degree of chlorination and 1260 a relatively high degree. Although the commercial mixtures contained some individual PCBs in common, they are different in composition. In addition, PCBs undergo chemical changes in the environment so that they differ from the original products. This is important because chemical structure is a major determinant of the hazardous properties of a compound. Aroclor 1260 was carcinogenic in animal tests, while Aroclor 1254 was negative, or at best slightly positive in an NCI bioassay, and Aroclor 1242 has not been subject to carcinogenicity testing. No set of environmental PCBs has been subject to cancer tests. However, under current procedures in the United States, data from Aroclor 1260, which yield the potency figure cited above, apply to all environmental samples.

The picture is as troublesome when one examines data from other countries. In Germany, examination of two commercial products, Clophen A 60 (close in composition to Aroclor 1260) and Clophen A 30 (similar to Aroclor 1242), gave strikingly different results when tested in the same rat strain, in the same laboratory and under otherwise similar experimental conditions. Clophen A 60 was teratogenic, with potency similar to that of Aroclor 1260, while Clophen A 30 yielded only benign tumours at about one-tenth the rate of Clophen A 60. All PCBs are regulated as if they were as carcinogenic as Aroclor 1260.

To determine an ADI in the vicinity of an excess lifetime cancer risk of 1 in 10^6, the upper limit on excess lifetime cancer risk can be estimated by multiplying the potency (risk per unit dose) by the dose people receive. Since the maximum lifetime risk level is considered 10^{-6} and the potency 7.7 per mg/kg/day,

$$10^{-6} = 7.7 \text{ per mg/kg/d} \times \text{negligible risk dose (mg/kg/day)}$$

and the 'negligible risk dose' equals $10^{-6}/7.7 = 1.3 \times 10^{-7}$ mg/kg bw/day. If this dose is ingested daily for a full lifetime, it should produce no more than 1 in 10^6 excess lifetime probability of cancer. This assumes, of course, that people exhibit, on average, the same response to PCBs as rats, that the PCBs people are exposed to are similar to Aroclor 1260, and that the linear, non-threshold model of cancer risk holds at all doses.

To estimate the PCB concentration in a food, like fish, that would produce average daily lifetime doses of no greater than 1.3×10^{-7} mg/kg/day, use is made of the most conservative estimate given in Table 13.3 for consumption of recreational fish from the Great Lake region, as fresh water bodies in those parts of the United States are known to contribute significantly to local exposure to PCBs (Ahmed, 1991). Assuming that recreational fishers consume 30 g/day of fish, then the tolerance (negligible risk dose × body weight/consumption rate) $= 3 \times 10^{-4}$ mg/kg, or 0.0003 ppm for PCBs (Rodricks and Rieth, 1991).

Enforcment of this limit for PCBs effectively translates into a ban on the consumption of a very large fraction of freshwater species and some marine species. Thus, some type of *balancing* of technical feasibility, economics (measured as losses of an important food source) and health risks was needed to arrive at an acceptable regulatory model. The FDA wisely established 2.00 ppm tolerance for Great Lakes fish (21 CFR 109.15 and 109.30). It must be recognized that the various laws governing the establishment of regulatory limits differ in the extent to which they allow departure from the 'risk-only' model (Rodricks, 1988).

GEMS/Food data on total PCBs are available from 23 countries, and cover important food commodities such as dairy milk, cereal, fish, fruits, oils and human milk (UNEP, 1992). Since fish generally contain higher levels of PCBs than other food category, diets containing higher amounts of fish will accordingly lead to higher intakes of PCBs (Ahmed, 1991; Jelinek, 1992).

National regulatory limits for PCBs in finfish and shellfish range from 500 to 5000 µg/kg, with various countries setting their limits depending on fishing grounds and species (WHO, 1987a). Generally, the median levels reported to GEMS/Food were mostly in the neighbourhood of 100 µg/kg, except for Denmark, Japan and the United States; It appears, however, that PCB levels in fish are declining (UNEP/FAO/WHO, 1988). Most locations around the world where fish contain high concentrations of PCBs are typically inland waters and enclosed seas. Very high levels of PCBs (up to 80 000 µg/kg) were reported in fish livers and liver oil in Sweden because PCBs tend to concentrate in that organ (Anderson *et al.*, 1984).

Only Japan and the United States reported intakes over several years. Both countries show a mean intake of < 0.05 µg/kg bw/day over an 8-year period (1980–88), but the US rates were an order of magnitude lower than in Japan, probably owing to lower consumption of fish in the US diet. Japanese consume on the average 90 g fish/person per day as compared to 20 g/person in the United States (UNEP/FAO/WHO, 1988). Intakes similar to those of the United States were reported from Australia and the United Kingdom, while those from Guatemala were somewhat higher (Table 13.6). Intakes from Finland were in the range 0.1–0.2 µg/kg bw/day. Substantially higher daily intakes in New Zealand, approaching 0.9 µg/kg bw/day for adults and 1.5 µg/kg bw/day for teen-aged males were attributed to higher intakes of PCBs from dairy products (Pickston *et al.*, 1985).

Table 13.6 Dietary intake of PCBs by adults[a]

Country	Year	Mean daily intake (μg/kg bw)
Australia[b]	1987	0.002
Finland[c]	1984	0.21
	1986	0.026
Guatemala	1988	0.012
Japan	1980	0.046
	1988	0.045
Netherlands	1984	0.2
New Zealand	1982	0.9
Switzerland[d]	1983	0.12
UK	1981	0.0005
	1985	ND[e]
USA	1980	0.0075
	1988	0.001

Modified from Jelinek (1992), with permission.
[a]GEMS/Food, unless otherwise referenced.
FDA suggested consumption maximum: 1 μg/kg bw/day.
[b]95th centile consumers.
[c]Moilanen et al. (1986).
[d]Wuthrich et al. (1985).
[e]Not detected.

National limits for meat vary from 200–300 μg/kg, and most PCB levels in meat range from 50 to 400 μg/kg in most countries, except for a few European countries during the 1970s, and after that time levels fell sharply to just above detection limits (UNEP, 1992).

PCBs were seldom detected in vegetables and vegetable oils, fruits, eggs, or cereals. However, high levels of PCB (up to 11 000 μg/kg) were reported in breakfast cereal in Sweden (WHO, 1976) and Mexico (Albert and Aldana, 1982), mainly due to contamination from packaging material.

Several countries established regulatory limits for PCBs in dairy milk ranging from 20 to 60 μg/kg (Van der Kolk, 1984). Median levels of PCBs in dairy milk as seen in GEMS/Food are generally below 20 μg/kg in Japan and are declining.

The average daily intake of PCBs for infants as estimated from levels in milk from eight countries is shown in Table 13.7. PCB intakes from Australia and the United States for infants and children were in the same range as those of adults. The estimated intakes for all countries are above FDA maximum ADI, except for India which restricts use of PCBs (Jani et al., 1988). A study of Eskimo Inuit women from the Hudson Bay region of Northern Quebec, who consume high amounts of fish and marine mammals reported a mean intake of abouts 13 μg/kg bw/day

Table 13.7 Dietary intake of PCBs by infants from human milk[a]

Country	Year	Mean daily intake (μg/kg bw)
Canada	1988	13.22[b]
	1988	3.12[c]
Denmark	1984	3.32
Finland	1992	1.92[d]
Germany	1982	10.5
	1983	11.9
Hong Kong	1985	2.19[e]
India	1982	ND[f]
Japan	1980	2.28
	1983	2.52
	1985	1.80
UK	1980	1.86

Modified from Jelinek (1992), with permission.
[a]GEMS/Food, unless otherwise referenced.
 FDA suggested consumption maximun: 1 μg/kg bw/day.
[b]Dewailly *et al.* (1989), for Inuit Eskimos.
[c]Dewailly *et al.* (1989), for Caucasians.
[d]Wickstrom *et al.* (1983), Helsinki.
[e]Ip and Phillips (1989) for ethnic Chinese.
[f]Not detected, Jani *et al.* (1988), Ahmedabad.

compared to 3.1 μg/kg in Southern Quebec near the St Lawrence river (Dewailly *et al.*, 1989). No difference was noted between various Inuit settlements. A major cause for the presence of PCBs in the Arctic is long-range atmospheric transport from industrialized populous regions (e.g. Great Lakes) to the south (Norstrom and Muir, 1988). Mean intake levels of approximately 10 μg/kg bw/day were estimated for PCB levels in fish and meat consumed in certain areas of Germany, referred to herein as FRG (Brunn *et al.*, 1985). Intakes of 2–3.5 μg/kg were estimated from Denmark, Finland (Wickstrom *et al.*, 1983), Hong Kong (Ip and Phillips, 1989), Japan and the United Kingdom. Whether such intakes are detrimental to children is not yet known (UNEP/FAO/WHO, 1988). Multilayer surveys from Japan indicate a general declining trend in recent years.

13.3.4.5 Polychlorinated dioxins and furans
Polychlorinated dibenzo-*p*-dioxins (PCDD) and furans (PCDF) contaminate the environment mainly as emission products of thermal processes of dioxins and furans in the presence of chlorine from industrial processes; from operations such as bleaching in paper mills; and by transformation from other industrial chemicals (e.g. chlorophenols,

poly(vinyl chloride) (PVC) monomers, phenoxy herbicides, chlorinated benzenes and chlorinated aliphatic compounds) (Hallikainen and Vartiainen, 1997). Humans can take these contaminants through inhalation, absorption through skin and consumption of food; the latter route being the most important method of intake (Travis and Hattemer-Frey, 1991). Most (> 90%) usually come from food (Furst, 1993). Other minor sources of contamination include coffee filters, milk cartons and cigarettes (Furst, 1993).

The EPA assessment of the risks of these compounds and congeners in the environment is based on the induction of the Ah receptor (Ahmed, 1995), and the use of toxicity equivalent factors (TEF). The toxicity of these mixtures is the product of the concentration of each chemical times its TEF value (EPA, 1994). Although these congeners are less toxic than TCDD, evidence from animal studies indicates altered plasma enzymes, increased hepatic and renal weights and accelerated hepatocellular degeneration, in addition to a decrease in cytochrome P450 count and decreased haemoglobin and serum albumin concentrations (Ecobichon, 1995).

International cooperation has existed on standardization of analytical methods and risk evaluation of PCDD/F in breast milk and food (Schimmel et al., 1994; WHO, 1989). Most studies of PCDD/F in food have dealt with either basic food components (e.g. milk, meat, fish, eggs and vegetables) or mixtures of food products or market basket surveys (Hallikainen and Vartiainen, 1997). Published surveys by government agencies in nine countries on total daily intake of these chemicals are illustrated in Table 13.8. There are no remarkable variations between intake data from various countries, but the main sources of exposure, milk and dairy products, fish or meat, can vary (Table 13.9). There are also local variations of food contaminants and variations among different consumer groups.

An ADI of 10 pg TEQ/kg bw/day is the value recommended for Europe (WHO, 1996), Canada (Birmingham et al., 1989) and Switzerland (Schmid and Schlatter, 1992) (Table 13.10). This value is based on a no-observable-effect level (NOEL) of 1000 pg/kg bw/day from results of carcinogenicity and reproductive studies in rats (Kociba et al., 1978; Murray et al., 1979) and a safety factor of 100. The Nordic countries and The Netherlands used a safety factor higher than WHO's (NORD, 1988). In the United States, the EPA and the FDA employed a more conservative approach by using a non-threshold linearized multistage model which gave ADIs 100–1000 times lower (Van den Huvel and Lucier, 1993). In Germany, three values have been proposed by the Federal Health Office: (1) < 1 pg indicates no effect on human health, and at a value of 1 pg precautions must be exercised; (2) a value between 1 and

Table 13.8 Daily intake of PCDD/F pg TEQ[a]/day

Country	Milk and dairy products	Meat and meat products	Fish and fish products	Eggs	Fruits and vegetables	Vegetable oil	Food industry items	Total daily intake	Total daily intake/ kg bw	Reference
Canada	48.8	24.8	16	17	1.3			108	1.8	Birmingham et al. (1989)
Finland	31	1.4	59.6	3				95	1.6	Hallikainen et al. (1995)
Germany	41.7	33.1	33.9	5.9	5.7	0.6		130	2.2	Beck et al. (1992)
Holland (median) (90th centile)	26	11	3			4 (+ fish oil)	16	60 118	1 2	Theelen et al. (1993)
Japan								63	1.3 (50 kg)	One et al. (1987)
Norway	9.9	5.4	12.7	4.3		17.1 (margarine)		49.4	0.8	SNT (1995)
Sweden	17–53	13.1	50–55	2.8	8.6	14.3 (+ margarine)		106–147	1.8–2.5	de Wit et al. (1997) Darnerud et al. (1995)
Switzerland	60									Schmid and Schlatter (1992)

Table 13.8 (Continued)

Country	Milk and dairy products	Meat and meat products	Fish and fish products	Eggs	Fruits and vegetables	Vegetable oil	Food industry items	Total daily intake	Total daily intake /kg bw	Reference
UK	35	38	7.7	4.6		10		125	1.8	MAFF (1992)
	26	10	3	3		6 (+ other fats)		69	1.0 (70 kg)	MAFF (1995)
USA	8.0	18.0	6.7	0.5				18–192 34.8 (TCDD)	0.3–3.0 (65 kg)	Schecter et al. (1994)

From Hallikainen and Vartiainen (1997). Used with permission.

[a] Refers to Toxic Equivalent (TEQ) method by which residues of the less toxic congeners of dioxins and furans are expressed in terms of the equivalent quantity of the most toxic congener (i.e TCDD). Residues of other congeners are multiplied by the weighting factor, referred to as the Toxic Equivalent Factor (TEF) to give the TEQ of each congener.

Table 13.9 Percentage exposure for PCDD/F from the most important food sources in different countries[a]

Country	Milk and dairy products	Meat and meat products	Fish and fish products	Other sources
Canada	45	23		
Finland	33		63	
Germany	32	26	26	
Holland	43			27% (food industry items)
Japan			27	27% (eggs)
Norway			26	35 (margarine)
Sweden	28		42	
UK	38			28% (bread and cereal products)

From Hallikainen and Vartiainen (1997). Used with permission.
[a]References as in Table 12.8; for UK, MAFF (1995).

Table 13.10 Tolerable daily intake values for PCDD/F in different countries

Country/agency	Tolerable Daily Intake (pg TEQ/kg bw/day)	Reference
WHO/EURO 1990	0	WHO (1996)
Canada/Ontario Ministry of Environment 1985	10	Birmingham et al. (1989)
Germany	1[a], 1–10[b], > 10[c]	Schulz (1993)
Holland/State Institute of National Health 1982	4	Theelen et al. (1993)
Nordic Council	5	NORD (1988)
Switzerland/Swiss Institute of Toxicology	10	Schmid and Schlatter (1992)
UK/MAFF	1–10	MAFF (1992)
USA/FDA 1983	0.0572	Van den Huvel and
USA/EPA 1986	0.006	Lucier (1993)

From Hallikainen and Vartiainen (1997). Used with permission.
[a] Value of precaution; [b] range of risk; [c] value for intervention.

10 pg implies no health consequences to humans, but requires efforts to minimize release of dioxin into the environment; and (3) > 10 pg for a long period implies a need for immediate regulatory action (Schulz, 1993).

Intake data for different age groups are shown in Table 13.11. Although the available data were insufficient to make a definitive conclusion about children's exposure to PCDD/F congeners, it was suspected that children are exposed to higher levels than adults because

Table 13.11 Intake estimates of PCDD/F for different age groups (pg TEQ/kg bw/day)

Country	Adult	Child	Infant	Neonate	Reference
Canada	0.49–2.0 (70 kg)	1.18–4.78 (33 kg)	2.6–10.7 (13 kg)	165 (5 kg)	Gilman and Newhook (1991)
Germany	2.2			27–418 147 (mean)	Beck et al. (1992)
Holland	1–2	>10 (+PCB)	>10 (+PCB)		Theelen et al. (1993)
USA	0.3–3.0			258–384	Schecter et al. (1994)

From Hallikainen and Vartiainen (1997). Used with permission.

milk, which contains higher concentration of these congeners, constitutes an important source of nutrition for children. Moreover, if TEQ of PCBs are considered along with those of PCDD/F, values for children will be close to those of adults, and intake values for children more than twice as high can be achieved (Hallikainen and Vartiainen, 1997).

Exposures to congeners of PCDD/F from food are generally low, and their significance to human health remains unknown. Setting limits in foods, or lowering consumption of milk and fish may reduce human exposure in the short run. In the long run, however, regulations to protect the environment may be the answer, as was seen in Germany, where regulatory measures that reduced the release of PCDD/F into the environment have led to reduction of emissions in central Europe by over 90% during the last 5 years (Basler, 1995).

13.3.4.6 Organochlorine insecticides

The chlorinated hydrocarbon (organochlorine) insecticides are a diverse group of agents belonging to three distinct chemical classes: (1) dichlorodiphenylethane (e.g. DDT and its metabolites, dicofols, perthane, methoxychlor, methlochlor); (2) chlorinated cyclodien (e.g. aldrin, dieldrin, heptachlor, chlordane, endosulfan); and (3) chlorinated benzene- and cyclohexane-related structures (e.g. hexa-chlorobenzene (HCB), hexachlorocyclohexane (HCH) and lindane) (Ecobichon, 1995). From the mid-1940s to the mid-1960s these chemicals were used extensively in various aspects of agriculture and forestry, in building and construction for protection against pests, and for human use to control a wide variety of pests. The properties that made these chemicals extremely effective insecticides (e.g. low volatility, chemical stability, lipid solubility, slow rate of biotransformation and degradation) also made them one of the major classes of environmental contaminants because of their persistence in the environment and their bioconcentration in fatty tissues and biomagnification within various food chains such as fish and wildlife, especially avian species (e.g. grebes, pelicans, falcons and eagles) that occupy the top trophic levels of their respective food chains (Colborn, 1991). Rachel Carson's book *Silent Spring* dramatized and drew worldwide attention to the plight of these avian species (Carson, 1962).

These chemicals persist primarily in fatty tissues, and are thus found in human milk and in dairy products, animal fat, fish and eggs. They interfere with membrane transport of sodium, potassium, calcium and chloride ions; inhibit selective enzymatic activities in the nervous system, display oestrogenic properties, and contribute to the release of, and/or persistence of, chemical transmitters at the terminal portion of nerve endings (Ecobichon, 1995; Murphy,1980). Many of these agents were shown to be carcinogenic in human and animal studies, and in short-term tests (NRC, 1989).

Even if these insecticides are used only in non-food applications, they will enter the food supply not only by direct drift onto crops but also by contaminating animal-derived foods via soil and water. These severe environmental problems led to bans in Europe and in North America. Nevertheless, they are still used extensively in developing countries because they are relatively inexpensive to manufacture and highly effective, and risk–benefit analyses favour their continued use for control of insects causing devastations to crops and human health (Forget et al., 1993).

GEMS/Food has collected data on pesticide residues in food since 1976. Organochlorine and organophosphate pesticides have been monitored in a wide variety of foods and human milk since 1980 by 22 countries for 13 pesticides over a 10-year period. Few countries have submitted data on all pesticides requested, mainly owing to non-usage or trivial usage, and fewer have reported them in all individual foods and food groups. Although developing countries are the major users of organochlorine pesticides, their representation was generally low. Only nine countries have submitted data on dietary intake of pesticides (Table 13.12) and no country submitted intake data for every year. DDT was the only pesticide for which all nine participating countries submitted data. The United States was the only country submitting intake data on all 13 pesticides and on residues of organochlorine pesticides through the entire monitoring period (Jelinek, 1992).

Aldrin and dieldrin. Aldrin was first used as an insecticide in the 1940s. It is toxic to warm-blooded animals and can be absorbed into the body by ingestion, by inhalation or through the skin. Aldrin in converted by plants and animals to dieldrin. Consequently, mainly dieldrin residues are detected in food (Ecobichon, 1995; Murphy, 1980).

Eight countries submitted data to GEMS/Food on mean daily adult dietary intake of aldrin/dieldrin (Table 13.12). In addition, the open literature contains data on Egypt (Abel-Gawaad and Shams El Dine, 1989) and Switzerland (Wuthrich et al., 1985). In most countries, the mean intake did not exceed 15% of the FAO/WHO JMPR ADI of 0.1 µg/kg bw, except for Egypt which showed a value of 1.36 µg/kg bw (~136%). Of the countries reporting data for more than one year, a definite downward trend occurred in intake of aldrin/dieldrin (expressed as percentage of ADI), which was below recommended dietary intake levels (Figure 13.1). This general downward trend is largely due to bans imposed on the use of aldrin and dieldrin (UNEP, 1992).

Dietary intakes for infants and children from human milk reported to GEMS/Food and published for Australia (NHMRC, 1990), Hong Kong

Table 13.12 Countries submitting data to GEMS/Food on daily intake of pesticides

	Australia	Finland	Guatemala	Japan	Netherlands	New Zealand	Thailand	USA	UK
Aldrin + dieldrin	×		×	×	×	×	×	×	×
DDT, total	×	×	×	×	×	×	×	×	×
Endosulfan	×		×					×	
Endrin			×					×	
HCB			×	×	×			×	×
HCH, total		×	×	×	×			×	×
Gamma-HCH		×	×	×			×	×	
Heptachlor + epoxide	×		×	×	×		×	×	
Diazinon			×		×		×		
Fenitrothion	×							×	
Malathion			×	×		×	×	×	
Parathion			×	×		×	×	×	
Parathion-methyl		×	×		×		×		

From Jelinek (1992). Used with permission.

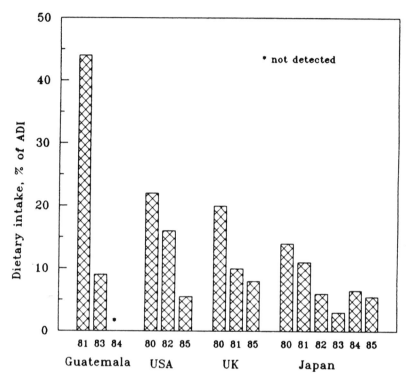

Figure 13.1 Median or mean dietary intake of aldrin plus dieldrin by adults composed for four countries during the period 1980–85. (From UNEP/FAO/WHO (1988). Used with permission.)

(Ip and Phillips, 1989), Iraq (Al-Omar *et al.*, 1986) Tunisia (Jemaa *et al.*, 1986) and the United States (Gartrell *et al.*, 1985, 1986; Gunderson, 1988; FDA, 1988a, 1989, 1990) were in the same range as for adults (Table 13.13). The average intakes for Australia, Germany, Hong Kong, Iraq, Japan, Tunisia and the United Kingdom were all above the ADI, with estimates from Iraq being the highest (2.28 µg/kg bw). Australia and developing countries, where pesticide-treated houses were studied, reported the highest ADI levels

In view of these findings, continued monitoring of aldrin/dieldrin in the diet and commercial foods, and in human milk, was recommended (Jelinek, 1992).

DDT. Dichlorodiphenyltrichloroethane-1,1,1-trichloro-2,2-bis[*p*-chloro-phenyl]ethane (DDT) was widely used from the 1940s to the 1960s in agriculture and in public health programmes to control diseases such as malaria, sleeping sickness and yellow fever. Many developed countries

Table 13.13 Dietary intake of aldrin/dieldrin by infants from human milk[a]

Country	Year	Daily intake (μg/kg bw)		
		Median	Mean	90th centile
Australia[b]	1981	1.56		
Denmark	1984	0.084		0.21
Germany	1982	0.155		0.33
	1983	0.168		0.32
Guatemala	1983	ND[c]		ND
Hong Kong[d]	1985	0.650	0.821	
Iraq[e]	1984		2.28	
Japan	1980	0.12		0.60
	1981	0.13		0.26
Tunisia[f]	1982		0.72	
UK	1980	0.24		0.60

Modified from Jelinek (1992), with permission.
[a]GEMS/Food, unless otherwise referenced.
 FAO/WHO ADI: 0.1 μg/kg bw/day.
[b]Stacey and Tatum (1985).
[c]Not detected.
[d]Ip and Phillips (1989): milk with 2.85% average fat content.
[e]Al-Omar et al. (1986).
[f]Jemaa et al. (1986).

have banned or restricted the use of DDT. It remains one of the main pesticides used in India (> 70% of total insecticide use during the mid-1980s (ICS, 1986) and in other developing countries to destroy mosquitoes and combat malaria (Murphy, 1980; UNEP, 1992). Although no confirmed ill effects of DDT have been reported in infants, some epidemiological studies have suggested that it may be associated with increased incidence of pancreatic cancer (Garabrant et al., 1992). DDT is most often found in food with a high fat content, particularly milk and dairy products. Data on adult daily intakes are presented in Table 13.14. The mean intake from Egypt (13.7 μg/kg bw/day) was about 70% of the ADI of 20 μg/kg bw established by JMPR. High levels of DDT contamination in cereals, eggs and meat has been reported in India (Kalra and Chawla, 1983; Kaphalia et al., 1985). The DDT intakes of other countries were far below the ADI (< 1%) with continued decline in some countries. Dietary intakes of DDT by infants and toddlers reported by Austria and the United States were similar to those for adults (Jelinek, 1992).

Intakes of DDT in human milk were reported to GEMS/Food by Denmark, Germany, Guatemala, Japan and the United Kingdom, together with studies carried out in Australia (Stacey and Tatum, 1985), Brazil (Sant' Ana et al., 1989), Finland (Wickstrom et al., 1983),

Table 13.14 Dietary intake of DDT by adults[a]

Country	Year	Mean/median daily intake	
		(µg/kg bw)	% ADI
Australia	1980	0.39	1.95
	1987	0.026	0.13
Egypt[b]	1988	13.7	68.5
Finland[c]	1984	0.041	0.21
	1986	0.026	0.13
Guatemala	1982	0.26	1.3
	1988	0.031	0.16
India	1981[d]	3.9	19.5
	1983[e]	3.6	18.0
Japan	1980	0.056	0.28
	1988	0.02	0.10
Netherlands	1985	0.004	0.02
New Zealand	1982	0.003	0.015
Switzerland[f]	1983	0.03	0.15
Thailand	1980	1.6	8.0
	1987	0.0008	0.004
UK	1980	0.05	0.25
	1985	0.05	0.25
US	1980	0.36	1.8
	1988	0.025	0.13

Modified from Jelinek (1992), with permission.
[a]GEMS/Food, unless otherwise referenced.
 FAO/WHO ADI: 20 µg/kg bw/day.
[b]Abel-Gawaad and Shams El Dine (1989).
[c]Moilanen et al. (1988).
[d]Singh and Chawla (1988), vegetarian to non-vegetarian.
[e]Kaphalia et al. (1985).
[f]Wuthrich et al. (1985).

Greece (Fytianos et al., 1985), Hong Kong (Ip and Phillips, 1989), India (Kalra and Chawla, 1983; Jani et al., 1988), Iraq (Al-Omar et al., 1986), Sweden (WHO, 1985) and Tunisia (Jemaa et al., 1986). The median or mean daily intake of DDT by infants was much higher than the adult intake, and intakes for Guatemala, Hong Kong, India, Iraq and Tunisia were 41.4, 47.2, 36.7, 19.2 and 17.4 µg/kg bw/day, respectively, which ranged from just below the ADI to substantially above. However, intakes reported from other countries were about 10–30% of the ADI (Table 13.15). In Hong Kong, vegetables sprayed with DDT accounted for about 40% of total food consumption, and seafood containing the insecticide was also widely consumed. The Indian samples were collected from Ahmedabad and reflected the continued use of DDT in that country. The high levels reported in Baghdad reflected the use of DDT

Table 13.15 Dietary intake of DDT by infants from human milk[a]

Country	Year	Daily intake (µg/kg)		
		Median	Mean	90th centile
Australia[b]	1981		5.0	
Brazil	1987[c]		4.2	
	1987[d]		2.0	
Denmark	1982	4.3		8.4
Finland[e]	1982		3.7	
Germany	1982	4.8		9.0
	1983	5.8		10.0
Greece[f]	1983		0.2	
Guatemala	1983	41.4		189.6
Hong Kong[g]	1985	40.0	47.2	
India	1980[h]	45.6		
	1982[i]	36.7		
Iraq[j]	1984		19.2	
Japan	1980	4.4		13.2
	1985	2.9		7.2
Sweden[k]	1983	5.6		14.3
Tunisia[l]	1982		17.4	
UK	1980	3.7		9.2
	1988	0.025	0.13	

Modified from Jelinek (1992), with permission.
[a]GEMS/Food, unless otherwise referenced.
  FAO/WHO ADI: 20 µg/kg bw/day.
[b]Stacey and Tatum (1985), Perth.
[c]Sant' Ana et al. (1989), Sao Paolo.
[d]Sant' Ana et al. (1989), rural Sao Paolo.
[e]Wickstrom et al. (1983), Helsinki.
[f]Fytianos et al. (1985), Northern Greece.
[g]Ip and Phillips (1989), ethnic Chinese.
[h]Kalra and Chawla (1983).
[i]Jani et al. (1988), Ahmedabad.
[j]Al-Omar et al. (1986), Baghdad.
[k]WHO (1985).
[l]Jemaa et al. (1986).

for malaria control programme in Iraq for more than 25 years, as well as other agriculture uses, before its ban in 1976.

Most countries that are major users of DDT have not submitted data, but published data suggest that produce is highly contaminated in some regions of the world where DDT is still in use, or where it has been banned only recently. Therefore, monitoring this insecticide in human milk and foods, including total diets, especially in developing countries, has been recommended (Jelinek, 1992).

HCH. Technical grade hexachlorocyclohexane (HCH) consists of a mixture of α-, β-, γ- and δ-isomers. It was banned or severely restricted for use as an insecticide in most developed countries, and some developing countries (e.g. China) have banned or severely restricted its use. In India, however, HCH is still one of the major pesticides used (UNEP/FAO/WHO, 1988). No ADI has been established by JMPR for the technical-grade product. Daily dietary intakes from 8 countries (Finland, Guatemala, Japan, The Netherlands, Switzerland, Thailand, the United Kingdom and the United States) showed mean intakes that were generally < 0.04 μg/kg bw. India showed a mean intake around 2 μg/kg bw in Punjab (Singh and Chawla, 1988) and 20 μg/kg bw in Lucknow (Kaphalia *et al.*, 1985). A general decline in intake was observed in Guatemala, Japan, the United Kingdom and the United States. Intakes for infants and children were about the same as for adults in the United States.

The estimated average daily dietary intakes of total HCH isomers by infants from human milk were generally higher than for adults, around 1–5 μg/kg bw (Table 13.16). In studies near Sao Paolo, Brazil, higher levels were found in rural areas compared to urban areas because mothers in rural areas worked in cotton, sugar cane and coffee plantations that frequently used HCH. Moreover, house spraying of HCH in rural areas was also conducted by public health authorities as part of a campaign to combat insects (Sant' Ana *et al.*, 1989). Much higher intakes of approximately 22 μg/kg wt/day in Punjab (Singh and Chawla, 1988), 27 μg kg bw/day in Ahmedabad (Jani *et al.*, 1988) and 61 μg/kg bw/day for ethnic Chinese in Hong Kong (Ip and Phillips, 1989) were reported.

In view of these high intakes, increased monitoring of HCH in diets and commercial foods was recommended for countries where HCH is still in use, or discontinued recently. More frequent monitoring of HCH in human milk was also recommended for countries where it has recently been employed for agricultural use or for public health purposes (Jelinek, 1992).

δ-HCH. Lindane, approximately 99% δ-HCH, is now restricted to use in residential buildings against household pests (UNEP/FAO/WHO, 1988). JMPR established an ADI of 8 μg/kg bw for this insecticide. Studies conducted in Guatemala, Japan, The Netherlands, Switzerland, the United Kingdom and the United States reported mean daily intakes around ≤ 0.1% of ADI. Studies in India reported intakes around 5% or 38% of the ADI (Singh and Chawla, 1988; Kaphalia *et al.*, 1985): the difference between these intakes is ascribed to either vegetarian or non-vegetarian diets. In Egypt, intakes about 125% of the ADI were reported (Abdel-Gawaad and Shams El Dine, 1989).

Table 13.16 Dietary intake of total HCH isomers by infants from human milk[a]

Country	Year	Daily intake (µg/kg bw)		
		Median	Mean	90th centile
Australia[b]	1981		0.12	
Brazil	1987[c]		1.73	
	1987[d]		5.56	
Denmark	1982		0.34	
Germany	1982	1.23		2.93
	1983	1.39		2.91
Hong Kong[e]	1985	60.9	56.4	
India	1981[f]	21.6		
	1982[g]	27.0		
Japan	1980	3.48		14.4
	1985	2.4		7.2
Tunisia[h]	1982		4.7	
UK	1980	0.42		2.0
Yugoslavia[i]	1981		1.2	

Modified from Jelinek (1992), with permission.
[a]GEMS/Food, unless otherwise referenced.
 No FAO/WHO ADI established.
[b]Stacey and Tatum (1985), Perth.
[c]Sant' Ana et al. (1989), Sao Paolo.
[d]Sant' Ana et al. (1989), rural Sao Paolo.
[e]Ip and Phillips (1989), ethnic Chinese.
[f]Kalra and Chawla (1983), Punjab.
[g]Jani et al. (1988), Ahmedabad .
[h]Jemaa et al. (1986).
[i]Jan (1983).

The median/mean estimated daily dietary intakes of lindane from human milk in studies conducted in Brazil, Denmark, Germany, Greece, Hong Kong, India, Japan and the United Kingdom were ≤ 2% of the ADI, whereas estimated intakes in Baghdad were around 45%. Thus, with the exception of Egypt, dietary intakes of lindane in reporting countries during the 1980s from both ordinary diets and human milk were low. Downward trends were also observed in several countries because lindane is a minor portion of commercial HCH, and was restricted in many countries (Jelinek et al., 1992).

HCB. Hexachlorobenzene was widely used as a fungicide on cereal grains until the 1970s, when its use had almost ceased. HCB emissions from chemical and waste disposal operations during the production of non-pesticidal organochlorine chemicals such as perchloroethylene were reported to contaminate foods. A temporary ADI of 0.6 µg/kg bw was withdrawn by JMPR in 1978 (UNEP/FAO/WHO, 1988).

Dietary intakes of HCB were reported to GEMS/Food by Guatemala, Japan, The Netherlands, the United Kingdom and the United States. In addition, intakes were reported in the open literature from Finland (Moilanen et al., 1986) and Switzerland (Wuthrich et al., 1985). All mean daily intakes for adults were < 0.025 µg/kg bw. Data from the United States for intakes by infants and young children were in the same low range as in adults.

Data on the levels of HCB in human milk were submitted to GEMS/Food by Denmark, Greece, Japan, the United Kingdom and the United States. In addition, data were published for Australia, Finland, Greece, Hong Kong, Israel, Tunisia and the former Yugoslavia (Table 13.17). Intake data for both regular diets and human milk were reported only by investigators from Finland, Japan and the United Kingdom; thus only limited comparisons can be made. The lowest estimated intakes from human milk were around 0.2–0.5 µg/kg bw, which were substantially higher than those reported in the literature. Much higher intakes in the range of 1–5 µg/kg bw were reported from Australia, Greece, Germany

Table 13.17 Dietary intake of HCB by infants from human milk[a]

Country	Year	Daily intake (µg/kg bw)		
		Median	Mean	90th centile
Australia[b]	1981		1.08	
Denmark	1982	0.5		0.8
Finland[c]	1982		0.28	
Germany	1982	2.73		5.32
	1983	3.16		5.62
Greece[d]	1983		2.73	
Hong Kong[e]	1985	0.14	0.17	
Israel[f]	1982	0.19	0.31	0.50
Japan	1980	0.22		0.43
	1982	0.24		0.44
Tunisia[g]	1982		3.6	
UK	1980	0.36		0.9
Yugoslavia[h]	1981		0.25	

Modified from Jelinek (1992), with permission.
[a]GEMS/Food, unless otherwise referenced.
 FAO/WHO ADI: 0.6 µg/kg bw/day (withdrawn).
[b]Stacey and Tatum (1985), Perth.
[c]Wickstrom et al. (1983), Helsinki.
[d]Fytianos et al. (1985), northern Greece.
[e]Ip and Phillips (1989), ethnic Chinese.
[f]Weisenberg (1986).
[g]Jemaa et al. (1986).
[h]Jan (1983).

and Tunisia. Because of the higher estimated dietary intakes from human milk, monitoring of HBC was recommended in both human milk and regular diets in countries where it was found in remarkable levels in foods of animal origin (Jelinek, 1992).

Heptachlor and its epoxide. Heptachlor was widely used for treating seed, foliage, fruits and cereals until the early 1970s, when restrictions on its use were imposed, primarily to reduce contamination of milk and animal products. Heptachlor is metabolized in the environment to its epoxide. The ADI for these products was previously set by JMPR at 0.5 µg/kg bw, but was subsequently lowered in 1991 to 0.1 µg/kg bw (Jelinek, 1992).

Dietary intakes for these products were reported to GEMS/Food by Australia, Guatemala, Japan, The Netherlands, Thailand and the United States, in addition to reports in the open literature from Finland (Moilanen *et al.*, 1986) and Switzerland (Wuthrich *et al.*, 1985). The highest mean daily dietary intake for adults was 0.031 µg/kg bw (about 30% of ADI), and most intakes were below this level. Intakes for infants and young children from Australia and the United States were also in the same range. Estimated average daily intakes by infants from human milk reported from Australia, Denmark, Germany, Greece, Iraq and Japan were generally below the ADI, except for Australia (Stacey and Tatum, 1985) and Iraq (Al-Omar *et al.*, 1986) which reported intakes of 0.6 and 4.32 µg/kg bw, respectively (Table 13.18). In the Australian study, the houses of some participants had been treated previously with heptachlor.

Table 13.18 Dietary intake of heptachlor and its epoxide by infants from human milk[a]

Country	Year	Daily intake (µg/kg bw)		
		Median	Mean	90th centile
Australia[b]	1981		0.6	
Denmark	1982	0.084		0.126
Germany	1982	0.046		0.185
	1983	0.050		0.147
Greece[c]	1983		0.013	
Iraq[d]	1984		4.32	
Japan	1980	ND		0.25
	1981	ND		0.23

Modified from Jelinek (1992), with permission.
[a]GEMS/Food, unless otherwise referenced.
 FAO/WHO ADI: 0.1 µg/kg bw/day.
[b]Stacey and Tatum (1985), Perth.
[c]Fytianos *et al.* (1985), northern Greece.
[d]Al-Omar *et al.* (1986), Baghdad.

Overall, the very low dietary intakes of heptachlor and its epoxide from regular diet reflect the restrictions or bans on its use. However, the higher intakes from human milk would require continued monitoring of this dietary source (Jelinek, 1992).

Endrin. Endrin is used as an insecticide mainly on cotton. Its FAO/WHO ADI is 0.2 µg/kg bw. Limited dietary intakes were reported to GEMS/Food by Guatemala and the United States, in addition to published data for Egypt (Abel-Gawaad and Shams El Dine, 1989), Switzerland (Wuthrich *et al.*, 1985), and Thailand (Vongbuddhapitak *et al.*, 1983). The highest mean dietary intake was from Egypt (2.4 µg/kg bw), and represents $\sim 1200\%$ of the ADI. Thailand reported a much lower intake of 0.035 µgkg bw ($\sim 17\%$ of the ADI), while intakes from Guatemala, Switzerland and the United States were much lower. Intakes for infants and children in the United States were of the same general level as for adults.

From the limited available data, it appears that dietary intake of endrin is low; however, because the potential exists for high levels to appear in foods, more countries should continue monitoring for residues of this insecticide (Jelinek, 1992).

Too limited dietary intake data were globally available for levels of other organochlorine pesticides such as chlordane, dicofol, methoxychlor, or toxaphene to permit reliable conclusions on their intake (UNEP/FAO/WHO, 1988).

13.3.4.7 *Organophosphorus pesticides*
In contrast to organochlorine compounds, organophosphorus pesticides are not generally stable in the environment, are extensively metabolized by animals and, consequently, have increasingly replaced organochlorine pesticides. They have been included in GEMS/Food since 1980. The residues are expected to occur mainly in raw crops (e.g. cereal grains, vegetables and fruits). When taken at higher concentrations, organophosphorus pesticides will pose a threat to human health because of their inhibitory effect on acetylcholinesterase (Ecobichon, 1995; Murphy, 1980).

Diazinon. Diazinon was first produced in the early 1950s to control pests of public health importance, and for use on fruits and vegetables, especially maize and alfalfa. The JMPR has established an ADI of 2 µg/kg bw for diazinon. Limited dietary intake data were reported to GEMS/Food for Japan, New Zealand and the United States, and in the open literature for Australia (NHMRC, 1990) and Switzerland (Wuthrich *et al.*, 1985). Mean daily dietary intakes for adults are generally < 0.01 µg/kg bw, except for Japan in 1983, when an intake of 0.16 µg/kg bw was reported due to inclusion of a heavily contaminated vegetable sample. Daily intakes by infants and children in the United States were also below 0.01 µg/kg.

Fenitrothion. Fenitrothion is a broad-spectrum insecticide with low acute mammalian toxicity. It has been used widely on fruits, vegetables, grains, tea, coffee and cotton, as well as in other public health uses to control pests. The JMPR ADI for fenitrothion is 5 µg/kg bw. Limited dietary intake data were reported to GEMS/Food by Australia for 1986 and 1987, and by the United States for 1988, in addition to published data for Japan (Matsumoto *et al.*, 1987). The mean daily dietary intakes for adult Japanese and Americans were < 0.03 µg/kg bw. Australia reported intakes for the 95th centile consumers of 2.2 µg/kg bw/day ($\sim$ 44% of the ADI), with grain and cereal products as the main source of this insecticide in the diet (NHMRC, 1990). Dietary intakes for infants and young children were in same range as adults in Australia, while in the United States the reported values were $\leq$ 0.003 µg/kg bw. Fenitrothion continued to be the most frequently detected organophosphorus residue in Australia, and 1987 intakes were similar to those of 1986 (NHMRC, 1990).

Malathion. Malathion was introduced commercially in 1950 as a broad-spectrum insecticide and acaricide, but is used mainly on grains and cereal crops. The FAO/WHO ADI for malathion is 20 µg/kg bw. Dietary intake data were reported to GEMS/Food by Guatemala, Japan, New Zealand, Thailand and the United States, and in the open literature for Australia and Switzerland. Adult intake values submitted to GEMS/Food were < 1% of the ADI, although the United States reported somewhat higher intakes than other countries.

Ethyl parathion. Ethyl parathion was introduced commercially in 1947 as a broad-spectrum insecticide in agriculture, but its use has recently restricted in favour of the less toxic methyl variety. JMPR established an ADI of 5 µg/kg bw. Data were submitted to GEMS/Food by Guatemala, Japan, New Zealand, Thailand and the United States, and appeared in the open literature for Australia. Mean daily dietary intakes of < 0.002 µg/kg bw were generally found, and the insecticide was not detected in surveys from Australia, Japan, or Switzerland. In the United States, the mean dietary intakes for infants and children were similar to those of adults (UNEP/FAO/WHO, 1988).

Methyl parathion. Methyl parathion is used as a broad-spectrum insecticide for a wide variety of agricultural applications, especially for control of cotton pests. The JMPR ADI for it is 20 µg/kg bw. Daily dietary intakes of < 0.002 µg/kg bw were reported by the United States for adults, infants and young children. It was also reported in one study in Guatemala and was not detected in diets from Japan and Thailand (UNEP/FAO/WHO, 1988).

13.3.4.8 Other organophosphorus pesticides
Data on mean adult daily dietary intakes of 18 other pesticides not included in GEMS food were reported in studies from Australia, Japan, Switzerland and the United States. The daily intakes for all of them were < 0.1 µg/kg bw, far below their respective FAO/WHO ADIs (Jelinek, 1992).

Most of the dietary intake data for organophosphorus pesticides submitted to GEMS/Food were from developed countries. Developing countries should strive to conduct surveys since use of these pesticides will be on the rise owing to gradual phasing out of organochlorine pesticides (Ecobichon, 1995; Jelinek, 1992).

13.3.4.9 Other pesticides
Other categories of pesticides for which dietary intake data have been reported in the literature include fungicides (11 reported by Australia, Japan, Switzerland and the United States), *N*-methylcarbamate (6 reported by Australia and the United States), pyrethroids (6 reported by Australia and the United States), substituted ureas (one reported by the United States) and 6 under 'other' categories reported by Japan, Switzerland and the United States. The highest intakes were < 1% of ADI for almost all of them (Jelinek, 1992).

13.3.4.10 Organic contaminants of potential toxicity to seafood in the United States

Distribution of organic contaminants. Few data exist for potential seafood contaminants such as the persistent chlorinated hydrocarbon pesticides. Minimal data also exist for industrial chemicals and by-products such as PCBs, B[*a*]Ps and dioxins as potential seafood contaminants and for the less frequently detected pesticides, e.g. chloropyrifos, dachtal (DCPA), diazinon, ethylene dibromide (EDB), malathion, mirex, toxaphene, omethoate, pentachloroaniline, tencnazene and trifluralin (FDA, 1988b; Gunderson, 1988). In specific circumstances, such as farm ponds in heavily agricultural areas, other chemicals (including those that are not known to bioconcentrate, e.g. atrazine) can be found in fish (Kansas DHE, 1988).

Some pesticides detected were specific to various regions. The carboxylic acid herbicide 2,4-(dichlorophenoxy)acetic acid (2,4-D) has been found in oysters from northern Chesapeake Bay and in Alaskan bivalves (NOAA, 1988). Fish from the Arroyo, Colorado and adjacent lower Laguna Madre in Texas contained measurable concentrations of pesticides such as ethion, carbophenothion, ethyl parathion and methyl parathion (NOAA, 1988). Most of these chemicals exhibit a majority of the following physical characteristics: low water solubility, high octanol/water ratio, low vapour pressure, lipophilicity, persistence and the tendency to accumulate and biomagnify in aquatic food resources. Alone or in combination, the chemicals pose risks for wildlife and humans (Ahmed, 1991; Colborn, 1991).

As in the case of inorganic contaminants, evidence on the distribution of organic contaminants confirms that finfish and shellfish from domestic freshwater and marine environments are contaminated with a number of organic contaminants that are potentially toxic to humans. The distribution of these contaminants also seems to follow a log-normal distribution. Contamination varies greatly with geographical location and species (Ahmed, 1991).

Estimation of carcinogenic risk. From the per capita consumption patterns (Table 13.3) and average concentration of contaminants calculated from available FDA surveillance and consumption data, a rough estimate (mg/kg bw/day) of the average amount of industrial organic chemicals and pesticides delivered to the American consumer via commercial seafood is presented in Table 13.19. These calculations assume a standard body weight of 70 kg. The table provides EPA upper confidence limit estimates of national aggregate lifetime cancer risk to carcinogenic organic chemicals. It can be seen that the overall estimated cancer risk is dominated by the (highly uncertain) estimate for PCBs, dieldrin, DDT and dioxins. Nevertheless, the overall risk is not negligible. The bulk of contamination, however, comes from a minor, localized and identifiable hazardous portion of the overall seafood commerce (Ahmed, 1991).

13.3.4.11   Unavoidable non-nutritive contaminants

In addition to the synthetic chemicals above, many other non-nutritive substances ($\sim 15\,000$) are unavoidably present in, formed in, or contaminate the diet. Evaluation of the health effects of such contaminants is complicated because no adequate studies have been performed on average and peak exposures, the potency of such substances, the quality of experimental and epidemiological data, the long-term health effects, and potential synergistic or antagonistic interaction among these substances in the diet (Ahmed, 1992c; NRC, 1989).

Table 13.19 Dietary exposures estimated from selected FDA surveillance data, 1984–88

Chemical	Estimated aggregate exposure (mg/kg bw/day)				EPA cancer potency (mg/kg bw/day)[a]	Indicated upper confidence limit cancer risk
	American finfish	American shellfish	All imported	Total		
Benzene hexachloride	2.9×10^{-8}	1.2×10^{-8}	1.0×10^{-8}	5.1×10^{-8}	6.3	3.2×10^{-7}
Chlordane	2.3×10^{-7}		5.1×10^{-8}	2.8×10^{-7}	1.3	3.6×10^{-7}
Dacthal (DCPA)	4.0×10^{-8}			4.0×10^{-8}		
tDDT	8.8×10^{-6}	2.1×10^{-8}	2.5×10^{-7}	9.0×10^{-6}	0.34	3.1×10^{-6}
Dieldrin	4.9×10^{-7}		2.5×10^{-8}	5.1×10^{-7}	16	8.2×10^{-6}
Endrin	1.7×10^{-8}			1.7×10^{-8}		
Heptachlor	4.3×10^{-8}		3.1×10^{-8}	7.4×10^{-8}	4.5	3.3×10^{-7}
Lindane			1.5×10^{-9}	1.5×10^{-9}	1.3	2.0×10^{-9}
Mirex	8.7×10^{-8}			8.7×10^{-8}		
Nonachlor	2.4×10^{-7}		1.5×10^{-8}	2.6×10^{-7}		
Octachlor	1.0×10^{-8}			1.0×10^{-7}		
Omethoate	1.2×10^{-8}			1.2×10^{-8}		
Pentachlorophenol	1.5×10^{-9}			1.5×10^{-9}		
Pentachloroaniline		2.0×10^{-9}		2.0×10^{-9}		
PCBs	6.9×10^{-6}	6.3×10^{-7}	3.7×10^{-7}	7.9×10^{-6}	7.7	6.0×10^{-5}
2,3,7,8-TCDD	1.5×10^{-11}			1.5×10^{-11}	1.6×10^{5}	2.3×10^{-6}
Tecnazene			1.5×10^{-9}	1.5×10^{-9}		
Total						$7.5 \times 10^{-5\,\text{b}}$

Modified from Ahmed (1991), with permission.

[a]This level is an upper estimate of the actual risk which may be as low as zero.

[b]Because these are 95% upper-confidence limit estimates, it is not strictly correct to add them. However, the statistical error in this case is not large compared to the other uncertainties of the analysis.

PAHs. Low levels of approximately 100 polycyclic aromatic hydrocarbons (or amines) have been identified as contaminating a variety of foods. About 20 of these PAHs have been shown to be carcinogenic in several species of laboratory animals, and some are potent mutagens. PAHs exert their toxic, carcinogenic and mutagenic effects only after metabolic activation, and their carcinogenic activity varies from very weak to potent. Of the five PAHs found to be carcinogenic and mutagenic, three (benzo[*a*]pyrene (B[*a*]P), dibenz[*a,h*]anthracene and benzanthracene) occur in the average US diet. B[*a*]P constitutes a significant portion of the total amount of carcinogenic PAHs in the environment; and levels as high as 50 µg/kg have been detected in charcoal-broiled steaks. Major factors that contribute to PAH contamination are smoking and char broiling of meat and fish leading to pyrolysis of fat drippings, use of curing smoke, and contact with petroleum and coal-tar products (Ahmed, 1991; NRC, 1989).

Foods smoked at home may contain higher levels of B[*a*]P than commercial foods treated with liquid smokes. Food packaging materials contaminated with PAHs are also a major dietary source. PAHs are not monitored by the FDA, and no ADI has been established for PAHs. The total daily intake of PAHs in the United States range from 1.6 to 16 µg (Santodonato *et al.*, 1981).

Various heterocyclic amines reported in cooked foods are shown in Table 13.20. All these compounds were highly mutagenic in the Ames test and other short-term tests, and several of these mutagenic pyrolysates were also carcinogenic in rat studies. The presence of a carcinogenic chemical in a pyrolysed amino acid or protein mixture does not, however, necessarily imply that these carcinogens will also be present in normally cooked, uncharred foods (NRC, 1989). Moreover, data on the quantities in food indicate that intakes of heterocyclic amines are negligible (1/5000 of the dose needed to develop cancer in 50% of animals fed carcinogens over their lifetime). Thus, these compounds at their present levels may not pose a serious cancer risk to humans (Sugimura *et al.*, 1986).

Although occupational studies show an association between PAHs and the incidence of skin and lung cancer, there are no definitive epidemiological studies that link consumption of food contaminated with PAHs and cancer in humans (NRC, 1989).

Nitrates, nitrites and N-*nitroso compounds.* Nitrites, and to a lesser extent nitrates, have been used as preservatives in many foods; for example, nitrites are used in cured meats to reduce risk of botulism, and they occur also naturally in food. Because they can be converted to *N*-nitroso compounds, which are strongly carcinogenic in many species under a variety of conditions, there has been much concern during the past three

Table 13.20 Aromatic heterocyclic amines in cooked food

Compound	Beef	Hamburger	Food-grade beef extract	Bacto beef extract	Sun-dried sardine	Sun-dried fish	Chicken
IQ[a]		0.02		20–40, 41.6 41–142	158		
MeIQ[b]					72		
MeIQx[c]	0.5 2.4	0.05 1.0	3.1 2.8	58.7, 112 142–157 200–300			
4,8-DiMeIQx[d]				10.0			
Trp-1[e]	53				13.3		
Trp-p-2[f]	1.6						
Glu-p-2[g]					280		
AαC[h]	650.8						180.4
MeAαC[i]	63.5						15.1

From Sugimura *et al.* (1986). Used with permission.
[a]2-Amino-3-methylimidazo[4,5-*f*]quinone.
[b]2-Amino-3,4-dimethylimidazo[4,5-*f*]quinoline.
[c]2-Amino-3,8-dimethylimidazo[4,5-*f*]quinoxaline.
[d]2-Amino-3,4,8-trimethylimidazo[4,5-*f*]quinoxaline.
[e]3-Amino-1,4-dimethyl-5-*H*-pyrido[4,3-*b*]indole.
[f]3-Amino-1-methyl-5-*H*-pyrido[4,3-*b*]indole.
[g]2-Aminodipyrido[1,2-a:3′,2′-*d*]imidazole.
[h]2-Amino-9*H*-pyrido[2,3-*b*]indole or 2-amino-α-carboline.
[i]2-Amino-3-methyl-9*H*-pyrido[2,3-*b*]indole, or 2-amino-3-methyl-α-carboline.

decades about their role in the aetiology of human cancer (Hotchkiss *et al.*, 1992).

Environmental sources of nitrate include well water and residues on vegetables due to certain agricultural pesticides (BIBRA, 1996). Although nitrate levels in food are generally about 21 µmol/l, levels up to 1600 µmol/l (100 mg/l) have been reported. However, western diets contain on the average 1–2 mmol nitrate/person/day (Hotchkiss *et al.*, 1992).

Nitrites are present in saliva and in urine of people with bladder infections. N-nitroso compounds can be formed in the stomach and bladder from the action of nitrites on ingested amines, which can be present naturally in foods, from residues of agricultural chemicals in food, and from drugs and medicines (Lijinsky, 1986). Environmentally, nitrite is formed from nitrate or ammonium ions by certain microorganisms in soil, water and sewage. *In vivo*, nitrite is formed from nitrate by microorganisms in the mouth and stomach, followed by nitrosation of secondary amines and amides in the diet. The average US diet provides

~ 0.8 mg nitrite/day (NRC, 1989). Table 13.21 shows sources of nitrate and nitrite in the diet. Many sources of nitrite are also sources of vitamin C. Nitrosation reactions can be inhibited by preferential, competitive neutralization of nitrite with naturally occurring and synthetic antioxidants such as vitamin C, vitamin E, sulphates, butylated hydroxyanisole (BHA), butylated hydroxytoluene (BHT), gallic acid or proteins (Hotchkiss et al., 1992).

The chief source of nitrosamines in the American diet was beer until recently, when the malting process was modified. Currently, the most important dietary source is cured meats, especially bacon, which may provide an average of 0.17 µg of nitropyrrolidine/person/day. This amount may be considerably lower if bacon is treated with antioxidants such as ascorbic acid. In the United States, the daily intake of nitrosamines from all dietary sources is estimated to be 1.1 µg/person/day (NRC, 1989).

Residual nitrites in cured meats and fish are an important source of nitrosating agents in the stomach since they provide concentrations of nitrite much higher than those in saliva. Many N-nitroso compounds can be formed in vivo from these sources and the carcinogenic effects of many of them are unknown (Lijinsky, 1986).

Table 13.21 Nitrate and nitrite contents of food

Vegetables	Nitrate (mg/kg)	Nitrite (mg/kg)	Meat	Nitrate (mg/kg)	Nitrite (mg/kg)
Artichoke	12	0.4	Unsmoked side bacon	134	12
Asparagus	44	0.6	Unsmoked back bacon	160	8
Green beans	340	0.6	Peameal bacon	16	21
Lima beans	54	1.1	Smoked bacon	52	7
Beets	2400	4	Corned beef	141	19
Broccoli	740	1	Cured corned beef	852	9
Brussels sprouts	120	1	Corned beef brisket	90	3
Cabbage	520	0.5	Pickled beef	70	23
Carrots	200	0.8	Canned corned beef	77	24
Cauliflower	480	1.1	Ham	105	17
Celery	2300	0.5	Smoked ham	138	50
Cron	45	2	Cured ham	767	35
Radish	1900	0.2	Belitalia (garlic)	247	5
Rhubarb	2100	NR[a]	Pepperoni (beef)	149	23
Spinach	1800	2.5	Summer sausage	135	7
Tomatoes	58	NR[a]	Ukranian sausage (Polish)	77	15
Turnip	390	NR[a]	German sausage	71	17
Turnip greens	6600	2.3			

From Hotchkiss et al. (1992). Used with permission.
[a]NR = not reported.

Poly(vinyl chloride). PVC is classified as an indirect food additive by the FDA, whereas the monomer, which may be present at low levels as a residue in PVC, is regarded as a contaminant. PVC is the parent compound for a series of copolymers used in food packaging materials. The monomer may migrate into foods, and PVC has been detected in a variety of alcoholic drinks (0.2–1.0 mg/l), in vinegars (~ 9 mg/l), and in products packed and stored in PVC containers, for example, edible oils (0.05 to 14.8 mg/l), margarine and butter (0.5 mg/kg), and finished drinking water in the United States (10.0 µg/l) (NRC, 1989). There are no estimated average exposures to PVC. Although no epidemiological studies on exposure to PVC as a food contaminant have been reported, several occupational exposure studies have linked PVC to cancer (NRC, 1989). IARC concluded that there is sufficient evidence that PVC is carcinogenic in humans (IARC, 1987).

Acrylonitrile. This chemical contaminates food through the migratory property of the monomer, which is present in small amounts in the polymer, from styrene packages (BIBRA, 1996). It was detected in the United Kingdom in margarine tubs and in food packaging films. In the United States, acrylonitrile was detected in margarine (13–45 µg/kg) and olive oils (38–50 µg/l), and in minute quantities in nuts (NRC, 1989). There are no estimates of average daily exposure to acrylonitrile in the United States, and the effects of human exposure to acrylonitrile from food packaging or drinking water have not been completely evaluated. However, a retrospective study of male employees exposed to acrylonitrile at a textile plant indicated a trend towards increased risk of cancer at all sites, especially the lung (O'Berg, 1980). This limited evidence, plus the finding that acrylonitrile is carcinogenic in rats upon ingestion or inhalation, and mutagenic in the Ames *Salmonella* and in *Escherichia coli* short-term tests, suggest that acrylonitrile, under certain conditions, might increase cancer risks in humans (NRC, 1989).

Diethylstilboestrol (DES). Among the approximately 20 growth hormones, commonly used in animal feed to promote weight in cattle, attention has focused mainly on the synthetic oestrogen DES following reports that it was carcinogenic in animals. In the United States, the use of DES in humans for various preventive and therapeutic applications terminated in 1978. Until 1979, DES was permitted as a growth promoter for cattle and sheep under certain conditions delineated by the FDA until the late 1970s but was then banned in the United States (NRC, 1985). There are no epidemiological reports on the health effects of DES residues in food. There is, however, sufficient evidence that therapeutic doses of DES during pregnancy produce vaginal and cervical cancers in the female offspring of treated women (Herbst and Cole, 1978). In animal studies it produced mainly mammary tumours in mice, rats and Syrian hamsters, and it also produced positive results in a number of short-term tests (NRC, 1989).

13.3.5 Comparative risks of food contaminants

13.3.5.1 Today's perception

The level of risk associated with consumption of food in the United States and most developed countries nowadays appears to be quite low. Although major advances could be made towards improving the wholesomeness of global supply, most improvements are needed in developing countries (UNEP, 1992). The perception of low risk from food is, however, not universally accepted. Although certain additives may contribute to cancer and death in the United States, life style (obesity, smoking, consumption of alcohol, lack of exercise, etc.) and major dietary components such as fats are considered to contribute to the most common causes of death due to chronic degenerative diseases such as heart diseases, stroke and cancer (Ahmed, 1992c; NRC, 1989). The dilemma of perception arose because the risk of developing chronic diseases as a result of food intake is uncertain, and cannot be precisely quantified or separated from other risk factors (NRC, 1989).

The present-day perception associating major risks with the food supply seems to be due to (1) current analytical methods allowing detection of traces of potentially hazardous additives in food; (2) food laws in the United States focusing attention and resources on food additives and contaminants at the expense of naturally occurring substances in food; and (3) considerable attention to potential hazards in foods drawn by several consumer groups and the natural health food industry (Clydesdale, 1982; Haas, 1991; Haas et al., 1986).

13.3.5.2 Comparison with other societal risks

Risks can be viewed as vital or non-vital. Vital risks are those that are potentially life-threatening (e.g. additives suspected of causing cancer), while non-vital risks (e.g. additive suspected of causing occasional hives) have a more limited impact. The ready availability of food in most developed countries is considered a vital benefit, although most consumers take it for granted. Consequently, the general perception is that food consumption should involve no vital and few non-vital risks. This goal of absolute safety, while commendable, is impossible to achieve in real life (Taylor and Sumner, 1990).

Risks can also be classified as either voluntary (e.g. cigarette smoking, drinking alcoholic beverages, motorcycling) or involuntary (e.g. floods, earthquakes, influenza) (Oser, 1978). All of these risks fall into the vital category. Table 13.22 presents the relative risks associated with some voluntary and involuntary activities. It is difficult to categorize eating as voluntary or involuntary, because humans need to eat and therefore voluntarily accept any associated risk. However, they can choose from among a variety of foods and, if informed, could make an educated risk

Table 13.22 Risk of death from voluntary and involuntary activities in the United States

	Risk of death per million persons per year
Voluntary activities	
Travel	
Motorcycle	20 000
Automobile	20–30
Airplane	8
Bicycle	10
Sports	
Car racing	1200
Rock climbing	1000
Skiing	170
Canoeing	400
Football playing	40
Eating or drinking	
Alcohol[a], one bottle of wine per day	75
Alcohol[a], one bottle of beer per day	20
Miscellaneous	
Smoking 20 cigarettes per day	5000
Contraceptive pills	20
Vaccination against smallpox	3
Involuntary activities	
Travel	
Run over by road vehicle	50
Eating or drinking	
Aflatoxin[a], 4 tablespoons peanut butter per day	40
Charcoal-broiled steak[a], $\frac{1}{2}$lb per week	0.4
Inhalation and ingestion of food[b]	10
Miscellaneous	
Floods	2.2
Influenza	200
Bites and stings of venomous animals and insects	0.2
Lightning	0.6
Earthquake (California)	1.7
Fires	29
Falls	71

From Taylor and Sumner (1990). Used with permission.
[a]Risk of cancer only.

choice. With today's awareness of risks inherent in eating foods in the United States, consumption of alcoholic beverages implies a willingness to accept that risk voluntarily. On the other hand, the presence of PAHs in charcoal-broiled steak may not be so apparent to the average consumer, and thus the decision to consume these foods without information about the degree of risk should be viewed as an involuntary one (Taylor and Sumner, 1990).

Most voluntary risks are avoidable, although some form of travel or walking seems essential. Involuntary risks, on the other hand, can be lessened by careful choices to avoid them. Comparative risk decisions allow individuals to make personal choices. For example, the risk of death from influenza must be weighed against the risks associated with vaccination against the virus before any personal decision can be made (Oser, 1978).

Although there are many procedures available for quantitative assessment of risks of food contaminants (see Chapter 12), procedures for quantifying benefits are irrelevant under the current FFDCA, and thus little attention has been given to developing them (Taylor and Sumner, 1990).

13.3.6 Impact of chemical contaminants on food

13.3.6.1 Health aspects
The most difficult area for risk evaluation of foods is that of chemical contaminants, because the health effects suspected do not take the form of obvious, distinctive and acute illness. The potential risks of concern (e.g. modest changes in the overall risk of cancer, subtle impairments of neurological development in fetuses and children) are generally quite difficult to measure directly in people exposed at levels that are common for consumers in developed countries. Immunocompromise may increase cancer risk (Hattis, 1986).

Inferences about the potential magnitude of these problems must be based on the level of specific chemicals present, on observations of human populations and experimental animals exposed at relatively high doses, on reasonable theories about the likely mechanisms of action of specific toxicants, and on the population distributions of sensitivity and human exposure. In nearly all cases, the current state of knowledge on these subjects must be regarded as quite tentative. Additionally, the number and variety of chemical contaminants are substantial, although a small minority constitute the bulk of risk that can be assessed quantitatively (Hattis et al., 1987). Some examples of risks that may be significant include reproductive effects from PCBs and methylmercury; carcinogenesis from selected congeners of PCBs, dioxins and dibenzofurans; and, possibly, parkinsonism in the elderly from long-term mercury exposure. Several other metallic and pesticide residues also exert significant health effects (Ahmed, 1991; Ecobichon, 1995; Goyer *et al.*, 1995). Examples of some of these hazards are presented in Table 13.23.

13.3.6.2 Economic aspects

Chemical contamination of food is of considerable economic importance for countries where revenues from export of foods represent a large proportion of gross national product (UNEP, 1992). As more countries institute control systems to prevent the importation of contaminated food, it becomes necessary for exporting countries to ascertain that their products meet the requirements laid down by the prospective importer. The recent bans by the EC of fishery products imported from some developed countries to member countries of the EU illustrates this point (FAO, 1997c). Trade agreements between countries address the question of a consistent approach in regulations and standards to protect all forms of life and to facilitate commerce between countries; this calls for the harmonization and equivalency of regulatory requirements and inspection procedures for all foods (Ahmed, 1991).

The United States issues Memoranda of Understanding (MOUs), either as formal agreements between a US government agency (e.g. the FDA) and another government agency (federal, state, local), or as informal agreements with a foreign government or another foreign institution exporting its products to the United States, to ensure that these countries have met certain minimum standards of safety of their export products (FDA, 1975). Foods not meeting the requirements of importing countries are usually refused entry or, after appropriate treatment, admitted as animal feed or used for industrial purposes, thus commanding much lower prices and leading to severe economic losses (UNEP/FAO/WHO, 1988). Similarly, a country that does not have

Table 13.23 Human health hazards associated with foods

Traditional toxicity	Irreversible or poorly reversible effects
Acute poisoning (natural intoxications, many teratogenic and reproductive effects, change in birth weight, neurological effects in infants)	Molecular (stochastic) processes (mutagenesis, carcinogenesis, and some teratogenesis)
Chronic toxicity (cholinesterase, inhibition of haem synthesis enzymes, neurological and kidney function effects)	Chronic cumulative effects (atherosclerosis, hypertension, depletion of mature oocytes, neurological Parkinsonism, Alzheimer)

Derived from Ahmed (1991).

systems for monitoring and/or control of risk of contaminated foods bears the risk of being used as a dumping ground for low-quality foods rejected by other countries (UNEP, 1992).

13.4   Role of governments, industry and consumer advocates in protecting public health

Chemical contamination of food can occur during growth, processing, transportation and storage of food. Therefore, measures to prevent and control food contamination in order to improve public health should include the setting of realistic standards to be met by food producers and processors; application of good agricultural and manufacturing practices to control sources of contamination; introduction of legislative measures to limit environmental contamination; actions to embargo, detain, seize, or recall to prevent the sale of contaminated food; issuance of advisories or warnings for certain products (e.g. seafood); and educating public safety agencies'staff, the food producers and consumers on measures to increase safety.

Prevention of food contamination is obviously preferred to control after occurrence; not only does prevention provide better protection, it also requires fewer resources in the long run (UNEP, 1992).

Accurate data on current national levels and trends in food contaminants are essential to governments when implementing regulations to prevent food contamination. Thus, a strong national monitoring programme will provide guidance for the prevention and control of chemical contamination of food by such measures as (1) defining the magnitude of the problem by establishing levels of contaminants and the risk they pose to public health; (2) identifying foods involved, and determining the source and/or the reason for their presence therein; (3) pointing out the need for control to both food producers and governments, and providing guidance for drafting regulations, if necessary; (4) specifying date(s) for the enforcement limits established by government; (5) ensuring that accurate measurements are made at the appropriate concentrations; (6) advising other regulatory/legislative bodies carrying out food and environmental policing; (7) providing access to international markets by ensuring the quality and safety of exported food; and (8) protecting countries from use as dumping grounds for substandard foods (Ahmed, 1991; UNEP, 1992).

The overuse of persistent environmental chemicals and pesticides for various purposes is an area of particular concern. To prevent or alleviate this problem, government agencies should inform farmers of good agricultural practices in the use of pesticides to protect the food supply, agricultural workers and the environment from potential contamination by these chemicals. Moreover, if the effect of food contamination by a widely used substance represents a serious risk to people's health, governments should restrict or prohibit its use, as was the case with DDT, aldrin and PCBs in most developed countries. However, the ban on manufacturing and use of such chemicals has not prevented their accumulation in the environment, or contamination of the food supply for a long time. Therefore, the presence of such chemicals in food must continue to be monitored (Ahmed, 1991; UNEP/FAO/WHO, 1988).

If measures to restrict common toxic substances at source are not effective in reducing their levels in food, then governments must introduce regulations that set safe levels of these contaminants in food, institute national monitoring programmes to observe their implementation, and set penalties if these measures are ignored. Such regulations will induce industry and agriculture to clean their waste themselves. A new systematic approach to the control of toxic environmental chemicals called the 'Sunset Process' has been tried recently. The premise of this process is that some chemicals, as well as processes and products associated with them, must be eliminated through ban, phase-out, use restriction or substitution. Identification of candidate chemicals would occur through a set of comprehensive criteria that include consideration of the potential of chemicals to reach the environment, contaminate food and pose threats to human health. Implementation of such a process could occur under either existing statutes, or by enactment of new legislation (Wahlstrom, 1989).

Government programmes must also exist for educating legislators, regulators, public health officials, industry personnel and the general public on regulatory and monitoring programmes currently in use. Appropriate government agencies should be able to provide information regarding good agricultural practices in such matters as the safe use of pesticides, the prevention and control of mycotoxins, hazards associated with consumption of seafood contaminated with persistent chemicals; and to provide information to the public concerning the importance of washing fruits and vegetables to remove residues of pesticides, the proper storage of opened canned food, and the role of lifestyle changes in reducing chronic degenerative diseases (Ahmed, 1991, 1992c; UNEP/FAO/WHO, 1988).

Industry also has major responsibilities in protecting public health against ill effects of food contaminants. These responsibilities are both individual (performed by individual producers, processors, distributors, food service operators or retailers), and collective (carried out by various industries together). Industry needs the cooperation and help of government to fulfil its public health responsibilities through measures such as the following.

(1) Enhancing the nutritional quality of food with the objective of minimizing the potential of both acute and chronic diseases that arise from consumption of foods, without undermining the abundance and affordability of the supply itself.

(2) Adherence to regulations that minimize the risk by operating according to standards enacted by various government regulatory agencies. However, industry must ensure the effectiveness of regulations, changing conditions of use, as well as changes in industrial practices, all of which can affect the efficacy of regulations.

(3) Monitoring the effectiveness of enforcement procedures of government regulations to ensure fair dealing and the integrity of the system. While the first priority of regulatory enforcement is for health-related issues, economic violations cannot be ignored, because lack of enforcement of economic fraud indicates a weakness in the regulatory mechanism, which leads to erosion of confidence in the system itself.

(4) Ensuring that regulations are specific and current.

(5) Educating the public as more is learned about specific risks from various factors, or the risk to particular sections of the population (e.g. pregnant women and immunosuppressed individuals) who are at increased risk from certain contaminants.

(6) Educating consumers about the importance of safe handling of foods to prevent cross-contamination.

(7) Funding of research on methods to identify and reduce potential risks from environmental contamination (Weddig, 1991).

Various advocate groups (or activists) in the United States monitor and criticize federal and state programmes on food safety, identify inadequacies in current government regulations, and demand stronger standards and enforcement programmes to ensure the safety of the nation's food supply (Haas, 1991; Haas et al., 1986). These groups exert formidable influence on public health through perceptions and choices they exert on their members, and on scientific institutions and governments. For example, scientists and regulators believe the major hazards associated with food to be in the following decreasing order of seriousness: (1) foodborne of microbial origin; (2) nutritional from either deficient or excess nutrient intake; (3) environmental contaminants; (4) naturally occurring toxicants (e.g. aflatoxins); (5) food and colouring additives. In contrast, public perception of the risks associated with foods is the reverse order of the above hazards. This discrepancy between what scientists/regulators rank and what the public perceives regarding food ingredients and contaminants arises probably because the public wants assurance that unnatural substances, which are not normal constituents of food, are completely safe. However, it is impossible to ensure the complete safety of any substance for all persons under all conditions of use. Thus, any uncertainty about safety results in the public's suspicion of a greater risk and potential cover-up, leading to lack of trust in scientific measures and in the regulatory process itself (Taylor and Sumner, 1990).

In the United States, many consumers are disenchanted with the role government agencies play in protecting public health. A measure of this is that public demands for direct consumer information programmes, such as California's Proposition 65, seem to have expanded in recent years (Ahmed, 1991). Moreover, consumers, who perceive themselves to be educated and generally informed about food safety, want a role in the decision-making processes of the safety of the food supply. This is exemplified by the public outcry on the ban of saccharin use by FDA in the 1970s, leading Congress to enact the Saccharin Study and Labeling Act (Public Law 95-203), which required warnings on foods and drinks containing saccharin. Although the labels alert consumers to the potential risks, they leave the decision regarding acceptance of the risks up to individual consumers as, of course, is the case with cigarette smoking.

Consumers should continue to be involved in establishing the priorities for maintaining and improving the overall safety of food. However, in order for them to have credible influence, they have to properly understand the significance of major hazards associated with food and how they rank in importance (Taylor and Sumner, 1990).

Acknowledgements

I express my thanks to Mr Carlos A. Lima dos Santos and Mr Alan W. Randell of FAO, Drs Gerald Moy and John L. Hermann of WHO and Dr Anja Hallikainen of the National Food Administration, Helsinki for the valuable material and information they provided; and to the editors of this book, Drs Colin Moffat and Kevin Whittle, for their suggestions and encouragement.

References

Abel-Gawaad, A.A.Z. and Shams El Dine, A. (1989) Insecticide residues in total diet samples. *Journal of the Egyptian Society Toxicology*, **4** 79-89.

Ahmed, F.E. (ed.) (1991) *Seafood Safety*, Institute of Medicine, US National Academy of Sciences, National Academy Press, Washington, DC.

Ahmed, F.E. (1992a) Evaluation of public health aspects of fishery products safety in the US. *Journal of Occupational Medicine and Toxicology*, **1** (1) 1-16.

Ahmed, F.E. (1992b) Assessment of cattle inspection in the United States from a public health perspective. *Journal of Occupational Medicine and Toxicology*, **1** (2) 209-222.

Ahmed, F.E. (1992c) Effect of diet on cancer and its development in humans. *Environmental Carcinogenesis Ecotoxicology Reviews*, **C10** (2) 141-180.

Ahmed, F.E. (1995) Applications of molecular biology to biomedicine and toxicology. *Environmental Carcinogenesis Ecotoxicology Reviews*, **C13** (1) 1-51.

Ahmed, F.E. and Anderson, R.D. (1994) Fishery resources, consumption and import trends, and biotechnology developments in the U.S.A. *Fisheries Research*, **19** 1-5.

Ahmed, F.E. and Thomas, D.B. (1992) Assessment of the carcinogenicity of the non-nutritive sweetener cyclamate. *CRC Critical Reviews in Toxicology*, **22** (2) 81-118.

Al-Omar, M.A., Abdul-Jalli, M.A., Al-Ogaily, F.H., Tawfik, S.J. and Al-Bassomy, M.A. (1986) A follow up of maternal milk contamination with organochlorine insecticide residues. *Environmental Pollution Series A*, **42** 79-91.

Anderson, O., Lender, C.E. and Vaz, R. (1984) Levels of organochlorine pesticides, PCBs and certain other organohalogen compounds in fish products in Sweden, 1976–1982. *Var Foeds*, **36** (Supplement 1) 1-59.

Basler, A. (1995) Dioxins and related compounds—status and regulatory aspects in Germany. *Environmental Sciences Pollution Research*, **2** (2) 117-121.

Beecher, G.R. and Vanderslice, J.T. (1984) Determination of nutrients in foods: factors that must be considered, in *Modern Methods of Food Analysis* (eds K.K. Stewart and J.R. Whitaker), Avi, Westport, CT, pp 29-55.

Beck, H., Drab, A. and Mathar, W. (1992) PCDDs, PCDFs and related contaminants in the German food supply. *Chemosphere*, **25** 1539-1550.

BIBRA (1996) *BIBRA Bulletin*, **35** (2) 35-63.

Birmingham, B., Thorpe, B., Frank, R., Clement, R., Tosine, H., Fleming, G., Ashman, J., Wheeler, J., Ripley, B.D. and Ryan, J.J. (1989) Dietary intake of PCDD and PCDF from food in Ontario, Canada. *Chemosphere*, **19** 507-512.

Brown, D.P. and Jones, M. (1981) Mortality and industrial hygiene study of workers exposed to polychlorinated biphenyls. *Archives of Environmental Health*, **36** 120-129.

Brunn, H., Berlich, H.D. and Mueller, F.J. (1985) Residues of pesticides and polychlorinated biphenyls in game animals. *Bulletin of Environmental Contamination and Toxicology*, **34** 527-532.

Carson, R. (1962) *Silent Spring*, Houghton Mifflin, Boston.

Clydesdale, F.M. (1982) Nutritional consequences of technology. *Journal of Food Protection*, **81** 816-820.

Colborn, T.E. (1991) Nontraditional evaluation of risk from fish contaminants, in *Proceedings of Symposium on Seafood Safety issues* (ed. F.E. Ahmed), Institute of Medicine, US National Academy of Sciences, Washington, DC, pp 119-122.

Costa, D.A. and Amdur, M. (1995) Air pollution, in *Casarett & Doull's Toxicology: The Basic Science of Poisons* (eds C.D. Klassen, M.O. Amdur and J. Doull), 5th edn, McGraw-Hill, New York, pp 95-122.

Crews, H.M. and Hanley, A.B. (eds) (1995) *Biomarkers in Food Chemical Risk Assessment*, Royal Society of Chemistry, Turpin Distribution Services, Letchworth.

Darnerud, P. O., Wicklund-Glynn, A., Andersson, O., Atumas, S., Johnsson, H., Linder, C.-E. and Becker, W. (1995) PCBs and dioxins in fish. *Vâr Fφda*, **2** 10-21.

Dewailly, E., Nantel, A., Weber, J.P. and Meyer, F. (1989) High levels of PCBs in breast milk of Inuit women from Arctic Quebec. *Bulletin of Environmental Contaminations and Toxicology*, **43** 641-646.

De Wit, C. Lexen, K. and Strandell, M. (1997) *The Swedish Dioxin Survey: Levels, Sources and Trends of Dioxins and Dioxin-Like Substances in the Swedish Environment*, Part I, in press.

DHEW (Department of Health, Education and Welfare) (1972) *Ten-State Nutrition Survey 1968–1970*, Vols I–VI, DHEW Publ. No. HSM-72-8130, 72-8130, 72-8131, 72-8132, 72-8133, 72-8134, Centers for Disease Control, Health Services and Mental Health Administration, Atlanta.

Eaton, B.S. and Konner, M. (1985) Paleolithic nutrition: a consideration of its nature and current implication. *New England Journal of Medicine*, **312** 283-289.

EC (European Commission) (1996a) *Assessment of Dietary Intake of Ochratoxin A by the Population of EU Member States*, Food Science and Techniques, Report of Experts Participating in Task 3.2.2, EUR 17523, Helsinki.

EC (European Commission) (1996b) *Dietary Exposure to Cadmium*, Food Science and Techniques, Report of Experts participating in Task 3.2.4, EUR 17527, Helsinki.

Ecobichon, D.J. (1995) Toxic effects of pesticides, in *Casarett & Doull's Toxicology: The Basic Science of Poisons* (eds C.D. Klassen, M.O. Amdur and J. Doull), 5th edn, McGraw-Hill, New York, pp 643-689.

EPA (Environmental Protection Agency) (1989) *Assessing Human Health Risks from Chemically Contaminated Fish and Shellfish: A Guidance Manual*, EPA-50318-89-002, Office of Marine and Estuarine Protection, Washington, DC.

EPA (US Environmental Protection Agency) (1990), *Exposure Factors Handbook, EPA/600/ 8-89/043*, March, Exposure Assessment Group, Office of Health and Environmental Assessment, EPA, Washington, DC.

EPA (US Environmental Protection Agency) (1991a) *Risk Assessment Guidance for Superfund, Volume I: Human Evaluation Manual; Supplemental Guidance*, 'Standard Default Exposure Factors', Interim final, OSWER Directive: 9285.6-03, 25 March, Office of Emergency and Remedial Responses, Toxic Integration Branch, EPA, Washington, DC.

EPA (US Environmental Protection Agency) (1994) *Health Assessment Document for 2,3,7,8-Tetrachlorodibenzo-p-dioxin (TCDD) and Related Compounds*. External Review Draft. EPA/600/BP-92/00/b, Office of Health and Environmental Assessment, Office of Research and Development, EPA, Washington, DC.

FAO (Food and Agriculture Organization) (1995) *Codex Alimentarius, General Requirements*, Joint FAO/WHO Food Standards Programme, Volume 1A, 2nd edn, Rome.

FAO (Food and Agriculture Organization) (1997a) *Codex Alimentarius Commission, Report of the Twenty-Eight Session of the Codex Commission on Food Additives and Contaminants*, Manilla, The Phillippines, 18–22 March, 1996, FAO, Rome.

FAO (Food and Agriculture Organization) (1997b) *Codex Alimentarius Commission, Report of the Twenty-Ninth Session of the Codex Committee on Food Additives and Contaminants*, The Hague, The Netherlands 17–21 March, 1997, FAO, Rome.

FAO (Food and Agriculture Organization) (1997c) *Fish Inspector*. Issue 38, 1–6, FAO, Rome.

FAO/WHO (Food Agriculture Organization/World Health Organization) (1988) *Analytical Quality Assurance Studies 1985–1987*, WHO, Geneva.

FDA (US Food and Drug Administration) (1975) Interagency agreements and memorandum of understanding, in *Staff Manual Guide IV.282d.1.GT No 75-24*, February 19, FDA, Washington, DC, p 4.

FDA (US Food and Drug Adminstration) (1988a) Food and Drug Administration pesticide program residues in food—1987. *Journal of the Association of Official Analytical Chemists*, **71** 156a-174a.

FDA (US Food and Drug Adminstration) (1988b) *Compliance Program Guidance Manual, FY 86 Pesticides and Industrial Chemicals in Domestic Food*, Program No. 7304-004, FDA, Bureau of Foods, Washington, DC.

FDA (US Food and Drug Adminstration) (1989) Food and Drug Administration pesticide program residues in food—1988. *Journal of the Association of Official Analytical Chemists*, **72** 133a-152a.

FDA (US Food and Drug Adminstration) (1990) Food and Drug Administration pesticide program residues in food—1988, *Journal of the Association of Official Analytical Chemists*, **73** 127a-146a.

Foran, J.A. (1991) Regulations playing roles in reducing chemical contaminants in fish and shellfish, in *Proceedings of Symposium on Seafood Safety Issues* (ed. F.E. Ahmed), Institute of Medicine, National Academy of Sciences, Washington, DC, pp 183-199.

Forget, G., Goodman, T. and de Villiers, A. (eds) (1993) *Impact of Pesticides on Health in Developing Countries*, International Development Research Center, Ottawa.

Furst, P. (1993) Contribution of different pathways to human exposure to PCDDs/PCDFs, 1993, *Dioxin '93, Organohalogen Compounds*, **13** 1-8.

Fytianos, K., Vasikiotis, G., Weil, L, Kavlendis, E. and Laskadidis, N. (1985) Preliminary status of organochlorine compounds in milk products, human milk and vegetables. *Bulletin of Environmental Contamination and Toxicology*, **34** 504-508.

Garabrant, D.H., Held, J., Langholz, B., Peters, J.M. and Mack, T. M. C. (1992) DDT and related compounds and risk of pancreatic cancer. *Journal of the National Cancer Institute*, **84** (10) 746-771.

Gartrell, M.C., Craun, J.C., Podrebarac, D.S. and Gunderson, E.L. (1985) Pesticides, selected elements and other chemicals in infant and toddler total diet samples, October, 1979–September, 1980. *Journal of the Association of Official Analytical Chemists*, **68** 1163-1183.

Gartrell, M.C., Craun, J.C., Podrebarac, D.S. and Gunderson, E.L. (1986) Pesticides, selected elements and other chemicals in infant and toddler total diet samples, October, 1980–March, 1982. *Journal of the Association of Official Analytical Chemists*, **69** 146-161.

Gaynor, K. (1977) The Toxic Substances Control Act: a regulatory morass. *Vanderbilt Law Review*, **30** 1149.

Gilman, A. and Newhook, R. (1991) An updated assessment of the exposure of Canadians to dioxins and furans. *Chemosphere*, **23** 1661-1667.

Goyer, R.A., Klassen, C.D. and Walkes, M.P. (1995) *Metal Toxicology*, Academic Press, Orlando, FL.

Gunderson, F.L. (1988) FDA Total Diet Study, April 1982–April 1984, Dietary intakes of pesticides, selected elements and other chemicals. *Journal of the Association of Official Analytical Chemists*, **71** 1200-1209.

Haas, E. (1991) Protecting consumer from unsafe seafood: education and beyond, in *Proceedings of Symposium on Seafood Safety issues* (ed. F.E. Ahmed), Institute of Medicine, US National Academy of Sciences, Washington, DC, pp 41-61.

Haas, E., Heiman, D., and Jones, M. (1986) *The Great American Fish Scandal: Health Risks Unchecked*, Public Voice for Food and Health Policy, Washington, DC.

Hallikainen, A. and Vartiainen, T. (1997) Food control surveys of polychlorinated dibenzo-*p*-dioxins and dibenzofurans and intake estimates. *Food Additives and Contaminants*, **14** (4) 355-366.

Hallikainen, A., Mustaniemi, A. and Vartiainen, T. (1995) *Dioxin Intake from Food, 1/1995*, National Food Administration, Helsinki.

Hampe, E.C., Jr, and Wittenberg, M. (1964) *The Lifeline of America, Development of Food Industry*, McGraw-Hill, New York.

Hattis, D. (1986) The promise of molecular epidemiology for quantitative risk assessment. *Risk Analysis*, **6** 181-193.

Hattis, D., Erdreich, L. and Ballew, M. (1987) Human variability in susceptibility to toxic chemicals—a preliminary analysis of pharmacokinetic data from normal volunteers. *Risk Analysis*, **7** 415-426.

Herbst, A.L. and Cole, P. (1978) Epidemiological and clinical aspects of clear cell adenocarcinoma in young women, in *Intrauterine Exposure to Diethylstilbesterol in the Human, Proceedings of a Symposium on DES*, 1977 (ed. A.L. Herbst), American College of Obstetricians and Gynecologists, Chicago.

Hotchkiss, J.H., Helser, M.A., Maragos, C.M. and Weng, Y.M. (1992) Nitrate, nitrite, and *N*-nitrosocompounds, in *Food Safety Assessment* (eds J.W. Finley, S.F. Robinson and D.J. Armstrong), American Chemical Society, Washington, DC, pp 400-418.

Hutt, P.B. and Hutt, P.B., II (1984) A history of government regulation of adulteration and misrepresentation of food. *Food and Drug Cosmetic Law Journal*, **39** 2-73.

IARC (1987) IARC Monograph on the Evaluation of Carcinogenic Risks of Chemicals to Humans, suppl. 7, Overall Evaluation of Carcinogenicity: an Updating of IARC Monographs, vols 1 to 42, International Agency for Research on Cancer, Lyon.

ICS (1986) Demand pattern of pesticides during the seventh five-year plan, in *Indian Chemical Statistics, 1986-87*, Ministry of Industry, Government of India, New Delhi, pp 136-139.

Ip, H.M.H. and Phillips, D.J.H. (1989) Organochlorine residues in human breast milk in Hong Kong. *Archives of Environmental Toxicology*, **18** 490-498.

Jacobson, J.L., Jacobson, S.W. and Humphrey, H.E. (1990) Effects of *in utero* exposure to polychlorinated biphenyls and related contaminants on cognitive functioning in young children. *Journal of Pediatrics*, **116** (1) 38-45.

James, W.P.T. (1997) *Food Standards Agency—An Interim Proposal*, Special Publication, 55 pp.

Jan, J. (1983) Chlorobenzene residues in human fat and milk. *Bulletin of Environmental Contamination and Toxicology*, **30** 595-599.

Jani, J.P., Patel, J.S., Shah, Z.M.P., Gupta, S.K. and Kashyap, S.K. (1988) Levels of organochlorine pesticides in human milk in Ahmedabad, India. *International Archives of Environmental Health*, **60** 111-113.

Jelinek, C.F. (1992) *Assessment of Dietary Intake of Chemical Contaminants*, United Nations Environmental Programme, Nairobi.

Jemaa, Z., Sabbah, S., Driss, M.R. and Bouguerra, M.L. (1986) Hexachlorobenzene in a Tunisian mother's milk, cord blood and foodstuff, in *Hexachlorobenzene: Proceedings of International Symposium* (eds C.P. Morris and J.R.P. Cabral), IARC Scientific Publication No. 77, International Agency for Research on Cancer, Lyon, pp 139-142.

Kalra, R.R. and Chawla, R.P. (1983) *Studies on Pesticide Residues and Monitors of Pesticide Pollution*, Final Technical Report PL-480 Project, Department of Entomology, Punjab Agricultural University, Ludhiana.

Kansas DHE (Department of Health and Environment) (1988) *A Survey of Pesticides in Tuttle Creek Lakes, its Tributaries and the Upper Kansas River*, Water Quality Assessment Section, Bureau of Water Protection, Kansas Department of Health and Environment, Topeka, Kansas.

Kaphalia, B.S., Siddiqui, F.S. and Seth, T.D. (1985) Contamination levels in different food items and dietary intake of organochlorine pesticide residues in India. *Indian Journal of Medical Research*, **81** 71-78.

Kilgore, W.W. and Li M.-Y. (1980) Food additives and contaminants, in *Casarett and Doull's Toxicology, The Basic Science of Poisons* (eds J. Doull, C.D. Klassen and M.O. Amdur), 2nd edn, Macmillan, New York, pp 593-607.

Kociba, R., Keyes, D., Beyer, J., Carreon, R., Wade, C., Dittenber, D., Kalnins, R. and Humiston, C. (1978) Result of a two-year chronic toxicity and oncogenicity study of 2,3,7,8-tetrachlorodibenzo-*p*-dioxin in rats. *Toxicology and Applied Pharmacology*, **46** 279-303.

Kotsonis, F.N., Burdock, G.A. and Flamm W.G. (1995) Food toxicology, in *Casarett & Doull's Toxicology: The Basic Science of Poisons* (eds C.D. Klassen, M.O. Amdur and J. Doull), 5th edn, McGraw-Hill, New York, pp 909-949.

Kuratsune, M. (1976) Epidemiologic studies on Yusho, in *Poisoning and Pollution* (ed. K. Higuchi), Kodansha, Tokyo, pp 9-23.

Lijinsky, W. (1986) The significance of *N*-nitroso compounds as environmental carcinogens. *Journal of Environmental Science and Health*, **4** 1-45.

Luckey, T.D. (1968) Introduction to food additives, in *Handbook of Food Additives* (ed. T.E. Furia), The Chemical Rubber Co, Cleveland, OH.

Madans, J.H., Kleinman, J.C., Cox C.S., Barbano, H.E., Feldman, J.J., Cohen, B., Finucane, F.F. and Cornoni-Hantley, J. (1986) 10 years after NHANES I: report of initial follow up, 1982–84. *Public Health Report*, **101** 465-473.

MAFF (Ministry of Agriculture, Fisheries and Food) (1992) Dioxins in food, *Food Surveillance Information Sheet, No. 31*, HMSO, London.

MAFF (Ministry of Agriculture, Fisheries and Food) (1995) Dioxins in food—UK dietary intakes, *Food Surveillance Information Sheet, No. 71*, HMSO, London.

MAFF (Ministry of Agriculture, Fisheries and Food) (1997) *Food Law*, Revised July 1997, HMSO, London.

Matsumoto, H., Murukami, Y., Kuwabara, K., Tanaka, R. and Kashimoto, T. (1987) Average daily intakes of pesticides and polychlorinated biphenyl in total diet samples in Osaka, Japan. *Bulletin of Environmental Contamination and Toxicology*, **38** 954-958.

McCabe, M. (1989) Pesticide law enforcement: a blueprint for state action. *National Environmental Enforcement Journal*, (June), 3-6.

Menzer R.E. and Nelson, J.O. (1980) Water and soil pollution, in *Casarett and Doull's Toxicology, The Basic Science of Poisons* (eds J. Doull, C.D. Klassen and M.O. Amdur), 2nd edn, Macmillan, New York, pp 632-658.

Miller, R.W. (1997) *This is Codex Alimentarius*, Food and Agriculture Organization of the United Nations, 5th edn, Rome.

Moilanen, R., Pyysalo, H. and Kimpulainen, J. (1986) Average total dietary intakes of organochlorine compounds from the Finnish diet. *Zeitschirift für Lebensmittel-Untersuchung und -Forschung*, **182** 484-488.

Murphy, S.D. (1980) Pesticides, in *Casarett and Doull's Toxicology, The Basic Science of Poisons*, (eds J. Doull, C.D. Klassen and M.O. Amdur), 2nd edn, Macmillan, New York, pp 357-408.

Murray, F., Smith, F., Nitche, K., Humiston, C., Kociba, R. and Schwetz, B. (1979) Three generation reproductive study of rats given 2,3,7,8-tetrachlorodibenzo-*p*-dioxin (TCDD) in the diet. *Toxicology and Applied Pharmacology*, **50** 241-252.

NHMRC (National Health and Medical Research Council) (1990) *The 1987 Market Basket Survey*, Australian Government Publishing Service, Canberra.

NMFS (National Marine Fisheries Service) (1990) *Fisheries Statistics of the United States—1989*, Current Fisheries Statistics No. 8900, NMFS, Government Printing Office, Washington, DC.

NOAA (National Oceanic and Atmospheric Administration) (1988) *PCB and Chlorinated Pesticide Contamination in US Fish and Shellfish: a Historical Assessment Report*, NOAA in Technical Memorandum, NOS OMA 39, Seattle, WA

NORD (1988) *Miljîrapport 1988/7, Nordisk Dioxinrisk Bedîmning*, Nordic Council, Copenhagen. pp 1-111.

Norstrom, R.J. and Muir, D.C.G. (1988) Long range transport of organochlorines in the Arctic and sub-Arctic: evidence from analysis of marine mammals and fish, in *Toxic Contamination in the Great Lakes*, Vol. I: Chronic Effects of Toxic Contaminants in Large Lakes (ed. N.W. Schmidtke), Lewis Publishing, Chelsea, MI, pp 119-137.

NRC (National Research Council) (1978), *Saccharin: Technical Assessment of Risks and Benefits. Part I: Saccharin and Its Impurities*, Assembly of Life Sciences, National Academy Press, Washington, DC.

NRC (National Research Council) (1980) *Drinking Water and Health*, Vol. 3, US National Academy of Sciences, National Academy Press, Washington, DC.

NRC (National Research Council) (1985) *Meat and Poultry Inspection: The Scientific Basis of the Nation's Program*, National Academy Press, Washington, DC.

NRC (National Research Council) (1989) *Diet and Health: Implications for Reducing Chronic Disease Risk*, National Academy Press, Washington, DC.

O'Berg, M.T. (1980) Epidemiologic study of workers exposed to acrylonitrile. *Journal of Occupational Medicine*, **22** 245-252.

Ono, M., Kashima, Y., Wakinomoto, T. and Tatsakawa, R. (1987) Daily intake of PCDDs and PCDFs by Japanese through food. *Chemosphere*, **16** 1823-1828.

Oser, B.L.(1978) Benefits/risks: whose? what? how much? *Food Technology*, **32** (8) 55-58.

Pao, E.M., Fleming, K.H., Guenther, P.M., and Nickle, F.J. (1982) *Foods Commonly Eaten by Individuals: Amounts Per Day and Per Eating Occasion*, Home Economics Records Report 44, US Department of Agriculture, Washington, DC.

Pickston, N.L., Brewerton, H.V., Drysdale, J.M., Hughes, J.T., Smith, J.M., Love, J.L., Sutcliffe, E.R. and Davidson, F. (1985) The New Zealand diet: a survey of elements, pesticides, colours and preservations. *New Zealand Journal of Technology*, **1** 81-89.

Rodricks, J.V. (1988) Origins of risk assessment in food safety decision making. *Journal American College of Toxicology*,**7** 539-542.

Rodricks, J.V. and Rieth, S.H. (1991) Assessing and managing risks associated with consumption of chemically contaminated seafood, in *Proceedings of Symposium on Seafood Safety Issues* (ed. F.E. Ahmed), Institute of Medicine, National Academy of Sciences, Washington, DC, 160-182.

Sant' Ana, L.S., Vassilieff, I. and Jokl, L. (1989) Levels of organochlorine insecticides in milk of mothers from urban and rural areas of Botucatu, SP, Brazil. *Bulletin of Environmental Contamination and Toxicology*, **42** 911-918.

Santodonato, J., Howard, P. and Basu, D. (1981) Health and ecological assessment of polynuclear aromatic hydrocarbons. *Journal of Environmental Pathology and Toxicology*, **5** 1-364.

Schecter, A., Startin, J., Wright, C., Kelly, M., Papke, O., Lis, A., Ball, M. and Olson, J. (1994) Congener-specific levels of dioxins and dibenzofurans in US food and estimated daily dioxin toxic equivalent intake. *Environmental Health Perspectives*, **102** 962-966.

Schimmel, H., Griepink, G., Maier, E.A., Kramer, G.N., Roos, A.H. and Tuinstra, L.G.M.T. (1994) Intercomparison study on milk powder fortified with PCDD and PCDF. *Fresenius Journal of Analytical Chemistry*, **348** 1-2, 37-46.

Schmid, P. and Schlatter, C.H. (1992) Polychlorinated dibenzo-*p*-dioxins (PCDDs) and polychlorinated dibenzofurans (PCDFs) in cow's milk from Switzerland. *Chemosphere*, **24** 1013-1030.

Schulz, D. (1993) PCDD/PCDF-German policy and measures to protect man and the environment. *Chemosphere*, **27** 501-507.

Singh, P.P. and Chawla, R.P. (1988) Insecticide residues in total diet samples in Punjab, India. *Science of the Total Environment*, **76** 139-146.

SNT (1995) *Fremmedstoffer og smitt estoffer-holder maten mdl?*, SNT, Norway, pp 1-39.

Stacey, C.I. and Tatum, T. (1985) House treatment with organochlorine pesticides and their levels in human milk—Perth, Western Australia. *Bulletin of Environmental Contamination and Toxicology*, **35** 202-208.

Sugimura, T., Sato, S., Ohgaki, H., Takayama, S., Nago, M. and Wakabayashi, K. (1986) Mutagens and carcinogens in cooked food, in *Genetic Toxicology of the Diet* (ed. I. Knudsen), Alan R. Liss, New York, pp 85-107.

Swain, W.R. (1988) Human health consequences of consumption of fish contaminated with organochlorine compounds, in *Aquatic Toxicology* (eds D.C. Malins and A. Jensen), Vol. 11, Elsevier Science, Amsterdam, pp 115-143.

Tannahill, R. (1973) *Food in History*, Stein and Day Publisher, Briarcliff Manor, New York.

Taylor, S.L. and Sumner, S.S. (1990) Risks and benefits of foods and food additives, in *Food Additives* (eds A.L. Branen, P. M. Davidson and S. Salminen), Marcel Dekker, New York, pp 663-685.

Theelen, R.M.C., Liem, A.K.D., Slob, W. and Van Wijnen, J.H. (1993) Intakes of 2,3,7,8-chlorine substituted dioxins, furans and planar PCBs from food in The Netherlands: medium and distribution. *Chemosphere*, **27** 1625-1635.

Travis, C. and Hattemer-Frey, H. (1991) Human exposure to dioxin. *Science of the Total Environment*, **104** 97-127.

UNEP (United Nations Environment Programme) (1992) *The Contaminants of Food*, Nairobi.

UNEP/FAO/WHO (1988) *Assessment of Chemical Contaminants in Food*, Report of the Results of United Nations Environment Programme, Food and Agriculture Organization and World Health Organization Programme on health-related environmental monitoring, Geneva.

USDA (US Department of Agriculture) (1984) *Nationwide Food Consumption Survey, Nutrient Intakes: Individuals in 48 States, Year 1977–78*, Report No. 1-2, Consumer Nutrition Division, Human Nutrition Information Service, Hyattsville, MD.

USDA (US Department of Agriculture) (1985) *Nationwide Food Consumption Survey, Continuing Survey of Food Intakes of Individuals, Women 19–50 Years and Their Children 1–5 Years, 1 Day, 1985*, Report No. 85-1, Nutrition Monitoring Division, Human Nutrition Information Service, Hyattsville, MD.

USDA (US Department of Agriculture) (1986a) *Nationwide Food Consumption Survey, Continuing Survey of Food Intakes of Individuals, Low Income Women 19–50 Years and Their Children 1–5 Years, 1 Day, 1985*, Report No. 85-2, Nutrition Monitoring Division, Human Nutrition Information Service, Hyattsville, MD.

USDA (US Department of Agriculture) (1986b) *Nationwide Food Consumption Surveys, Continuing Survey of Food Intakes of Individuals, Men 19–50 Years, 1 Day, 1985*, Report No. 85-3, Nutrition Monitoring Division, Human Nutrition Information Service, Hyattsville, MD.

USDA (US Department of Agriculture) (1987a) *Nationwide Food Consumption Surveys, Continuing Survey of Food Intakes of Individuals, Women 19–50 Years and Their Children, 1986*, Report No. 86-1, Nutrition Monitoring Division, Human Nutrition Information Service, Hyattsville, MD.

USDA (US Department of Agriculture) (1987b) *Nationwide Food Consumption Survey, Continuing Survey of Food Intakes of Individuals, Low Income Women 19–50 Years and Their Children 1–5 Years, 1 Day, 1986*, Report No. 86-2, Nutrition Monitoring Division, Human Nutrition Information Service, Hyattsville, MD.

USDA (US Department of Agriculture) (1987c) *Nationwide Food Consumption Surveys, Continuing Survey of Food Intakes of Individuals, Women 19–50 Years and Their Children 1–5 Years, 4 Days, 1985*, Report No. 85-4, Nutrition Monitoring Division, Human Nutrition Information Service, Hyattsville, MD.

USDA (US Department of Agriculture) (1988) *Nationwide Food Consumption Surveys, Continuing Survey of Food Intakes of Individuals, Low Income Women 19–50 Years and Their Children 1–5 Years, 4 Days, 1985*, Report No. 85-5, Nutrition Monitoring Division, Human Nutrition Information Service, Hyattsville, MD.

USDA (US Department of Agriculture) (1990) *USDA Agricultural Handbook No. 8, Consumption of Foods—Raw, Processed, and Prepared*, Nutrient Data Research, Human Nutrition Information Service, Hyattsville, MD.

Van den Huvel, J.P. and Lucier, G. (1993) Environmental toxicology of polychlorinated dibenzo-*p*-dioxins and polychlorinated dibenzofurans. *Environmental Health Perspectives*, **100** 189-200.

Van der Kolk (1984) *Consideration of a Codex Approach to Contamination of Foodstuffs with Polychlorinated biphenyls (PCBs)*, Codex Committee of Pesticide Residues, CX/PR 84/10, FAO, Rome.

Vongbuddhapitak, A., Sungwaranond, B., Thoophom, G. and Atisook, K. (1983) Pesticide residues intake in Bangkok, *The Bulletin of the Department of Medical Sciences, B.E. 2526*, **25** 131-141

Wahlstrom, B. (1989) Sunset for dangerous chemicals. *Nature*, **341** 276.

Weddig, L.J. (1991) Responsibility of industry in protecting public health, in *Proceedings of Symposium on Seafood Safety Issues* (ed. F.E. Ahmed), Institute of Medicine, National Academy of Sciences, Washington, DC, pp 33-40.

Weisenberg, E. (1986) Hexachlorobenzene in human milk: a polyhalogenated risk, in *Hexachlorobenzene: Proceedings of an International Symbosium* (eds C.R. Morris and J.R.P. Cabral), IARC Scientific Publication No. 77, International Agency for Research on Cancer, Lyon, pp 193-200.

WHO (UN World Health Organization) (1976) *Polychlorinated Biphenyl and Terphenyl*, Environmental Health Criteria No. 2, Geneva.

WHO (UN World Health Organization) (1985) *The Quantity and Quality of Breast Milk*, United Nations, Geneva.

WHO (UN World Health Organization) (1987a) *PCBs, PCDDs and PCDFs: Prevention and Control of Accidental and Environmental Exposures*, WHO Regional Office for Europe, Copenhagen.

WHO (UN World Health Organization) (1987b) *Evaluation of Certain Food Additives and Contaminants*, Technical Report Series 759, Geneva.

WHO (UN World Health Organization) (1989) *Levels of PCBs, PCDDs and PCDFs in Breast Milk: Results of WHO Coordinated Interlaboratory Quality Control Studies and Analytical Field Studies*, Environmental Health Series, No. 34, Geneva.

WHO (UN World Health Organization) (1996) *Levels of PCBs, PCDDs and PCDFs in Human Milk, Second Round of WHO-Coordinated Exposure Study*, Environmental Health in Europe No. 3, WHO European Center for Environment and Health, Bilthoven, Copenhagen, Nancy, Rome.

WHO/FAO (World Health Organization/Food Agriculture Organization) (1995) *Application of Risk Analysis to Food Standards Issues*, Report of the Joint FAO/WHO Expert Consultation, Geneva, Switzerland, 13–17 March 1995, Geneva.

Wickstrom, K., Pyysalo, H. and Simes, M.A. (1983) Levels of chlordane, hexachlorobenzene and DDT compounds in Finnish human milk in 1982. *Bulletin of Environmental Contamination and Toxicology*, **31** 251-256.

Worobec, M.D. and Ordway, G. (1989) *Toxic Substances Control Guide*, Bureau of National Affairs, Washington, DC.

Woteki, C.E. (1986) Dietary survey data: sources and limits to interpretation. *National Reviews*, Supplement, **44** 204-213.

Wuthrich, C., Mueller, F., Blaser, O. and Marer, B. (1985) Pesticides and other chemical residues in Swiss diet samples. *Mitteilungen aus dem Gebiete der Lebensmitteluntersuchung und Hygiene*, **76** 260-276.

Yu, M.L., Hsu, B.C., Bladen, B.C. and Rogan, W.J. (1991) *In utero* PCD/PCDF: relation of developmental delay to dysmorphology and dose. *Neurotoxicology and Teratology*, **13** (2) 195-202.

Index